PLANT METABOLITES AND VEGETABLES AS NUTRACEUTICALS

AAP Advances in Nutraceuticals

PLANT METABOLITES AND VEGETABLES AS NUTRACEUTICALS

Edited by

Raj K. Keservani, PhD

Bui Thanh Tung, PhD

Rajesh K. Kesharwani, PhD

Eknath D. Ahire, MS Pharm

AAP | APPLE ACADEMIC PRESS

First edition published 2025

Apple Academic Press Inc.
1265 Goldenrod Circle, NE,
Palm Bay, FL 32905 USA

760 Laurentian Drive, Unit 19,
Burlington, ON L7N 0A4, CANADA

CRC Press
2385 NW Executive Center Drive,
Suite 320, Boca Raton FL 33431

4 Park Square, Milton Park,
Abingdon, Oxon, OX14 4RN UK

Library and Archives Canada Cataloguing in Publication

Title: Plant metabolites and vegetables as nutraceuticals / edited by Raj K. Keservani, PhD, Bui Thanh Tung, PhD, Rajesh K. Kesharwani, PhD, Eknath D. Ahire, MS Pharm.
Names: Keservani, Raj K., 1981- editor. | Bui, Tung Thanh, editor. | Kesharwani, Rajesh Kumar, 1978- editor. | Ahire, Eknath D., editor.
Description: First edition. | Includes bibliographical references and index.
Identifiers: Canadiana (print) 20240350367 | Canadiana (ebook) 20240350375 | ISBN 9781774915448 (hardcover) | ISBN 9781774915455 (softcover) | ISBN 9781032680125 (ebook)
Subjects: LCSH: Functional foods. | LCSH: Plant metabolites. | LCSH: Vegetables. | LCSH: Plant metabolites—Therapeutic use. | LCSH: Vegetables—Therapeutic use.
Classification: LCC QP144.F85 P53 2024 | DDC 612.3—dc23

Library of Congress Cataloging-in-Publication Data

CIP data on file with US Library of Congress

ISBN: 978-1-77491-544-8 (hbk)
ISBN: 978-1-77491-545-5 (pbk)
ISBN: 978-1-03268-012-5 (ebk)

ABOUT THE BOOK SERIES: AAP ADVANCES IN NUTRACEUTICALS

Series Editor:
Raj K. Keservani, PhD, MPharm
Faculty of B. Pharmacy, CSM Group of Institutions, Allahabad, India

Rajesh K. Kesharwani, PhD
Nehru Gram Bharati (Deemed to be University), Prayagraj, India

Anil K. Sharma, PhD, MPharm
Department of Pharmacy, School of Medical and Allied Sciences, GD Goenka University, Gurugram, India

In the modern era, mankind has witnessed a paradigm shift with respect to fundamental eating behavior. The lack of physical workouts and busy schedules at offices and in households have promoted the consumption of junk foods, which eventually results in numerous diseases and disorders since the nutritional content from fast food is inadequate. Plumpness and obesity have become a global health threat. The leading causes of death from most developing nations are noncommunicable diseases such as cardiovascular diseases, cancer, arthritis, osteoporosis, and liver toxicity. Patients suffering from such lifestyle ailments are a bit apprehensive towards prolonged use of costly modern therapeutics, which encourages instead the use of alternative approaches for management of such diseases and disorders.

The emerging sector of the nutraceutical industry encompasses products derived from nature, dietary supplements, and functional foods. Nutraceuticals are used for treatment and prevention of a broad range of diseases, such as the common cold, arthritis, sleep-related disorders, cancers, cardiovascular complications, metabolic disorders, and others.

The research on nutraceuticals is increasing day by day, considering the beneficial effects of food or food supplements in the management of diverse diseases. The issue of paramount concern is standardization and establishment of clinical efficacy of nutraceuticals, which is in fact a challenge for researchers around the globe.

This book series aims at realizing the significance of the variety of nutraceuticals for human well-being. The books in the series will also emphasize the role of dieticians and nutritionists for the prescription of judicious eating habits. The key food components such as carbohydrates, proteins, and lipids as well as micronutrients (vitamins, minerals) are demonstrated to maintain good health, obviating the need for medicines. Thus, nutraceuticals that are indeed derived from food ingredients are believed to be potential alternative therapeutics.

Asymptomatic diseases/disorders necessitate proper diagnosis. Stressed circumstances have been reported to cause weight loss. Workplace stress is a key variable in the genesis and propagation of metabolic disorders. The nuances of the underlying mechanisms are still to be deciphered. Poor lifestyle trends have been attributed along with the consumption of junk foods in several such instances. Sugar-rich carbonated beverages, the use of contraband drugs, and the consumption of liquor in excess are assigned as risk factors. This book series will cover advances and applications in addition to providing the basics of nutraceuticals.

The books in this series will be valuable resources for industry professionals to design and develop quality products for end users. In addition, the book series will be fortified by ideas of innovation, concept building, manufacturing aspects, quality control, and regulatory status of nutraceuticals.

Ostensibly, business has seen a rise in the nutraceutical consumption in recent years. Pure, safe, efficacious products are the need of the hour globally. There are a number of books on nutraceuticals in the market, yet these are only a few books presenting certain aspects of nutraceuticals therein. This book series aims to offer a holistic view of this promising strategy, which is picking up pace with time. This book series will be a unique endeavor to bring manufacturing, research and development, and marketing strategies under a single umbrella.

CURRENT BOOKS IN THE SERIES

- **Micronutrients and Macronutrients as Nutraceuticals**
  Editors: Prakash Chandra Gupta, PhD, Sayan Bhattacharyya, MBBS, MD, Nisha Sharma, PhD, MPharm, Rajesh K. Kesharwani, PhD, and Raj K. Keservani, PhD, MPharm

- **Formulations, Regulations, and Challenges of Nutraceuticals**
  Editors: Tingirikari Jagan Mohan Rao, PhD, Rajesh K. Kesharwani, PhD, Raj. K. Keservani, PhD, and Anil K. Sharma, PhD

- **Food Supplements and Dietary Fibers in Health and Disease**
  Editors: Bhushan R. Rane, PhD, MPharm, Raj K. Keservani, PhD, MPharm, Durgesh Singh, DPhil, Nayan A. Gujarathi, PhD, and Ashish S. Jain, PhD
- **Nutraceuticals in Cancer Prevention, Management, and Treatment**
  Editors: Raj K. Keservani, PhD, MPharm Bui Thanh Tung, PhD, Sippy Singh, DPhil, and Rajesh K. Kesharwani, PhD
- **Advances in Flavonoids for Human Health and Prevention of Diseases**
  Editors: Nisha Sharma, PhD, Deepika Saini, PhD, Rajesh K. Kesharwani, PhD, Prakash Chandra Gupta, and Raj. K. Keservani, PhD
- **The Flavonoids: Extraction and Applications**
  Editors: Deepika Saini, PhD, Rajesh K. Kesharwani, PhD, and Raj K. Keservani, PhD, MPharm
- **Immune-Boosting Nutraceuticals for Better Human Health: Novel Applications**
  Editors: Urmila Jarouliya, PhD, Raj K. Keservani, PhD, MPharm Rajesh K. Kesharwani, PhD, Virendra K. Patel, PhD, and Adi D. Bharti, MPharm
- **Applications of Functional Foods in Disease Prevention**
  Editors: Raj K. Keservani, PhD, MPharm, and Eknath D. Ahire, MSPharm
- **Plant Metabolites and Vegetables as Nutraceuticals**
  Editors: Raj K. Keservani, PhD, MPharm, Bui Thanh Tung, PhD, Rajesh K. Kesharwani, PhD, MPharm, and Eknath D. Ahire, MSPharm
- **Preventive and Therapeutic Role of Vitamins as Nutraceuticals**
  Editors: Khemchand R. Surana, Eknath D. Ahire, MSPharm, Raj K. Keservani, PhD, MPharm, and Rajesh K. Kesharwani, PhD
- **Nutraceuticals and Bone Health**
  Editors: Deepak Sharma, PhD, Madan Mohan Gupta, PhD, Anil K. Sharma, PhD, MPharm, Raj K. Keservani, PhD, MPharm, and Rajesh K. Kesharwani, PhD
- **Flavonoids as Nutraceuticals**
  Editors: Rajesh K. Kesharwani, PhD, MTech, Deepika Saini, PhD, Raj K. Keservani, PhD, MPharm, and Anil Kumar Sharma, PhD
- **Herbals as Nutraceuticals: Their Role in Healthcare**
  Editors: Raj K. Keservani, PhD, MPharm, Rajeshwar Kamal Kant Arya, PhD, and Rajesh K. Kesharwani, PhD

- **Nutraceutical Fruits: Overview and Disease Prevention**
  Editors: Raj K. Keservani, PhD, MPharm
- **Nutrigenomics and Nutraceuticals**
  Editors: Raj K. Keservani, PhD, MPharm, Eknath D. Ahire, MSPharm, Shubham J. Khainar, PhD, Sanjay J. Kshirsagar, PhD, and Rajesh K. Kesharwani, PhD
- **Nutraceuticals in Cardiac Health Management**
  Editors: Raj K. Keservani, PhD, MPharm, Rajesh K. Kesharwani, PhD, Praveen Kumar Jain, PhD, Bhushan R. Rane, PhD, and Nayan A. Gujarathi, PhD
- **Nutraceuticals for the Treatment and Prevention of Sexual Disorders**
  Editors: Raj K. Keservani, PhD, MPharm, Sharangouda J. Patil, PhD, and Ivan Aranha, PhD
- **Nutraceuticals in Obesity Management and Control**
  Editors: Raj K. Keservani, PhD, MPharm, and Alka Lohani, PhD
- **Antioxidants as Nutraceuticals**
  Editors: Nayan A. Gujarathi, PhD, Raj K. Keservani, PhD, MPharm, Rajesh K. Kesharwani, PhD, Bhushan R. Rane, PhD, and Yogeeta Sameer Goyal, PhD
- **The Nature of Nutraceuticals: History, Properties, Sources, and Nanotechnology**
  Editors: Rajesh K. Kesharwani, PhD, Prashant Kumar, PhD, and Raj K. Keservani, PhD, MPharm
- **Nutraceuticals in Insomnia and Sleep Problems**
  Editors: Raj K. Keservani, PhD, MPharm, Sayan Bhattacharyya, PhD, and Rajesh K. Kesharwani, PhD
- **Nutraceuticals in Respiratory and Pulmonary Diseases**
  Editors: Deepika Saini, PhD, Rajesh K. Kesharwani, PhD, and Raj K. Keservani, PhD, MPharm
- **Nutraceuticals in Arthritis and Psoriasis**
  Editors: Meenakshi Jaiswal, PhD, Raj K. Keservani, PhD, MPharm Rajesh K. Kesharwani, PhD, and Swati G. Talele, PhD

ABOUT THE EDITORS

Raj K. Keservani, PhD, MPharm
Associate Professor, Faculty of B. Pharmacy, CSM Group of Institutions, Prayagraj, Uttar Pradesh, India

Raj K. Keservani, PhD, MPharm, is working as an Associate Professor in the Faculty of B. Pharmacy at the CSM Group of Institutions, Prayagraj, India. He has more than 12 years of academic (teaching) experience from various institutes of India in pharmaceutical education. He has supervised more than 25 undergraduate students for their research work. He has published over 50 peer-reviewed papers in the field of pharmaceutical sciences in national and international journals, six patents, over 70 book chapters, three co-authored books, and over 40 edited books. He is also active as a reviewer for several international scientific journals. His research interests include nutraceutical and functional foods, novel drug delivery systems (NDDS), transdermal drug delivery/drug delivery, health science, cancer biology, and neurobiology. Dr. Keservani graduated with a pharmacy degree from the Department of Pharmacy, Kumaun University, Nainital, Uttarakhand, India. He received his Master of Pharmacy (MPharm) (specialization in pharmaceutics) from the School of Pharmaceutical Sciences, Rajiv Gandhi Proudyogiki Vishwavidyalaya, Bhopal, India. He earned his PhD degree from IKG Punjab Technical University, Jalandhar, Punjab, India. Dr. Keservani is a life member of the Society of Pharmaceutical Education and Research (SPER).

Bui Thanh Tung, PhD
Associate Professor, University of Medicine and Pharmacy, Vietnam National University, Hanoi, Vietnam

Bui Thanh Tung, PhD, is an Assistant Professor working at the University of Medicine and Pharmacy, Vietnam National University, Hanoi, Vietnam. His main research focuses on medicinal plants for the treatment of various diseases such as diabetes, Alzheimer's disease, and cancer. His special interest is in the development of new drugs from Vietnamese natural resources. He writes and presents widely on issues of natural products, bioactive compounds, and their pharmacological activities. Dr. Tung received his PhD in Pharmacology from Seville University, Spain

Rajesh K. Kesharwani, PhD, MTech
Associate Professor, Department of Computer Application, Nehru Gram Bharati (Deemed to be University), Prayagraj, Uttar Pradesh, India

Rajesh K. Kesharwani, PhD, MTech, is an Associate Professor, Department of Computer Application, Nehru Gram Bharati (Deemed to be University), Prayagraj, India. He has more than 11 years of research and nine years of teaching experience at various institutes of India, imparting bioinformatics and biotechnology education. He has received several awards, including the NASI-Swarna Jayanti Puruskar by the National Academy of Sciences of India. He has supervised one PhD and more than 20 undergraduate and graduate students for their research work. Dr. Kesharwani has authored over 49 peer-reviewed articles, 49 book chapters, and 35 edited books with international publishers. He has been a member of many scientific communities as well as a reviewer for many international journals. He has presented many papers at various national and international conferences. Dr. Kesharwani received his PhD from the Indian Institute of Information Technology, Allahabad, and worked at NIT Warangal for two semesters. He has been a recipient of a Ministry of Human Resource Development (India) Fellowship and a Senior Research Fellowship from the Indian Council of Medical Research, India. His research fields of interest are medical informatics, protein structure and function prediction, computer-aided drug designing, structural biology, drug delivery, cancer biology, nanobiotechnology, and biomedical sciences.

Eknath D. Ahire, MS Pharm
Assistant Professor, Department of Pharmaceutics, MET's, Institute of Pharmacy, Bhujbal Knowledge City, Adgaon, Nashik, Maharashtra, India

Eknath D. Ahire, MS Pharrm, is currently working as an Assistant Professor in the Department of Pharmaceutics at MET's, Institute of Pharmacy, Bhujbal Knowledge City, Adgoan, Nashik, (Affiliated to Savitribai Phule Pune University), Pune. He has obtained his graduation (BPharm) degree from Savitribai Phule Pune University. He obtained his master's degree (MS Pharm) in Pharmaceutics from the National Institute of Pharmaceutical

Education and Research-Ahmedabad (NIPER-A). Currently, he is also a PhD scholar with Savitribai Phule, Pune University, Maharashtra, and India. He was awarded with NFST (RGNF) PhD Research Scholar fellowship, by the Government of India. He is working for QCS Labs LLP, Mumbai, as an inspector for Nigeria's regulatory market pre-shipment inspection under the Earn and Learn scheme. He has a total of five years of experience, working as Formulation Scientist II at Torrent Research Center – Gandhinagar, Gujrat, India, as well as five years of academic experience. He is working as a reviewer for several reputed international journals. He is a life member of the Association of Pharmaceutical Teachers of India (APTI). Having more than 60 publications (20 researches/reviews and 40 book chapters) published in different national and international well-reputed and indexed journals with six textbooks and three edited books. He has participated in and attended several national and international conferences and workshops. He has worked on nanotechnology-based novel drug delivery systems, nutraceuticals, and immunology.

CONTENTS

CONTRIBUTORS

Eknath D. Ahire
Department of Pharmaceutics, METs, Institute of Pharmacy, BKC, Adgaon, Nashik, Maharashtra, India

Ankit Awasthi
Department of Applied Agriculture, Central University of Punjab, Ghudda, Bathinda, Punjab, India

Amruta Balekundri
Department of Pharmacognosy, KLE College of Pharmacy Belagavi, KLE University of Higher Education and Research, Belagavi, Karnataka, India

Mayur S. Bidkar
Department of Pharmaceutics, Sandip Institute of Pharmaceutical Sciences, Nashik, Maharashtra, India

Navneet Bithel
Professor, Department of Botany and Microbiology, Gurukula Kangri (Deemed to be University), Haridwar, Uttarakhand, India

Dinesh B. Chaudhari
Department of Pharmaceutics, Sandip Institute of Pharmaceutical Sciences, Nashik, Maharashtra, India

Darcia D'mello
Department of Biotechnology, School of Life Sciences, Garden City University, Bangalore, Karnataka, India

Myrene Roselyn D'souza
Department of Biochemistry, Mount Carmel College Autonomous, Karnataka, India

Deepa Devi
Department of Botany and Microbiology, Gurukula Kangri (Deemed to be University), Haridwar, Uttarakhand, India

N. Eramma
Department of Microbiology, Acharya Bangalore B-School, Bangalore, Karnataka, India

Akshay U. Ghorpade
Department of Pharmaceutics, Sandip Institute of Pharmaceutical Sciences, Nashik, Maharashtra, India

Aparajita Gupta
Department of Microbiology, Dolphin PG College of Science and Agriculture, Chunni Kalan, Fatehgarh Sahib, Punjab, India

Anil G. Jadhav
Department of Pharmaceutics, Sandip Institute of Pharmaceutical Sciences, Nashik, Maharashtra, India

S. Renuka Jyothi
Department of Life Science, School of Sciences, Jain University, Bangalore, Karnataka, India

Raj K. Keservani
Faculty of B. Pharmacy, CSM Group of Institutions, Prayagraj, Uttar Pradesh, India

Nguyen Bao Kim
VNU University of Medicine and Pharmacy, Vietnam National University, Hanoi, Vietnam

Manish Kumar
Department of Pharmacy, Madhav University, Pindwara, Rajasthan, India

Shailendra Kumar
Institute of Microbiology, Dr. Ram Manohar Lohia Avadh University, Ayodhya, Uttar Pradesh, India

Tushar Lokhande
Mahatma Gandhi Vidyamandir's, Pharmacy College, Panchavati, Nashik, Maharashtra, India

Rahul Kumar Mishra
Department of Pharmaceutical Science and Technology, Madan Mohan Malaviya University of Technology, Gorakhpur, Uttar Pradesh, India

Vidisha Mishra
Department of Home Science, R.B.B.M. College, B.R.A. Bihar University, Muzaffarpur, Bihar, India

Raghu Ningegowda
Department of Chemistry, Jyoti Nivas College, Bangalore, Karnataka, India

Sharangouda J. Patil
Department of Zoology, NMKRV College for Women (Autonomous), Bangalore, Karnataka, India

Devangna Paul
Department of Biochemistry, Mount Carmel College Autonomous, Karnataka, India

S. O. Sadashiv
Department of Studies in Food Technology, Shivagangothri, Davangere University, Davanagere, Karnataka, India

Shalni Satya
Department of Environmental Science, Central University of Rajasthan, Rajasthan, India

Kamal Kant Sharma
Research Scholar, Department of Botany and Microbiology, Gurukula Kangri (Deemed to be University), Haridwar, Uttarakhand, India

Munish Sharma
Research Scholar, Department of Plant Sciences, Central University of Himachal Pradesh, Shahpur, Kangra, Himachal Pradesh, India

Munit Sharma
Research Scholar, Department of Botany and Microbiology, Gurukula Kangri (Deemed to be University), Haridwar, Uttarakhand, India

Ashwini V. Shelke
Sandip Institute of Pharmaceutical Sciences, Mahirawani, Nashik, Maharashtra, India;
Mahatma Gandhi Vidyamandir's, Pharmacy College, Panchavati, Nashik, Maharashtra, India

Ajay Kumar Shukla
Institute of Pharmacy, Dr. Ram Manohar Lohia Avadh University, Ayodhya, Uttar Pradesh, India

Shivam Singh
Scientist B, Government District Hospital Cum Trauma Center, Waidhan, Singrauli, Madhya Pradesh, India

Nguyen Nhu Son
VNU University of Medicine and Pharmacy, Vietnam National University, Hanoi, Vietnam

Ankur Srivastav
Institute of Pharmacy, Dr. Ram Manohar Lohia Avadh University, Ayodhya, Uttar Pradesh, India

Yashi Srivastava
Department of Applied Agriculture, Central University of Punjab, Ghudda, Bathinda, Punjab, India

Swati G. Talele
Department of Pharmaceutics, Sandip Institute of Pharmaceutical Sciences, Nashik, Maharashtra, India

Bui Thanh Tung
VNU University of Medicine and Pharmacy, Vietnam National University, Hanoi, Vietnam

Madhuri Verma
Assistant Professor, Naraina Group of Institutions of Pharmacy, Kanpur, Dr. A.P.J. Abdul Kalam Technical University, Uttar Pradesh, India

ABBREVIATIONS

1C	one-carbon
6-OHDA	oxidopamine or 6-hydroxydopamine or 2,4,5-trihydroxydopamine
AD	Alzheimer's disease
ADCD	autophagy-dependent cell death
ADH	alcohol dehydrogenase enzyme
Ah	aryl hydrocarbon
AIH	autoimmune hepatitis
ALD	alcoholic liver disease
ALF	acute liver failure
ALS	amyotrophic lateral sclerosis
AMPK	5′-monophosphate-activated protein kinase
anti-LKM-1	anti-liver kidney microsomal
Aβ	amyloid β
BBB	blood-brain barrier
BBR	berberine
BD	Behcet's disease
BMI	body mass index
BuOH fr.	butanolic fraction
CA1	cornu ammonis 1
CAD	coronary heart disease
CCl_4	carbon tetrachloride
CD1	cluster of differentiation 1
CDA	Canadian Diabetes Association
CDK	cyclin-dependent kinase
CEE	crude ethanol extract
CNS	central nervous system
COX-2	cyclooxygenase-2
CPP	calcium pyrophosphate
CSF	cerebro spinal fluid
CVD	cardiovascular disease
CYLD	cylindromatosis
DA	dopaminergic
DALY	disability-adjusted life year
DAPI	4,6-diamidino-2-phenylindole
DHS	dihydroxy-trans-stilbene
DIM	diindolylmethane

DPPH	2,2-diphenyl-1-picryldrazyl
DSSE	defatted sesame seed extracts
EGCG	epigallocatechin-3-gallate
EMT	epithelial-mesenchymal transition
EO	essential oil
ER	epigenome-regulated
ESLD	end-stage of liver disease
ETSBAS	electron transfer signal-blocking activities
FADD	Fas-associated death domain
FDA	Food and Drug Administration
FLD	fatty liver disease
FMF	familial Mediterranean fever
FMHG	fast shifting healthcare items
GFAP	glial fibrillary acidic protein
GHDx	global health data exchange
Grn	glucoraphanin
GS	glucosinolate
GSE	grapeseed extract
GSH	glutathione
HBsAg	hepatitis B surface antigen
HCC	hepatocellular carcinoma
HD	Huntington's disease
HDL	high-density lipoprotein
HF	heart failure
HFE	high iron Fe
HIV	human immunodeficiency virus
HMP	hexose monophosphate shunt
HO-1	heme oxygenase-1
Htt	huntingtin
IBD	inflammatory bowel disease
IC_{50}	half-maximal inhibitory concentration
IgG	immunoglobulin G
IL-1α	interleukin-1α
IL-1β	interleukin-1β
IUCN	International Union for Conservation of Nature and Natural Resources
JNK	C-Jun N-terminal kinase
LDL	low-density lipoprotein
LDL-C	low-density lipoprotein cholesterol
LID	L-dopa-induced dyskinesia
LOOH	lipid hydroperoxide
LPS	lipopolysaccharide
MAPK	mitogen-activated protein kinase

MCP-1	monocyte chemoattractant protein-1
mGLUT9	mouse-glucose transporter 9
MI	myocardial infarction
mi-RNA	micro-RNA
MMP	mitochondrial membrane potential
MND	motor neuron disease
MPP	methyl-4-phenylpyridinium
MPT	multipurpose tree species
MPTP	1-methyl-4-phenyl-1,2,3,6-tetrahydropyridine
mRNA	messenger ribonucleic acid
MSNs	medium spiny neurons
MSU	monosodium urate
MVA	mevalonic acid
NAFLD	non-alcoholic fatty liver disease
NASH	non-alcoholic steatohepatitis
NCCMNIH	National Center for Complementary Medicine of the National Institutes of Health
ND	neurodegenerative disease
nfvPPA	non-fluent/agrammatic variant primary progressive aphasia
NF-κB	nuclear factor kappa-B
NIH	National Institute of Health
NLRP3	nucleotide-binding domain-like receptor protein 3
NO	nitric oxide
Nrf2	nuclear factor erythroid 2-related factor 2
NSAIDs	nonsteroidal anti-inflammatory drugs
NSCLC	non-small cell lung cancer cells
OAT1	organic anion transport 1
PAD	peripheral artery disease
PBR	peripheral type benzodiazepine receptor
PD	Parkinson's disease
PEP	postencephalitic parkinsonism
PGE2	prostaglandin E2
PI3K/Akt	phosphatidylinositol-3-kinase and protein kinase B
PM2.5	particulate matter
PO	potassium oxonate
PSC	primary sclerosing cholangitis
RES	resveratrol
RHIMs	RIP homotypic interaction motif
RIP	kinase receptor proteins
ROS	reactive oxygen species
SAMC	S-allyl mercapto cysteine
SCA	spinocerebellar ataxia
SDI	socio-demographic index

SIRT1	Sirtuin (silent mating type information regulation 2 homolog) 1
SMA	spinal muscular atrophy
SMN	survival motor neuron
SN	substantia nigra
SOD	superoxide dismutase
svPPA	semantic variant PPA
THP-1	Tamm-Horsfall protein 1
TLR	toll-like receptor
TNF	tumor necrotic factors
TNF-α	tumor necrosis factor-alpha
TNF-β	tissue necrosis factor-beta
TSE	transmissible spongiform encephalopathies
TXNIP	thioredoxin interacting protein
UA	uric acid
URAT1	urate transporter
USDA	United States Diabetes Association
UTI	urinary tract infections
VCI	vascular cognitive impairment
VEGF	vascular endothelial growth factor
WHO	World Health Organization
XOD	xanthine oxidase

PREFACE

Vegetables and plant metabolites as nutraceuticals are nutritionally important for maintaining the body's energy balance and are promising as highly effective treatments for a number of ailments. Nutraceuticals may also have chemopreventive activities, as already suggested by traditional medicine. Today, nutraceuticals are included in nutritional supplements sold at counters without prescription, and self-prescription of these products has significantly increased over the years.

This book illustrates the role of *Plant Metabolites and Vegetables as Nutraceuticals* in the prevention and/or treatment of human diseases. The present book strives to provide beneficial information to its readers relevant to plant metabolites and vegetables as nutraceuticals. The text of this book is written by highly skilled, experienced, and renowned scientists and researchers around the globe with up-to-date information to offer insights on herbal medicines to readers, researchers, academicians, scientists, and industrialists worldwide.

This book *Plant Metabolites and Vegetables as Nutraceuticals* comprises 14 chapters in three parts, which describe the use of plant metabolites and vegetables for improving health and their mechanisms.

PART I: NUTRACEUTICAL POTENTIAL AND HEALTH PROMOTION OF PLANT METABOLITES AND VEGETABLES

Information on health promotion in *Plant Metabolites and Vegetables as Nutraceuticals* has been presented in Chapter 1: Potentials of Vegetables and Plant Metabolites in Healthcare, written by Munit Sharma and colleagues. The authors have given an introduction to phytochemicals and their functions to protect plants against hazards faced in the environment, such as moisture, drought, and extreme salt conditions. They also describe the inverse relationship between eating fruits and vegetables and the incidence of serious and chronic diseases.

Chapter 2: Nutraceutical Potential of Selected Vegetables, written by Sharangouda J. Patil and colleagues, focuses on various plants in Ayurvedic formulation and advanced organics in modern natural medicine in global scenarios recognized as important food. The presence of biological molecules in these vegetable extracts explores various pharmacological and nutraceutical properties on many levels.

PART II: NUTRACEUTICAL BENEFITS OF BIOACTIVE AND FLAVONOIDS

Plants contain naturally occurring chemical constituents known as phytochemicals which are produced through a series of different metabolic reactions.

Chapter 3: Phytochemicals Present in Vegetables for Health Promotion, written by Kamal Kant Sharma and colleagues, gives a general introduction to phytochemicals proven to have potential health benefits for humans. The usage of health-promoting food has risen dramatically as an outcome of changing lifestyles, higher purchasing power, and better-informed customers.

Chapter 4: Flavonoids in Vegetables as Nutraceuticals, written by Mayur S. Bidkar, illustrates how flavonoids provide nutritional qualities to fruits and vegetables and provide a better understanding of their structures and biological activities, specifying their potential as therapeutic compounds and also for forecasting and managing food standard.

Chapter 5: Nutraceutical Bioactive Compounds in Fruits and Vegetables, written by Ajay Kumar Shukla and his co-worker, discusses the bioactive compounds of fruits and vegetables and their impact on various diseases.

PART III: PLANT METABOLITES AND VEGETABLES IN PREVENTION AND TREATMENT OF DISEASES

Plants have been used to prevent and treat various diseases since their first discovery. Many plant metabolites have been reported to exhibit anti-cancer activity through various metabolic/biological pathways. Their consumption has been associated with reducing the risks of development of different types of cancers. Chapter 6: Effect of Plant and Vegetable-Based Secondary Metabolites on Cancer Treatment, written by Ankit Awasthi and Yashi Srivastava, provides the latest knowledge and updates on the anti-cancer potential of several vegetables.

Gout is the most common form of inflammatory arthropathy globally and is caused by the deposition of monosodium urate (MSU) crystals. People with persistent hyperuricemia can have chronic gouty arthritis, tophi, irreversible structural joint damage, and even disability. Chapter 7: Plant Metabolites and Vegetables for Gout Prevention and Treatment, written by Bui Thanh Tung and colleagues, discusses about the pathogenesis and main therapeutic targets of the herbs and herbal compounds in the treatment of gouty arthritis. They also introduce a number of potential herbal plants and isolated phytochemicals that are used and studied to prevent and reduce symptoms of this disease.

Cardiovascular disease (CVD) has surpassed cancer as the top cause of mortality worldwide and is considered a significant public health problem. The recent increase in popularity of natural products and alternative medicine has renewed interest in traditional heart disease treatments. Certain dietary habits have been linked to better cardiovascular health in epidemiological research. Chapter 8: Plant Metabolites and Vegetables for Heart Disease

Prevention and Treatment, written by Ashwini V. Shelke and colleagues, discusses about plant-based metabolites and foods for Heart Disease Prevention and Treatment.

The liver is a major organ in the human body, which plays an eminent role in the regulation of metabolism, detoxifying toxic substances, and synthesizing useful biomolecules. Plants serve as a huge repository of bioactive secondary metabolites such as phenols, flavonoids, alkaloids, carotenoids, etc., and large amounts of these chemicals in the diet appear to have a significant effect in the prevention and treatment of liver disorders. Chapter 9: Plant Metabolites and Vegetables for Liver Disease Prevention and Treatment, written by Sharangouda J. Patil and *colleagues,* provides an overview of the current progress in research on plant-derived metabolites that serve as potential new therapeutic agents for the treatment of various liver diseases.

Chapter 10: Plant Metabolites and Vegetables for Brain Disease Prevention and Treatment, written by Madhuri Verma, gives information that secondary metabolites of vegetables and herbs are promising in preventing and managing neurodegenerative disorders. These metabolites elucidate their activity on mitochondrial changes in neurodegenerative disease (ND), also have antioxidant, anti-aggregation, memory enhancing, revert neuronal function, and neuroprotective activity in animal model and clinical trials.

Bioactive compounds play an important role in the production of nutraceutical food. Nutraceutical foods have a valuable role in improving human health. These foods have proved valuable because they protect the body from chronic disorders. Different fruits and vegetables contain different types of biological properties for the treatment and prevention of gastrointestinal problems. Chapter 11: Plant Metabolites and Vegetables for Gastrointestinal Disease Prevention and Treatment, written by Vidisha Mishra, reviews the plant metabolites and phenolic compounds of vegetables against gastrointestinal disease.

Chapter 12: Plant Metabolites and Vegetables for Diabetes Prevention and Treatment, written by Amruta Balekundri and Eknath D. Ahire, provides an overview of the nutraceutical composition of plants and vegetables, their health benefits, and their applications in diabetes prevention and management.

Worldwide the consumption of plant and vegetable-based secondary metabolites in kidney disease treatment is enhanced due to their range of health benefits like the prevention of kidney disease, discussed in Chapter 13: Plant Metabolites and Vegetables for Kidney Diseases Prevention and Treatment, written by Amruta Balekundri and coworker.

Devangna Paul and Myrene Roselyn D'souza report on how phytochemicals may have a role in the prevention and moderation of some of the symptoms of Parkinson's disease (PD) in Chapter 14: Delaying the Onset of Parkinson's Disease Through Phytotherapy.

PART I

Nutraceutical Potential and Health Promotion of Plant Metabolites and Vegetables

CHAPTER 1

POTENTIAL OF VEGETABLES AND PLANT METABOLITES IN HEALTHCARE

MUNISH SHARMA,[1] NAVNEET BITHEL,[2] KAMAL KANT SHARMA,[3] and MUNIT SHARMA[3]

[1]*Research Scholar, Department of Plant Sciences, Central University of Himachal Pradesh, Shahpur, Kangra, Himachal Pradesh, India*

[2]*Professor, Department of Botany and Microbiology, Gurukula Kangri (Deemed to be University), Haridwar, Uttarakhand, India*

[3]*Research Scholar, Department of Botany and Microbiology, Gurukula Kangri (Deemed to be University), Haridwar, Uttarakhand, India*

ABSTRACT

Balanced diets include vegetables as an essential component. In order to lower the risk of a variety of diseases including cancer and diet-related problems, the World Health Organization (WHO) recommends five to eight servings (400–600 grams) a day. Amongst dietary supplements, vegetables attain a significant stage due to their low calorific value, and high phytochemical and fiber content, which places them at a considerable level of functional food. Vegetables have long been used as a source of nutrition including fruits and vegetables, pigments, juices, and nutraceuticals in your diet will reduce the risk of cancer and various disorders like hypertension and diabetes, etc. A variety of phytochemicals possess strong antioxidant properties on their own. Phytochemicals in fruits and vegetables like lycopene, beta-carotene, and vitamin C play an important role in the prevention of various diseases. For example, lycopene has been shown to protect against carcinogens that cause liver, cervix, and prostate cancer. There are plenty of foods that contain

Plant Metabolites and Vegetables as Nutraceuticals.
Raj K. Keservani, Bui Thanh Tung, Rajesh K. Kesharwani, and Eknath D. Ahire (Eds.)

lycopene, including tomatoes, watermelon, pink grapefruit, apricots, and pink guavas. Among its many functions, vitamin C plays an essential role in wound healing, and immunity protection. The main sources of vitamin C are Indian gooseberry, citrus fruits such as orange, lemon, potatoes, kiwifruit, and broccoli. This chapter summarizes the awareness of the health benefits of vegetables and plant metabolites' role in human health care. Importantly, because several phytochemicals influence the same genes and pathways that pharmaceuticals target, vegetable-rich diets combined with medicinal therapy are being explored as innovative treatments. As a result, phytochemicals found in vegetables may be a promising approach to disease prevention.

1.1 INTRODUCTION

Phytochemicals are physiologically active and naturally occurring chemical molecules that improve human health and illness prevention. An extensive diversity of natural pigments which serve a variety of biological functions and contribute to human health (Sharma et al., 2021). It has been estimated that about 10,000 different phytochemicals are found in leaves, roots, stems, fruits, flowers, legumes, fungi, and herbs while some remain unknown (Saxena et al., 2013). A phytochemical can be classified by its chemical structure into the following categories: phenolic acid, flavonoids, flavanols, and anthocyanidins (Sharma et al., 2020; Keservani & Sharma, 2014). The levels of these pigments, however, varied from plant to plant depending on climatic conditions (Zhang & Gan, 2015). The human body requires dietary active components to function effectively. Nutraceuticals are fruits that contain bioactive chemicals that minimize the risk of a variety of ailments. Veggies are considered preventive foods due to their low-calorie content and high vitamin and mineral content (Keservani et al., 2010a, b, 2017, 2020; Sharma et al., 2021). Vegetables may help alleviate disorders caused by a lack of nutrients including omega-3 fatty acids, and antioxidants (vitamin. E, vitamin C, carotene, zinc, cryptoxanthin, lycopene, and minerals). Because of the presence of bioactive components such as phytochemicals, vegetables offer nutraceutical potential. Nutraceutical-rich vegetables have been demonstrated to benefit gastrointestinal disorders, cardiovascular disease (CVD), cancer, diabetes, and other diseases (Keservani et al., 2020). Vegetables can not only reduce illness progression, but they can also treat numerous diseases and improve the immune system (Sharma et al., 2020a). Phytochemicals are thought to provide much of their potential for disease prevention, the most well-documented of which is antioxidant preventive activity. Nonetheless, due to additive and synergistic effects, there is a strong understanding that vegetable consumption is more essential in giving health advantages than consumption of just one of their contents. It is believed that phytochemicals play a major role in their disease prevention ability, the best-documented of which is antioxidant protection (Sharma et al., 2020b). The use of isolated phytochemicals (Keservani et al., 2016a) as supplements should be approached with caution because different meta-studies have produced diverse and often conflicting conclusions about their effects on health. That is because dosages may exceed the recommended daily intake (Aune et al., 2012). Furthermore, there is a strong consensus that consuming whole fruits and vegetables is more important than taking just one of its parts for health purposes related to additive and synergistic effects (Boeing et al., 2012). Phytonutrients are sourced from a variety of food groups such as cereal grains, vegetables, fruits (Keservani et al., 2016b), legumes, nuts tea, etc. These phytonutrients contain biological and chemical components such as glucose, protein (Singh et al., 2017b), lipids, water, fat-soluble vitamins (Singh et al., 2017a), and minerals, all of them are important for human oxidative metabolism. Plant-derived metabolites include flavonoids, carotenoids, phytoestrogens, phytates, isoflavones, isothiocyanates, indoles, saponins, and terpenes. These secondary plant metabolites are referred to as accessory health factors. Cereals, legumes, nuts, and seeds also may contribute to disease mitigation (Bhaskarachary et al., 2015). Secondary metabolites in plants are natural sources of chemically bioactive components that have been utilized in a healthy diet, traditional medicine, and a number of different applications. The goal was to improve PSM production is to attain maximum yields suitable for commercial applications. A genetic, environmental, and agronomic factor influences secondary plant metabolite content (Juric et al., 2020). Today's economy has led to a significant intake of caloric intake processed bottled and packaged meals. Due to the excessive consumption of these energy-dense foods, we are losing a significant amount of nutrients from our diets. Various chronic diseases, such as overweight, obesity, CVD, diabetes, stroke, hypertension, and some kinds of cancer, are caused by unhealthy diets, which cause a significant nutritional gap (Swallah et al., 2020). The normal diet involves the ingestion of certain plant secondary metabolites that are thought to have health-promoting qualities against various metabolic syndrome events (Cisneros-Zevallos, 2021). To improve the concept of nutritionally functional food, it is critical to know biologically active molecules, identify prevention and protection strategies, and confirm the efficacy of these molecules in a large-scale human study. While determining the bioavailability of nutrients, whether supplied as nutraceuticals or enriched foods, is not specified, it is recommended that potential harmful components, such as allergens, be identified. Several studies suggest that the antioxidant activities of fruits vary significantly based on genotype, species, and growth stage, which can determine their nutritional value (Lavechhia et al., 2013). The phytochemical composition of fruits and vegetables is the first step in identifying their possible health benefits. Fresh veggies, like fruits, are appealing due to their high concentration of bioactive components (Sharma et al., 2020c). Due to the general quantity and intensity of their bioactive components, vegetables from the Brassicaceae family have acquired considerable interest in recent years, and this attention has been justified. It was expected that consumption of 150–200 g of these vegetables from conventional and organic agriculture would provide the same amount of Iron, Copper, Fe, Mn, and Zn recommended in the dietary guidelines due to the fact that organic agriculture does not reduce heavy metal contamination or improve nutritional quality of Brassicaceae (Sharma et al., 2020d). Only slight variations in heavy metal (Cd, Ni, and Pb) permissible intakes were identified in between two cropping methods (mainly for Pb). There are several factors, including genetic source, environmental conditions, and vegetable cultivation methods that can affect the amount of these chemicals in Brassica plants (e.g., phenols, vitamin C, minerals, glucosinolates (GSs)). Although the nutrient quality of fruit and vegetables depends on a variety of factors, the product's storage after harvest has the greatest impact on loss and degradation of phytochemical ingredients (Ezzat et al., 2020). Fruits and vegetables have been found to be nutritionally beneficial in several studies published recently. In addition, fruit and vegetable-rich diets combined with medical therapy are being studied as potential treatment methods, since phytochemicals activate genes and pathways targeted by pharmaceuticals. It is therefore possible that phytochemicals found in fruits and vegetables could treat a wide range of diseases. The chapter summarizes some of the latest findings on the phytochemical composition of fruits and vegetables, as well as their health-promoting effects and mechanisms in various models, including animal and human trials.

1.2 SECONDARY METABOLITES EXPRESSION

Climate/season, edaphic conditions, and the interaction of many other plants as well as other living organisms may influence the presence or absence of specific secondary metabolites in herbal medicines (Rane et al., 2020a–c). The association between plants and insect flora is another factor that can influence secondary metabolite biosynthesis in plants. Plants that produce a fragrant perfume (essential oils (EOs)) to attract insects and birds and produce nectar and pollen as food for these pollinators (Sharma et al., 2021). Many flowers contain honey or pollen, both of which are common foods for insects and birds. Bees and butterflies visit flowers to flowers, extracting honey or pollen or both. They also carry pollen on their parts of the body during this activity, which benefits in pollination when visiting other plants. Numerous flowers have structurally modified floral sections to regulate insect pollination. These insects can also spawn on plants. They deposit millions of eggs, and the larvae that hatch from these eggs feed on the plant's leaves, sometimes completely ruining the plants due to overfeeding. During the course of evolution, plants began to produce some toxic materials, causing a huge percentage of the feeding larvae to be killed. In contrary, the bug began to develop resistance, allowing many larvae to survive (Sharma et al., 2021). It is producing a growing number of harmful substances. This became identical to the interaction between plants and insects, which resulting in the evolution of numerous chemical substances, mostly secondary metabolites, in both plants and insects. The evolutionary basis of crossbreeding was certainly nature's mechanism for reshuffling genes in order to produce new varieties. Similarly, abiotic factors had an effect on the plants, and the plants responded by producing numerous substances (Sharma et al., 2020c). The desert pressures the plants in extreme drought conditions, and the plants respond by producing substances that help them protect themselves from the desert's stress. The presence of EO components, mainly monoterpenes and sesquiterpenes, generates the aroma of sandalwood. The distinctive fragrance components are only fully formed in the sandalwood trees that grow in certain Karnataka Forest areas. The aroma of sandalwood grown in various parts of India or around the world differs from the ingredients in the relevant chemicals (Mohiuddin, 2019).

1.2.1 ROLE OF SECONDARY METABOLITES IN PLANTS AND IN HUMANS

There are a number of plant secondary metabolites that have bioactive properties and are part of the plant's system that adapts to its environment and responds to potential biotic and abiotic problems. Abiotic stress includes wounding, light, temperature, gas composition, and chemical elicitors, while biotic stress includes challenges from microorganisms, insects, and animals in general. These compounds fall into many categories, with the most prevalent being phenolics, terpenoids, and alkaloids, depending on the plant species. Some rather compounds are contributed to the formation through a broad range of biological pathways that are popularly controlled by a signal transduction pathway that involves, among several other things, signals, transposable elements, the expression of genes associated with the enzymes that regulate the pathways, potential genes that directly influence the expression of other genes, and cross-talks between these pathways at the molecular and physiological levels. Plants have evolved a highly regulated mechanism for synthesizing these bioactive substances. When the plant is not stressed, and is said to be in homeostasis, and redox status appears to be one of the important signal components that allows the company to function properly. Although the preceding explanation makes it quite clear that plants produce active molecules for their own purposes, it is also well known that humans have been using these same secondary metabolites to boost their health for ages (Zevallos et al., 2017). Certain bioactive compounds play a significant role in the odor, color, and taste of plant components. These chemical molecules were previously assumed to be functionally insignificant, therefore plant biologists paid no attention to them. However, organic chemists have been investigating their chemical structures and properties since the 1850s. Secondary metabolites are now known to play an active and vital role in possible defense mechanisms, particularly in chemical agents amongst plants and their pathogens. Due to their potential applications in human nutrition, cosmetics, and medications, as well as their role in plant defense, secondary metabolites have attracted the attention of scientists in recent decades. This significant increase in interest in secondary metabolites is both intellectual and commercial in character (Ahmed et al., 2017).

1.2.2 METABOLITES IN SOIL

Soil quality refers to the ability of soil to support plant and animal life both below and above ground, as well as ecosystem health. There is a close correlation between the physiochemical and eco-biological properties of soil. Physical structure influences the chemical composition and soil microbiome, which in turn dictates biological activity. In physical properties, cation exchange capacity, pH, salinity, and macro- and micronutrients are taken into account, and in chemical properties, cation-exchange capacity, texture, and porosity are taken into consideration. Plant biodiversity decreases when agricultural systems reduce plant biochemical variety and soil biological diversity. Plants influence their surroundings in a variety of ways, including interactions with the rhizosphere soil microbiota, soil nutrient cycling, allelopathy, and herbivore defenses. Plants expel specific PSMs that boost their ability to absorb nutrients from the soil. The seeds and root of Alfalfa (*Medicago sativa*) contain flavonoids that suppress the growth of *Sinorhizobium meliloti*, an N-fixing gram-negative bacterium (Clemenson et al., 2020).

1.2.3 ROLE OF PLANT SECONDARY METABOLITES IN COMBATING PEST INDUCED STRESS IN BRINJAL

Phenolic acids, flavonoids, lignans, stilbenes, and tannins are all members of the phenolics class, and their complexity ranges from monomeric phenolic acids to complex polymers like tannin. The most prevalent type of antioxidant supplement in our diet is phenolics found in fruits and vegetables. Secondary metabolites can take many different forms. Terpenoids and steroids are a few examples, as are alkaloids, fatty acid-derived substances and polyketides, non-ribosomal polypeptides, and enzyme cofactors. Many natural antioxidants, including polyphenols, are abundant in brinjal, the most prevalent of which is acid (CGA). There is also crypto chlorogenic acid (4-O-caffeoylquinic acid) and neochlorogenic acid (3-O-caffeoylquinic acid), albeit in lower concentrations. Apart from its antioxidant capabilities. CGA has anti-inflammatory, anti-mutagenic, and anti-proliferative properties. One of the most significant antioxidants in our diet is GA. Brinjal is well-known as the primary vegetable source of CGA, accounting for more than 75% of total brinjal phenolics (Talukder et al., 2021).

1.2.4 PLANT POLYPHENOLS

Polyphenolic compounds are rich in bioactive compounds that protect against UV radiation and pathogenic infection. Polyphenols are naturally present chemicals included in a wide range of food products, including fruits, vegetables, cereals, and beverages. Grapes, apples, pears, cherries, and berries, for example, provide 200–300 mg polyphenols per 100 gm fresh weight. In food, polyphenols can assist in bitterness, pungency, color, flavor, odor, and oxidative stability. Clinical studies and related meta-analyzes towards the end of the 20th century clearly indicated that long-term intake of plant polyphenol-rich diets gave some protection against the development of malignancies, cardiovascular illnesses, diabetes, osteoporosis, and neurological disorders. Polyphenolic compounds and other dietary phenolics are receiving attention from the scientific community due to their remarkable therapeutic properties (Pandey & Rizvi, 2009). In terms of health-promoting properties and functions, phenolic compounds are among the most common phytochemical classes. Phenolic compounds are secondary metabolites found across the plant kingdom that have complicated structures and activities. It is considered to be the most important and abundant chemical class in the plant world. The presence of several antioxidants helps the protective mechanisms by these fruits against diseases. Recent research has shown that polyphenolic compounds are preferably the phytochemicals with both functional and health-promoting qualities found in the plant world, and a lack of them may increase the risk of disease. Phytonutrients are powerful antioxidants that can influence cellular metabolism as well as carcinogenic detoxification. In addition to aromatic rings and one or more hydroxyl methyl groups, phenolic or polyphenolic compounds also produce functional by-products (esters, methyl ethers, glycosides, etc.). Their molecular structures may differ from composite high-molecular-mass polymers. A number of studies support the belief that eating fruits on a daily basis contributes significantly to a healthy diet, and eating fewer may increase the risk of some chronic illnesses such as cancer, heart disease, and stroke. Vegetables are important sources of minerals, thiamine, vitamins, niacin, pyridoxine, folic acid, iron, dietary fiber, magnesium, malic acid, tartaric acid, calories, and chlorogenic acids. Some fruits and their products have a significant number of antioxidants in their phytochemical components (phytonutrients), which might affect metabolic activities and detoxify carcinogens, potentially regulating the reasons that influence cell tumors (Swallah et al., 2020). In comparison to other forms of plant bioactive molecules, polyphenols in the diet have also been shown to minimize endothelial dysfunction, lower blood pressure and cholesterol, and regulate energy metabolism (Sreenavasulu & Fernie, 2022).

1.2.5 UV-B-INDUCED SECONDARY METABOLITES

Many ultraviolet investigations have concentrated on measuring secondary plant metabolite quantities in plants subjected to larger concentrations of ultraviolet Rays hitting the earth's surface as a function of stratospheric ozone depletion. Experiments have shown even low, ecologically relevant UV-B levels produce significant changes in the plant's biosynthetic pathways, resulting in a concentration of predominantly phenolic compounds, carotenoids, and GSs. As the general populace has grown more aware and concerned about food quality, one interesting idea is to exploit UV-B-induced metabolic changes in fruits, vegetables, and herbs to meet the customer interest in natural health-promoting packaged foods. UV-B also activates wound and defense signaling pathways, which could also result in the accumulation of secondary plant metabolites participating in the plant anti-herbivore defense system, such as GSs and phenolic compounds (Mackerness et al., 1999; Shreiner et al., 2012).

1.2.6 FLAVONOIDS

Flavonoids are naturally occurring phenolic substances found in fruits, vegetables, cereals, bark, roots, stems, flowers, tea, and wine. Even when flavonoids were extracted as useful chemicals, these naturally occurring substances were recognized for their therapeutic properties. Following that, the association between flavonoid ingestion and long-term consequences on mortality was examined, and it was discovered that flavonoid intake is negatively connected with risk of coronary heart disease (CAD). Flavones have a planar structure due to a double bond in the central aromatic ring. Probably one of the best flavonoids is quercetin. Quercetin levels are very high in onions, apples, broccoli, and berries. The second group is composed of flavanones, which are predominantly observed in citrus fruit. Naringin is an example of a flavonoid from any of these classes. Catechin-containing flavonoids could be found in green and black tea, as well as red wine, whereas anthocyanins can be found in strawberries and other berries, grapes, wine, and tea. Antiviral, anti-inflammatory, anticancer, anti-thrombogenic, and antiosteoporosis activities are among treatment outcomes (Nijveldt et al., 2001).

1.2.7 TERPENOIDS

Terpenoids (terpenes) are compounds found in plant resin glands, or trichomes, that are responsible for the scent of a plant. Their interaction with cannabinoids, known as the "entourage effect," contributes to a cannabis strain's specific effect. The phytoconstituents of plants around the world have already been identified, and cannabis naturally contains 250 non-cannabinoid terpenoids and 145–150 cannabinoid terpenoids. Approximately 50% of marijuana smoke and vapor consists of terpenes, while 10–20% is cannabinoids and 10–30% is other (less potent) terpenes. The bulk of commonly occurring terpenoids was found in many different plants. ß-Caryophyllene, for example, can be found in plants as diverse as black pepper, *Cannabis sativa* strains, and rosemary (Singh et al., 2018). Terpenoids have been identified to protect cells from both biotic and abiotic stresses. Terpenoids, which stimulating photorespiration, can also protect plants from UV-induced and oxidative damage. Carotenoids are the most well-known terpenoids involved in antioxidant protection. Protective terpenes in plants include triterpene glycosides (or saponins) such as – tomatine in tomato fruits and leaves and avenacin in oat (Avena sativa) roots. Avenacin and -tomatine are phytoanticipins, which are pre-formed active compounds that prevent against microbial attack (Yeshi et al., 2020).

1.2.8 LIGNIN

Lignin is among the most extremely important metabolites produced by the phenylalanine/tyrosine biosynthetic pathway in plant cells. It is also the second most prevalent natural polymer, providing for 30% of the biosphere's organic carbon concentration. Lignin and its metabolites play a vital role in plant growth and development. Lignin, a synthetic phenolic polymer, improves the mechanical properties and surface chemistry of plant cell walls while significantly increasing mineral transport through plant vascular tissue. Furthermore, lignin is a major factor against pests and pathogens. Lignin synthesis might even have a function in plant lodging resilience and stress response. Lignin can alternatively be employed as a resource in the energy or medicinal industries. Lignin development in the cell wall enhances the mechanical resilience of plant stems significantly. It has significant consequences for crop lodging resistance (Liu et al., 2018).

1.2.9 SAPONINS

Several Saponins have physiological and pharmacological effects that include cardio protection, immunomodulation, antifatigue, and hepatoprotection. Saponins' antifungal activity is largely attributed to their capacity to combine with sterols in fungal membranes, producing hole development and membrane loss. They even have an influence on permeability. Dietary saponins have been shown to lower cholesterol, activate the immune system, and limit tumor growth. Saponins limit active transport by increasing overall enterocyte permeability. Saponins can form insoluble compounds with micronutrients such as zinc and iron (Poiroux-Gonord, 2010).

1.3 BRASSICALES

The *Brassicales* is classified by GSs, a type of secondary plant metabolite. *Brassicales* plants produce GSs, which are important in the global market due to their overall health benefits and the cancer-preventive characteristics of their metabolic by-products. When a plant cell is injured, the natural enzyme myrosinase enzymatically eliminates GSs, creating a variety of oxidation products such as nitriles, epithionitriles, and/or isothiocyanates. It is worth noting that isothiocyanates have been connected to the pungency of these veggies and have been demonstrated to have anti-cancerogenic, anti-inflammatory, and anti-diabetic properties (Neugart et al., 2018).

1.3.1 ALKALOIDS

Alkaloids are a diverse collection of compounds present in Flora And fauna all over the world. All of them seem to be secondary compounds and assemblage of various elements and macromolecules formed from amino acids or bond formation. The amino acids that provide their nitrogen atom and structure are being used to categorize this chemical manufacturing category. Alkaloids with the same origin or basic nucleus may have different biosynthetic pathways and biological activities. Alkaloids are chemical compounds that occur naturally and usually contain basic nitrogen atoms. They could contain a chemical that is neutral or even somewhat acidic. A few synthetic chemicals are also categorized as alkaloids. Alkaloids may contain sulfur, along with carbon, nitrogen, or hydrogen, and, in exceptional instances, bromine, phosphorus, or chlorine. Carl F. W. Meissner, a German scientist, coined the concept "alkaloid" in 1819 and derived it from the Arabic name 'al-qali,' which is associated with the plant from which soda was first extracted. Alkaloids are low-molecular-weight molecules that amounts to approximately 20% of plant-based secondary metabolites. So far, over 12,000 alkaloids have been isolated from various plant genera (Dey et al., 2020) (Table 1.1).

TABLE 1.1 β-Carotene, Lycopene, and Ascorbic Acid and Their Pharmacological Potential

Sl. No.	Plant Name	Part Used	Active Constituents	Activity	References
1.	Araza fruit (*Eugenia stipitata* mcvaugh)	Pulp	Vitamin C amount 44 mg/100 gm.	Strong antioxidant activity.	Khomdram & Devi (2010)
2.	Amla (*Emblica officinalis*)	Pulp	Vitamin C content 379.7 mg/100 gm	Strong antioxidant activity.	Khomdram & Devi (2010)
3.	Beet root (*Beta vulgaris*)	Root	Total carotenoid 0.56 μg/g.	Suppress the breast cancer cell lines (MCF 7).	Choo & Sin (2012)
4.	Carrot (*Daucus carota* L.)	Root	Total carotenoid 0.96 μg/g.	Anticancer activity against cell lines of human breast cancer (MCF 7).	Choo & Sin (2012)
5.	Mango (*Mangifera indica* L.)	Pulp	β-carotene from 661.27–2,220 μg/100 g, ascorbic acid – 9.79 to 77.71 mg/100.	Antioxidant	Ancilla & Sumathi (2016)
6.	Green spinach (*Sauropus androgynous* L.)	Whole green leafy vegetable.	Total carotenoid 0.96 μg/g.	Human breast cancer cell lines (MCF 7).	Choo & Sin (2012)
7.	Red spinach (*Amaranthus dubius* Mart. Ex Thell.)	Red spinach	Total carotenoid 0.88 μg/g.	Anticancer activity against Human breast cancer cell lines (MCF 7).	Choo & Sin (2012)
8.	Red fleshed watermelon (*Citrullus lanatus*)	Whole fruit	Ascorbic acid content (86.32 mg/kg) and lycopene (9.50 mg/kg) contents.	The anti-radical power and ferrous ion chelating activity.	Sun et al. (2012)
9.	Yellow fleshed watermelon (*Citrullus lanatus*)	Whole fruit	Ascorbic acid content (52.05 mg/kg) and lycopene (0.04 mg/kg).	Antioxidant activity.	Sun et al. (2012)
10.	Tomato (*Solanum lycopersicum* L.)	Whole fruit	Lycopene content (3.05–11.77), ascorbic acid (12.72–20.41).	Antioxidant and anticarcinogen against Cervical HeLa, Metastatic prostate DU 145.	Meng (2016)

TABLE 1.1 *(Continued)*

Sl. No.	Plant Name	Part Used	Active Constituents	Activity	References
11.	Yellow guava (*Psidium guajava*)	Pulp, leaves	Vitamin C content (228.3 mg), β-carotene (374 mg).	*Ex vivo* myeloid leukemia blasts, Cervical HeLa, Metastatic prostate DU 145, Human mouth KB.	Ooe et al. (2016)
12.	Japanese blue tomato (*Solanum lycopersicum*)	Peel, seed, pulp	Lycopene (1.13, 0.05, 0.68 mg/gm) and β-carotene in seed 0.02 mg/gm.	Antioxidant activity and prevention of eye diseases.	Khan et al. (2011)
13.	Red berries (*Rivina humilis* L.)	Whole fruit	Lycopene	Strong antioxidant activity.	Abou-Elella & Mourad (2015)
14.	Pumpkin (*Cucurbita maxima*)	Pulp	β-carotene 3,100 mg, vitamin C content 9 mg.	Strong antioxidant, cardiovascular disorder, anti-diabetic, Hepatoprotective, breast cancer MCF-7, colon cancer CT-26, prostate cancer RM-1.	Garzon et al. (2012)
15.	Mausami (*Citrus medica* L.)	Pulp	Vitamin C content 11.7 mg/100 gm.	Strong antioxidant activity.	Khomdram & Devi (2010)

1.4 FRUITS AND VEGETABLES HAVING INCREASED HEALTH POSSIBILITIES

Although several factors might support the development and prevalence of the disease, the contribution of some fruits and vegetables should indeed be evaluated.

1.4.1 BROCCOLI (BRASSICA OLERACEA)

Broccoli, like cauliflower, Brussels sprouts, kohlrabi, cabbage, and mustard, is a cruciferous vegetable that belongs to the Brassicaceae (previously known as Cruciferae) family. Before the flower buds develop, the flower heads (also known as broccoli, broccoli florets, or broccoli heads) are consumed (Buck, 1956). Broccoli's primary components are GSs, which are included in all members of the Brassicaceae family. Hydrogenation of GSs produces a number of biologically active products, including isothiocyanates, thiocyanates, nitriles, and oxazolidine-2-thiones. These products have always had an impact on various biological processes via influencing transcription factor levels, signaling pathways, the cell cycle, and apoptosis. Broccoli is enriched in isothiocyanate sulforaphane and other GSs derivatives considered health benefits (Zhang et al., 1992). Broccoli also contains a number of plant secondary metabolites, particularly flavonoids, 2-caffeoylquinic acid derivatives, and cinnamic acid derivatives. Ascorbic acid, xanthophylls (lutein, zeaxanthin), carotenoids, tocopherols, proteins (Keservani et al., 2015; Jarouliya & Keservani, 2019b), carbohydrates, and chlorophyll is also present. Broccoli is also high in minerals like selenium and calcium, as well as fiber, vitamins, and amino acids. Broccoli has been widely recognized not just for its flavor, but also for its therapeutic properties (Latte et al., 2011). Sulforaphane, an isothiocyanate pharmacologically active glucoraphanin (Grn) metabolite, has been recognized as the compound essential for many of broccoli's beneficial properties. Grn, also described as sulforaphane GS, is a long-acting antioxidant and the most efficient occurs naturally inducer of phase 2 detoxification enzymes (Zhang et al., 1992). Moreover, investigations have already shown that broccoli phytochemicals can protect the eyes from retinitis pigmentosa and fight inflammation, synovial hyperplasia, and bone deterioration in animals with rheumatoid arthritis (Tanito et al., 2005; Kong et al., 2010).

1.4.2 DANDELION (*TARAXACUM OFFICINALE*)

Dandelion is a perennial herbaceous plant that is associated with artichokes and the Composite family. The leaves of this plant are eaten fresh as a salad, whereas the roots are roasted and being used as a coffee alternative (Schutz et al., 2005). Flower extracts are however employed as artificial sweeteners in a wide range of food products, including carbonated beverages, frozen dessert desserts, baked goods, puddings, and cheese. Dandelion has traditionally been used as herbal medicine to treat gastrointestinal problems, hepatobiliary abnormalities, hyperlipidemia, dyspepsia, rheumatism, and renal illnesses, in addition to being used as food. The high phenolic content of dandelion, particularly hydroxycinnamic acid derivatives, and flavonoids such as chlorogenic and chicoric acids, luteolin, and quercetin glycosides are connected to its health-promoting potential. Besides from all these secondary metabolites, dandelion contains inulin, a Compositae carbohydrate, minerals, as well as certain amino acids (Williams et al., 1996; Schutz et al., 2005; Gonzalez-Castejon et al., 2012).

1.4.2 GARLIC (*ALLIUM SATIVUM*)

Garlic is a member of the Liliaceae family. Onions, shallots, leeks, and chive are close relatives. Many communities have used garlic for therapeutic and diagnostic purposes for thousands of years (Simonetti, 1990). It contains additional sulfur compounds than any other Allium species, particularly diallyl sulfate, alliin, ajoene, and allicin, as well as several enzymes, amino acids, and minerals like selenium (Newall et al., 1996). Sulfur compounds are considered for garlic's pungent odor along with several of its medicinal properties (Block, 2010). Fresh garlic has been shown in clinical trials to have substantial antimicrobial properties (Sohn et al., 2009). Moreover, research has indicated that garlic has anti-inflammatory properties, anti-cancer qualities, and positive health benefits on numerous cardiovascular risks (Block, 2010; Sobenin et al., 2010). Numerous epidemiological studies have found a link between increased garlic consumption and a lower risk of various types of cancer (Fleischauer & Arab, 2001). An analysis of data from seven population studies shows that the more garlic taken, the lower the risk of stomach, esophagus, pancreatitis, and colorectal cancer (Tanaka et al., 2004; Chan et al., 2005).

1.4.3 COCOA (THEOBROMA CACAO)

Coffee beans (Kesharwani et al., 2016) and their compounds, such as chocolate, are full of antioxidants, especially catechins, epicatechins, and the polymer procyanidins. In combination with these phytochemicals, cocoa includes flavonoids and xanthine, theobromine is perhaps the most prevalent of the latter, which both possess vasodilatory qualities (Hooper et al., 2012). Cocoa as a plant and chocolate as food may have beneficial health effects against oxidative stress and inflammatory processes, as well as rectifying cardiovascular disorders, cancer, as well as other pathological conditions, due to the high composition of this succession of phytochemicals with a variety of biological processes (Ramos-Romero et al., 2012). Many clinical and experimental research suggests that consuming cacao frequently reduces the risk of CVD and many of its risk factors (Corti et al., 2009). Dietary chocolate has been shown to enhance hypertensive, inflammatory, atherogenic, and thrombotic health problems, as well as insulin sensitivity and vascular endothelial function (Buijsse et al., 2010). According to a recent meta-analysis, higher levels of chocolate consumption are correlated with a one-third reduction in the incidence of cardiometabolic diseases (Desch et al., 2010). This beneficial relationship was considerable for any CVD (37% reduction), diabetes (31% reduction), and stroke (29% reduction). There are very few observational studies and minimal information that correlate the incidence of cancer to cocoa used. Although some intervention studies have reported changes for the better in biomarkers evaluating antioxidant capacity despite their short duration, there is very little information related to cancer markers. However, these metabolic investigations demonstrated that regular consumption of cocoa and its derivatives increase plasma antioxidant levels, which may be a favorable effect because it protects against reactive oxygen species (ROS) (Weisburger, 2001).

1.4.4 OLIVES (OLEA EUROPAEA)

A variety of medicinal and gastronomic uses are associated with the olive tree. *Olea europaea*, native to parts of the Mediterranean and Central Asia. Olives have traditionally been used as moisturizers, laxatives, nutrients, sedatives, and tonics (Belarbi et al., 2011). Traditional therapies include colic, baldness, paralysis, rheumatic pain, sciatica, and hypertension. Because raw olives are incredibly bitter and inedible, they must be prepared before consumption (Waterman & Lockwood, 2007). They can be collected when it was still unripe (green olives) or even when fully ripe (black olives). To manufacture olive oil, ripe olives are processed. *In vitro* and animal research have showed that olives have significant antioxidant activity, which may account for several of the Mediterranean diet's recognized beneficial qualities (Kawaguchi et al., 2011). Although the structure of olives is complicated, the primary groups of chemicals thought to contribute to the observed health benefits include oleic acid, phenols, and squalene (Owen et al., 2000), all of which have been proven to reduce oxidative stress (Visioli & Bernardini, 2011). Significant epidemiological studies have suggested that daily consumption of olive and derivative products minimizes the chances of some illnesses, including breast, skin, and colon cancer (Owen et al., 2000). The prevention of DNA damage, which inhibits mutagenesis and carcinogenesis, is one mechanism associated with the anti-cancer benefits of hydroxytyrosol and oleuropein. Moreover, oleuropein has been reported to always have hepatoprotective characteristics (Domitrovic et al., 2012).

1.4.5 GRAPES (VITIS VINIFERA L.)

The health effects of grapes on CVD risk, notably endothelial function, LDL oxidation, atherosclerosis development, and oxidative reducing stress, have been intensively investigated (Weseler et al., 2011). Moderate grape eating has been shown to lower mortality (Vislocky & Fernandez, 2010). Numerous *in vitro*, animal, and human research have shown that grape polyphenols are beneficial to cardiovascular disorders such as heart disease and diabetes (Chuang & McIntosh, 2011). Several investigations also have demonstrated whether chemoprevention is beneficial. Data on skin cancer prevention are extremely powerful (Neves et al., 2012). Grapes have also been shown to have beneficial effects on oral health, immunological function (Keservani et al., 2015a), anti-inflammatory, and anti-allergic (Lee et al., 2012) systems. Further studies suggest that grape phenolic compounds possess antibacterial and antiviral activities. New and interesting results from independent research showed that grapes can assist with other deadly diseases including Alzheimer's and age-related cognitive decline (Wang et al., 2012). Numerous studies also have suggested whether prevention and treatment are beneficial. Data on skin cancer prevention are particularly strong (Neves et al., 2012). Grapes have been shown to have beneficial impacts on oral health, immunological function, and anti-inflammatory, and anti-allergic (Lee et al., 2012) systems. Recent research indicates that grape phenolic compounds possess antimicrobial and antioxidant properties. New and exciting data from individual research show that grapes can assist with other devastating diseases including Alzheimer's and age-related cognitive decline (Wang et al., 2012).

1.5 VEGETABLES

There were only a few investigations on vegetable antioxidant activity. Garlic, broccoli mushroom, white cabbage and cauliflower, kidney, pinto beans, beans, beet, and corn have been shown to have high antioxidant activity (Kahkonen et al., 1999). There are also many other vegetables that are rich in antioxidants, such as spinach, kale, broccoli, beets, red bell peppers, onions, corn, eggplant, cauliflower, and cucumbers (Prior & Cao, 2000). Onion, kale, tomato, and certain lettuce cultivars have large concentrations of quercetin (Hertog et al., 1992). Kaempferol levels remain elevated in kale, broccoli, and endive. Garlic extracts are also being investigated as possible cardiovascular and anticancer medications. When garlic cloves are mashed, allicin is formed, which gradually decomposes to form sulfur-containing molecules possessing chemo-preventive potential. Mushrooms, white cabbage, cauliflower, and garlic have been demonstrated to have potent anti-disease properties (Gazzani et al., 1998). Flavonoids have been associated with the greatest numbers in Allium vegetables. Onion contains natural flavanols, which have been demonstrated to scavenge free radicals (Suh et al., 1999). Because it is deficient in antioxidants to commence with, the brain may be more vulnerable to the harmful effects of free radicals. A diet rich in vegetables may assist with age-related neurodegenerative diseases (NDs) such as Parkinson's and Alzheimer's (Okuda et al., 1992). There has been a recent study in the United States that has demonstrated that rats given blueberry, strawberry, or spinach extracts developed fewer motor abnormalities as they aged, and they showed better memory performance when compared to their study counterparts. Furthermore, the groups that received these supplements exhibited indicators of vitamin E present in their brains. Spinach is believed to be essential for preventing memory loss and Alzheimer's disease (AD), but whether the results will remain constant in humans remains to be determined. The phytochemicals in these extracts are thought to have properties that enhance cell membrane permeability, allowing vital nutrients and chemical signals to move in and out of the cell and so lowering inflammatory processes in the tissues (Clarke, 1999). Recently, special attention has been paid to edible vegetables that are substantial in plant secondary metabolites that prompt antioxidant properties enzymes (e.g., glutathione-S-transferase, quinone reductase, and epoxide hydrolase), which inactivate reactive carcinogens by destroying their reactive centers or conjugating them with endogenous ligands, triggering their elimination from the body (Talalay, 1992). This inducer activity seems to be mostly due to the highly reactive thiocyanates in crucifers (e.g., broccoli, cabbage, cauliflower, etc.). In a rat mammary tumor model, isothiocyanate, a sulforaphane present in broccoli, has been shown to have an anticarcinogenic effect (Zhang et al., 1994). GSs, which are believed to be anticarcinogenic, are very stable precursors of isothiocyanates, which are commonly found in very high quantities in crucifers, and their degradation by myrosinase is necessary for the reported effects (Williamson et al., 1998). Other phytochemicals in green leafy vegetables that have already been associated with antioxidant properties include fat-soluble antioxidants such vitamin E, coenzyme Q10, lycopene, b-carotene, and a-carotene, as well as water-soluble antioxidants including vitamin C.

1.5.1 CAROTENOIDS, VITAMIN E, AND VITAMIN C

Several carotenoids, particularly b-carotene, lycopene, lutein, and zeaxanthin, were shown to have antioxidant activity, but b-carotene has gained the most interest. The antioxidant vitamins are vitamin C, vitamin E, and beta-carotene combined simultaneously (Table 1.1). All function together again and independently to prevent or delay oxidative responses that contribute to degenerative diseases such as cancer, CVD, cataracts, and many others (Elliot, 1999). A significant proportion of the world's population does not consume enough fruits and vegetables to consume sufficient levels of antioxidant vitamins to avoid oxidative damage. Some studies have found that smokers who consume a high-carotenoid diet have a lower risk of developing the disease than those who consume a low carotenoid diet. There has also been evidence that diets high in carotenoids may lower breast cancer risk. Moreover, vitamin C has been shown to inhibit the development of N-nitroso compounds, which are cancer-causing chemicals derived from nitrates and nitrites found in conserved foodstuffs and some drinking water.

1.5.2 KAEMPFEROL, QUERCITIN, AND LUTEOLIN

The DPPH (2,2-diphenyl-1-picryldrazyl) and Lipid peroxidation assays indicate that the leaf has the highest antioxidant capacity. The DPPH assay reports an IC_{50} (half-maximal inhibitory concentration) value of 49.86 g/ml.

Lipid peroxidation LPO activity measured in the plant's leaf was 23.715 mols MDA g-1DW14. (iii) *Duranta erecta*: The GC MS analysis of *Duranta erecta* reveals the presence of 14 compounds that are found in all parts, including Benzene 1,4-diol, Trans-cinnamic acid, 1,2-Benzene-dicarboxylic acid, and many compounds that are similar to flavonoids, including 2H-1-benzopyran-2-one,6,7-dimethoxy-, 2(4H Fruits and vegetables are high in natural antioxidants and have high levels of physiologically active components, which provide therapeutic benefits beyond simple nourishment (Oomah & Mazza, 2000). They are a significant source of dietary antioxidants, which enhance plasma antioxidant capacity and therefore also prevent atherosclerosis-related disorders in humans (Cao et al., 1998). Intake of fruits and vegetables has therefore been linked with lower cancer incidence and mortality in various human groups and case-control studies for all prominent cancers. Experiments with cells revealed that vegetables have antitumorigenic properties. Total fruit and vegetable intake, cardio- and cerebrovascular diseases, and death all have a significant unfavorable connection. Vegetarians and nonvegetarians who eat a lot of fruits and vegetables have lower blood pressure. D-limonene, epigallocatechin gallate, soya protein, isoflavones, vitamins A, B, C, E, tocopherols, calcium, selenium, chlorophyllin, alipharin, sulfides, catechin, tetrahydro curcumin, seas aminol, glutathione (GSH), uric acid (UA). These chemicals may operate as anticancer or cardioprotective agents independently or in combination through a variety of mechanisms. The antioxidant compounds mentioned above are some of the most important natural antioxidants found in fruits and vegetables that have been proved to have an essential role in disease prevention.

1.6 PROSPECTS AND SCOPE

The protective effects of phytochemicals, as well as their potential to be included in foods or dietary additives as nutraceuticals, have already had a substantial effect on the food industry, according to a recent study (Dillard & German, 2000). A Traditional meal is now providing a significant challenge to the nutrition and agriculture industries. The food industry can help to increase nutritional density by providing designer foods that include phytonutrients as well as regular nutrients (proteins, fats, and carbohydrates). The concept of an antioxidant defense system in processed foods is fast spreading and quickly becoming an important factor for determining product quality. With the development of industrialization, competition in the market among worldwide firms will soon secure the antioxidant capacity parameter's place in nutritional branding. The development of a practical way of measuring antioxidant activity for industrial applications will be critical in this context. This will enhance the production of fruits and vegetables as well as the production of functional ingredients and refreshments. The quantity of information currently available is recommended to assess natural antioxidant content or examine its capacity to slow lipid oxidation in accelerated testing. They do not provide a comprehensive view of a meal's overall antioxidant capacity as a result of chemical action via several processes such as chain breaking, oxygen scavenging, and metal chelation. Overall antioxidant activity in foods should thus be determined by monitoring these properties as well as their relative importance in the food product. This will benefit immensely scientists, food processors, and consumers, as the emerging developments are toward functional foods with specific health benefits. A better understanding of the role and fate of natural and process-induced antioxidants, as selected by processing methods, on both food security and public health is required in research. Additionally, in order to draw conclusive results, research is needed to examine the beneficial impact of phytochemicals against neurodegenerative illnesses must be expanded to be included *in vivo* investigations in addition to *in vitro* studies.

1.7 CONCLUSIONS

The impact of nutrition on general health status has been known since ancient times when pathologic diseases were treated with plants and natural foods thought to have medicinal properties. Currently, epidemiological evidence has confirmed the association between vegetable and fruit diets and a lower risk of chronic diseases such as cancer and CVD. Increasing vegetable intake products also reduces the risk of obesity and enhances diabetes control. Moreover, data suggest that diets rich in vegetables decrease the risk of premature death. Besides, it being rich in vitamins and minerals, minerals, and fiber, fruits, and vegetables also contain a broad range of phytochemicals having strong bioactivity, the most well-known of which are polyphenols. Recently, consumers have been told that phytochemicals with antioxidant properties found in some foods may inhibit some of the processes involved in cancer growth. Epidemiologic studies on the impact of phytochemicals on human health, on either side, have been inconclusive. Even if there is significant proof, results must always be thoroughly examined and communicated in connection to human health benefits, because phytochemicals may have low bioavailability and may also be extensively metabolized. Furthermore, some phytochemicals can be toxic and mutagenic in some cell culture procedures. Furthermore, fruits and vegetables include a diverse range of metabolites that cannot be duplicated by highly purified phytochemicals. In this light, and until more is learned about phytochemical activity, absorption, and diffusion in the body, it appears reasonable to recommend fruits and vegetables in their entirety (minimum of 400 g/day) as dietary supplements rather than separate components.

KEYWORDS

- **alkaloids**
- ***Brassicales***
- **flavonoids**
- **lignin**
- **metabolites**
- **plant secondary metabolites**
- **saponins**

REFERENCES

Abou-Elella, F., & Mourad, R., (2015). Anticancer and antioxidant potentials of ethanolic extracts of *Phoenix dactylifera, Musa acuminata* and *Cucurbita maxima*. *Research Journal of Pharmaceutical, Biological and Chemical Sciences, 6*(1), 710.

Ahmed, E., Arshad, M., Khan, M. Z., Amjad, M. S., Sadaf, H. M., Riaz, I., & Ahmad, N., (2017). Secondary metabolites and their multidimensional prospective in plant life. *Journal of Pharmacognosy and Phytochemistry, 6*(2), 205–214.

Ancilla, S. B., & Sumathy, V. J., (2016). Anticancer activity of crude extract and carotenoid pigments from fruits. *European Journal of Pharmaceutical and Medical Research, 4*(5), 276–280.

Aune, D., Chan, D. S., Vieira, A. R., Rosenblatt, D. A., Vieira, R., Greenwood, D. C., & Norat, T., (2012). Fruits, vegetables, and breast cancer risk: A systematic review and meta-analysis of prospective studies. *Breast Cancer Res. Treat., 134*(2), 479–493.

Belarbi, M., Bendimerad, S., Sour, S., Soualem, Z., Baghdad, C., Hmimed, S., Chemat, F., & Visioli, F., (2011). Oleaster oil positively modulates plasma lipids in humans. *J. Agric. Food Chem., 59*(16), 8667–8669.

Bender, C., Straßmann, S., & Heidrich, P., (2021). Cellular antioxidant effects and bioavailability of food supplements rich in hydroxytyrosol. *Appl. Sci.*, *11*, 4763.
Bhaskarachary, K., Naveena, N., & Polasa, K., (2015). Potential benefits of plant metabolites for human health. *The Indian Journal of Nutrition and Dietetics*, *52*(2), 213–225.
Biondi, F., Balducci, F., Capocasa, F., Visciglio, M., Mei, E., Vagnoni, M., Mezzetti, B., & Mazzoni, L., (2021). Environmental conditions and agronomical factors influencing the levels of phytochemicals in *Brassica* vegetables responsible for nutritional and sensorial properties. *Appl. Sci.*, *11*, 1927.
Block, E., (2010). *Garlic and Other Alliums: The Lore and the Science* (1st edn.). Royal Society of Chemistry, Cambridge.
Boeing, H., Bechtold, A., Bub, A., Ellinger, S., Haller, D., Kroke, A., Leschik-Bonnet, E., et al., (2012). Critical review: Vegetables and fruit in the prevention of chronic diseases. *Eur. J. Nutr.*, *51*(6), 637–663.
Buck, P. A., (1956). Origin and taxonomy of broccoli. *Economic Botany, 10*(3), 250–253.
Buijsse, B., Weikert, C., Drogan, D., Bergmann, M., & Boeing, H., (2010). Chocolate consumption in relation to blood pressure and risk of cardiovascular disease in German adults. *Eur. Heart J. 31*(13), 1616.
Camara-Martos, F., Sevillano-Morales, J., Rubio-Pedraza, L., Bonilla-Herrera, J., & De Haro-Bailon, A., (2021). Comparative effects of organic and conventional cropping systems on trace elements contents in vegetable *Brassicaceae*: Risk assessment. *Appl. Sci.*, *11*, 707.
Cao, G., Booth, S. L., Sadowski, J. A., & Prior, R. L., (1998). Increase in human plasma antioxidant capacity after consumption of controlled diets high in fruits and vegetables. *American Journal of Clinical Nutrition*, *68*, 1081–1087.
Chan, J. M., Wang, F., & Holly, E. A., (2005). Vegetable and fruit intake and pancreatic cancer in a population-based case-control study in the San Francisco Bay area. *Cancer Epidemiol. Biomarkers Prev., 14*(9), 2093–2097.
Chatzopoulou, E., Carocho, M., Di Gioia, F., & Petropoulos, S., (2020). The beneficial health effects of vegetables and wild edible greens: The case of the mediterranean diet and its sustainability. *Appl. Sci.*, *10*, 9144.
Choo, W. S., & Sin, W. Y., (2012). Ascorbic acid, lycopene, and antioxidant activities of red-fleshed and yellow-fleshed watermelons. *Advances in Applied Science Research*, *3*(5), 2779–2784.
Chuang, C. C., & McIntosh, M. K., (2011). Potential mechanisms by which polyphenol-rich grapes prevent obesity-mediated inflammation and metabolic diseases. *Annu. Rev. Nutr., 31*, 155–176.
Cisneros-Zevallos, L., (2021). The power of plants: How fruit and vegetables work as source of nutraceuticals and supplements. *International Journal of Food Sciences and Nutrition*, *72*(5), 660–664.
Clarke, M. P., (1999). *Can Food Forestall Ageing*. Agriculture Research USDA.
Clemensen, A. K., Provenza, F. D., Hendrickson, J. R., & Grusak, M. A., (2020). Ecological implications of plant secondary metabolites-phytochemical diversity can enhance agricultural sustainability. *Frontiers in Sustainable Food Systems*, *4*, 547–826.
Clemenson, G. D., Stark, S. M., Rutledge, S., & Stark, C. E. L., (2020). Enhancing hippocampal-based memory in older adults through video games. *Behav. Brain Res.* 10.1016/j.bbr.2020.112667.
Corti, R., Flammer, A. J., Hollenberg, N. K., & Luscher, T. F., (2009). Cocoa and cardiovascular health. *Circulation, 119*(10), 1433–1442.

Desch, S., Schmidt, J., Kobler, D., Sonnabend, M., Eitel, I., Sareban, M., Rahimi, K., Schuler, G., & Thiele, H., (2010). Effect of cocoa products on blood pressure: Systematic review and meta-analysis. *Am. J. Hypertens*, *23*(1), 97–103.

Dey, P., Kundu, A., Kumar, A., Gupta, M., Lee, B. M., Bhakta, T., & Kim, H. S., (2020). Analysis of alkaloids (indole alkaloids, is quinoline alkaloids, tropane alkaloids). In: *Recent Advances in Natural Products Analysis* (pp. 505–567). Elsevier.

Dillard, C. J., & German, J. B., (2000). Phytochemicals: Nutraceuticals and human health. *Journal of the Science of Food and Agriculture*, *80*, 1744–1756.

Domitrovic, R., Jakovac, H., Marchesi, V. V., Sain, I., Romic, Z., & Rahelic, D., (2012). Preventive and therapeutic effects of oleuropein against carbon tetrachloride-induced liver damage in mice. *Pharmacol. Res.*, *65*(4), 451–464.

Elliot, J. G., (1999). Application of antioxidant vitamins in foods and beverages. *Food Technology*, *53*, 46–48.

Ezzat, A., Heged us, A., Szabo, S., Ammar, A., Szabo, Z., Nyeki, J., Molnar, B., & Holb, I. J., (2020). Temporal changes and correlations between quality loss parameters, antioxidant properties and enzyme activities in apricot fruit treated with methyl jasmonate and salicylic acid during cold storage and shelf-life. *Appl. Sci.*, *10*, 8071.

Fleischauer, A. T., & Arab, L. (2001). Garlic and cancer: A critical review of the epidemiologic literature. *J. Nutr.,* (3s), 1032–1040.

Folts, J. D., (2002). Potential health benefits from the flavonoids in grape products on vascular disease. *Adv. Exp. Med. Biol.*, *505*, 95–111.

Garzon, G. A., et al., (2012). *Journal of. Agricultural Food Chemistry, 60*(18), 4709–4717.

Gazzani, G., Papetti, A., Massolini, G., & Daglia, M., (1998). Antioxidative and pro-oxidant activity of water-soluble components of some common diet vegetables and thermal treatment. *Journal of Food Chemistry, 6*, 4118–4122.

Gonzalez-Castejon, M., Visioli, F., & Rodriguez-Casado, A., (2012). Diverse biological activities of dandelion. *Nutr. Rev.*, *70*(9), 534–547.

Gracian-Alcaide, C., Maldonado-Lobon, J., Ortiz-Tikkakoski, E., Gomez-Vilchez, A., Fonolla, J., Lopez-Larramendi, J., Olivares, M., & Blanco-Rojo, R., (2020). Effects of a combination of elderberry and reishi extracts on the duration and severity of respiratory tract infections in elderly subjects: A randomized controlled trial. *Appl. Sci.*, *10*, 8259.

Hertog, M. G. L., Hollman, P. C. H., & Katan, M. B., (1992). Content of potentially anticarcinogenic flavonoids of 28 vegetables and 9 fruits commonly consumed in the Netherlands. *Journal of Agricultural Food Chemistry, 40*, 2379–2383.

Hooper, L., Kay, C., Abdelhamid, A., Kroon, P. A., Cohn, J. S., Rimm, E. B., & Cassidy, A., (2012). Effects of chocolate, cocoa, and flavan-3-ols on cardiovascular health: A systematic review and meta-analysis of randomized trials. *Am. J. Clin. Nutr.*, *95*(3), 740–751.

Hsia, C. H., Tung, Y. T., Yeh, Y. S., & Chien, Y. W., (2021). Effects of *Gynura bicolor* on glycemic control and antioxidant ability in prediabetes. *Appl. Sci.*, *11*, 5066.

Jarouliya, U., & Keservani, R. K., (2019a). Pathways leading to child obesity; Chapter 12. In: Debasis, B., (eds.), *Global Perspectives on Childhood Obesity: Current Status, Consequences and Prevention* (2nd edn., pp. 137–146). Academic Press, Elsevier. ISBN: 9780128128404.

Jarouliya, U., & Keservani, R. K., (2019b). Protein function as cell surface and nuclear receptor in human diseases; Chapter 1. In: Sanjay, B., & Debarshi, K. M., (eds.), *Medicinal Chemistry with Pharmaceutical Product Development*, (pp. 1–32). Apple Academic Press, CRC Press, Taylor & Francis Group. ISBN: 9781771887106.

Juric, S., Sopko, S. K., Krol-Kilinska, Z., Zutic, I., Uher, S. F., Dermic, E., & Vincekovic, M., (2020). The enhancement of plant secondary metabolites content in *Lactuca sativa* L. by encapsulated bioactive agents. *Scientific Reports, 10*(1), 1–12.

Kahkonen, M. P., Hopia, A. I., Vuorela, H. J., et al., (1999). Antioxidant activity of plant extracts containing phenolic compounds. *Journal of Agricultural Food Chemistry, 47*, 3954–3962.

Kawaguchi, K., Matsumoto, T., & Kumazawa, Y., (2011). Effects of antioxidant polyphenols on TNF-a-related diseases. *Curr. Top. Med. Chem., 11*(14), 1767–1779.

Keservani, R. K., & Sharma, A. K., (2014). Flavonoids: Emerging trends and potential health benefits. *Journal of Chinese Pharmaceutical Sciences, 23*(12), 815.

Keservani, R. K., Kesharwani, R. K., Sharma, A. K., & Jarouliya, U., (2015a). Dietary supplements, nutraceutical, and functional foods in immune response (immunomodulators); Chapter 20. In: Debasis, B., Harry, G. P., & Anand, S., (eds.), *Nutraceutical and Functional Foods in Human Life and Disease Prevention* (343–358). Boca Raton, Fl. CRC Press, Taylor and Francis. ISBN: 9781482237214.

Keservani, R. K., Kesharwani, R. K., Sharma, A. K., Vyas, N., & Chadoker, A., (2010b). Nutritional supplements: An overview. *International Journal of Current Pharmaceutical Review and Research, 1*(1), 59–75.

Keservani, R. K., Kesharwani, R. K., Vyas, N., Jain, S., Raghuvanshi, R., & Sharma, A. K., (2010a). Nutraceutical and functional food as future food: A review. *Der Pharmacia Lettre, 2*(1), 106–116.

Keservani, R. K., Sharma, A. K., & Jarouliya, U., (2015). Protein and peptide in drug targeting and its therapeutic approach. *Ars Pharmaceutica (Internet), 56*(3), 165–177.

Keservani, R. K., Sharma, A. K., & Kesharwani, R. K., (2016a). Nutraceutical and functional foods for cardiovascular health. In: *Food Process Engineering* (pp. 257–278, 291–312). Apple Academic Press, CRC Press. ISBN: 9781771884020.

Keservani, R. K., Sharma, A. K., & Kesharwani, R. K., (2016b). Medicinal effect of nutraceutical fruits for the cognition and brain health. *Scientifica*, 10. Article ID 3109254. doi:10.1155/2016/3109254.

Keservani, R. K., Sharma, A. K., & Kesharwani, R. K., (2017). An overview and therapeutic applications of nutraceutical and functional foods. *Recent Advances in Drug Delivery Technology,* 160–201. ISBN:9781522507543.

Keservani, R. K., Sharma, A. K., & Kesharwani, R. K., (2020). *Nutraceuticals and Dietary Supplements: Applications in Health Improvement and Disease Management*. CRC Press. ISBN: 9781771888738.

Kesharwani, R. K., Singh, P., & Keservani, R. K., (2016). Green coffee bean extract and chlorogenic acids: Chemistry and novel antioxidant benefits; Chapter 1. In: Bagchi, D., Moriyama, H., & Swaroop, A., (eds.), *Green Coffee Bean Extract in Human Health* (pp. 1–17). Boca Raton, Taylor and Francis, USA. ISBN: 9781498716376.

Khan, M. I., et al., (2011). Acute, subacute and sub chronic safety assessment of betalains rich *Rivina humilis* L. Berry juice in rats. *Food and Chemical Toxicology, 49*(12), 3154–3157.

Khomdram, S., & Devi, G. S., (2010). Determination of antioxidant activity and vitamin C of some wild fruits of manipuurasuhma. *The Bioscan, 5*(3), 501–504.

Kong, J. S., Yoo, S. A., Kim, H. S., Kim, H. A., Yea, K., Ryu, S. H., Chung, Y. J., Cho, C. S., & Kim, W. U., (2010). Inhibition of synovial hyperplasia, rheumatoid T cell activation, and experimental arthritis in mice by sulforaphane, a naturally occurring isothiocyanate. *Arthritis Rheum, 62*(1), 159–170.

Latte, K. P., Appel, K. E., & Lampen, A., (2011). Health benefits and possible risks of broccoli—An overview. *Food Chem. Toxicol.*, *49*(12), 3287–3309.

Lavecchia, T., Rea, G., Antonacci, A., & Giardi, M. T., (2013). Healthy and adverse effects of plant-derived functional metabolites: The need of revealing their content and bioactivity in a complex food matrix. *Critical Reviews in Food Science and Nutrition*, *53*(2), 198–213.

Lee, T., Kwon, H. S., Bang, B. R., Lee, Y. S., Park, M. Y., Moon, K. A., Kim, T. B., et al., (2012). Grape seed proanthocyanidin extract attenuates allergic inflammation in murine models of asthma. *J. Clin. Immunol., 32*(6), 1292–1304.

Li, H., Subbiah, V., Barrow, C. J., Dunshea, F. R., & Suleria, H. A., (2021). Phenolic profiling of five different australian grown apples. *Appl. Sci.*, *11*, 2421.

Liu, Q., Luo, L., & Zheng, L., (2018). Lignin's: Biosynthesis and biological functions in plants. *International Journal of Molecular Sciences*, *19*(2), 335.

Mackerness, S. A. H., & Jordan, B. R., (1999). Changes in gene expression in response to ultraviolet B-induced stress. In: Pessarakli, M., (ed.), *Handbook of Plant and Crop Stress* (2nd edn, pp. 749–768). Marcel Dekker, New York.

Meng, C., (2016). Suppression of tomato SlNAC1 transcription factor delays fruit ripening. *Journal of Plant and Physiology*, *193*, 88–96.

Mohiuddin, A. K., (2019). Impact of various environmental factors on secondary metabolism of medicinal plants. *J. Pharmacol. Clin. Res.*, *7*(1), 1–10.

Neugart, S., Baldermann, S., Hanschen, F. S., Klopsch, R., Wiesner-Reinhold, M., & Schreiner, M., (2018). The intrinsic quality of brassicaceous vegetables: How secondary plant metabolites are affected by genetic, environmental, and agronomic factors. *Scientia Horticulturae*, *233*, 460–478.

Neves, A. R., Lucio, M., Lima, J. L., & Reis, S., (2012). Resveratrol in medicinal chemistry: A critical review of its pharmacokinetics, drug-delivery, and membrane interactions. *Curr. Med. Chem.*, *19*(11), 1663–1681.

Newall, C. A., Anderson, L. A., & Phillipson, J. D., (1996). Herbal medicines: A guide for health-care professionals. Pharmaceutical Press, London.

Nijveldt, R. J., Van, N. E. L. S., Van, H. D. E., Boelens, P. G., Van, N. K., & Van, L. P. A., (2001). Flavonoids: A review of probable mechanisms of action and potential applications. *The American Journal of Clinical Nutrition*, *74*(4), 418–425.

Okuda, T., Yoshida, T. & Hatano, T., (1992). Antioxidant effects of tannins and related phenols. In: Huang, M. T., Ho, C. T., & Lee, C. Y., (eds.), *Phenolic Compounds in Foods and Their Effect on Health* (pp. 87–97). Washington, DC: American Chemical Society.

Ooe, E., et al., (2016). Analysis and characterization of anthocyanins and carotenoids in Japanese blue tomato. *Bioscience Biotechnology Biochemistry, 80*(2), 341–349.

Oomah, B. D., & Mazza, G., (2000). Functional foods. In: Francis, F. J., (ed.). *The Wiley Encyclopaedia of Science & Technology* (2nd edn., Vol. 2, pp. 1176–1182). New York, NY: Wiley.

Owen, R. W., Giacosa, A., Hull, W. E., Haubner, R., Wurtele, G., Spiegelhalder, B., & Bartsch, H., (2000). Olive oil consumption and health: The possible role of antioxidants. *Lancet Oncol.*, *1*(2), 107–112.

Pandey, K. B., & Rizvi, S. I., (2009). Plant polyphenols as dietary antioxidants in human health and disease. *Oxid. Med. Cell Longev., 2*(5), 270–278. doi: 10.4161/oxim.2.5.9498. PMID: 20716914; PMCID: PMC2835915.

Poiroux-Gonord, F., Bidel, L. P., Fanciullino, A. L., Gautier, H., Lauri-Lopez, F., & Urban, L., (2010). Health benefits of vitamins and secondary metabolites of fruits and vegetables and prospects to increase their concentrations by agronomic approaches. *Journal of Agricultural and Food Chemistry*, *58*(23), 12065–12082.

Prior, R. L., & Cao, G., (2000). Antioxidant phytochemicals in fruits and vegetables: Diet and health implications. *Horticulture Science*, *35*, 588±592.

Ramos-Romero, S., Perez-Cano, F. J., Ramiro-Puig, E., Franch, A., & Castell, M., (2012). Cocoa intake attenuates oxidative stress associated with rat adjuvant arthritis. *Pharmacol. Res.*, *66*(3), 207–212.

Rane, B. R., Bharath, M. S., Patil, R. R., Keservani, R. K., & Jain, A. S., (2020b). Novel approaches in nutraceuticals; Chapter 11. In: Rajesh, K. K., Raj, K. K., & Anil, K. S., (eds.), *Enhancing the Therapeutic Efficacy of Herbal Formulations Through Novel Drug Delivery Systems* (pp. 241–266). IGI Global International Publisher, Pennsylvania, USA. ISBN13: 9781799844532.

Rane, B. R., Patil, A. S., Keservani, R. K., & Jain, A. S., (2020a). Novel approaches in herbal formulation; Chapter 2. In: Rajesh, K. K., Raj, K. K., & Anil, K. S., (eds.), *Enhancing the Therapeutic Efficacy of Herbal Formulations Through Novel Drug Delivery Systems* (pp. 43–68). IGI Global International Publisher, Pennsylvania, USA. ISBN13: 9781799844532.

Rane, B. R., Tadavi, S. A., & Keservani, R. K., (2020c). Naturopathy; Chapter 12. In: Anil, K. S., Raj, K. K., & Surya, P. G., (eds.), *Herbal Product Development* (pp. 321–347). Apple Academic Press, CRC Press, Taylor & Francis Group. ISBN: 9781771888776.

Redondo-Calvo, F. J., Montenegro, O., Padilla-Valverde, D., Villarejo, P., Baladron, V., Bejarano-Ramirez, N., Galan, R., Gomez, L. A., Villasanti, N., Illescas, S., et al., (2021). Thiosulfate-enriched *Allium sativum* extract as an adjunct to antibiotic treatment of sepsis in a rat peritonitis model. *Appl. Sci.*, *11*, 4760.

Saxena, M., et al., (2013). Phytochemistry of medicinal plants. *Journal of Pharmacognosy and Phytochemistry*, *1*(6), 168–182.

Schreiner, B., et al., (2012). Role of the AAA protease Yme1 in folding of proteins in the intermembrane space of mitochondria. *Mol. Biol. Cell*, 23(22), 4335–4346.

Schreiner, M., Mewis, I., Huyskens-Keil, S., Jansen, M. A. K., Zrenner, R., Winkler, J. B., & Krumbein, A., (2012). UV-B-induced secondary plant metabolites-potential benefits for plant and human health. *Critical Reviews in Plant Sciences*, *31*(3), 229–240.

Schutz, K., Kammerer, D. R., Carle, R., & Schieber, A., (2005). Characterization of phenolic acids and flavonoids in dandelion (*Taraxacum officinale* WEB. ex WIGG.) root and herb by high-performance liquid chromatography/electrospray ionization mass spectrometry. *Rapid Commun. Mass. Spectrom*, *19*(2), 179–186.

Sharma, A. K., Sharma, A. K., Sharma, M., & Sharma, M., (2021a). Medicinal and aromatic plants: A potential source of novel bioactive compounds showing antiviral efficacy against coronavirus (SARS-CoV-2). *Indian Journal of Ecology*, *48*(14), 7–16.

Sharma, M., Sharma, A. K., & Sharma, M., (2021b). Ethno-botanical study of medicinal plants from unexplored area of District Ramban (J&K) India. *Indian Journal of Agricultural Research*, *55*, 702–708.

Sharma, M., Sharma, A. K., Thakur, R., & Sharma, M. (2020). Dynamics of traditional information of medicinal plants from Hilly Terrains of Ramban (J&K) India. *Indian Journal of Ecology*, *47*(4), 1009–1013.

Sharma, M., Thakur, R., & Sharma, M., (2020b). Ethnomedicinal, phytochemical and pharmacological properties of *Crocus sativus* (Saffron). *Journal of Indian Botanical Society*, *99*(3, 4), 46–55.

Sharma, M., Thakur, R., & Sharma, M., (2020c). Ethnobotanical survey of medicinal plants of unexplored hilly areas of district Ramban (J&K). *International Journal of Botany Studies, 5*(3), 55–63.

Sharma, M., Thakur, R., Sharma, M., Sharma, A. K., & Sharma, A. K., (2020d). Changing SCENARIO of medicinal plants diversity in relation to climate changes: A review. *Plant Archives, 20*(2), 4389–4400.

Simonetti, G., (1990). In: Schuler, S., (ed.), *Simon & Schuster's Guide to Herbs and Spices (Nature Guide Series)* (pp. 261, 262). Inc, New York.

Singh, P., et al., (2018). Nuclear mRNA surveillance mechanisms: Function and links to human disease. *J. Mol. Biol., 430*(14), 1993–2013.

Singh, P., Kesharwani, R. K., & Keservani, R. K., (2017a). Antioxidants and vitamins: Roles in cellular function and metabolism. In: *Sustained Energy for Enhanced Human Functions and Activity* (pp. 385–407). Academic Press. ISBN:9780128054130.

Singh, P., Kesharwani, R. K., & Keservani, R. K., (2017b). Protein, carbohydrates, and fats: Energy metabolism. In: *Sustained Energy for Enhanced Human Functions and Activity* (pp. 103–115). Academic Press. ISBN:9780128054130.

Singh, S., Thakur, A., & Singh, S. K., (2018). *Secondary Metabolite in Fruit Crop: A Mini Review, 5*(4), 794–803.

Sobenin, I. A., Pryanishnikov, V. V., Kunnova, L. M., Rabinovich, Y. A., Martirosyan, D. M., & Orekhov, A. N., (2010). The effects of time-released garlic powder tablets on multifunctional cardiovascular risk in patients with coronary artery disease. *Lipids Health Dis., 9*, 119.

Sohn, D. W., Han, C. H., Jung, Y. S., Kim, S. I., Kim, S. W., & Cho, Y. H., (2009). Anti-inflammatory and antimicrobial effects of garlic and synergistic effect between garlic and ciprofloxacin in a chronic bacterial prostatitis rat model. *Int. J. Antimicrob. Agents, 34*(3), 215–219.

Sreenivasulu, N., & Fernie, A. R., (2022). Diversity: Current and prospective secondary metabolites for nutrition and medicine. *Curr. Opin. Biotechnol., 74*, 164–170. doi: 10.1016/j.copbio.2021.11.010. Epub 2021 Dec 20. PMID: 34942505.

Suh, H. J., Lee, J. M., Cho, J. S., Kim, Y. S., & Chung, S. H., (1999). Radical scavenging compounds in onion skin. *Food Research International, 32*, 659–664.

Sun, Y. D., et al., (2012). Dynamic QTL analysis for fruit lycopene content and total soluble solid content in a *Solanum lycopersicum* S. *pimpinellifolium* cross. *Genetics and Molecular Research, 11*(4), 3696–3710.

Swallah, M. S., Sun, H., Affoh, R., Fu, H., & Yu, H., (2020). Antioxidant potential overviews of secondary metabolites (polyphenols) in fruits. *International Journal of Food Science.*

Talalay, P., (1992). The role of enzyme induction in protection against carcinogenesis. In: Wattenberg, L., Lipkin, M., Boone, G. W., & Kello, G. J., (eds.). *Cancer Chemoprevention* (pp. 469–478). Boca Raton, FL: CRC Press.

Talukder, P., Dutta, D., Ghosh, E., Bose, I., & Bhattacharjee, S., (2021). Role of plant secondary metabolites in combating pest induced stress in brinjal (*Solanum melongena* L.). *Journal of Environmental Engineering and Landscape Management, 29*(4), 449–453.

Tanaka, S., Haruma, K., Kunihiro, M., Nagata, S., Kitadai, Y., Manabe, N., Sumii, M., et al., (2004). Effects of aged garlic extract (AGE) on colorectal adenomas: A double-blinded study. Hiroshima *J. Med. Sci., 53*(3, 4), 39–45.

Tanito, M., Masutani, H., Kim, Y. C., Nishikawa, M., Ohira, A., & Yodoi, J., (2005). Sulforaphane induces thioredoxin through the antioxidant responsive element and attenuates retinal light damage in mice. *Invest. Ophthalmology. Vis. Sci., 46*(3), 979–987.

Visioli, F., & Bernardini, E., (2011). Extra virgin olive oil's polyphenols: Biological activities. *Curr. Pharm. Des., 17*(8), 786–804.

Vislocky, L. M., & Fernandez, M. L., (2010). Biomedical effects of grape products. *Nutr. Rev., 68*(11), 656–670.

Wang, J., Ferruzzi, M. G., Ho, L., Blount, J., Janle, E. M., Gong, B., Pan, Y., et al., (2012). Brain-targeted proanthocyanin metabolites for Alzheimer's disease treatment. *J. Neurosci., 32*(15), 5144–5150.

Waterman, E., & Lockwood, B., (2007). Active components and clinical applications of olive oil. *Altern. Med. Rev., 12*(4), 331–342.

Weisburger, J. H., (2001). Chemo preventive effects of cocoa polyphenols on chronic diseases. *Exp. Biol. Med. (Maywood), 226*(10), 891–897.

Weseler, A. R., Ruijters, E. J., Drittij-Reijnders, M. J., Reesink, K. D., Haenen, G. R., & Bast, A., (2011). Pleiotropic benefit of monomeric and oligomeric flavanols on vascular health: A randomized controlled clinical pilot study. *PLoS One, 6*(12), e28460.

Williams, C. A., Goldstone, F., & Greenham, J., (1996). Flavonoids, cinnamic acids and coumarins from the different tissues and medicinal preparations of *Taraxacum officinale*. *Phytochemistry, 42*(1), 121–127.

Williamson, G., Faulkner, K., & Plumb, G. W., (1998). Glucosinolates and phenolics as antioxidants from plant foods. *European Journal of Cancer Prevention, 7*, 17–21.

Yeshi, K., Crayn, D., Ritmejeryte, E., & Wangchuk, P., (2022). Plant secondary metabolites produced in response to abiotic stresses has potential application in pharmaceutical product development. *Molecules, 27*(1), 313.

Yeshi, K., Ruscher, R., Hunter, L., Daly, N. L., Loukas, A., & Wangchuk, P., (2020). Revisiting inflammatory bowel disease: Pathology, treatments, challenges, and emerging therapeutics including drug leads from natural products. *Journal of Clinical Medicine, 9*, 1273. https://doi. org/10.3390/jcm9051273.

Zevallos, V. F., Raker, V., Tenzer, S., Jimenez-Calvente, C., Ashfaq-Khan, M., Ruessel, N., Pickert, G., et al., (2017). Nutritional wheat amylase-trypsin inhibitors promote intestinal inflammation via activation of myeloid cells. *Gastroenterology, 152*(5), 1100–1113.e1112. https://doi.org/10.1053/j.gastro.2016.12.006.

Zhang, Y. J., & Gan, R. Y., (2015). Antioxidant phytochemicals for the prevention and treatment of chronic diseases. *Molecules, 20*(12), 21138–21156.

Zhang, Y., Kenski, T. W., Cho, C. G., Posner, G. S., & Talalay, P., (1994a). Anticarcinogenic activities of sulforaphane and structurally related synthetic norbornyl isothiocyanates. *Proceeding of National Academy of Sciences, USA, 91*, 2147–2150.

Zhang, Y., Talalay, P., Cho, C. G., & Posner, G. H., (1992). A major inducer of anti-carcinogenic protective enzymes from broccoli: Isolation and elucidation of structure. *Proc. Natl. Acad. Sci. (USA), 89*(6), 2399–2403.

CHAPTER 2

NUTRACEUTICAL POTENTIAL OF SELECTED VEGETABLES

N. ERAMMA,[1] S. O. SADASHIV,[2] RAGHU NINGEGOWDA,[3] S. RENUKA JYOTHI,[4] and SHARANGOUDA J. PATIL[5]

[1]*Department of Microbiology, Acharya Bangalore B-School, Bangalore, Karnataka, India*

[2]*Department of Studies in Food Technology, Shivagangothri, Davangere University, Davanagere, Karnataka, India*

[3]*Department of Chemistry, Jyoti Nivas College, Bangalore, Karnataka, India*

[4]*Department of Life Science, School of Sciences, Jain University, Bangalore, Karnataka, India*

[5]*Department of Zoology, NMKRV College for Women (Autonomous), Bangalore, Karnataka, India*

ABSTRACT

The consumption of vegetables has various benefits, and it has been practiced worldwide throughout history. Vegetables are classified as plant parts that are used for food preparation by humans to consume or feed domesticated animals. The overall meaning is to balance the nutrition in our diet with vegetables which are applied vegetable plants referred it as edible plants due to their principal nutraceutical agents, namely the barks, bulbs, flowers, fruits, leaves, legumes, nuts, rhizomes, roots, stems, seeds, etc. The biological compounds are responsible for nutraceutical therapeutic benefits by these parts of foods and their products due to their secondary metabolism

Plant Metabolites and Vegetables as Nutraceuticals.
Raj K. Keservani, Bui Thanh Tung, Rajesh K. Kesharwani, and Eknath D. Ahire (Eds.)

aids in preventing physiological stress or free radicals' potential and protect from oxidative damage. These active principles of biological molecules are scattered in all the parts of edible and non-edible vegetable foods. The aim of the chapter emphasizes addressing on nutraceutical potential of selected vegetable plants to describe their nutraceutical and pharmaceutical significance in the narrative research literature by collecting data from PubMed, Springer, Elsevier, NISCAIR, MDPI, etc., of 12 years. The potential vegetable plants possess all the metabolites and are rich in amino acids, fats, flavonoids, glycosides, phenolics, proteins, and tannins concentrations and they act as antioxidants or therapeutic nutrition for all age groups. In this context reviewed 5 vegetables extensively for their phytochemistry and nutritional composition of *Coriandrum sativum* L., *Mentha arvensis, Momordica charantia* L., *Moringa oleifera, Raphanus sativus* L. plant's potential in their parts such as three leafy, one un-ripened fruit and one tuber. These plants and products are considered for various properties by being applied in a wide range of activities in Ayurvedic formulation and advanced organics in modern natural medicine in global scenarios recognized as an important food. The presence of biological molecules in these vegetable extracts explored various pharmacological and nutraceutical properties on many levels. Depending upon the constituent concentrations biological properties like antiallergic, anticancer, antidiabetic, antidyslipidemic, anti-inflammatory, antimicrobial, antimutagenic, antioxidant, antipyretic, cardioprotective, hypotensive, immunomodulatory, neurological, ulcerogenic activities were reported in pre-clinical studies.

2.1 INTRODUCTION

The consumption of edible foods in raw and processing methods for health benefits has been a day-to-day practice for centuries. The choice and alternative principal food products show the richness of bioactive components in diet (Yasmeen et al., 2017). Vegetables are the primary food ingredients and possess dietary nutraceutical (Keservani et al., 2010a, b, 2017, 2020) molecules like flavonoids (Keservani & Sharma, 2014; Keservani et al., 2016a, b), glucosinolates (GSs), polyphenols, phytoestrogens, terpenes, etc., and exhibit potential role in the combatting degenerative or chronic health issues (Baiano & Previtali, 2018). These bioactive nutraceutical compounds are commonly dispersed in all vegetables; it is widely distributed in edible and non-edible portions of vegetable plants to consider as food molecules as well as bio-wastes. Hence, their intake of food is considered mainly a regular diet and as supplements in the form of therapeutic food also obtained as a functional food by the nutraceutical industry with needful action against various health ailments.

Vegetables are parts and products (barks, bulbs, flowers, fruits, leaves, legumes, nuts, rhizomes, roots, stems, seeds) of plants, used raw, and processed as per their convenient method to increase the value and sensory characteristics of processed food diets. These foods provide organic compounds as primary and secondary metabolites which are responsible for the biological function of organisms. Behind the any properties of vegetables depend upon their phytoconstituents such as alkaloids, amino acids, carbohydrates, flavonoids (Keservani & Sharma, 2014), glycosides, proteins (Singh et al., 2017b), saponins, steroids, tannins, and triterpenoids, etc., are the designed derivatives pharmacologically useful in the modern and traditional medicine of system for the of various disease treatments (Haleshappa et al., 2020, 2021, 2022; Opara, 2014). Most of the useful application of vegetable health benefits depends on antioxidant potential, in that traditional vegetables, are highly potent in nutraceutical agents, and in literature, many reports were emphasized with horseradish, ginger, turmeric, and wasabi due to the richness of antioxidant activities depend upon their active ingredients like allyl isothiocyanate, curcuminoids, and gingerols (shogaols, paradols, and zingerone), polyphenols, etc. (Nahak et al., 2014).

In this context, chapter revealed extensively on five vegetables for their phytochemistry to know their biomolecules and their therapeutic action along with nutritional composition of *Coriandrum sativum* L., *Mentha arvensis, Momordica charantia* L., *Moringa oleifera, Raphanus sativus* L. plant's potential as an food ingredient in the edible parts and products such as three leafy, one un-ripened fruit and one tuber selected as nutraceutical vegetables.

2.2 *CORIANDRUM SATIVUM* L.

Coriander (*Coriandrum sativum* L.) is belonging to family Umbelliferae and genus Coriandrum very common food garnishing plant in food preparation, possesses medicinal essential oils (EOs) as a spice as well as therapeutic medicinal plants (Mandal & Mandal, 2015). It is simple herbaceous plant characteristics as branched, glabrous, slender cultivated throughout India in all days, it is having special characteristic feature lime aroma when it gets expose to air or rubbed (Mourya et al., 2021). Coriander has different vernacular names in different languages, i.e., English (coriander), Urdu (Dhania), Arabic (Kuzbara), Hindi (Dhania), Chinese (Yuan sui), Greek (korion). The "coriander," is consequential from Greek word for "bedbug" due to its spanking smell in the fresh foliage and it is resembled also to that of bug plague-ridden bed line (Nadeem et al., 2014). Coriander is indigenous medicinal plants of the Mediterranean and is also distributed widely in Russia, Europe (Central), Africa (North) as well in Asian countries (Sahib, 2013). Coriander was used in time-honored Greek medicines by Hippocrates (460–377 BC). The Egyptians called this herb as "spice of happiness" (Nadeem et al., 2014).

Coriander plants containing EO in its leaves, stems, flowers, fruits, and in seeds. The plant leaves, shoots, and seeds are commonly used in food preparation and decoration, in traditional approach as a folk medicine also used as season specific spice during food preparation (Mandal & Mandal, 2015). The fruits of coriander, also known as the seeds, are globular in structure, aromatic, bittersweet, and spicy taste in nature. Coriander seeds are recognized essential ingredient in curry powder preparation; it is most important in the preparation of meat dishes with the mincing and stews process (Sahib, 2013). In the context of Indian traditional medicine, coriander is used as medicament in many of the disorders such as digestive issue, respiratory problems, and urinary system complications, along with that it is having properties of carminative, diaphoretic, diuretic, and stimulant. In the Iranian folklore as a traditional medicine, it is reported for their application in dyspeptic complaints and loss of appetite as common medical problems (Mourya et al., 2021).

2.2.1 PHYTOCHEMISTRY OF *CORIANDRUM SATIVUM* L.

The fragrance characteristics like odor and flavor are considered such plant as an aromatic plant, its dried seeds, mature fruits, fresh herbage are key component for the phytochemistry. The aliphatic aldehyde like C_{10} to C_{16} are main component as aromatic in this plant with the aroma like fetid in nature, it is observed that fresh oil of this herb produces aromatic EO, as per the literature reported isolated major components as monoterpenes and monoterpene hydrocarbons in the oil, in the form of oxygenated condition (Mourya et al., 2021). The most important constituents of different parts of coriander plant are the EO, fatty oil, and other crude oils are presented in Table 2.1. The fatty oil is light yellow in color and has a characteristic of smell. The oil is unique in that it contains high amount of petroselinic acid (C_{18}: 1n–12) (Sahib, 2013). The EO composition in the content of ripe, dried fruits or seeds weight are in between 0.03% and 2.6%, and in the fatty oil content varies in the range of 9.9 to 27.7% in that 1% of S-(+)-linalool around 60–70% as a major component, in that many minor fraction of active constituents also present in the EO such as borneol, camphor, citronellol, geraniol, geraniol acetate, limpnene, α-pinene, γ-terpinene, p-cymene of monoterpene hydrocarbons, some compounds of organic form such as coriandrin, coriandrons A-E, dihydrocoriandrin, digustilide, flavonoids, furan, isocoumarins, neochidilide, phenolic acids, pthlides, pyrazine, pyridine, sterols, tetrahudrofuran, thiazole as a heterocyclic main components in the oil (Nimish et al., 2011). Over the last decade, EOs have gained popularity as a source of bioactives, with many potential health benefits (Sahib, 2013) (Figure 2.1).

TABLE 2.1 Composition Variation of *C. sativum* Essential Oil from Different Parts of the Plant

Source	Metabolites	References
Flowers	Benzofuran, 2,3-dihydro (15%), hexadecanoic acid, methyl ester (10.32%), 2,4a-epoxy-3,4,5,6,7,8,-hexahydro-2,5,5,8,a-tetramethyl-2h-l-benzofuran (9.35%), and dodecanoic acid (5%).	Mandal & Mandal (2015)
Leaves	Decanal (19.09%), trans-2-decenal (17.54%), dodecanal (4.1%), dodecan-1-ol (3.13%), cyclodecane (12.15%), cis-2-dodecena (10.72%), and 2-decen-1-ol (12.33%).	
Seeds	Camphor (3%–5.1%), geranyl acetate (0.2%–5.4%), p-cymene (0.1%–8.1%), α-pinene (0.2%–10.9%), γ-terpinine (0.3%–11.2%), and linalool (58.0–80.3%).	

2.2.2 NUTRITIONAL COMPOSITION OF C. SATIVUM

The coriander fresh leaves contain 87.9% of moisture, 3.3% of protein, 6.5% of carbohydrates, 1.7% of total ash as an proximate components, 0.14% of calcium, 0.06% of phosphorus, 0.01% of iron, 60 mg/100 g of vitamin B2, 0.8 mg/100 g of niacin, 135 mg/100 g of vitamin C trace amount of vitamin A as an vital inorganic nutrients. In the fresh basis if coriander seeds in 100 g reported for the 11 g of starch, 20 g of fat, 11 g of protein, and around 30 g of crude fiber as organic constituents (Mandal & Mandal, 2015). The dried seed of coriander contains 11.37% of water, 11.49% of crude protein, 19.15% of fat, 28.43% of crude fiber, 10.53% of starch, 10.29% of pentosans, 1.92% of sugar, 4.98% mineral constituents and 0.84% of EO are also reported. In the EO major compounds such as linalool (67.7%), α-pinene (10.5%), β-terpinene (9.0%), geranyl acetate (4.0%), camphor (3.0%) and geraniol (1.9%) are characterized and elucidate in 100% of the oil (Nadeem et al., 2014) (Table 2.2).

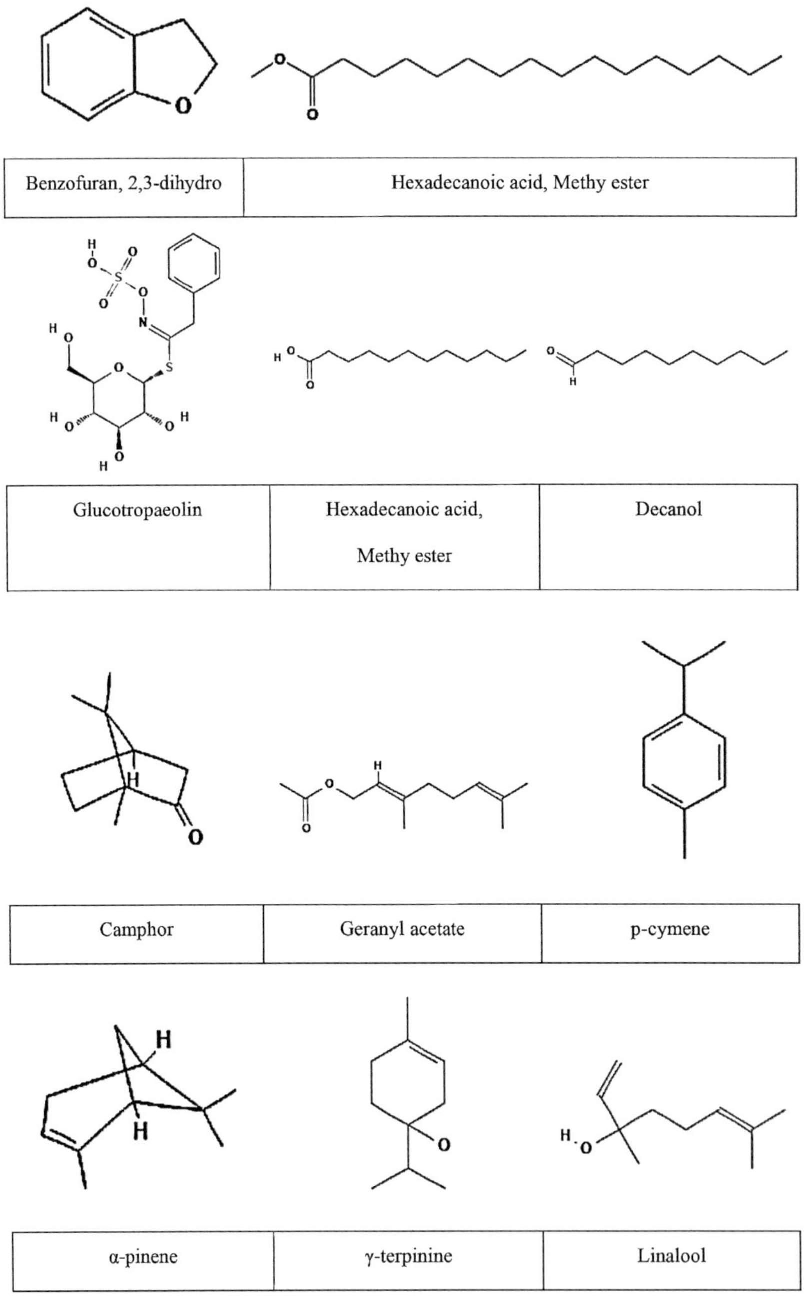

FIGURE 2.1 Molecular name and structures of phytoconstituents of different parts in *C. sativum* plant.

TABLE 2.2 Nutrient Compositions of Coriander Leaf and Seeds as per USDA

Nutrient Composition	Coriander Leaf Per 100 gm	Coriander Seed Per 100 gm	References
Water	7.30 g	8.86 g	Bhat et al. (2014)
Energy	279 kcal	298 kcal	
Protein	21.93 g	12.37 g	
Carbohydrate	52.10 g	54.99 g	
Total fat	4.78 g	17.77 g	
Dietary fiber	10.40 g	41.9 g	
Iron	42.46 g	16.32 g	
Calcium	1,246 mg	709 mg	
Magnesium	694 mg	330 mg	
Phosphorus	481 mg	409 mg	
Sodium	211 mg	35 mg	
Potassium	4,466 mg	1,267 mg	
Zinc	4.72 mg	4.70 mg	
Vitamin C	566.7 mg	21.0 mg	
Riboflavin	1.500 mg	0.290 mg	
Niacin	10.707 mg	2.130 mg	
Thiamine	1.252 mg	0.239 mg	
Vitamin B12	0.00 ug	0.00 ug	
Vitamin A	293 ug	0.00 ug	
Vitamin D (D2 + D3)	0.00 ug	0.0 ug	
Total saturated fatty acids	0.115 g	0.990 g	
Total monounsaturated fatty acids	2.232 g	13.580 g	
Total polyunsaturated fatty acids	0.328 g	1.750 g	
Cholesterol	0.00 mg	0 mg	

Source: National Nutrition Data Base (2013).

2.2.3 *MEDICINAL PROPERTIES OF C. SATIVUM*

Coriander plant and their seeds have been used as an edible product and medicines from the ancient time. The parts of coriander plant give many products such as leaves, flower, fruit, and seed exhibit many biological properties like antioxidant, diuretic, ant-diabetic, sedative, anti-microbial activity, anti-convulsant, hypnotic, anthelmintic, and anti-mutagenic in any parts of the plant (Nadeem et al., 2014). Coriander plant extracts or compounds demonstrated in research studies to prove their antioxidant potential, and it is due to the presence of strong polyphenols, various types of vitamins (Singh et al., 2017a) and sterol constituents (Mahleyuddin et al., 2021). Coriander has one of the major Iranian traditional medicines in the treatment of anticonvulsants, anti-depressant, and for its nerve soothing, sedative, and anxiolytic properties (Sahib, 2013). Coriander oil is considered as EO due to the properties of antimicrobial, along with that useful application like carminative, stomachic, and spasmolytic. Coriander products are also useful in diarrhea, gastritis issues, dyspepsia, and their relative diverse complications like antibilious, stimulation in digestion and stomachic properties (Nadeem et al., 2014). The coriander fruits produced a momentous reduction in all cholesterol associated lipids, whereas extracts compact high-density lipoprotein (HDL) cholesterol (Naeem, 2022). A cardioprotective effect also reported due to its richness of polyphenol bioactive components of *C. sativum,* it is having rich antioxidant potential hence it is associated in the problems of myocardial infarction (MI) prevention due to the cellular injury in myofibrillar area by the induction of isoproterenol and it is severe cardiac damage observed in the pre-clinical studies in male Wistar albino rats (Mahleyuddin et al., 2021) (Table 2.3).

TABLE 2.3 Coriander Compounds Showing Various Applications

Compounds Involved	Functionally Active	References
Coriander oil	Antimicrobial activity	Bhat et al. (2014)
Coriander extract	Antianxiety effect	
Aqueous and seed extracts	Antiglycemic activity	
BHA	Inhibition of oxidative products	
Terpenoids	Inhibition of oxidative products	
Flavonoids	Oxidative damage protection	
Phenolics	Antioxidant activity	
Hydroxyl groups	Metal detoxification	
Other constituents	Hypolipidemic effect	

2.3 *MENTHA ARVENSIS* L.

A perennial herb of Europe known as *Mentha arvensis* L., or peppermint, has naturalized in the northern United States and Canada and is grown all over the world (Kapp et al., 2013). The Lamiaceae family includes the genus Mentha, which has a taxonomy with roughly 42 species, 15 hybrids, hundreds of subspecies, and thousands of cultivars. From the species *M. arvensis* L., *M. aquatica* L., *M. spicata* L., *M. longifolia* L., and *M. suaveolens* Ehrh, 11 naturally occurring hybrids have been created; the majority of hybrids are infertile but can proliferate due to their very invasive rhizome (Tafrihi et al., 2021). *Mentha arvensis*, often known as wild mint, corn mint, or field mint, is well-known around the world. Padina, phutya, or puttiya are frequently used as local names for *Mentha arvensis*. It has a very long history. It was thought that the Greek mythical nymph Minthe was where the word "mentha" first appeared (Nazim et al., 2020). It is one of the oldest medicinal plants in use today and is employed in both Eastern and Western cultures. The herb was utilized in cooking and healing in the ancient Greek, Roman, and Egyptian societies. The volatile oils found in peppermint, which are employed in the flavor, fragrance, and pharmaceutical industries; make it one of the most economically significant aromatic and therapeutic plants right now. About 8,000 tons of peppermint oil are produced annually in the world (Balakrishnan, 2015).

2.3.1 PHYTOCHEMISTRY OF MENTHA ARVENSIS L.

Mentha arvensis produces a powerful aroma as a result of the phenolic and flavonoid levels in the oil. Caffeic acid and its derivatives caftaric acid, cinnamic acid, ferulic acid, and oleanolic acid are also present in Mentha species. Polyphenol levels are high in Mentha species. These plants also contain flavonoids such luteolin and its derivatives apigenin, acacetin, diosmin, and salvigenin, which make up roughly 10–70 of the total phenolics, as well as flavanols like catechin, epicatechin, and coumarins like esculetin and scopoletin. The main focus of mint compositions is the EOs. In Mentha species, they are primarily composed of alcohols, ketones, esters, ethers, and oxides and are colorless, pale yellow or greenish yellow (Tafrihi et al., 2021). Menthol is the primary component of the EO of peppermint and is mostly responsible for the agent's anti-spasmolytic effects. Various constituents of peppermint oil are limonene (1.0–5.0%), cineole (3.5–14.0%), menthone (14.0–32.0%), menthofuran (1.0–9.0%), isomenthone (1.5–10.0%), menthyl acetate (2.8–10.0%), isopulegol (0.2%), menthol (55.0%), pulegone (4.0%) and carvone (max. 1.0%) (Balakrishnan, 2015) (Table 2.4 and Figure 2.2).

TABLE 2.4 The Most Abundant Active Compounds of *Mentha* spp.

Compounds	Percentage (%)	IUPAC Name	References
Menthone	14–32	(2S,5R)-2-Isopropyl-5-methylcyclohexanone	Loolaie et al. (2017)
Menthofuran	1–9	3,6-Dimethyl-45,6,7-tetrahydro-1-benzofuran	
Limonene	1–5	1-Methyl-4-(1-methylethenyl)-cyclohexane	
Cineole	3.5–14	1,3,3-Trimrthyl-2-oxabicyclo[2,2,2,] octane	
Isomenthone	1.5–10	(2R,5R)-5-methyl-2-propan-2-2ylcyclohexane-1-one	
Menthyl acetate	2.8–10	Acetic acid (1R,2S,5R)-2-isopropyl-5-methylcyclohexyl]ester	
Isopulegol	0.2	5-methyl-2-p2rop-1en-2-ylcyclo-hexan-1-ol	
Menthol	30–35	(1R, 2S, 5R)-2-Isopropyl-5-methyl cyclohexanol	
Pulegone	4	p-Methyl-4(8)-en-3-one	
Carvone	1	2-Methyl-5-(prop-1-en-2-yl) cyclohex-2-en-1-one.	

2.3.2 MEDICINAL PROPERTIES OF *MENTHA ARVENSIS* L.

Due to its aroma and flavoring capabilities, this herbal (Sharma et al., 2020; Rane et al., 2020a–c) is mostly utilized in cosmetics, pharmaceuticals, and personal care items. It has antinociceptive, antibacterial, anti-inflammatory, and antioxidant characteristics that are used to stop the inflammation process, diabetes, cardiovascular disease (CVD), and other chronic degenerative diseases. It is employed in the production of pesticides, and antibiotics deter the Dipterian vector that spreads sickness (Sharma & Gautam, 2022). Topical vasodilators like menthol help other topical skin medicines absorb better. The absorption of cortisone, mannitol, indomethacin, morphine hydrochloride, and propranolol are believed to be improved by menthol. It serves as the primary component of ayurvedic drugs such Daburs "Pudin Hara." Menthol, the primary component in cough syrups and ointments like Vicks Vaporub, etc., can be found naturally in the oil. (i) Peppermint leaf and oil are utilized all throughout the world in traditional medicine, as flavorings, and in cosmetic and medicinal products. (ii) Of all the volatile oils, peppermint oil is the one that is used most frequently. (iii) Herbalists

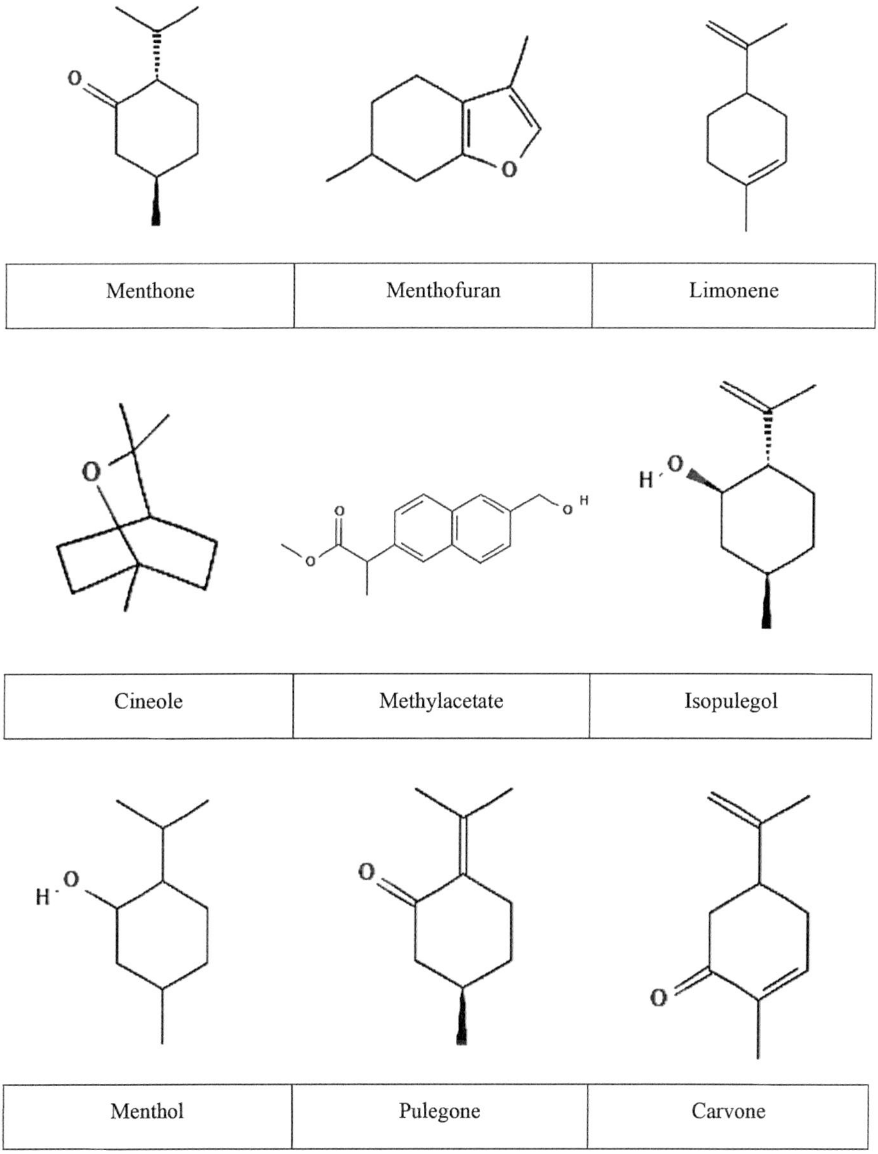

FIGURE 2.2 Structure of active compounds in *Mentha* spp.

classify peppermint as an analgesic, anticatarrhal, antimicrobial, rubefacient, stimulant, and emmenagogue in addition to being an astringent, antiseptic, antipruritic, antispasmodic, antiemetic, carminative, diaphoretic, somewhat bitter. And (iv) to treat respiratory congestion, peppermint oil vapor was inhaled. Tea with peppermint oil added was a traditional cure for throat and mouth mucosal inflammation, bronchitis, and coughs (Balakrishnan, 2015) (Figure 2.3).

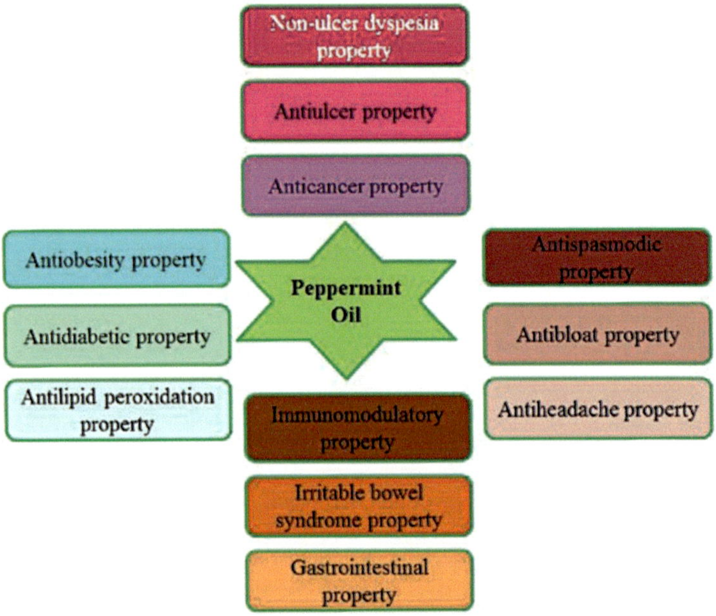

FIGURE 2.3 The major health benefits of peppermint oil of *M. arvensis.*

2.4 *MOMORDICA CHARANTIA* L.

The Cucurbitaceae family includes *Momordica charantia*, often known as bitter gourd, balsam pear, bitter melon, kugua, or karela. *Momordica charantia* (a member of the Momordica genus) has a long history of use as a vegetable and a remedy (Jia et al., 2017). In South Asia, bitter gourd is a well-liked vegetable. The generic name "Momordica" is derived from the Latin word 'Mordica,' which means "to bite," alluding to the serrated edges of the leaves that resemble biting marks. The plant, often referred to as "bitter melon" or "bitter gourd," lives up to its common name because the fruit and all other parts of the plant are quite bitter (Krishnendu &

Nandini, 2016). It is believed to have its origin in India and moved to China in the 14th century. This climber thrives in tropical and subtropical environments. It is having different vernacular names, i.e., in Hindi is known as Karela, Sanskrit is Karavelli, Marathi is Karli, Gujarati is Karelo, Bangali is Baramasiya, Kannada is Karali, Malayalam is Kaypa, Tamil is Pakar, and Telugu is Kakara. The bitter gourd is one of nature's most abundant gifts, but due to its bitter taste, which is caused by the presence of a bitter substance called momordicin, it is also one of the vegetables that people discard the most frequently (Thakur & Sharma, 2016). According to the World Health Organization (WHO), 150 different types of medicinal plants are utilized commercially for therapeutic purposes (WHO).

2.4.1 PHYTOCHEMISTRY OF MOMORDICA CHARANTIA

Vegetables that include bitter gourds are considered as credible sources of a variety of bioactive compounds. Various bioactive compounds in bitter gourd vegetables have already been discovered through numerous studies (Sorifa, 2018). Among the bioactive components present in *M. charantia* fruit are carbohydrates, proteins, lipids, triterpenoids, saponins, polypeptides, flavonoids, alkaloids, and sterols (Jia et al., 2017). Because it contains a large number of phytochemicals and proteins, bitter gourd has enormously beneficial health effects (Sorifa, 2018) (Table 2.5).

TABLE 2.5 Major Bioactive Components of *M. charantia* and Their Related Functions

Plant Parts	Major Metabolites	Functions	References
Pericarp, fruit	Sterols	Antimicrobial	Jia et al. (2017)
Fruit, pericarp, seed	Phenolics	Anti-inflammation, antioxidant, immune enhancement	
Seed flesh	Lipids	Antitumor antioxidant	
Seed	Peptides and proteins	Antimicrobial, anti-tumor, immune suppression, RNA N-glycosidase, polynucleotide adenosine, DNase-like, glycosidase (PAG), phospholipase, superoxide dismutase (SOD)	
Stem, leave, fruit	Terpenoids	Antidiabetic, anticancer, antioxidant, hypoglycemic, chemoprevention	
Fruit, root, seed	Saponins	Antiviral, antihyperglycemic, hypolipidmic	

TABLE 2.6 Physicochemical and Nutritional Composition of Bitter Gourd Vegetable

Parameters (%)	Amount	References
Moisture	93.8	Sorifa (2018)
Protein (g)	0.9	
Carbohydrates (g)	0.2	
Fat (g)	0.1	
Dietary fiber (g)	3.3	
Organic acids (g)	0.1	
IDF (% dry basis)	30.2	
SDF (% dry basis)	14.4	
Acidity (%)	0.03	
TSS (Brix)	3.2	
Reducing sugars (%)	2.5	
Total sugars (%)	2.8	
Chlorophyll (mg/L)	10.9	
Vitamin C (mg)	50.0	
Vitamin A (mg)	0.04	
Thiamin (mg)	0.05	
Riboflavin (mg)	0.03	
Niacin (mg)	0.4	
Phosphorus (mg/100 g)	69.0	
Calcium	22.0	
Potassium	26.0	
Magnesium	16.0	
Iron	0.9	
Sodium	3.0	
Zinc	0.1	
Ash (%)	0.8	

2.4.2 *NUTRITIONAL COMPOSITION OF MOMORDICA CHARANTIA*

M. charantia is a very healthy vegetable since it contains a significant amount of protein, ascorbic acid, calcium, iron, and phosphorus. The total fat content of whole fruit ranged from 2.9 to 6.4% of dry matter. Bitter gourd is rich in vitamin A, vitamin B1, vitamin B2, vitamin B3, and vitamin B9. There are also plenty of inorganic minerals, including phosphorus, potassium, calcium, magnesium, salt, iron, and zinc (Krishnendu & Nandini, 2016). Various types of amino acids are also present in it, including glutamine, asparagines, glycine, lysine, alanine, leucine, valine, arginine, proline, serine, isoleucine, phenylalanine, tryptophan, histidine, threonine, and methionine (Sorifa, 2018). Fruits are rich in a range of nutrients that are crucial for a healthy diet for humans, including proteins, minerals, vitamins, and many more (Krishnendu & Nandini, 2016). It possesses bilious properties, antidotal, and antipyretic effects. Its fruits and leaves are a rich resource of B vitamins, especially thiamine (4%), riboflavin (4%), niacin (2%), vitamin B6 (3%) and folate (13%). Its leaves are nutrient-dense and have been reported to be source of calcium (1%), magnesium (4%), potassium (7%), phosphorus (5%), and iron (3%). With 88 mg of vitamin C per 100 g, bitter gourd fruits are a good source of vitamins and minerals (Thakur & Sharma, 2016) (Table 2.6).

2.4.3 NUTRACEUTICAL PROPERTIES OF MOMORDICA CHARANTIA

In most underdeveloped nations, particularly in Asia and Africa, bitter melon is used as a food, herb, and medicinal. Unripe bitter melon fruit and seeds have undergone substantial investigation into their phytochemical and therapeutic properties (Valyaie et al., 2021). Ayurvedic and Chinese medicine have long employed *Momordica charantia* to treat diabetes and its complications. Studies on the bitter gourd's role as anti-diabetic, anti-ulcerogenic, antimutagenic, antioxidant, anti-tumor, anti-lipolytic, abortifacient, antiviral, hypoglycemic, and immunomodulatory properties (Thakur & Sharma, 2016). *M. charantia's* fruits, vines, leaves, and even roots have all been used in traditional medicine to cure diseases like diabetes, diarrhea, carbuncles, and diabetes-related toothaches. Items related to *M. charantia* are currently quite popular; one such item is bitter gourd tea, also known as gohyah or herbal tea produced from dried slices is mostly used for therapeutic purposes (Jia et al., 2017). Bitter gourd contains astringent compounds like charantin, vicine, glycosides, and karavilosides as well as polypeptide-p, a plant insulin that is hypoglycemic and lowers blood sugar levels by boosting beta cells, increasing glucose uptake, and increasing glycogen synthesis in the liver, muscles, and fat cells (Thakur & Sharma, 2016; Sorifa, 2018). *In vitro* studies have demonstrated that *M. charantia* proteins (momorcharin) have an inhibitory effect on the human immunodeficiency virus (HIV). Furthermore, its extract can be used as a broad-spectrum antibacterial drug to treat infections (Jia et al., 2017).

2.5 *MORINGA OLEIFERA*

The scientific name for the moringa tree, also referred to as the drumstick tree or horseradish tree, is *Moringa oleifera*. Due to its tolerance to a variety of soil and climatic conditions, this plant, which is native to India, thrives in tropical and subtropical regions all over the world (Gopalakrishnan et al., 2016). It is one of the most popular veggies in developing nations. A member of the Moringaceae family is *Moringa oleifera* (Mani & Roy, 2020). There are 13 species in the genus, including *M. oleifera*, *M. arborea*, *M. borziana*, *M. concanensis*, *M. drouhardii*, *M. hildebrandtii*, *M. longituba*, *M. ovalifolia*, *M. peregrine*, *M. pygmaea*, *M. rivae*, *M. ruspoliana*, and *M. sten.* The sale of fruits, leaves, fuel wood, seeds oil, seedcake as fertilizer, and bio-pesticide made by the multipurpose tree species (MPT) moringa aids in alleviating poverty in rural areas. Because of the valuable nutritional benefit, it provides in times of drought, Ethiopians refer to *Moringa stenopetala* as the "Mother of Farmers." It has been dubbed "Miracle Tree" or "Wonder Tree," "Natural Nutrition for Tropics," and "Mother's Best Friend" due to its numerous nutraceutical characteristics and suitability for agroforestry systems due to its numerous nutritional, medicinal, and industrial applications (Prabhakar & Rao, 2018; Kachhawa & Chawla, 2017). Different plant portions are rich in proteins, lipids, carbohydrates, amino acids, vitamins, and minerals. As a result, it is utilized as a nutrient source and dietary supplement in the tropics, particularly for infants and newborns, where malnutrition is a major problem (Singh et al., 2020). Each Indian household regularly prepares and consumes traditional dishes like chapatti, pulav, poha, parantha, dalia, poori, potato, vegetables, etc. (Kachhawa & Chawla, 2017).

2.5.1 PHYTOCHEMISTRY OF *MORINGA OLIFERA*

An edible plant termed *Moringa oleifera* is rapidly gaining recognition for its ability to treat and prevent malnutrition, particularly health issues linked to hidden hunger (Olusanya et al., 2020). Its phytochemicals GSs, isothiocyanates, glycoside compounds, and glycerol-1-9-octadecanoate, as well as terpenoids, tannins, saponins, anthraquinones, flavonoids, alkaloids, and reducing sugar, are anti-cancerous agents reported in this plant (Mani & Roy, 2020). In the stems, leaves, flowers, pods, and seeds, bioactive secondary metabolites have been found that have nutritional and medicinal perks, such as flavonoids (such as quercetin, kaempferol, and vanillin, among others), phenolic acids (ellagic acid, gallic acid, ferulic acid, and other phenolic acids in trace amounts), glycosides, and GSs (Singh et al., 2020). Increased amounts

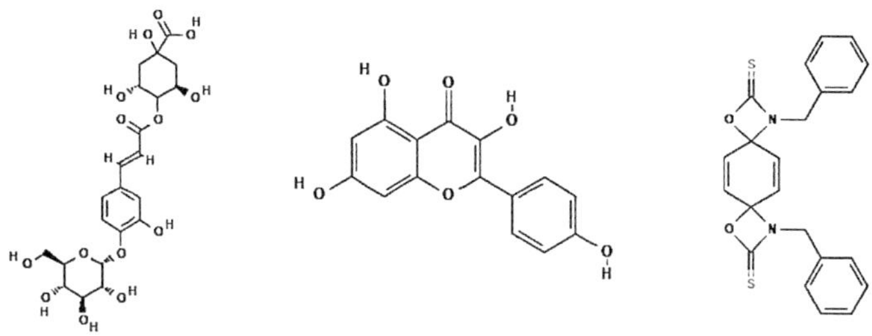

4-O-glucopyranosyl-caffeoyl quinic acid	Kaempferol	Pterygospermin

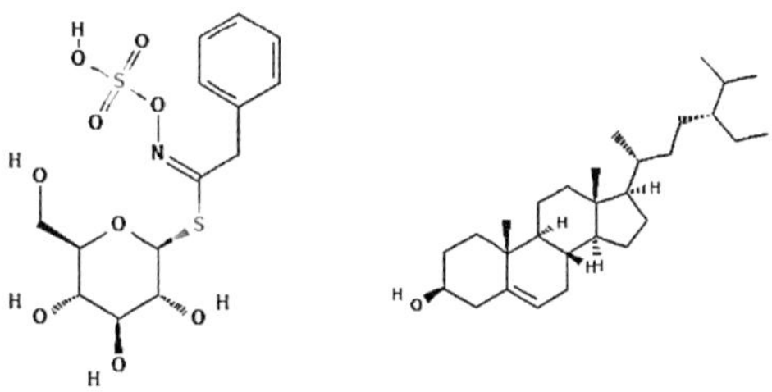

-α-L-rhamnopyranosyloxy benzyl glucosinolate	β-sitosterol

FIGURE 2.4 Structure of phytoconstituents of *M. oleifera*.

of nutrients and phytochemicals can be found in *Moringa oleifera* leaves. They contain a significant amount of protein, a range of minerals, including B vitamins, and an amino acid profile that is well-balanced. Quercetin and kaempferol in their 3-O-glycoside forms are two examples of the phenolic chemicals, which make up the other class of phytoconstituents discovered in plant leaves. Strong antioxidant properties can be found in quercetin and kaempferol. In studies using obese Zucker rats, the 3-O-glycoside derivative of quercetin known as isoquercitrin or isotrifolin was found to have hypotensive, antidyslipidemic, and anti-diabetic effects (Olusanya et al., 2020). The GSs O-ethyl-4-(-Lrhamnosyloxy) benzyl carbamate, 4-(-L-rhamnopyranosyloxy) benzylglucosinolate, and 4-(-L-rhamnosyloxy) benzylglucosinolate have been discovered to be present in every part of the *M. oleifera* plant. These compounds have well-known hypotensive, chemopreventive, and bactericidal effects (Singh et al., 2020) (Table 2.7 and Figure 2.4).

TABLE 2.7 Phytoconstituents Reported in Different Parts of *M. oleifera*

Plant Parts	Phytoconstituents	References
Leaves	4-O-glucopyranosyl-caffeoyl quinic acid	Singh et al. (2020)
Stem roots leaves and seeds	Kaempferol	
Roots and flowers	Pterygospermin	
Leaves, stem, and roots	-α-L-rhamnopyranosyloxy benzyl glucosinolate	
Seeds	β-sitosterol	

2.5.2 *NUTRITIONAL COMPOSITION OF MORINGA OLIFERA*

Moringa oleifera contains more than 90 nutritionally significant chemical components, including protein, fat, vitamins, minerals, crabs, and dietary fibers (Singh et al., 2020). Calcium, one of the most important nutrients for human growth, is one of the many components found in moringa that are essential for human growth and development. While 8 ounces of milk only has 300–400 mg per serving, moringa leaves have 1,000 mg per serving and a low caloric content, making them a better choice for diets for obese persons. More than 4,000 mg of moringa powder can be obtained. Since moringa powder can be taken in place of iron supplements, it functions well enough cure anemia. While beef only has 2 mg of iron, moringa leaf powder has 28 mg. According to reports, moringa provides more iron than spinach. The fibrous pods can be used to treat colon cancer and gastrointestinal issues. Around 30% of the pods, 44% of the leaves, and 31% of the flowers are amino acid rich. For sperm cells to grow normally in the diet, zinc is necessary. The synthesis of DNA and RNA also requires it. Zinc content in *M. oleifera* leaves ranges from 25.5 to 31.03 mg/kg, which is adequate for a daily diet (Gopalakrishnan et al., 2016; Mani & Roy, 2020). Argenine and histidine, which are found in moringa, are particularly important for newborns since they cannot produce enough protein to satisfy their growing needs. According to experts, protein deficits affect 30% of children in Sub-Saharan Africa. The moringa plant has the potential to be a very nourishing food.

2.5.3 *NUTRACEUTICAL PROPERTIES OF MORINGA OLEIFERA*

Moringa oleifera preparations with antibacterial, hypotensive, hypoglycemic, immunomodulatory, and anti-inflammatory properties. The seeds, for example, have potent antibacterial properties against *Staphylococcus aureus* and *Pseudomonas aeruginosa.* Flavonoids, quercetin, -sitosterol, and zeatin are all recognized to be natural antioxidants found in the leaves. Antispasmodic action has been discovered in *Moringa oleifera* roots and leaves. Anti-inflammatory action was found in hot water infusions of the leaves, seeds, flowers, roots, and bark. Anticancer, antiallergic, antioxidant, anti-inflammatory, immunomodulatory, antifungal, antibacterial, and hepatoprotective properties are all found in certain components of *M. oleifera* preparations (Table 2.8).

TABLE 2.8 Nutraceutical Properties of Different Parts of *M. olifera*

Plant Parts	Medicinal Properties	References
Pods/fruits	Treat malnutrition.	Pandey et al. (2019)
Bark	Promote digestion.	
Roots	Antilithic, antifertility, anti-inflammatory, promote digestion, beneficial against constipation, rheumatism, back pain, renal complications.	
Seeds	To treat abdominal tumors and antibacterial.	
Leaves	Useful in treating tumors, piles, eye, and ear infection, hypocholesterolemic, reduce glandular swelling, headaches, promote digestion.	
Flowers	Reduce cholesterol level, spleen enlargement, remedy for tumors, muscle disease.	

2.6 *RAPHANUS SATIVUS* L.

Plant-based diets have several health benefits for humans. It lowers the risk of cancer, CVDs, neurological disorders, and aging-related issues, among other things. Plants produce a variety of pharmacologically significant secondary metabolites. Due to their superior nutritional and medicinal potentials, vegetables from the cruciferous plant family have attracted a wide variety of dietary attention. GSs, phenolic compounds, tocopherols, carotenoids, and ascorbic acid are all found in cruciferous vegetables, according to several studies (Manivannan et al., 2019). The root vegetable crop *Raphanus sativus L.*, usually known as radish, is a well-known Brassicaceae family member. *R. sativus* L. is a plant that originated in Europe and Asia. It thrives in mild climates. It grows to a height of 30–90 cm, with thick roots of varied sizes, shapes, and colors. They have a potential flavor and use it as edible in food preparation. Traditional Japanese cuisine includes salted radish roots

(Takuan), which are consumed in the amount of roughly 5,00,000 tons per year in Japan. The roots of salted radish have a characteristic yellow color that develops during storage (Gutiérrez & Perez, 2014). Radishes are very different in colors as per the soil and species specific, ranging from bright white in Asia to blood red in Europe, purple, green, and black, while its flesh is white in most European and Asian countries (Gamba et al., 2021).

R. sativus has edible roots, leaves, and fruits. This plant's components have a wide range of ethnomedical applications. This plant has anticancer, antimicrobial, antidiabetic, diuretic, antifertility, hypertensive, antimicrobial, nephroprotective, gastroprotective, and hepatoprotective properties, to name a few. It is also used to treat gynecological issues and jaundice. Its seeds have been proven to be beneficial in the treatment of urinary problems in the Bechar province of Algeria's southwest. It has also been proven to have phytoremediation effects in contaminated soils (Agarwal & Varma, 2014).

2.6.1 *PHYTOCHEMISTRY OF RAPHANUS SATIVUS L.*

Several studies reported that cruciferous vegetables possess GSs, phenolic compounds, tocopherols, carotenoids, and ascorbic acid. Roots and leaves of radish contain many active principles like coumarins, alkaloids, nitrogenous compounds, gibberellins, glycosinolate, organic acids, phenolic compounds, sulfur pigments and sulforaphane. It contains antioxidant compounds (Jaafar et al., 2020). Calcium, iron, and ascorbic acid are abundant in radish leaves. When eaten with rice, it offers a good supply of calcium. Rice contains phytin that assists in calcification (Singh & Singh, 2013). Because of their potent secondary metabolites, radish root and leaf extracts are effective against antimicrobial microbes (Jaafar et al., 2020). It comprises water-soluble vitamins (B_1, B_2, B_3, B_5, B_6, B_9, and C) as well as minerals such as calcium, iron, magnesium, manganese, phosphorus, potassium, zinc, etc. Pungent smell compounds called GSs and their derivatives such as glucobrassicin, glucoerucin, glucoraphanin (Grn), glucoraphasatin, 4-hydroxyglucobrassicin, 4-methioxyglucobrassicin, neoglucobrassicin, etc., and potent principal bioactive component isothiocyanates are also identified in radish (e.g., sulforaphene, sulforaphane, and indole-3-carbinol) (Banihani, 2017). Sufloraphene, Beta-Sitosterol, Stearic acid, Sinapine, Castasterone, and Brassinolide are all found in the seeds. P-coumaric, ferulic, and caffeic acids are found in the leaves (Sreelekshmi & Pratibha, 2015) (Table 2.9).

TABLE 2.9 Chemical Compositions of *Raphanus sativus* L.

Nutritional Value/100 g	Radish Fruit	Radish Top	Radish Root	References
Moisture (%)	90.5	90.3	94	Singh & Singh (2013)
Edible matter (%)	88	–	99	
Fiber (%)	1.4	0.9	0.8	
Protein (%)	2.3	2.7	0.7	
Fat (%)	0.3	0.6	0.1	
Carbohydrates	4.7	3.4	3.4	
Minerals (%)	0.8	2.1	0.6	
Calcium (mg/100 g)	80	310	50.0	
Iron (mg/100 g)	2.8	16.1	0.4	
Phosphorus (mg/100 g)	100	60	22	
Vitamin A (I.U./100 g)	50	18.6	5.0	
Riboflavin (mg/100 g)	0.05	0.16	0.02	
Thiamine (mg/100 g)	0.07	0.03	0.06	
Vitamin C (mg/100 g)	69	103	15.0	
Nicotinic acid (mg/100 g)	0.2	0.3	0.5	

2.6.2 NUTRACEUTICAL PROPERTIES OF *RAPHANUS SATIVUS* L.

Radish is employed as a traditional cure for a range of ailments in traditional Indian medicine, Unani, Greeko-Arab, including jaundice, gallstones, liver issues, rectal prolapse, digestion issues and abdominal complications (Banihani, 2017). It is utilized for various biological properties such as antimicrobial, antiparasitic, antiscorbutic, aphrodisiac, carminative, cholagogue, diuretic, laxative, lithotripsic, stimulant, stomachic, etc. The juices from the fresh leaves are diuretic and laxative. Roots are used to address urinary problems and syphilitic illness, as well as piles and gastrodynic cramps (Ibrahim & Abdel-Mogib, 2019). Radish root is beneficial for troubles with the liver and gall bladder. It is used in homeopathy to treat headaches, insomnia, and persistent diarrhea. Butyl crotonylisothiocyanate sulfide, a major component, contributes to the characteristic radish odor. Methyl mercaptan contributes unpleasant odor to the radish (Singh & Singh, 2013). Radishes are diuretic, which means they increase urine production. It's a good detoxifier, and it quickens the healing of piles symptoms. It has the ability to relax the excretory system, improve water retention, and ease constipation (Sreelekshmi & Pratibha, 2015). The seed from different varieties of radish contains allyl, methyl, and isopropyl as glycosidically bounded oil derivatives in that isothiocyanates, and sulforaphene are more useful components for various applications in food industries. 4-methyl sulfenyl-3-butenyl cyanide compound corresponds as sulforaphane nitrile and it is a cleaved natural product of isolated glycosides. Radish seeds are also said to contain machrolysin, a broad-spectrum antibiotic that is specific for *Mycobacterium tuberculosis*. The non-drying fatty oil extracted from radish seeds is useful for soap manufacturing, illuminating, and culinary use (Singh & Singh, 2013).

2.7 CONCLUSION AND FUTURE PROSPECTS

Nutraceutical properties of any vegetable exhibit fundamental sources of potential phytoconstituents such as antioxidants, and anti-inflammatory activities due to its phenolic compounds as a key ingredient in the food and pharmaceutical industries. Vegetable plants and their products are highly significant due to the presence of a higher concentration of phenolics in all the edible components. In this scenario, vegetables gained world attention due to their higher value in nutrition and various applications in disease management, hence therapeutic molecules produced as nutraceutical and pharmaceutical products to mitigate problems as a key ingredient of food. The cost effect, availability, palatability, production rate, ecological conditions and economic importance are highly potential and interesting in vegetable production and consumption all over the world with and without meat food. In addition, the processing technology and application are very important in the usage of these vegetables in order to enhance the property to give more efficacies in any disease or disorder.

The future perspective of such work continued to explore the key ingredients which show the nutraceutical property to overcome any disease and also support like food, hence to fulfill the wide range of applications on any vegetables more research work required and to identify the active principles and should consider as a therapeutic molecule in support of traditional as well as modern medicine. It also suggests that in our diet at least any two such vegetables should be included with or without processing to impact its natural therapeutic action as immune boosters naturally by the diet and detoxifying various toxic elements.

KEYWORDS

- **antioxidants**
- **nutraceutical property**
- **organics**
- **pharmacological properties**
- **phytochemistry**
- **traditional medicine**
- **vegetables**

REFERENCES

Agarwal, K., & Varma, R., (2014). Radical scavenging ability and biochemical screening of a common Asian vegetable – *Raphanus sativus* L. *Int. J. Pharm. Sci. Rev. Res., 27*(1), 127–134.

Balakrishnan, A., (2015). Therapeutic uses of peppermint: A review. *J Pharmaceut. Sci Res., 7*(7), 474–476.

Banihani, S. A., (2017). Radish (*Raphanus sativus*) and diabetes. *Nutrients, 9*, 1–9.

Barua, R., UddinTalukder, M. E., Islam, M. S., Yesmin, F., Chakma, K., Kabir, M. G., & Bhuiyan, R. H., (2020). Nutritional analysis and phytochemical evaluation of bitter gourd (*Momordica charantia*) from Bangladesh. *Asian J. Agri. Food Sci., 8*(2), 11–17.

Bhat, S., Kaushal, P., Kaur, M., & Sharma, H. K., (2014). Coriander (*Coriandrum sativum* L.): Processing, nutritional and functional aspects. *Afr. J. Plant Sci., 8*(1), 25–33.

Gamba, M., Asllanaj, E., Raguindin, P. F., Glisic, M., Franco, O. H., Minder, B., et al., (2021). Nutritional and phytochemical characterization of radish (*Raphanus sativus*): A systematic review. *Trend Food Sci. Technol., 113*, 205–218.

Gopalakrishnan, L., Doriy, K., & Santhosh, K. D., (2016). *Moringa oleifera*: A review on nutritive importance and its medicinal application. *Food Sci. Human Wellness, 5*, 49–56.

Haleshappa, R., Patil, S. J., Usha, T., & Siddalinga, M. K. R., (2020). Phytochemicals, antioxidant profile and GCMS analysis of ethanol extract of *Simarouba glauca* seeds. *Asian J. Biolog. Life Sci., 9*, 379–385.

Haleshappa, R., Patil, S. J., Usha, T., & Siddalinga, M. K. R., (2021). Phytochemical analysis, *in vitro* evaluation of antioxidant and free radical scavenging activity of *Simarouba glauca* seeds. *Adv. Pharmacol. Pharm., 9*, 1–8.

Haleshappa, R., Sajeeda, N., Kolgi, R. R., Patil, S. J., & Siddalinga, M. K. R., (2022). Phytochemicals, anti-nutritional factors, and proximate analysis of *Simarouba glauca* seeds. *Int. Adv. Res. J. Sci. Engg. Technol., 9*(3), 218–227.

Jaafar, N. A., Ahmed, A. S., & Al-Sandooq, D. L., (2020). Detection of active compounds in radish *Raphanus sativus* L. and their various biological effects. *Plant Arch., 20*(2), 1647–1650.

Jia, S., Shen, M., Zhang, F., & Xie, J., (2017). Recent advances in *Momordica charantia:* Functional components and biological activities. *Int. J. Mol. Sci., 18*, 1–25.

Kachhawa, K., & Chawla, P., (2017). Improvement in nutritional quality of commonly consumed products with supplementation of drumstick leaves (*Moringa oleifera* L.). *Int J Home Sci., 3*(2), 478–485.

Kapp, K., Hakala, E., Püssa, T., Orav, A., Raal, A., & Hanski, L., (2013). Commercial peppermint (*Mentha piperita* L.) teas: Antichlamydial effect and polyphenolic composition. *Planta Medica*, 79. 10.1055/s-0033-1351813.

Keservani, R. K., & Sharma, A. K., (2014). Flavonoids: Emerging trends and potential health benefits. *Journal of Chinese Pharmaceutical Sciences, 23*(12), 815.

Keservani, R. K., Kesharwani, R. K., Sharma, A. K., Vyas, N., & Chadoker, A., (2010b). Nutritional supplements: An overview. *International Journal of Current Pharmaceutical Review and Research, 1*(1), 59–75.

Keservani, R. K., Kesharwani, R. K., Vyas, N., Jain, S., Raghuvanshi, R., & Sharma, A. K., (2010a). Nutraceutical and functional food as future food: A review. *Der Pharmacia Lettre, 2*(1), 106–116.

Keservani, R. K., Sharma, A. K., & Kesharwani, R. K., (2016a). Nutraceutical and functional foods for cardiovascular health. In: *Food Process Engineering* (pp. 257–278, 291–312). Apple Academic Press, CRC Press. ISBN: 9781771884020.

Keservani, R. K., Sharma, A. K., & Kesharwani, R. K., (2016b). Medicinal effect of nutraceutical fruits for the cognition and brain health. *Scientifica*, 10. Article ID 3109254. doi:10.1155/2016/3109254.

Keservani, R. K., Sharma, A. K., & Kesharwani, R. K., (2017). An overview and therapeutic applications of nutraceutical and functional foods. *Recent Advances in Drug Delivery Technology*, 160–201. ISBN:9781522507543.

Keservani, R. K., Sharma, A. K., & Kesharwani, R. K., (2020). *Nutraceuticals and Dietary Supplements: Applications in Health Improvement and Disease Management*. CRC Press. ISBN: 9781771888738.

Krishnendu, J. R., & Nandini, P. V., (2016). Nutritional composition of bitter gourd types *(Momordica charantia L.)*. *Int. J. Adv. Engg. Res. Sci., 3*(10), 96–104.

Loolaie, M., Moasefi, N., Rasouli, H., & Adibi, H., (2017). Peppermint, and its functionality: A review. *Arch. Clin. Microbiol., 8*(4), 54.

Mahleyuddin, N. N., Moshawih, S., Ming, L. C., Zulkifly, H. H., Kifli, N., Loy, M. J., et al., (2022). *Coriandrum sativum* L.: A review on ethnopharmacology, phytochemistry, and cardiovascular benefits. *Molecules, 27*(209), 1–19.

Mandal, S., & Mandal, M., (2015). Coriander (*Coriandrum sativum* L.) essential oil: Chemistry and biological activity. *Asian Pac. J. Trop. Biomed., 5*(6), 421–428.

Mani, A., & Roy, S., (2020). Nutritional importance and medicinal properties of drumstick. *Food Bever Processing, 6*(8), 52, 53.

Manivannan, A., Kim, J. H., Kim, D. S., Lee, E. S., & Lee, H. E., (2019). Deciphering the nutraceutical potential of *Raphanus sativus*: A comprehensive overview. *Nutrients, 11*(402), 1–15.

Maurya, U., Yadav, P., Maurya, S., Maurya, R., & Yadav, P., (2021). *Coriandrum sativum* L.: A nutritional food and phytomedicine. *Int. J. Crea. Res. Thoughts, 9*(1), 1730–1740.

Nadeem, M., Anjum, F. M., Issa, K. M., Tehseen, S., El-Ghorab, A., & Sultan, J. I., (2014). Nutritional and medicinal aspects of coriander (*Coriandrum sativum* L.): A review. *Brit. Food J., 115*(5), 743–755.

Naeem, M. Y. M., Kızıldeniz, T., Kaddour, A. A., Tursun, A. O., & Movila, M., (2022). *Coriander (Coriandrum sativum L.) a Nutritional, Aromatic and Medicinal Herb* (pp. 1–96). IKSAD Publishing House. Turkey.

Nayak, P., Kumar, T., Gupta, A. K., & Joshi, N. U., (2020). Peppermint a medicinal herb and treasure of health: A review. *J. Pharmacog. Phytochem., 9*(3), 1519–1528.

Nazim, M., Sadiq, Q. U., Nawaz, A., Anjum, S., Ali, M., & Maryam, H., (2020). *Mentha arvensis*, a medicinal and aromatic plant, has high nutritional value and several-uses: A review. *Buletin Agroteknologi., 1*(2), 37–49.

Nimish, P. L., Sanjay, K. B., Nayna, B. M., & Jaimik, R. D., (2011). Phytopharmacological properties of *Coriander sativum* as a potential medicinal tree: An overview. *J. Appl. Pharmaceut Sci., 01*(04), 20–25.

Olusanya, R. N., Kolanisi, U., Van, O. A., Ngobese, N. Z., & Siwela, M., (2020). Nutritional composition and consumer acceptability of *Moringa oleifera* leaf powder (MOLP)-supplemented mahewu. *South Afr. J. Bot., 129*, 175–180.

Opara, E. I., & Chohan, M., (2014). Culinary herbs and spices: Their bioactive properties, the contribution of polyphenols and the challenges in deducing their true health benefits. *Cul. Int. J. Mol. Sci., 15*(10), 19183–19202.

Pandey, V., Chauhan, V., Pandey, V. S., Upadhyaya, P., & Kopp, O., (2019). *Moringa olifera* Lam: A bifunctional edible plant from India, phytochemistry and medicinal properties. *J. Plant Stud., 8*(1), 10–19.

Prabhakar, B. N., & Rao, A. M., (2013). Drumstick: A miracle and nutraceutical tree–as source of income for rural livelihood. *Conference Proceedings: International Conference on Agriculture & Animal Sciences* (pp. 1–5). Srilanka.

Preeti, S., & Singh, J., (2013). Medicinal and therapeutic utilities of *Raphanus sativus*. *Int. J. Plant, Animal Environ. Sci., 3*(2), 103–105.

Rane, B. R., Bharath, M. S., Patil, R. R., Keservani, R. K., & Jain, A. S., (2020b). Novel approaches in nutraceuticals; Chapter 11. In: Rajesh, K. K., Raj, K. K., & Anil, K. S., (eds.), *Enhancing the Therapeutic Efficacy of Herbal Formulations Through Novel Drug Delivery Systems* (pp. 241–266). Edited by IGI Global International Publisher, Pennsylvania, USA. ISBN13: 9781799844532.

Rane, B. R., Patil, A. S., Keservani, R. K., & Jain, A. S., (2020a). Novel approaches in herbal formulation; Chapter 2. In: Rajesh, K. K., Raj, K. K., & Anil, K. S., (eds.), *Enhancing the Therapeutic Efficacy of Herbal Formulations Through Novel Drug Delivery Systems* (pp. 43–68). IGI Global International Publisher, Pennsylvania, USA. ISBN13: 9781799844532.

Rane, B. R., Tadavi, S. A., & Keservani, R. K., (2020c). Naturopathy; Chapter 12. In: Anil, K. S., Raj, K. K., & Surya, P. G., (eds.), *Herbal Product Development* (pp. 321–347). Apple Academic Press, CRC Press, Taylor & Francis Group. ISBN: 9781771888776.

Sahib, N. G., Anwar, F., Gilani, A. S., Hamid, A. A., Saari, N., & Alkharfy, K. M., (2013). Coriander (*Coriandrum sativum* L.): A potential source of high-value components for functional foods and nutraceuticals: A review. *Phytother Res., 27*, 1439–1456.

Sayed, I. H. Y. I., & Abdel-Mogib, M., (2019). Chemical constituents of radish, *Raphanus sativus* Linn. (*Brassicaceae*) leaves extracts and their insecticidal activity against mealybug, *Solenopsistinsley* (*Hemiptera: Pseudococcidae*). *Int. J. Cur. Adv. Res., 8*(10 B), 0179–20184.

Sharma, A. K., Keservani, R. K., & Gautam, S. P., (2020). *Herbal Product Development*. Apple Academic Press, CRC Press, Taylor & Francis Group. ISBN: 9781771888776.

Sharma, M., & Gautam, D., (2022). Phytoconstituents and medicinal value of *Mentha piperita*. *Mod. Phytomorphol., 15*, 156–160.

Singh, A. K., Rana, H. K., Tshabalala, T., Kumar, R., Gupta, A., Ndhlala, A. R., & Pandey, A. K., (2020). Phytochemical, nutraceutical, and pharmacological attributes of a functional crop *Moringa oleifera* Lam: An overview. *South Afr. J. Bot., 129*, 209–220.

Singh, P., Kesharwani, R. K., & Keservani, R. K., (2017a). Antioxidants and vitamins: Roles in cellular function and metabolism. In: *Sustained Energy for Enhanced Human Functions and Activity* (pp. 385–407). Academic Press. ISBN:9780128054130.

Singh, P., Kesharwani, R. K., & Keservani, R. K., (2017b). Protein, carbohydrates, and fats: Energy metabolism. In: *Sustained Energy for Enhanced Human Functions and Activity* (pp. 103–115). Academic Press. ISBN:9780128054130.

Sorifa, A. M., (2018). Nutritional compositions, health promoting phytochemicals and value-added products of bitter gourd: A review. *Int Food Res J., 25*(5), 1763–1772.

Sreelekshmi, G., & Pratibha, K., (2015). A safe body through moolaka –*Raphanus sativus* Linn. *Int Ayurvedic Med J., 3*(10), 3090–3096.

Tafrihi, M., Imran, M., Tufail, T., Gondal, T. G., Caruso, G., Sharma, S., et al., (2021). The wonderful activities of the genus Mentha: Not only antioxidant properties. *Molecule, 26*(1118), 1–22.

Thakur, A., & Sharma, R., (2016). Health promoting phytochemicals in vegetables: A mini review. *Intl J Food Ferment Technol., 8*(2), 107–117.

Valyaie, A., Azizi, M., Kashi, A., Sathasivam, R., Un Park, S., Akifumi, S. A., et al., (2021). Evaluation of growth, yield, and biochemical attributes of bitter gourd (*Momordica charantia* L.) cultivars under Karaj conditions in Iran. *Plants., 10*(1370), 1–19.

www.TFLJournal.org (accessed on 15 November 2023).

Yasmeen, R., Fukugawa, N. K., & Wang, T. T. Y., (2017). Establishing health benefits of bioactive food components: A basic research scientist's perspective. *Curr. Opin. Biotechnol., 44*, 109–114.

PART II
Nutraceutical Benefits of Bioactive and Flavonoids

CHAPTER 3

PHYTOCHEMICALS PRESENT IN VEGETABLES FOR HEALTH PROMOTION

KAMAL KANT SHARMA,[1] MUNIT SHARMA,[1] DEEPA DEVI,[1] APARAJITA GUPTA,[2] and SHIVAM SINGH[3]

[1]*Department of Botany and Microbiology, Gurukula Kangri (Deemed to be University), Haridwar, Uttarakhand, India*

[2]*Department of Microbiology, Dolphin PG College of Science and Agriculture, Chunni Kalan, Fatehgarh Sahib, Punjab, India*

[3]*Scientist B, Government District Hospital Cum Trauma Center, Waidhan, Singrauli, Madhya Pradesh, India*

ABSTRACT

Currently, several plant-based natural products has been emerged recently as a key source of nutraceuticals or functional foods that are playing an important role in providing therapeutic assistance through diet. They have received significant consideration in diet due to their safe and efficient nutritive as well as therapeutic values. Amongst dietary supplements, vegetables attain a significant stage due to its low calorific value, high phytochemical, and fiber content, which places it at a significant level of functional food. A portion of the common nutraceuticals, for instance, carotenoids from bell pepper, carrot, and melon and flavonoids as rutin, kaempferol, and myricetin are generally predominant in onion, hot pepper, broccoli, snap bean, kale, and lettuce like vegetables. These functional foods along with others play a vital role in disease inhibition and/or prevention of disease through promoting

Plant Metabolites and Vegetables as Nutraceuticals.
Raj K. Keservani, Bui Thanh Tung, Rajesh K. Kesharwani, and Eknath D. Ahire (Eds.)

antioxidant activity, regulating cell signaling or by other unspecified pathways. In several cases, this health benefit is not mainly due to the action of a single group of phytocomponent but is essentially the overall effect of multiple components interacting with each other at a time. Functional aspects of several traditional vegetables are being revealed, while new one (as dandelion) are still being added in diet having accompanying nutraceutical values.

3.1 INTRODUCTION

The demand for immunity enhancing nutraceuticals as well as functional foods is elevated significantly during the COVID-19 pandemic. Moreover, from the time, a dearth in the use of meat, poultry, and marine foodstuffs is noticed across the globe (Belton et al., 2021). The demand for plant-based nutraceutical in India is expected to cross $18 billion by 2025 according to 'The developing nutraceuticals market in India' a study accompanied by Invest India (Anumala et al., 2021). In Indian folklore and old stories, food grown from the earth has typically been regarded as being rich in some basic dietary micronutrients and fiber. However, in recent years they have been perceived as a significant source of phytochemicals that may contribute to well-being individually or in combination (Sarkar et al., 2015). In this manner, at present greater part of individuals have pointed products of the soil under the subtitle of "functional foods" (Alkhatib et al., 2017). There are numerous organically conceivable explanations behind this possibly defensive affiliation, including the way that a significant number of the phytochemicals act as antioxidants (Yahia et al., 2017). Phytochemicals present in vegetables are exceptionally different, for example, carotenoids, flavonoids, and phenolic compounds (Keservani et al., 2016a, b; Keservani & Sharma, 2014; Percival & Turner, 2002; Devkota, 2022). *In vitro*, plant phenolics exhibit significant antioxidative activity, which has significant implications for human welfare in the vast majority of cases. Rich sources are tea wine, organic products, and vegetables (Duthie & Crozier, 2000; Shahidi & Ambigaipalan, 2015).

In terms of horticulture, vegetables are generally herbaceous plant ingested raw or cooked as whole (such as cabbage and cauliflower), or its part (such as fruit, seeds, tubers, stems, leaves, or flower, etc.), usually eaten raw or cooked as part of a meal. They are considered practical food nutraceuticals in light of the fact that they give minerals and supplements which are wellbeing advancing. They have a significant impact in the human eating regimen and are a significant wellspring of naturally dynamic nutraceuticals (Pennington, 2003). In addition, variety of vegetable items has previously entered the human eating routine and are as yet entering like *Centella asiatica*, *Urtica dioica*, and *Taraxacum officinale*, as enhancements, yet due to the absence of appealing surface and savoring taste, their utilization in ordinary eating regimen is still under rated. Some models among vegetables are green leafy vegetables like spinach, and others are garlic, onion, and broccoli, ginger, root harvests, artichoke, and peppergrass (Chandrika & Kumara, 2015; Malik et al., 2021; Anumala et al., 2021).

The medicinal value of such vegetables is supplementary to its nutraceutical value which reveals its dual purpose to serve mankind (Anumala et al., 2021). The entire members of the *Cinchona* comprise a well-known alkaloids Quinine which is said to be responsible for malaria, sore throat, nervous ergonomic, anemia, and convalescence (Nair, 2021). Likewise, members of *Alliaceae* family like Garlic and Parsnip contains sulfur containing phytochemicals which amplify the glutathione (GSH) redox cycle and triggers specific immunity. Bioactivity of these phytochemicals shows an antioxidant, antimicrobial, anticarcinogenic, immune-modulating, and hepato-protective activity. Recent evidence proves that members of *Alliaceae* family (as Garlic) checks several cardiac complications viz atherosclerosis, hypercholesterolemia, and hypertension and cancer. It said to be an effective medication against various inflammations escalating due to bacterial and viral infections (Proust, 2008; Hassan, 2020). Some species of Genus *Origanum* commonly known as thyme (mostly *Thymus vulgaris*), is used as spice with antiseptic and antimicrobial properties as utilized in several traditional medicinal system of Asia against all kinds of infections (Patil et al., 2021). Moreover, several variations in physical and biochemical parameters are found among the members of *Cruciferae* family as Turnip, Cabbage, Broccoli, Cauliflower, and the Brussels sprouts which have been reported to contain anti-radical activity revealed by their remarkable contents in polyphenols, flavones, and condensed tannins (Yerou et al., 2022; Muneeb et al., 2022). Flavonoids are bioactive substances derived from several members of bell pepper and tea that have anti-inflammatory activity, and fortify the human body against allergies, viruses, and the factors inducing tumors (Pinto et al., 2021; Kim et al., 2021).

Recent evidence proves that naturally occurring phytochemicals can also act as nutraceuticals (Keservani et al., 2010a, b) which can retain anticarcinogenic activity and other associated effects (as anti-inflammatory), which are collectively referred to as chemo preventers (Calvani et al., 2020; Gupta

& Prakash, 2022). Evidence demonstrates that the intake of vegetable as entire might be more than the amount of the parts seems to act synergistically where the impact of basically not many of these is added substance (Singh & Rawat, 2021).

3.2 CLASSIFICATION OF PHYTOCHEMICALS

For the most part, phytochemicals have been named important metabolites, subjective to their activities. These secondary metabolites contain normal sugars, amino acids, proteins, purines, and pyrimidines of nucleic acids including different mixtures as chlorophylls, and some more (Koche et al., 2016). Besides, the investigation of these phytochemicals is additionally perplexing because of the broad variety even inside similar gathering of mixtures, and the metabolic deprivation or change that might happen during squashing or handling of plants (for example for Allium and Brassica components), thusly expands the intricacy of the blend (Tiwari et al., 2015).

Moreover, plants contain previously mentioned combinations of nutraceuticals dynamic phytochemicals, for the most part optional metabolites, having very factor gatherings of phytochemicals regarding their number, primary heterogeneity, and conveyance, with significant chance for association (Rowland, 1999; Tiwari et al., 2015; Shah, 2009).

Additionally, these nutraceutical (Keservani et al., 2017, 2020) constituents can be categorized on the basis of their biological sources, therapeutic conditions, or chemical characteristics of the bio-based products (Table 3.1).

TABLE 3.1 Classification of Nutraceuticals According to Their Chemical Groups

SL. No.	Class	Examples
1.	Inorganic mineral	Minerals as phosphorous and iron.
2.	Vitamins	Vitamins as vitamins A and B.
3.	Digestive enzymes	Enzymes as lipase and protease.
4.	Probiotics	Bacteria as *Lactobacillus* and *Bifidobacterium*.
5.	Prebiotic	Fibers as asparagus and chickpea.
6.	Dietary fibers	Fibers as carrots and broccoli.
7.	Cereals and grains	Mostly whole cereals and grains.
8.	Health drinks	Mostly fiber containing juice and soup.
9.	Antioxidants	Natural antioxidants as tomato and lemon.
10.	Phytochemicals	–
11.	Herbs and spices	–

3.3 NUTRACEUTICALS AND PHYTOCHEMICALS: PERCEPTION OF ITS BIOAVAILABILITY

The term nutraceutical term was authored by Dr. Stephen DE Felice in 1989, which as per him is a blend of nourishment and drug. Such mix might go from segregated supplements, dietary enhancements, and explicit weight control plans to hereditarily designed fashioner food varieties, home grown items and handled food sources like oats, soups, and refreshments (Arya et al., 2019; Anumala et al., 2021). A nutraceutical is determined to have a few physiological advantages as giving security against transferable and non-communicable diseases or stopping cell multiplication by restraining cell flagging or modifying the biochemical pathway (Daliu et al., 2018; Mao et al., 2018). At present, various nutraceutical food stuffs are famous (nonalcoholic beverages, zest combinations, sauces, and extraordinarily pre-arranged items) which remember for their recipes explicit parts or normal spices having a restorative impact to be specific tonic, stimulative, stomach related, defender or cancer prevention agent, which are extricated from the most popular and very much valued developed plants or from the unconstrained greenery (Fernandes et al., 2019; Egbuna & Dable-Tupas, 2020). Phytochemicals confined or refined from natural food stuffs are by and large sold in restorative structures (as container and syrup) yet not straightforwardly connected with food as entire are alluded to as nutraceuticals (Egbuna & Dable-Tupas, 2020). Though it seems easy to develop a pharmaceutical form (capsule, tablets, suppositories, etc.) containing bioactive food, it is relatively challenging to attain inadequate bioavailability for such nutraceuticals. The bioavailability is often imperiled by the reduced stability, solubility, and/or permeability of the bioactive constituent in the gastrointestinal tract (Kesharwani et al., 2010; Dima et al., 2020). For example, in the case of rutin the major disadvantage associated with its reduced bioavailability is primarily caused by its low aqueous solubility, poor stability and restricted membrane permeability, on other hand, another bioactive component curcumin agonizes restricted serum and tissue dispersal, relatively prompt metabolism, and diminutive circulation half-life (Kesharwani et al., 2010; Gullon et al., 2017). In addition, the solubilization of bioactive mixtures relies upon their chemical structure and the states of processing climate. Subsequently, the hydrophilic mixtures are by and large solvent in the gastrointestinal liquids as per pH and ionic strength, though the hydrophobic mixtures are basically solubilized in the digestive fluid, in the presence of the lipolytic proteins (Sarabandi et al., 2019; Dima et al., 2020). On other hand, the solubilization of hydrophobic

it reorganizes the central metabolism to suppress the metabolic dysregulation caused by excess glucose and that it identifies glucose biosynthesis and glucose one-carbon (1C) metabolism as critical mechanistic pathways (Bernuzzi, 2021). Another investigation, done *in vivo* and *in vitro* model systems, demonstrated the aforementioned compound's antihypertensive, cardioprotective, and hypocholesterolemia potential at various doses (Tebay et al., 2015; Qin et al., 2015).

3.5.4 PLANT-BASED STEROLS AND STANOLS

These comprise natural plant-based natural components such as phytosterols and phytosterols that are essentially incorporated into the human diet largely from vegetable oils, cereals, and fruits. Phytosterols like β-sitostanol and campestanol, have less abundance in nature, accounting for around 2% of the total sterols present in common sources such as corn, wheat, rye, and rice (Pascal et al., 2009; Gangopadhyay, 2021). Sitosterol, campesterol, sitostanol, and campestanol are the most abundant phytosterols and phytosterols in the human diet (Kardong et al., 2013; Harborne et al., 2013). Both of these sterol derivatives supply no energy to the body but aid in cholesterol reduction, cancer prevention, immunological modulation, and skin protection (Brufau et al., 2008; Patel et al., 2017). Both of these substances have been shown to be helpful in lowering serum cholesterol concentrations by competing for intestinal absorption with dietary and biliary cholesterol (Chang-Mo, 2020). Furthermore, they can also have antioxidant, antiulcer, immunomodulatory, antibacterial, antifungal, and wound healing activities, as well as prevent platelet aggregation and control inflammation (Dutta, 2003; Ogbe et al., 2015; Joshi et al., 2022). According to current research, these plant-based sterol/stanol derivatives are physiologically active chemicals with a variety of health applications (Schiepers et al., 2009; Poudel et al., 2019).

On the other hand, a recent meta-analysis study specified that although phytosterol supplementation affects only the body mass index (BMI), but not the other anthropometric indices of the study (Ghaedi et al., 2019). Moreover, some research has shown that not all phytosterols supplementation may be beneficial, and those individuals with sitosterolemia may be susceptible to overload. Also, limited evidence supports a reduction in fat-soluble vitamins and carotenoid concentrations through the decline in the number of available lipoprotein carriers by their consumption (Ghaedi et al., 2019; Chang-Mo, 2020). In some current research, it has been shown that not all phytosterols show beneficial effects for individuals having sitosterolemia. Also, less

3.3 NUTRACEUTICALS AND PHYTOCHEMICALS: PERCEPTION OF ITS BIOAVAILABILITY

The term nutraceutical term was authored by Dr. Stephen DE Felice in 1989, which as per him is a blend of nourishment and drug. Such mix might go from segregated supplements, dietary enhancements, and explicit weight control plans to hereditarily designed fashioner food varieties, home grown items and handled food sources like oats, soups, and refreshments (Arya et al., 2019; Anumala et al., 2021). A nutraceutical is determined to have a few physiological advantages as giving security against transferable and non-communicable diseases or stopping cell multiplication by restraining cell flagging or modifying the biochemical pathway (Daliu et al., 2018; Mao et al., 2018). At present, various nutraceutical food stuffs are famous (nonalcoholic beverages, zest combinations, sauces, and extraordinarily pre-arranged items) which remember for their recipes explicit parts or normal spices having a restorative impact to be specific tonic, stimulative, stomach related, defender or cancer prevention agent, which are extricated from the most popular and very much valued developed plants or from the unconstrained greenery (Fernandes et al., 2019; Egbuna & Dable-Tupas, 2020). Phytochemicals confined or refined from natural food stuffs are by and large sold in restorative structures (as container and syrup) yet not straightforwardly connected with food as entire are alluded to as nutraceuticals (Egbuna & Dable-Tupas, 2020). Though it seems easy to develop a pharmaceutical form (capsule, tablets, suppositories, etc.) containing bioactive food, it is relatively challenging to attain inadequate bioavailability for such nutraceuticals. The bioavailability is often imperiled by the reduced stability, solubility, and/or permeability of the bioactive constituent in the gastrointestinal tract (Kesharwani et al., 2010; Dima et al., 2020). For example, in the case of rutin the major disadvantage associated with its reduced bioavailability is primarily caused by its low aqueous solubility, poor stability and restricted membrane permeability, on other hand, another bioactive component curcumin agonizes restricted serum and tissue dispersal, relatively prompt metabolism, and diminutive circulation half-life (Kesharwani et al., 2010; Gullon et al., 2017). In addition, the solubilization of bioactive mixtures relies upon their chemical structure and the states of processing climate. Subsequently, the hydrophilic mixtures are by and large solvent in the gastrointestinal liquids as per pH and ionic strength, though the hydrophobic mixtures are basically solubilized in the digestive fluid, in the presence of the lipolytic proteins (Sarabandi et al., 2019; Dima et al., 2020). On other hand, the solubilization of hydrophobic

TABLE 3.2 List of Some Essential Macro- and Micro-Nutrients and Their Vegetable Source

SL. No.	Nutrient	General Functions	Vegetable Source	Daily Requirement for Age 18–50 Years	
				Male	Female
1.	Calcium	Bone strength and growth, blood clotting, muscle contraction.	Agathi, amaranth leaves celery, Chinese broccoli, French beans, okra, and turnip.	1,000 milligrams per day.	1,200 milligrams per day.
2.	Choline	Cell structure and signaling.	Brussels sprout, broccoli, mushroom, and peanut.	550 milligrams per day.	425 milligrams per day.
3.	Chromium	Regulates blood sugar levels.	Broccoli, green bean, potato, and tomato.	35 micrograms per day.	25 micrograms per day.
4.	Copper	Blood iron (Heme) regulation.	Asparagus, peanuts, shiitake mushroom, sweet potato, sweet potatoes, lima bean, turnip green, beet green, spinach, and soybean sprout.	900 micrograms per day.	1,000 micro-grams per day.
5.	Fluoride	Bone growth, reduce teeth cavity.	Potato, spinach, and tea.	4 milligrams per day.	3 milligrams per day.
6.	Folic acid (folate).	Prevent birth defect, improves cardiac health.	Collards, turnip, lettuce, and spinach.	400 micrograms per day.	400 micrograms per day.
7.	Iodine	Regulates release of thyroid hormones.	Broccoli, brussels sprouts, cabbage, cauliflower, spirulina, and soybean.	150 micrograms per day.	150 micrograms per day.
8.	Iron (Heme)	Required for red blood cells, act as cofactor for enzymes.	Spinach, beans, lentils, and soybean.	8 milligrams per day.	18 milligrams per day.
9.	Magnesium	Helps with heart rhythm, muscle, and nerve function, bone strength.	Collard green, kale, mustard, potato, soybean, spinach, turnip, and quinoa.	400 milligrams per day.	410 milligrams per day.
10.	Manganese	Helps in bones growth, act as cofactor and coenzyme for a few enzymes.	Chickpea, wheat germ, sweet potatoes, nuts, spinach, lima beans, and other legumes.	2.3 milligrams per da.	1.8 milligrams per day.

3.3 NUTRACEUTICALS AND PHYTOCHEMICALS: PERCEPTION OF ITS BIOAVAILABILITY

The term nutraceutical term was authored by Dr. Stephen DE Felice in 1989, which as per him is a blend of nourishment and drug. Such mix might go from segregated supplements, dietary enhancements, and explicit weight control plans to hereditarily designed fashioner food varieties, home grown items and handled food sources like oats, soups, and refreshments (Arya et al., 2019; Anumala et al., 2021). A nutraceutical is determined to have a few physiological advantages as giving security against transferable and non-communicable diseases or stopping cell multiplication by restraining cell flagging or modifying the biochemical pathway (Daliu et al., 2018; Mao et al., 2018). At present, various nutraceutical food stuffs are famous (nonalcoholic beverages, zest combinations, sauces, and extraordinarily pre-arranged items) which remember for their recipes explicit parts or normal spices having a restorative impact to be specific tonic, stimulative, stomach related, defender or cancer prevention agent, which are extricated from the most popular and very much valued developed plants or from the unconstrained greenery (Fernandes et al., 2019; Egbuna & Dable-Tupas, 2020). Phytochemicals confined or refined from natural food stuffs are by and large sold in restorative structures (as container and syrup) yet not straightforwardly connected with food as entire are alluded to as nutraceuticals (Egbuna & Dable-Tupas, 2020). Though it seems easy to develop a pharmaceutical form (capsule, tablets, suppositories, etc.) containing bioactive food, it is relatively challenging to attain inadequate bioavailability for such nutraceuticals. The bioavailability is often imperiled by the reduced stability, solubility, and/or permeability of the bioactive constituent in the gastrointestinal tract (Kesharwani et al., 2010; Dima et al., 2020). For example, in the case of rutin the major disadvantage associated with its reduced bioavailability is primarily caused by its low aqueous solubility, poor stability and restricted membrane permeability, on other hand, another bioactive component curcumin agonizes restricted serum and tissue dispersal, relatively prompt metabolism, and diminutive circulation half-life (Kesharwani et al., 2010; Gullon et al., 2017). In addition, the solubilization of bioactive mixtures relies upon their chemical structure and the states of processing climate. Subsequently, the hydrophilic mixtures are by and large solvent in the gastrointestinal liquids as per pH and ionic strength, though the hydrophobic mixtures are basically solubilized in the digestive fluid, in the presence of the lipolytic proteins (Sarabandi et al., 2019; Dima et al., 2020). On other hand, the solubilization of hydrophobic

mixtures takes place through the consideration of their atoms into blended micelle phase comprising of bile salts, free unsaturated fats, phospholipids, along with other related factors (Dima et al., 2020). The "bio-accessibility" is for the most part considered as the underlying step of the bioavailability and is made sense of as the small amount of consumed nutraceutical that becomes available for retention through the digestive epithelial film (Jafari & McClements, 2017).

3.4 VEGETABLES: FROM PHYTOCHEMICAL TO NUTRITION

Already, a range of vegetable products has been touched the human diet and are still arriving (viz nettle and dandelion), as supplements with a supporting role in nourishment and sometimes as therapeutic (Anumala et al., 2021). Due to the dearth in attractive texture and savoring taste, most of the population do not consume it directly, but ingest them by cooking it with spices or other additives which increases its taste (Raghavan, 2006; Anumala et al., 2021). Some examples of commonly ingested vegetables are spinach, dill, thyme, lemon grass (leafy vegetables), bell peppers, okra, pumpkin, tomatoes (fruit vegetables), artichoke, caper, cauliflower, clove (flower bud vegetable), courgetti flowers, courgetti flowers, squash blossoms (flower vegetable) including asparagus, bamboo shoots, kohlrabi (stem vegetables) including carrot, potato, turnip, beet root (root vegetables) and Muchmore (Tomas-Barberan & Robins, 1997; Radovich, 2011).

As per latest recommendation of WHO, it is advised to intake a minimum 300–400 g of vegetables and fruit per day, excluding starchy root stuffs as potato. Further, they advised to intake variety of seasonal vegetables and fruits in meals and also as snacks which can reduce the risk of a range of non-communicable diseases including malnutrition and also helps to ensure an adequate day-to-day intake of dietary fiber. A wide assortment of vegetables gives a scope of supplements and different bioactive compounds (Septembre-Malaterre et al., 2018).

The medicinal values of these vegetables are adjuvant to their nutraceutical value and hence their use in diet in sufficient amount could provide desired therapeutic implications (Ali et al., 2021). Likewise, the members of *Alliaceae* family (garlic, onion, parsnip) synthesize sulfur compounds which increases the redox cycle of GSH and activate specific immunity (Iciek et al., 2009). The bioactivity of aforementioned compounds exhibits a significant antioxidant, antimicrobial, anticarcinogenic, immune modulating and hepatoprotective potential (Iciek et al., 2009; Ali et al., 2021). There is vast diversity among the species of broccoli, brussels, sprouts, cabbage, and cauliflower that have free radical scavenging activity proven by their remarkable contents of flavones, polyphenols, and condensed tannins (Septembre-Malaterre et al., 2018; Anumala et al., 2021). The glycosidic flavanols and flavones isolated from vegetables as carrot, chive, lettuce, onion, and spinach that have anti-inflammatory activity, and fortify the human body against allergies, viruses, and the factors inducing various cancer and tumor (Huber et al., 2009; Hostetler et al., 2017). An enormous quantity of vegetables which had been utilized in diet since time have been displayed to assume an urgent part in the prevention of different illness (Jagdale et al., 2021). Notwithstanding the full scale and miniature nutrients necessary for typical digestion-a plant-based diet contains numerous non-nutritive phytoconstituents which may also assume a significant part in wellbeing upgrade (Grivetti & Ogle, 2000; Taylor et al., 2016) (Table 3.2).

As of late, a few vegetables have gotten substantial consideration in diet since they are protected, solid, and have likely healthy benefit as well as remedial impacts to control and forestall various infections (Nasri et al., 2014). Among regular dietary enhancements, vegetables being low in calories are loaded with nutrients, minerals, cancer prevention agents and phytochemicals viz lycopene from tomato, carotenoids from carrot and so on. Such phytochemicals assume a crucial part in sickness counteraction and a reduction in illness risk factors most probably through their cell reinforcement action (Lee et al., 2017). Practical traits of numerous conventional vegetables are presently being found, along with the presentation of a few new vegetable items that contains extra nutraceutical benefits (Gokhale, 2021).

3.5 MAJOR BIOACTIVE PHYTOCHEMICALS OF VEGETABLES

Foods such as vegetables that are derived from plants contain bioactive phytochemicals that can provide health benefits beyond fundamental nutrition by decreasing the risk of chronic diseases spreading (Chandra et al., 2012). The medical benefits of bioactive mixtures in whole food varieties cannot be fully understood until these phytochemicals can be confined and distinguished (Chandra et al., 2012; Liu, 2013). In addition, modern *in vitro* research suggests that beyond antioxidant activity these bioactive phytochemicals present in vegetables have multiple mechanisms of action

TABLE 3.2 List of Some Essential Macro- and Micro-Nutrients and Their Vegetable Source

SL. No.	Nutrient	General Functions	Vegetable Source	Daily Requirement for Age 18–50 Years	
				Male	Female
1.	Calcium	Bone strength and growth, blood clotting, muscle contraction.	Agathi, amaranth leaves celery, Chinese broccoli, French beans, okra, and turnip.	1,000 milligrams per day.	1,200 milligrams per day.
2.	Choline	Cell structure and signaling.	Brussels sprout, broccoli, mushroom, and peanut.	550 milligrams per day.	425 milligrams per day.
3.	Chromium	Regulates blood sugar levels.	Broccoli, green bean, potato, and tomato.	35 micrograms per day.	25 micrograms per day.
4.	Copper	Blood iron (Heme) regulation.	Asparagus, peanuts, shiitake mushroom, sweet potato, sweet potatoes, lima bean, turnip green, beet green, spinach, and soybean sprout.	900 micrograms per day.	1,000 micro-grams per day.
5.	Fluoride	Bone growth, reduce teeth cavity.	Potato, spinach, and tea.	4 milligrams per day.	3 milligrams per day.
6.	Folic acid (folate).	Prevent birth defect, improves cardiac health.	Collards, turnip, lettuce, and spinach.	400 micrograms per day.	400 micrograms per day.
7.	Iodine	Regulates release of thyroid hormones.	Broccoli, brussels sprouts, cabbage, cauliflower, spirulina, and soybean.	150 micrograms per day.	150 micrograms per day.
8.	Iron (Heme)	Required for red blood cells, act as cofactor for enzymes.	Spinach, beans, lentils, and soybean.	8 milligrams per day.	18 milligrams per day.
9.	Magnesium	Helps with heart rhythm, muscle, and nerve function, bone strength.	Collard green, kale, mustard, potato, soybean, spinach, turnip, and quinoa.	400 milligrams per day.	410 milligrams per day.
10.	Manganese	Helps in bones growth, act as cofactor and coenzyme for a few enzymes.	Chickpea, wheat germ, sweet potatoes, nuts, spinach, lima beans, and other legumes.	2.3 milligrams per da.	1.8 milligrams per day.

TABLE 3.2 *(Continued)*

SL. No.	Nutrient	General Functions	Vegetable Source	Daily Requirement for Age 18–50 Years	
				Male	Female
11.	Molybdenum	Cofactor for enzymes.	Banana, black-eyed pea, lima bean, potatoes, and nuts.	109 micrograms per day.	76 micrograms per day.
12.	Phosphorus	Required for bone growth and cell signaling.	Carrot, celery, cucumber, lettuce, peas, mushroom, onions, pepper, radish, and tomato.	1,596 milligrams per day.	1,189 milligrams per day.
13.	Potassium	Control blood pressure, reduces possibility of kidney stones.	Avocado, bean, banana, beet green, lentil, potato, and soybean.	3,400 milligrams per day.	2,600 milligrams per day.
14.	Selenium	Protects cells from damage. Helps manage thyroid hormone.	Asparagus, Brazil nuts, broccoli, plantain, Brussel sprout, spinach, and whole soybean.	55 micrograms per day.	55 micrograms per day.
15.	Sodium	Regulates body fluid balance.	Celery, carrot, beet, and spinach.	2,300 milligrams per day.	2,300 milligrams per day.
16.	Zinc	Supports your immune system and nerve function. Also important for reproduction.	Corn, broccoli, garlic, green pea, kale, mushroom, lima bean, lentil sprout and spinach.	11 milligrams per day.	8 milligrams per day.

interacting with each other (Doughari, 2012; van Breda & de Kok, 2018). Because these bioactivities differ extensively according to presence of respective phytochemical groups from one to another vegetable source (Zhang et al., 2015; Phan et al., 2018).

Vegetables are rich source of bioactive phytochemicals such as flavonoids, carotenoids, anthocyanins, vitamins, and other polyphenolics. Such components play a huge job in sickness counteraction/lessen illness risk factors essentially through cancer prevention agent activity probably interacting with other cell action/systems (Zhang et al., 2015; van Breda & de Kok, 2018). Novel breeding strategies resulted the development of artificially cultivated edible vegetables as bananas, broccoli, kale, and potatoes, that have been able to boost their phytochemicals content resulting to upraise their nutritional standards (Hounsome et al., 2008). Along with several phytochemicals, almost all vegetables contain sufficient amount of dietary fiber which also contribute to the overall health benefit (Dias, 2012). The important phytochemicals groups present in vegetables which are of nutraceuticals importance are discussed underneath.

3.5.1 CAROTENOIDS

Carotenoids are red, yellow, and orange natural, fat-soluble pigments, tracked down in Chloroplast and Chromoplast of plants. Carotenoids are the precursor of Vitamin A and are strong antioxidants that aides in forestalling a few types of malignant growth and heart sicknesses (Fiedor & Burda, 2014; Langi et al., 2018). Carotenoids are a class of hydrocarbon compounds comprising tetraterpenes (40 carbon iotas) and broadly formed a twofold security framework that regulates several capabilities including its color (Thakur & Sharma, 2018). Primarily, the majority of carotenoids are made out of eight isoprenoid units, such as β carotene. Hydrocarbons, such as α and β carotene, tracked down in carrots, are alluded to as carotenes and oxygenated subsidiaries, such as lutein and β cryptoxanthin, tracked down prevalently in corn and tangerines, separately, are alluded to as xanthophylls (Namitha & Negi, 2010; Bhatt & Patel, 2020). The most trademark element of carotenoids is their formed twofold bond chain that structures a chromophore liable for the trademark tones going from dreary (phytoene), to yellow (lutein), orange (βcarotene), and red (lycopene). This construction permits carotenoids to act in the energy-move responses of photosynthesis and as cell reinforcements (Namitha & Negi, 2010; Zia-Ul-Haq, 2021). Presently, most extreme nourishment research has been centered around these carotenoids and they have been widely studied up for their likely impacts and relationship to a few illness processes (Krinsky & Johnson, 2005). Out of these carotenoids, α and β carotene, along with β cryptoxanthin, can be utilized to retinol (vitamin A) in the body and consequently these are alluded to as provitamins (Harrison, 2012; Olmedilla-Alonso et al., 2020). Both α carotene and β cryptoxanthin can likewise supply vitamin A, but since just 50% of every particle has the important unsubstituted β ionone ring, they hypothetically have just around half the movement of β carotene (Muller & Bohm, 2011). β-carotene is essentially found in vegetables like apricot, carrots, red peppers, yams, and pumpkins, alongside green vegetables like broccoli, chard, and kale (Harrison, 2012). Additionally, β-cryptoxanthin is the guideline carotenoid is found in banana, ringer pepper, cucumber, tomato, and lettuce. That's what epidemiological examinations recommend β-cryptoxanthin may assume a part in security from fiery illnesses like angina, joint pain, and osteoporosis (Tanumihardjo, 2012). Recently, it was concentrated on that β-cryptoxanthin levels in blood can be essentially related with the decreased gamble of metabolic condition, for example, hypertension, hyperglycemia, and dyslipidemia, proposing that β-cryptoxanthin offers hostile to weight bioactivity (Jiao et al., 2019; Matsumoto et al., 2020). Additionally, expanded β-cryptoxanthin fixation in blood can reduce risk of different malignant growth types, including renal disease, bosom disease, oral and pharyngeal disease, skin tumorigenesis, colon malignant growth, stomach disease, as well as lung cancer (Jiao et al., 2019).

3.5.2 FLAVONOIDS AND THEIR DERIVATIVES

More than 8,000 types of flavonoids have been recognized today, making them one of the most ubiquitous plants phenolics. Usually, flavonoids found in plants are glucose or rhamnose. In terms of structure, this complex group of polyphenols has a basic C6–C3–C6 structure that can be divided into several subclasses, such as flavones, flavanones, flavanols, flavan-3-ols, anthocyanidins, and isoflavones (Gonzalez-Aguilar et al., 2009; Carpena et al., 2021). Out of these, flavones have the highest concentrations (5,320 mg 100 g^{-1}) as noted in chamomile and parsley, while in moderate amounts it is present in chickpea, millets, kumquat, and sorghum (Verma & Pratap, 2010; Hostetler et al., 2017). Likewise, flavanones are also found in moderate concentrations in chickpea, cumin, hawthorn berry, licorice, peppermint, and tomatoes and in high concentrations only in citrus fruits (Nakanishi et al., 2013).

Flavonoids have become widely recognized as effective health-promoting compounds due to their powerful antioxidant abilities. The structural properties of such compounds are used to determine their antioxidant capability (Gonzalez-Aguilar et al., 2009). The health assistances of such antioxidants present in natural resources are associated with their role in the inhibition of numerous complaints named "oxidative stress pathologies" (Forman & Zhang, 2021).

These structurally complex flavonoids and subtypes reveal a variety of biological activities *in vitro*, which may indicate their potential health-protective potency as anti-inflammatory, anticancer, and cardio protectant agents (Magozwi et al., 2021). A number of flavonoids have demonstrated inhibitory effects on ACE, which raises blood pressure, and cyclooxygenase, which creates prostaglandins and estrogen-producing enzymes. *In vitro* inhibitory effects of flavonoids may reduce platelet aggregation, lowering the risk of heart disease, and inhibit estrogen synthase, which binds estrogen to receptors in tissues, lowering the risk of cancer-related to estrogen (Hodek et al., 2002). So far, several bioactive properties of dietary flavonoids have been identified, including antiproliferative, cardiovascular, anti-inflammatory, antiviral, cardiovascular, free radical scavenging, lipid peroxidase inhibition, and inhibition of hydrolytic and oxidative enzymes (Kumar & Pandey, 2013).

Because of its structural properties, quercetin, the most prevalent flavonoid found in higher plants and an abundant dietary flavanol, functions as an antioxidant, anti-inflammatory, antibacterial, antiviral, antihepatotoxic, and mutagenesis agent (Kumar & Pandey, 2013).

Another, antimutagenic compound luteolin which is found in high quantities in sage, artichokes, parsley, wild carrots, thyme, and peanut hulls, inhibits macrophage activation and hinders inflammation as studied in endothelial cells. Additional research showed that luteolin blocked niacin-induced flushing, and amplified insulin sensitivity in endothelial cells (Ou et al., 2019; Xu et al., 2019). Another cocoa flavonoid, Piscataquis, has been described as a crucial contribution to the increase in plasma antioxidant capacity caused by cocoa ingestion, whereas quercetin acts almost as a key protective agent against LDL oxidation (Storniolo et al., 2020). Ample studies have revealed that the above-mentioned photo components provide health-protective assistance through their free radical scavenging capacity, eicosanoid synthesis, and platelet aggregation inhibition activity

(Nichenametla et al., 2006). Other flavonoids, such as curcumin, catechin, epicatechin-gallate, etc., found in variable amounts in vegetables were also reported by several authors to contain antiviral, antimicrobial, cardioprotective, and other health-promoting properties.

3.5.3 GLUCOSINOLATES (GSS)

Glucosides containing sulfur are abundant in green leafy vegetables, particularly Brassica spp. They are also found in relatively high concentrations in oilseeds like rapeseed and condiments like mustard seed, which is why they have a bitter taste and a strong smell (Thakur & Sharma, 2018; Miękus et al., 2020; Sikorska-Zimny & Beneduce, 2021). Glucosinolates (GSs) in certain plants are decomposed during food preparation and ingestion to synthesize biologically active chemicals that have been extensively researched for their medicinal properties (Conaway et al., 2000; Hayes et al., 2008).

These GSs have traditionally been classified as aromatic GSs. Moreover, this arrangement has subsequently been called into question, with the type of final product being recommended as a more useful classification technique (Blazevic et al., 2020). Sinigrin, glucobrassicin, and glucoiberin are the primary GSs identified in cabbage and kale (Nilsson et al., 2006; Cartea et al., 2008). Turnip (*B. rapa* var. rapa) contains chemicals similar to gluconapin and gluco brassica napin. Furthermore, its roots are high in progoitrin and gluconasturtiin (Krumbein et al., 2005; Cartea & Velasco, 2008).

Although several studies reveal that natural as well as artificial isothiocyanates have been shown to check various chronic diseases/disorders (such as cancer) in animal models in which sulforaphane is isolated from broccoli, and metabolites in the human body through the mercapturic acid pathway (Fahey et al., 2015; Angelino et al., 2015).

Sulforaphane is a natural treatment for a variety of chronic conditions, including Helicobacter pylori-induced intestinal inflammation, type 2 diabetes, and autistic spectrum disorder (Amjad et al., 2015; Axelsson et al., 2017; Yanaka, 2017).

Furthermore, via activating the Nrf2/Keap1 pathway, it reduces oxidative stress and inflammation caused by fine particulate matter (PM2.5) in human bronchial epithelial cells (Wang et al., 2022). It has been shown in clinical studies that patients with T2DM who consume broccoli sprouts with a high sulforaphane content have higher plasma and total antioxidant capacity (Bahadoran et al., 2013). A recent study on sulforaphane discovered that it reorganizes the central metabolism to suppress the metabolic dysregulation caused by excess glucose and that it identifies glucose biosynthesis and glucose one-carbon (1C) metabolism as critical mechanistic pathways (Bernuzzi, 2021). Another investigation, done *in vivo* and *in vitro* model systems, demonstrated the aforementioned compound's antihypertensive, cardioprotective, and hypocholesterolemia potential at various doses (Tebay et al., 2015; Qin et al., 2015).

3.5.4 PLANT-BASED STEROLS AND STANOLS

These comprise natural plant-based natural components such as phytosterols and phytosterols that are essentially incorporated into the human diet largely from vegetable oils, cereals, and fruits. Phytosterols like β-sitostanol and campestanol, have less abundance in nature, accounting for around 2% of the total sterols present in common sources such as corn, wheat, rye, and rice (Pascal et al., 2009; Gangopadhyay, 2021). Sitosterol, campesterol, sitostanol, and campestanol are the most abundant phytosterols and phytosterols in the human diet (Kardong et al., 2013; Harborne et al., 2013). Both of these sterol derivatives supply no energy to the body but aid in cholesterol reduction, cancer prevention, immunological modulation, and skin protection (Brufau et al., 2008; Patel et al., 2017). Both of these substances have been shown to be helpful in lowering serum cholesterol concentrations by competing for intestinal absorption with dietary and biliary cholesterol (Chang-Mo, 2020). Furthermore, they can also have antioxidant, antiulcer, immunomodulatory, antibacterial, antifungal, and wound healing activities, as well as prevent platelet aggregation and control inflammation (Dutta, 2003; Ogbe et al., 2015; Joshi et al., 2022). According to current research, these plant-based sterol/stanol derivatives are physiologically active chemicals with a variety of health applications (Schiepers et al., 2009; Poudel et al., 2019).

On the other hand, a recent meta-analysis study specified that although phytosterol supplementation affects only the body mass index (BMI), but not the other anthropometric indices of the study (Ghaedi et al., 2019). Moreover, some research has shown that not all phytosterols supplementation may be beneficial, and those individuals with sitosterolemia may be susceptible to overload. Also, limited evidence supports a reduction in fat-soluble vitamins and carotenoid concentrations through the decline in the number of available lipoprotein carriers by their consumption (Ghaedi et al., 2019; Chang-Mo, 2020). In some current research, it has been shown that not all phytosterols show beneficial effects for individuals having sitosterolemia. Also, less evidence supports a decline in fat-soluble vitamins and carotenoid concentrations through the reduction in the number of existing lipoprotein carriers by phytosterols ingestion. Thus, normal healthy populations taking an adequate diet not exceeding their daily intake of such sterols should be of no concern. The number of phytosterols and stanols intake allied with the therapeutic effects, is reported to be about 1.5~2.5 g/day which has been acclaimed to diminish plasma low-density lipoprotein cholesterol (LDL-C) without any side effects, as reported in pertinent research on phytosterols. In addition, there was no difference in cholesterol-lowering effect between phytosterol-enriched foods and phytosterol supplements provided by capsule or tablet (Chang-Mo, 2020). In conclusion, it is important to consider the appropriate intake of phytosterols/stanols in the human diet, that can reduce the aforementioned diseases without side effects as produced by regular drugs.

3.5.5 PLANT-BASED PHENOLICS

Polyphenols are naturally occurring compounds found in foods such as vegetables, fruits, cereals, and drinks. In higher plants, these compounds are widely distributed as secondary metabolites that are usually involved in protecting plants from UV radiation or pathogen (Sharma & Prakash, 2014). There are several classes of phenolic compounds based on the number of phenol rings and the structural components that bind the rings together. The most prevalent forms of hydrolysable tannins are gallotannins (Andres-Lacueva et al., 2010).

The expanding research on the absorption of food-borne polyphenols can be used to address the issues connected with the bioavailability of such chemicals. Most human bioavailability studies used dosages in the hundreds of milligram range, resulting in peak plasma concentrations in the 10–50 nmol/L range (Tiwari et al., 2015). In the stomach, bacterial metabolism of polyphenols includes ring dissociation, which results in the generation of a diverse spectrum of metabolites such as aldehydes and phenolic acids. These components may also exert local anti-inflammatory activity in the gut lumen that may be crucial for maintaining mucosal homeostasis and health through passive absorption across the colon (Larrosa et al., 2009; Tiwari et al., 2015). Traditionally, it was assumed that these compounds performed as antinutrients since most polyphenols, specifically tannins, had antagonistic effects on health and having other harmful qualities (Emire et al., 2013; Sharma et al., 2021). Because of their distinct water solubility and intermolecular complexation, these polyphenols are classed as condensed proanthocyanins or tannins (Shetty, 2007). The discovery of antioxidant activity in various polyphenols has provided a unified perspective on the health benefits afforded by many of these substances (Sharma & Prakash, 2014) It has also been shown that polyphenols have antioxidant properties and can reduce the risk of coronary heart disease (CAD), diabetes, hypertension, and several types of cancer in humans (Dillard & German, 2000; Preedy et al., 2013).

Previous investigations have found an opposite relationship between the risk of degenerative illnesses and consuming a polyphenol-rich diet. Further, a polyphenol-rich diet is associated with a lower prevalence of degenerative disorders, according to only a few epidemiological studies. An important source of dietary polyphenols is cereals, legumes (barley, beans, sorghum, maize, and pulses), oil seeds (rape, canola, flax, and olive seeds), fruits and vegetables, and beverages (tea, coffee, cocoa, beer, and wine) (Prakash & Kumar, 2011; Gassara et al., 2014; Salanța, 2020). Among fruits and vegetables, 200–6,300 mg of polyphenols can be found in apples, grapes, pear, cherries, and berries, as well as vegetables. Around 1 g of polyphenols is consumed by the average person every day 1, which is about 10 times more than vitamin C and 100 times more than vitamin E (Packer & Sies, 2001; Neveu et al., 2010). It is known that polyphenols such as catechins, epicatechins, epigallocatechin's from tea, isoflavones from soy, and tannic acid, as well as glycyrrhizin from licorice root, chlorogenic acid, and saponins, affect intestinal glucose transport (Pandey & Rizvi, 2009). A number of polyphenolic anti-aging substances, including anthocyanin from grapefruits have been shown to have anti-aging effects (Shukitt-Hale et al., 2008; Khare et al., 2020). Regular consumption of polyphenols like flavonoids has been shown to reduce the risk of some age-related illnesses such as Alzheimer's, Parkinson's, and dementia disease. These elements combat neurodegenerative illness by protecting neurons, increasing existing neural functions, and stimulating neuronal regeneration (Ataie et al., 2016). *Ginkgo biloba* extracts, combined with flavanone tangeretin (derived from lemon), have shown substantial neuroprotective activities and are promising agents against Parkinson's disease (PD) (Bagchi et al., 2015).

3.5.6 PLANT-BASED TERPENES AND TERE DERIVATIVES

Terpenes are the most common type of naturally occurring Phyto compound found in plants and some lower animals (Singh & Sharma, 2015). Terpenoids with connected functional groups such as alcohols, aldehydes, or ketones in their chemical structure can also be described (Prado-Audelo et al., 2021). Terpenes are traditionally defined as all-natural chemicals composed of isoprene subunits and derived primarily from plants. These chemicals are found in insects, microorganisms or marine organisms, mosses, lichens, as well as higher plants or their parts (roots, rhizomes, stems, leaves, blooms, fruits, seeds) that have a distinctive odor, flavor, or pharmacological activity (Breitmaier, 2006). Terpenoids are classified according to their carbon units as monoterpenes, sesquiterpenes, diterpenes, sesquiterpenes, and triterpenes (Perveen & Al-Taweel, 2018). Essential oils (EOs) contain mono and sesquiterpenes as their primary constituents. Tetraterpenoids, on the other hand, belong to a separate group of chemicals known as carotenoids (Croteau et al., 2000; Patil, 2020). Currently, a wide variety of terpenes are being used to aid in human health care (Dillard & German, 2000). Tocopherols and tocotrienols, which are naturally found in corn, mustard, broccoli, asparagus, and spinach, are the best-studied terpene antioxidants. Tomatoes (paste and canned form) and spinach (frozen and microwaved form) had the greatest combined mean, with roughly 4.987 (0.24 mg/100 g edible weight) and 4.227 (0.44 mg/100 g edible weight) respectively (Chun et al., 2006). The effect of such a vegetable diet on the decrease of various health issues (such as cancer) may be explained by the *in vivo* activities of terpenes (Sharma et al., 2017). Additionally, several terpenes or their derivative compounds have been shown to have antibacterial, anti-inflammatory, and immunomodulatory effects in numerous *in vitro* and *in vivo* studies (Paduch et al., 2007; Dambolena et al., 2016; Masyita et al., 2022). Terpenes are also utilized to improve skin penetration, and its derivative tocotrienol is an excellent apoptotic inducer for human breast cancer cells (Lee et al., 2003; Gupta & Prakash, 2022). Furthermore, the impact of a vegetable-rich diet on diabetes and cancer risk reduction may be explained by terpene activities *in vivo* (Prakash et al., 2012; Panigrahy et al., 2021; Gupta & Prakash, 2022). Several terpenes have also been employed as medications to promote human health, such as artemisinin, which is used to treat malaria (Willcox et al., 2004). Terpenes, in addition to these characteristics, enable versatility in route of drug administration and side effect prevention (Cox-Georgian et al., 2019).

3.6 RISK ASSOCIATED WITH USE OF PHYTOCHEMICALS AS FUNCTIONAL FOOD OR NUTRACEUTICAL

Despite wide acceptance of phytochemicals as nutraceuticals, minimal study has been done on the safety and efficacy of most individual phytochemicals in prevention of cancer (Yeung et al., 2008). Several studies have undoubtedly suggested that various phytochemicals, such as epigallocatechin gallate or resveratrol (RES), have beneficial benefits in the prevention or treatment of certain diseases like as cancer, but these findings have yet to be validated in randomized human clinical trials (Miller & Snyder, 2012; Bode & Dong, 2015). Concurrently, men taking carotene had a greater incidence of lung cancer (Kushi et al., 2012; Bode & Dong, 2015). Along with the risk described above, the quality and safety of such phytochemicals/nutraceuticals is a significant consideration.

In addition to the risks outlined above, these plant-based biochemicals, known as polyphenols, can bind to food antigens and trigger a strong immunological reaction (or allergy) in humans (Bessa et al., 2021; Huang et al., 2014). Food allergy is estimated to affect 4% of children and 1% of adults worldwide, with an increase in frequency over the last two decades (Elghoudi & Narchi, 2022). Indeed, the buildup of proteins and fortification of phytocompounds with minerals/chemicals or other substances anticipated for the purposes of increasing rapid development, production, and dispersion of such phytonutrients rapidly exacerbates the situation (Jackson et al., 2013).

With the exception of well-known food allergens such as milk and wheat, other foods, notably vegetables and herbals (Sharma et al., 2020; Rane et al., 2020a–c), can cause unexpected, severe adverse health effects such as allergies (Haddad et al., 2020). In the case of peanuts, the potential for adverse health consequences from any of their basic constituents persists due to their varying chemical properties deriving from climates (Greenhawt et al., 2020). Cross-reactivity is an immunological phenomenon that has implications for the diagnosis and management of some food allergies. According to the aforementioned factors, around half of latex allergy patients are allergic to cross-reactive plant-based foods (Radauer et al., 2008, 2014). Furthermore, as respiratory allergies have increased over the last few decades and become the most common dietary allergens among adults (Pawankar, 2014). Similarly, the latex allergic syndrome shares many similarities with the pollen-food allergy syndrome, and several plant foods, including potatoes, tomatoes, bell peppers, kiwis, chestnuts, peaches, avocados, and bananas, have been related to this illness (Nowak-Wegrzyn, 2013). Another study found that when *Brassica rapa* vegetables were exposed to salicylic acid and ethephon, two typical compounds that stimulate the plant defense response, the number of IgE-reactive antigens rose up to tenfold (Shahali & Dadar, 2018). As well as with sensitivities, food-borne chemicals, and impurities in processed or uncooked foods can cause serious health problems. Heavy metals, microbiological contamination, pesticide residues, molds, and mycotoxins are the most frequent types of pollutants, all of which can be both natural and/or man-made (Abdolshahi & Yancheshmeh, 2020). Natural or synthetic pesticides and herbicides are of particular importance since, due to their constant use in functional foods. Organophosphates, carbamates, organochlorides, and pyrethroids are the common synthetic pesticides used in the food business and they are all neuroactive and poisonous at high levels (Abreu-Villaca & Levin, 2017). Biopesticides and herbicides generated from natural plant, animal, or microbial components, on the other hand, are used to control specific target organisms. Xenoestrogens (e.g., bisphenol A) are one type of biopesticide formed from food dispensation (Ibarreta et al., 2001). Many reports show a synthetic xenoestrogens have a long half-life and the ability to significantly boost estrogen levels in animals, including humans, by causing hormonal abnormalities in both people and animals (Ibarreta et al., 2001; Sharma et al., 2021). Organic products (such as vegetables) cultivated without the utilization of synthetic herbicides and pesticides usually contains residues than conventionally farmed ones. Proper food handling, including thorough washing, peeling/skinning, and cooking, is the most straightforward and cost-effective technique to dramatically decrease pesticide/herbicide contamination (Bagchi et al., 2015). Similarly, molds from the Aspergillus and Penicillium genera are employed to benefit humans in fermented foods such as soy sauce. They are used to produce a variety of enzymes, including citric acid and lactic acid, which are essential and naturally present in plant-based foods, but their presence is undesirable and has the potential to cause significant health effects (Blandino et al., 2003; Nout & Aidoo, 2011). Mycotoxins have a complex chemical composition and thrive in damp and aerated environments. Among the most frequent mycotoxins, aflatoxins, which are produced by *Aspergillus flavus* and *Aspergillus parasiticus*, have a well-documented link to a variety of liver and kidney diseases, including cancer (Pessu et al., 2011). These aflatoxins infect a wide range of dietary plants, including ground nuts, parsnips, carrots, parsley, and celery, as well as grain crops such as rice and corn. Aflatoxin contamination is a major worry in developing countries (Awuchi & Ogueke, 2021; Beier & Nigg, 2019).

It is rare for food-borne diseases to be fatal, and they usually resolve within 7 days. Bacteria and viruses cause mild to severe acute gastrointestinal pain, nausea, vomiting, diarrhea, loss of appetite, severe abdominal cramps, and fever. Symptoms in adults with chronic illnesses may progress to more hazardous shock, depending on the infection (Krug et al., 2020).

Viruses have also been blamed for approximately 70% of food-related diseases, with growing transmission via food and drink (Havelaar et al., 2015). Norovirus is the pathogen mostly associated with influenza, accounting for more than 60% of all enteroviral etiologies (Hardstaff et al., 2018; Mattison et al., 2018). Consumption of contaminated foods is a common method of norovirus transmission (Hardstaff et al., 2018).

A different type of proteinaceous agent Prions, which can be spread through food that cause aggregation of certain cellular proteins. When these vulnerable proteins fold incorrectly, they can cause neurodegenerative spongiform encephalopathies. The disease, identified in humans as Creutzfeldt-Jakob disease can be transmitted through intake of infected animal products. Creutzfeldt-Jakob disease progresses quickly and fatally after a long period of dormancy. Although human instances of classic and variant Creutzfeldt-Jakob disease are uncommon, bovine spongiform encephalopathy has been documented in Japan, the Netherlands, and Denmark. Because prions cannot be removed by cooking, the prevalence of prions in several nations has resulted in changes in consumer dietary choices (Hilton, 2006; Watson et al., 2021).

3.7 VARIOUS APPROACHES OF VEGETABLE-BASED PHYTOCHEMICALS/NUTRACEUTICALS IN HEALTH AND DISEASES

Scientific evidence for chronic illnesses indicates micronutrients is crucial in preventing the development of these diseases. Nutritionists have been investigating the phytochemical components within fruits and vegetables (phytochemicals) that may protect against these diseases based on possible associations between high vegetable consumption and a lower prevalence (Reboredo-Rodrguez et al., 2018; Howes & Simmonds, 2014). Following the COVID-19 pandemic, the quick development of apparent natural health foods, now usually characterized by the market as functional food or, at times, nutraceuticals, has had a significant impact on consumers (Ayseli et al., 2020). Capsaicin, a spicy chemical found in several pepper kinds and ginger in varying amounts, has anticarcinogenic and antimutagenic properties (Shah et al., 2020). Curcumin, another phytochemical natural product, possesses anti-inflammatory and cancer-prevention characteristics (Abd El Hack et al., 2021). One of two key iso-flavonoids found in leguminous vegetable soybean is genistein (5,7,4-trihydroxyisoflavone). Genistein inhibits mitogen-stimulated proliferation in human breast cancer cells grown *in vitro*.

Furthermore, these soy iso-flavonoid conjugates have chemo-preventive action in carcinogen-induced breast cancer rat models (Boutas et al., 2022). As a result, it has been used in treating inflammatory metabolic diseases, neoplastic neurodegenerative diseases (NDs), and disorders of the central nervous system (CNS) (Ginwala et al., 2019; Sahin et al., 2019). In the United States and Europe, cardiovascular disorders are the leading cause of death among postmenopausal women. In addition to lowering total and LDL cholesterol, iso-flavonoids found in soy products, soy protein, and flaxseed can also increase HDL cholesterol, lowering heart disease risk (Houston, 2022). There is proof to substantiate the concept that phytoestrogen usage helps vegetarians have a lower incidence of cardiovascular disease (CVD) and may also operate as a cardio protectant (Yamagata, 2019; Wolters et al., 2020). Several studies have found that those who eat a lot of iso-flavonoids had a lower risk of acquiring cancers (Sikdar et al., 2019; Mali et al., 2019). Similar links have been discovered between iso-flavonoid and lignan intakes and thyroid, ovarian, and breast malignancies in women (Sharma & Prakash, 2014). When compared to cancer mortality rates, phytoestrogen intake has a negative link with breast, ovarian, prostate, and colon cancer (Cherdshewasart et al., 2007; Al-Azzawi & Wahab, 2002). Finally, a diet rich in such vegetable-based phytonutrients may provide a variety of pharmacological effects, such as isoflavones, RES, lignans, and so on. As a result of consuming such foods in the diet, females are more likely to suffer from estrogen-related ailments such as osteoporosis, cognitive decline, and CVD, as well as a higher likelihood of breast cancer (Cherdshewasart et al., 2009; Al-Azzawi & Wahab, 2010).

3.8 CONCLUSION

Several studies have lately suggested that vegetable-based phytochemicals or nutraceuticals play a vital role in the prevention of a variety of cancer diseases. New products and components based on these phytochemicals, known as nutraceuticals or functional foods, have enormous growth potential in both the food and pharmaceutical industries. The most well-known vegetables include broccoli, which contains GSs that have been found to reduce cancer risk, and dandelion, which has anti-inflammatory properties; A study found that garlic contains allyl sulfides, which inhibit cell growth in cancer cells; cocoa is associated with a lower risk of CVD; and olives contain phytochemicals that promote health. A number of edible herbs have been used for cognitive enhancement, including ginkgo (*Ginkgo biloba*), ginseng (*Panax ginseng*), and saw palmetto (*Serenoa repens*) for benign prostatic hyperplasia. Some genuinely researched commercial items derived from such plant-based natural products contain unique elements that can provide health benefits faster than consuming conventionally healthy meals alone. The bioavailability, metabolizing capacity, safety, and quality of these goods eaten in natural/artificial (fortified) form are other issues that need to be investigated thoroughly.

KEYWORDS

- **bioavailability**
- **calorific value**
- **carotenoid**
- **flavonoid**
- **functional food**
- **nutraceuticals**
- **phytochemicals**

REFERENCES

Abd. El-Hack, M. E., El-Saadony, M. T., Swelum, A. A., Arif, M., Abo, G. M. M., Shukry, M., Noreldin, A., et al., (2021). Curcumin, the active substance of turmeric: Its effects on health and ways to improve its bioavailability. *J Sci Food Agric., 101*(14), 5747–5762.

Abdolshahi, A., & Yancheshmeh, B. S., (2020). Food contamination. In: *Mycotoxins and Food Safety*. Intech Open.

Abreu-Villaca, Y., & Levi, E. D., (2017). Developmental neurotoxicity of succeeding generations of insecticides. *Environ Int.*, *99*, 55–77.

Al-Azzawi, F., & Wahab, M., (2002). Estrogen and colon cancer: Current issues. *Climacteric.*, *5*(1), 3–14.

Alcocer, F., Whitley, D., Salazar-Gonzalez, J. F., Jordan, W. D., Sellers, M. T., Eckhoff, D. E., Suzuki, K., et al., (2002). Quercetin inhibits human vascular smooth muscle cell proliferation and migration. *Surgery*, *131*(2), 198–204.

Ali, S., Chouhan, R., Sultan, P., Hassan, Q. P., & Gandhi, S. G., (2021). A comprehensive review of phytochemistry, pharmacology and toxicology of the genus *Aconitum* L. *Adv. Trad. Med.*, 1–22.

Alkhatib, A., Tsang, C., Tiss, A., Bahorun, T., Arefanian, H., Barake, R., Khadir, A., & Tuomilehto, J., (2017). Functional foods and lifestyle approaches for diabetes prevention and management. *Nutrients*, *9*(12), 1310.

Amjad, A. I., Parikh, R. A., Appleman, L. J., Hahm, E. R., Singh, K., & Singh, S. V., (2015). Broccoli-derived sulforaphane and chemoprevention of prostate cancer: From bench to bedside. *Curr. Pharmacol. Rep.*, *1*(6), 382–390.

Andrés-Lacueva, C., Medina-Remon, A., Llorach, R., Urpi-Sarda, M., Khan, N., Chiva-Blanch, G., & Lamuela-Raventos, R. M., (2010). Phenolic compounds: Chemistry and occurrence in fruits and vegetables. *Fruit and Vegetable Phytochemicals*, *2010*, 53–88.

Angelino, D., Dosz, E. B., Sun, J., Hoeflinger, J. L., Van, T. M. L., Chen, P., Harnly, J. M., Miller, M. J., & Jeffery, E. H., (2015). Myrosinase-dependent and–independent formation and control of isothiocyanate products of glucosinolate hydrolysis. *Front Plant Sci.*, *6*, 831.

Anumala, V., Sarma, P., & Phurailatpam, A., (2021). *Fruits and Vegetables as Nutraceutical: Nature's Medicine.* United States: Taylor & Francis.

Arya, M. S., Reshma, U. R., Syama, S. T., & Anaswara, S. J., (2019). Nutraceuticals in vegetables: New breeding approaches for nutrition, food and health: A. *J. Pharmacogn Phytochem.*, *8*(1), 677–682.

Ataie, A., Shadifar, M., & Ataee, R., (2016). Polyphenolic antioxidants and neuronal regeneration. *Basic Clin. Neurosci.*, *7*(2), 81.

Awuchi, C. G., & Ogueke, C. (2021). Mycotoxin Levels and Impacts on Chemical and Functional Properties of some Grains Sold in Owerri, Imo State, Nigeria. Retrieved March, 7.

Axelsson, A. S., Tubbs, E., Mecham, B., Chacko, S., Nenonen, H. A., Tang, Y., Fahey, J. W., et al., (2017). Sulforaphane reduces hepatic glucose production and improves glucose control in patients with type 2 diabetes. *Sci. Transl. Med.*, *9*(394), eaah4477.

Ayseli, Y. I., Aytekin, N., Buyukkayhan, D., Aslan, I., & Ayseli, M. T., (2020). Food policy, nutrition and nutraceuticals in the prevention and management of COVID-19: Advice for healthcare professionals. *Trends Food Sci. Technol.*, *105*, 186–199.

Bagchi, D., Preuss, H. G., & Swaroop, A., (2015). *Nutraceuticals and Functional Foods in Human Health and Disease Prevention.* United Kingdom: CRC Press.

Bahadoran, Z., Mirmiran, P., & Azizi, F., (2013). Potential efficacy of broccoli sprouts as a unique supplement for management of type 2 diabetes and its complications. *J Med Food.*, *16*(5), 375–382.

Beier, R. C., & Nigg, H. N., (2019). Toxicology of naturally occurring chemicals in food. In: *Food Borne Disease Handbook* (pp. 37–186). CRC Press.

Belton, B., Rosen, L., Middleton, L., Ghazali, S., Mamun, A. A., Shieh, J., Noronha, H. S., Dhar, G., Ilyas, M., Price, C., & Nasr-Allah, A., et al., (2021). COVID-19 impacts and adaptations in Asia and Africa's aquatic food value chains. *Mar. Policy*, *129*, 104523.

Benameur, T., Soleti, R., Panaro, M. A., La Torre, M. E., Monda, V., Messina, G., & Porro, C., (2021). Curcumin as prospective anti-aging natural compound: Focus on brain. *Molecules, 26*(16), 4794.

Bernuzzi, F., (2021). *Understanding the Role of Dietary Sulforaphane in Regulating Metabolic Pathways Using Genomic Approaches*. Doctoral dissertation, University of East Anglia.

Bessa, C., Francisco, T., Dias, R., Mateus, N., Freitas, V. D., & Pérez-Gregorio, R., (2021). Use of polyphenols as modulators of food allergies. From chemistry to biological implications. *Front Sustain Food Syst.*, *5*, 623611.

Bhatt, T., & Patel, K., (2020). Carotenoids: Potent to prevent diseases review. *Nat. Prod. Bioprospect.*, *10*(3), 109–117.

Blandino, A., Al-Aseeri, M. E., Pandiella, S. S., Cantero, D., & Webb, C., (2003). Cereal-based fermented foods, and beverages. *Food Res. Int.*, *36*(6), 527–543.

Blažević, I., Montaut, S., Burcul, F., Olsen, C. E., Burow, M., Rollin, P., & Agerbirk, N., (2020). Glucosinolate structural diversity, identification, chemical synthesis, and metabolism in plants. *Phytochemistry*, *169*, 112100.

Bode, A. M., & Dong, Z., (2015). Toxic phytochemicals and their potential risks for human cancer. *Cancer Prev Res.*, *8*(1), 1–8.

Boutas, I., Kontogeorgi, A., Dimitrakakis, C., & Kalantaridou, S. N., (2022). Soy isoflavones and breast cancer risk: A meta-analysis. *In Vivo*, *36*(2), 556–562.

Breitmaier, E., (2006). *Terpenes: Flavors, Fragrances, Pharmaca, Pheromones.* John Wiley & Sons.

Brufau, G., Canela, M. A., & Rafecas, M., (2008). Phytosterols: Physiologic and metabolic aspects related to cholesterol-lowering properties. *Nutr. Res.*, *28*(4), 217–225.

Calvani, M., Pasha, A., & Favre, C., (2020). Nutraceutical boom in cancer: Inside the labyrinth of reactive oxygen species. *Int. J. Mol. Sci.*, *21*(6), 1936.

Carpena, M., Caleja, C., Nuñez-Estevez, B., Pereira, E., Fraga-Corral, M., Reis, F. S., Simal-Gandara, J., et al., (2021). Flavonoids: A group of potential food additives with beneficial health effects. In: Lage, M. Á. Á. P., & Otero, P., (eds.), *Food Additives*. IntechOpen.

Cartea, M. E., & Velasco, P., (2008). Glucosinolates in *Brassica* foods: Bioavailability in food and significance for human health. *Phytochem Rev.*, *7*(2), 213–229.

Chandra, S., Sah, K., Bagewadi, A., Keluskar, V., Shetty, A., Ammanagi, R., & Naik, Z., (2012). Additive and synergistic effect of phytochemicals in prevention of oral cancer. *European J Gen Dent.*, *1*(3), 142–147.

Chandrika, U. G., & Kumara, P. A. P., (2015). Gotu kola (*Centella asiatica*): Nutritional properties and plausible health benefits. *Advances in Food and Nutrition Research, 76*, 125–157.

Chang-Mo, C., (2020). A study on phytosterols or phytostanols consumption and the protection from cardiovascular disease. *Protection Convergence, 5*, 22–31.

Cherdshewasart, W., Panriansaen, R., & Picha, P., (2007). Pretreatment with phytoestrogen-rich plant decreases breast tumor incidence and exhibits lower profile of mammary ERα and ERβ. *Maturitas*, *58*(2), 174–181.

Chun, J., Lee, J., Ye, L., Exler, J., & Eitenmiller, R. R., (2006). Tocopherol and tocotrienol contents of raw and processed fruits and vegetables in the United States diet. *J. Food Compost. Anal.*, *19*(2, 3), 196–204.

Conaway, C. C., Getahun, S. M., Liebes, L. L., Pusateri, D. J., Topham, D. K., Botero-Omary, M., & Chung, F. L., (2000). Disposition of glucosinolates and sulforaphane in humans after ingestion of steamed and fresh broccoli. *Nutr Cancer*, *38*(2), 168–178.

Cox-Georgian, D., Ramadoss, N., Dona, C., & Basu, C., (2019). Therapeutic and medicinal uses of terpenes. In: *Medicinal Plants* (pp. 333–359). Springer, Cham.

Croteau, R., Kutchan, T. M., & Lewis, N. G., (2000). Natural products (secondary metabolites). *Biochemistry and Molecular Biology of Plants, 24*, 1250–1319.

Daliu, P., Santini, A., & Novellino, E., (2018). A decade of nutraceutical patents: Where are we now in 2018. *Expert Opin. Ther. Pat.*, *28*(12), 875–882.

Dambolena, J. S., Zunino, M. P., Herrera, J. M., Pizzolitto, R. P., Areco, V. A., & Zygadlo, J. A., (2016). Terpenes: Natural products for controlling insects of importance to human health—A structure-activity relationship study. *Psyche.*

De Ancos, B., Colina-Coca, C., González-Peña, D., & Sánchez-Moreno, C., (2015). Bioactive compounds from vegetable and fruit by-products. *Biotechnol. Bioact. Compd. Sources Appl,* 1–36.

Demmig-Adams, B., López-Pozo, M., Stewart, J. J., & Adams, W. W., (2020). Zeaxanthin and lutein: Photoprotectors, anti-inflammatories, and brain food. *Molecules, 25*(16), 3607.

Devkota, H. P., (2022). Nutraceutical compounds, classification, biosynthesis, and function. In: *Nutraceuticals Production from Plant Cell Factory* (pp. 3–28). Springer, Singapore.

Dias, J. S., (2012). Major classes of phytonutriceuticals in vegetables and health benefits: A review. *J. Nutr Ther., 1*(1), 31–62.

Dillard, C. J., & German, J. B., (2000). Phytochemicals: Nutraceuticals and human health. *J. Sci. Food Agric., 80*(12), 1744–1756.

Dima, C., Assadpour, E., Dima, S., & Jafari, S. M., (2020). Bioavailability of nutraceuticals: Role of the food matrix, processing conditions, the gastrointestinal tract, and nano delivery systems. *Compr. Rev. Food Sci., 19*(3), 954–994.

Doughari, J. H., (2012). *Phytochemicals: Extraction Methods, Basic Structures, and Mode of Action as Potential Chemotherapeutic Agents* (pp. 1–33). Rijeka, Croatia: INTECH Open Access Publisher.

Duthie, G., & Crozier, A., (2000). Plant-derived phenolic antioxidants. *Curr. Opin. Lipidol., 11*(1), 43–47.

Dutta, P. C., (2003). *Phytosterols as Functional Food Components and Nutraceuticals*. CRC Press.

Egbuna, C., & Dable-Tupas, G., (2020). *Functional Foods and Nutraceuticals* (Vol. 1, pp. 1–632). Springer Nature Switzerland AG.

Elghoudi, A., & Narchi, H., (2022). Food allergy in children-the current status and the way forward. *World J. Clin. Pediatr., 11*(3), 253.

Emire, S. A., Jha, Y. K., & Mekam, F., (2013). Role of anti-nutritional factors in food industry. *Beverage & Food World, 2*, 23–28.

Fahey, J. W., Holtzclaw, W. D., Wehage, S. L., Wade, K. L., Stephenson, K. K., & Talalay, P., (2015). Sulforaphane bioavailability from glucoraphanin-rich broccoli: Control by active endogenous myrosinase. *PloS One, 10*(11), e0140963.

Fernandes, S. S., Coelho, M. S., & De Las Mercedes Salas-Mellado, M., (2019). Bioactive compounds as ingredients of functional foods: Polyphenols, carotenoids, peptides from animal and plant sources new. In: *Bioactive Compounds* (129–142). Woodhead Publishing.

Fiedor, J., & Burda, K., (2014). Potential role of carotenoids as antioxidants in human health and disease. *Nutrients, 6*(2), 466–488.

Forman, H. J., & Zhang, H., (2021). Targeting oxidative stress in disease: Promise and limitations of antioxidant therapy. *Nat. Rev. Drug Discov., 20*(9), 689–709.

Gangopadhyay, N., (2021). Oils extracted from nuts and grains. *Oil and Oilseed Processing: Opportunities and Challenges*, 61–79.

Gassara, F., Brar, S. K., & Verma, M., (2014). Phenolics as nutraceuticals and functional foods. *Nutraceuticals and Functional Foods*, 155.

Ghaedi, E., Varkaneh, H. K., Rahmani, J., Mousavi, S. M., Mohammadi, H., Fatahi, S., Pantovic, A., et al., (2019). Possible anti-obesity effects of phytosterols and phytostanols supplementation in humans: A systematic review and dose–response meta-analysis of randomized controlled trials. *Phytother. Res.*, *33*(5), 1246–1257.

Ginwala, R., Bhavsar, R., Chigbu, D. G. I., Jain, P., & Khan, Z. K., (2019). Potential role of flavonoids in treating chronic inflammatory diseases with a special focus on the anti-inflammatory activity of apigenin. *Antioxidants, 8*(2), 35.

Gokhale, J. S., Lele, S. S., & Ananthanarayan, L., (2021). Indian traditional foods and diets: Combining traditional wisdom with modern science of nutraceuticals and functional foods. In: *Nutrition, Food and Diet in Ageing and Longevity* (pp. 357–392). Springer, Cham.

Gonzalez-Aguilar, G. A., Rosa De Carrillo, L. A., & Alvarez-Parrilla, E., (2009). *Fruit and Vegetable Phytochemicals: Chemistry, Nutritional Value and Stability.* Germany: Wiley.

Greenhawt, M., Shaker, M., Wang, J., Oppenheimer, J. J., Sicherer, S., Keet, C., Wallace, D., et al., (2020). Peanut allergy diagnosis: A 2020 practice parameter update, systematic review, and grade analysis. *J. Allergy Clin. Immunol., 146*(6), 1302–1334.

Grivetti, L. E., & Ogle, B. M., (2000). Value of traditional foods in meeting macro-and micronutrient needs: The wild plant connection. *Nutr Res Rev.*, *13*(1), 31–46.

Gullón, B., Lú-Chau, T. A., Moreira, M. T., Lema, J. M., & Eibes, G., (2017). Rutin: A review on extraction, identification and purification methods, biological activities, and approaches to enhance its bioavailability. *Trends Food Sci Technol.*, *67*, 220–235.

Gupta, C., & Prakash, D., (2022). Phytochemicals as the source of natural immunomodulator and their role in cancer chemoprevention. In: *Immunomodulators and Human Health* (pp. 209–241). Springer, Singapore.

Haddad, M. A., Dmour, H., Al-Khazaleh, J. F. M., Obeidat, M., Al-Abbadi, A., Al-Shadaideh, A. N., et al., (2020). Herbs and medicinal plants in Jordan. *J. AOAC Int.*, *103*(4), 925–929.

Harbourne, N., Marete, E., Jacquier, J. C., & O'Riordan, D., (2013). Conventional extraction techniques for phytochemicals. *Handbook of Plant Food Phytochemicals: Sources, Stability, and Extraction*, 397–411.

Hardstaff, J. L., Clough, H. E., Lutje, V., McIntyre, K. M., Harris, J. P., Garner, P., & O'Brien, S. J., (2018). Foodborne and food-handler norovirus outbreaks: A systematic review. *Foodborne Pathog. Dis.*, *15*(10), 589–597.

Harrison, E. H., (2012). Mechanisms involved in the intestinal absorption of dietary vitamin A and provitamin A carotenoid. *Biochim. Biophys. Acta Mol. Cell Biol. Lipids BBA-Mol. Cell Biol. L.*, *1821*(1), 70–77.

Hassan, B. A. R., (2020). *Medicinal Plants: Use in Prevention and Treatment of Diseases.* United Kingdom: Intech Open.

Havelaar, A. H., Kirk, M. D., Torgerson, P. R., Gibb, H. J., Hald, T., Lake, R. J., et al., (2015). World Health Organization global estimates and regional comparisons of the burden of foodborne disease in 2010. *PLoS Medicine*, *12*(12), e1001923. World Health Organization Foodborne Disease Burden Epidemiology Reference Group..

Hayes, J. D., Kelleher, M. O., & Eggleston, I. M., (2008). The cancer chemopreventive actions of phytochemicals derived from glucosinolates. *Eur. J. Nutr.*, *47*(2), 73–88.

Hilton, D. A., (2006). Pathogenesis and prevalence of variant Creutzfeldt–Jakob disease. *J. Path.*, *208*(2), 134–141.

Ho, M. H. K., Wong, W. H. S., & Chang, C., (2014). Clinical spectrum of food allergies: A comprehensive review. *Clin Rev Allergy Immunol.*, *46*(3), 225–240.

Hodek, P., Trefil, P., & Stiborová, M., (2002). Flavonoids-potent and versatile biologically active compounds interacting with cytochromes P450. *Chem Biol Interact.*, *139*(1), 1–21.

Hostetler, G. L., Ralston, R. A., & Schwartz, S. J., (2017). Flavones: Food sources, bioavailability, metabolism, and bioactivity. *Adv Nutr.*, *8*(3), 423–435.

Hounsome, N., Hounsome, B., Tomos, D., & Edwards-Jones, G., (2008). Plant metabolites and nutritional quality of vegetables. *J. Food Sci.*, *73*(4), R48–R65.

Houston, M. C., (2022). Nutrition and nutritional supplements in the management of dyslipidemia and dyslipidemia-induced cardiovascular disease. *Nutritional and Integrative Strategies in Cardiovascular Medicine* (pp. 97–130). CRC Press.

Howes, M. J. R., & Simmonds, M. S., (2014). The role of phytochemicals as micronutrients in health and disease. *Curr Opin Clin Nutr Metab Care.*, *17*(6), 558–566.

Huang, H. W., Hsu, C. P., Yang, B. B., & Wang, C. Y., (2014). Potential utility of high-pressure processing to address the risk of food allergen concerns. *Compr Rev Food Sci.*, *13*(1), 78–90.

Huber, L. S., Hoffmann-Ribani, R., & Rodriguez-Amaya, D. B., (2009). Quantitative variation in Brazilian vegetable sources of flavonols and flavones. *Food Chem.*, *113*(4), 1278–1282.

Ibarreta, D., Daxenberger, A., & Meyer, H. H., (2001). Possible health impact of phytoestrogens and xenoestrogens in food. *APMIS*, *109*(S103), S402–S425.

Iciek, M., Kwiecień, I., & Włodek, L., (2009). Biological properties of garlic and garlic-derived organosulfur compounds. *Environmental and Molecular Mutagenesis*, *50*(3), 247–265.

Jackson, J. C., Weatherspoon, L., Nnyepi, M., Malete, L., Mokgatlhe, L., Lyoka, P., & Bennink, M., (2013). Sorghum bean composite porridge nutritional quality and acceptability. *Nutr. Food Sci.*

Jafari, S. M., & McClements, D. J., (2017). Nanotechnology approaches for increasing nutrient bioavailability. *Adv Food Nutr Res.*, *81*, 1–30.

Jagdale, Y. D., Mahale, S. V., Zohra, B., Nayik, G. A., Dar, A. H., Khan, K. A., et al., (2021). Nutritional profile and potential health benefits of super foods: A review. *Sustainability*, *13*(16), 9240.

Jiao, Y., Reuss, L., & Wang, Y., (2019). β-Cryptoxanthin: Chemistry, occurrence, and potential health benefits. *Curr. Pharmacol. Rep.*, *5*(1), 20–34.

Joshi, T., Patni, K., Belwal, T., Maurya, H., & Kumar, A., (2022). Sterols: Benificial or detrimental for human health. In: *Antioxidants Effects in Health* (pp. 475–484). Elsevier.

Keservani, R. K., & Sharma, A. K., (2014). Flavonoids: Emerging trends and potential health benefits. *Journal of Chinese Pharmaceutical Sciences*, *23*(12), 815.

Keservani, R. K., Kesharwani, R. K., Sharma, A. K., Vyas, N., & Chadoker, A., (2010b). Nutritional supplements: An overview. *International Journal of Current Pharmaceutical Review and Research*, *1*(1), 59–75.

Keservani, R. K., Kesharwani, R. K., Vyas, N., Jain, S., Raghuvanshi, R., & Sharma, A. K., (2010). Nutraceutical and functional food as future food: A review. *Der Pharmacia Lettre.*, *2*(1), 106–116.

Keservani, R. K., Kesharwani, R. K., Vyas, N., Jain, S., Raghuvanshi, R., & Sharma, A. K., (2010a). Nutraceutical and functional food as future food: A review. *Der Pharmacia Lettre*, *2*(1), 106–116.

Keservani, R. K., Sharma, A. K., & Kesharwani, R. K., (2016a). Nutraceutical and functional foods for cardiovascular health. In: *Food Process Engineering* (pp. 257–278, 291–312). Apple Academic Press, CRC Press. ISBN: 9781771884020.

Keservani, R. K., Sharma, A. K., & Kesharwani, R. K., (2016b). Medicinal effect of nutraceutical fruits for the cognition and brain health. *Scientifica, 10*. Article ID 3109254. doi:10.1155/2016/3109254.

Keservani, R. K., Sharma, A. K., & Kesharwani, R. K., (2017). An overview and therapeutic applications of nutraceutical and functional foods. *Recent Advances in Drug Delivery Technology*, 160–201. ISBN:9781522507543.

Keservani, R. K., Sharma, A. K., & Kesharwani, R. K., (2020). *Nutraceuticals and Dietary Supplements: Applications in Health Improvement and Disease Management*. CRC Press. ISBN: 9781771888738.

Khare, P., Maurya, R., Bhatia, R., Mangal, P., Singh, J., Podili, K., et al., (2020). Polyphenol rich extracts of finger millet and kodo millet ameliorate high fat diet-induced metabolic alterations. *Food Funct., 11*(11), 9833–9847.

Kim, M. J., Kim, D. W., Kim, J. G., Shin, Y., Jung, S. K., & Kim, Y. J., (2021). Analysis of the chemical, antioxidant, and anti-inflammatory properties of pink pepper (*Schinus molle* l.). *Antioxidants*, *10*(7), 1062.

Koche, D., Shirsat, R., & Kawale, M., (2016). An overerview of major classes of phytochemicals: Their types and role in disease prevention. *Hislopia Journal, 9*(1, 2), 0976–2124.

Krinsky, N. I., & Johnson, E. J., (2005). Carotenoid actions and their relation to health and disease. *Mol. Asp. Med.*, *26*(6), 459–516.

Krug, M., Chapin, T., Danyluk, M., Goodrich-Schneider, R., Schneider, K., Harris, L., & Worobo, R., (2020). *Outbreaks of Foodborne Disease Associated with Fruit and Vegetable Juices*, 1922–2019: FSHN12-04/FS188, rev. 6/2020. EDIS, 2020(5).

Krüger, C., & Malleyeck, I., (2009). Diagnosing possible infantile cow's milk protein allergy in rural Africa, when history and physical examination are the only tools: A case report. *Cases J., 2*(1), 1–4.

Krumbein, A., Schonhof, I., & Schreiner, M., (2005). Composition and contents of phytochemicals (glucosinolates, carotenoids and chlorophylls) and ascorbic acid in selected *Brassica* species (*B. juncea*, *B. rapa* subsp. nipposinica var. chinoleifera, *B. rapa* subsp. chinensis and *B. rapa* subsp. rapa). *J. Appl. Bot. Food Qual.*, *79*(3), 168–174.

Kumar, S., & Pandey, A. K., (2013). Chemistry and biological activities of flavonoids: An overview. *Sci. World J.*

Kushi, L. H., Doyle, C., McCullough, M., Rock, C. L., Demark-Wahnefried, W., Bandera, E. V., Gapstur, S., Patel, A. V., Andrews, K., Gansler, T., & American Cancer Society 2010 Nutrition and Physical Activity Guidelines Advisory Committee, (2012). American Cancer Society Guidelines on nutrition and physical activity for cancer prevention: Reducing the risk of cancer with healthy food choices and physical activity. *CA Cancer J. Clin.*, *62*(1), 30–67.

Langi, P., Kiokias, S., Varzakas, T., & Proestos, C., (2018). Carotenoids: From plants to food and feed industries. *Microbial Carotenoids*, 57–71.

Larrosa, M., Luceri, C., Vivoli, E., Pagliuca, C., Lodovici, M., Moneti, G., & Dolara, P., (2009). Polyphenol metabolites from colonic microbiota exert anti-inflammatory activity on different inflammation models. *Mol. Nutr. Food Res.*, *53*(8), 1044–1054.

Lee, M. T., Lin, W. C., Yu, B., & Lee, T. T., (2017). Antioxidant capacity of phytochemicals and their potential effects on oxidative status in animals—A review. *Asian-australas J Anim Sci.*, *30*(3), 299.

Liu, R. H., (2003). Health benefits of fruit and vegetables are from additive and synergistic combinations of phytochemicals. *Am J Clin Nutr.*, *78*(3), 517S–520S.

Magozwi, D. K., Dinala, M., Mokwana, N., Siwe-Noundou, X., Krause, R. W., Sonopo, M., McGaw, L. J., et al., (2021). Flavonoids from the genus euphorbia: Isolation, structure, pharmacological activities and structure–activity relationships. *Pharmaceuticals*, *14*(5), 428.

Mahmood, A. R., Abdullah, M. R., & Khalaf, H. S., (2022). Use of flavonoids and green tea extracts as antioxidants induced by oxidative stress: A review article. *Galore Int. J. Health. Sci. Res.*, *7*(2).

Mali, A. V., Padhye, S. B., Anant, S., Hegde, M. V., & Kadam, S. S., (2019). Anticancer and antimetastatic potential of enterolactone: Clinical, preclinical, and mechanistic perspectives. *Eur. J. Pharmacol.*, *852*, 107–124.

Malik, G., Singh, D. B., Malik, A. A., & Gulzar, A., (2021). Underutilized vegetable species of Indian Himalayas. In: *Neglected and Underutilized Crops-Towards Nutritional Security and Sustainability* (pp. 113–132). Springer, Singapore.

Mao, X. Y., Jin, M. Z., Chen, J. F., Zhou, H. H., & Jin, W. L., (2018). Live or let die: Neuroprotective and anti-cancer effects of nutraceutical antioxidants. *Pharmacol Ther.*, *183*, 137–151.

Mares, J., (2016). Lutein and zeaxanthin isomers in eye health and disease. *Annu. Rev. Nutr.*, *36*, 571.

Masyita, A., Sari, R. M., Astuti, A. D., Yasir, B., Rumata, N. R., Emran, T. B., Nainu, F., & Simal-Gandara, J., (2022). Terpenes and terpenoids as main bioactive compounds of essential oils, their roles in human health and potential application as natural food preservatives. *Food Chem.: X.*, 100217.

Matsumoto, M., Waki, N., Suganuma, H., Takahashi, I., Kurauchi, S., Sawada, K., Tokuda, I., et al., (2020). Association between biomarkers of cardiovascular diseases and the blood concentration of carotenoids among the general population without apparent illness. *Nutrients*, *12*(8), 2310.

Mattison, C. P., Cardemil, C. V., & Hall, A. J., (2018). Progress on norovirus vaccine research: Public health considerations and future directions. *Expert. Rev. Vaccines*, *17*(9), 773–784.

Międkus, N., Marszałek, K., Podlacha, M., Iqbal, A., Puchalski, C., & Świergiel, A. H., (2020). Health benefits of plant-derived sulfur compounds, glucosinolates, and organosulfur compounds. *Molecules*, *25*(17), 3804.

Miller, P. E., & Snyder, D. C., (2012). Phytochemicals and cancer risk: A review of the epidemiological evidence. *Nutr. Clin. Pract.*, *27*(5), 599–612.

Muller, L., & Bohm, V., (2011). Antioxidant activity of β-carotene compounds in different *in vitro* assays. *Molecules*, *16*(2), 1055–1069.

Muneeb, U. R., & Mubashir, H. M., (2022). *Edible Plants in Health and Diseases: Cultural, Practical and Economic Value* (Vol. 1). Singapore: Springer Nature Singapore.

Nair, K. P., (2021). Cinchona (*Cinchona* sp.). In: *Tree Crops* (pp. 129–151). Springer, Cham.

Nakanishi, K., Goto, T., & Itô, S., (2013). *Natural Products Chemistry* (Vol. 1). Academic press.

Namitha, K. K., & Negi, P. S., (2010). Chemistry and biotechnology of carotenoids. *Crit. Rev. Food Sci. Nutr.*, *50*(8), 728–760.

Nasri, H., Baradaran, A., Shirzad, H., & Rafieian-Kopaei, M., (2014). New concepts in nutraceuticals as alternative for pharmaceuticals. *Int. J. Prev Med.*, *5*(12), 1487.

Neveu, V., Perez-Jiménez, J., Vos, F., Crespy, V., du Chaffaut, L., Mennen, L., ... & Scalbert, A. (2010). *Phenol-Explorer: An Online Comprehensive Database on Polyphenol Contents in Foods*. Database, 2010, bap024.

Nichenametla, S. N., Taruscio, T. G., Barney, D. L., & Exon, J. H., (2006). A review of the effects and mechanisms of polyphenolics in cancer. *Crit. Rev. Food Sci. Nutr.*, *46*(2), 161–183.

Nilsson, J., Olsson, K., Engqvist, G., Ekvall, J., Olsson, M., Nyman, M., & Åkesson, B., (2006). Variation in the content of glucosinolates, hydroxycinnamic acids, carotenoids, total antioxidant capacity and low-molecular-weight carbohydrates in *Brassica* vegetables. *J. Sci. Food Agric.*, *86*(4), 528–538.

Nout, M. J., & Aidoo, K. E., (2011). Asian fungal fermented food. In: *Industrial Applications* (pp. 29–58). Springer, Berlin, Heidelberg.

Nowak-Wegrzyn, A. (2013). *Clinical Manifestations and Diagnosis of Oral Allergy Syndrome (Pollen-Food Allergy Syndrome)*. Retrieved September, 11, 2016.

Ogbe, R. J., Ochalefu, D. O., Mafulul, S. G., & Olaniru, O. B., (2015). A review on dietary phytosterols: Their occurrence, metabolism, and health benefits. *Asian J. Plant Sci Res.*, *5*(4), 10–21.

Olmedilla-Alonso, B., Rodríguez-Rodríguez, E., Beltrán-de-Miguel, B., & Estévez-Santiago, R., (2020). Dietary β-cryptoxanthin and α-carotene have greater apparent bioavailability than β-carotene in subjects from countries with different dietary patterns. *Nutrients*, *12*(9), 2639.

Ou, H. C., Pandey, S., Hung, M. Y., Huang, S. H., Hsu, P. T., Day, C. H., Pai, P., et al., (2019). Luteolin: A natural flavonoid enhances the survival of HUVECs against oxidative stress by modulating AMPK/PKC pathway. *Am. J. Chinese Med.*, *47*(03), 541–557.

Packer, L., & Sies, H., (2001). *Flavonoids and Other Polyphenols*. Elsevier.

Paduch, R., Kandefer-Szerszeń, M., Trytek, M., & Fiedurek, J., (2007). Terpenes: Substances useful in human healthcare. *Arch. Immunol Ther.*, *55*(5), 315–327.

Pandey, K. B., & Rizvi, S. I., (2009). Plant polyphenols as dietary antioxidants in human health and disease. *Oxid. Med. Cell. Longev.*, *2*(5), 270–278.

Panigrahy, S. K., Bhatt, R., & Kumar, A., (2021). Targeting type II diabetes with plant terpenes: The new and promising antidiabetic therapeutics. *Biologia.*, *76*(1), 241–254.

Pascal, L. S., Segal, R., Dima, S., & Constantin, O., (2009). Phytosterol emulsions for functional foods: Preparation and stability. *Lucrări Științifice-Universitatea de Științe Agronomice Și Medicină Veterinară București. Seria F, Biotehnologii.*, (Special Volume) (pp. 256–260).

Patel, K., Kumar, V., Verma, A., Rahman, M., & Patel, D. K., (2017). β-sitosterol: Bioactive compounds in foods, their role in health promotion and disease prevention "a concise report of its phytopharmaceutical importance." *Curr. Tradit. Med.*, *3*(3), 168–177.

Patil, A. S., (2020). *Plant Secondary Metabolites: Isolation, Characterization & Biological Properties*. Studera Press.

Patil, S. M., Ramu, R., Shirahatti, P. S., Shivamallu, C., & Amachawadi, R. G., (2021). A systematic review on ethnopharmacology, phytochemistry and pharmacological aspects of *Thymus vulgaris* Linn. *Heliyon.*, *7*(5), e07054.

Pawankar, R., (2014). Allergic diseases and asthma: A global public health concern and a call to action. *World Allergy Organ J.*, *7*(1), 1–3.

Pennington, J. A. T., (2003). Definitions and classifications for fruit and vegetables. *IARC Handbooks of Cancer Prevention*, *8*, 1–21.

Percival, S. S., & Turner, R. E., (2002). Applications of herbs to functional foods. In: *Handbook of Nutraceuticals and Functional Foods* (pp. 415–428). CRC Press.

Perveen, S., & Al-Taweel, A., (2018). *Terpenes and Terpenoids*. BoD–Books on Demand.

Pessu, P. O., Agoda, S., Isong, I. U., Adekalu, O. A., Echendu, M. A., & Falade, T. C., (2011). Fungi and mycotoxins in stored foods. *Afr. J. Microbiol. Res.*, *5*(25), 4373–4382.

Phan, M. A. T., Paterson, J., Bucknall, M., & Arcot, J., (2018). Interactions between phytochemicals from fruits and vegetables: Effects on bioactivities and bioavailability. *Crit. Rev. Food Sci. Nutr.*, *58*(8), 1310–1329.

Pinto, T., Aires, A., Cosme, F., Bacelar, E., Morais, M. C., Oliveira, I., Ferreira-Cardoso, J., et al., (2021). Bioactive (poly) phenols, volatile compounds from vegetables, medicinal and aromatic plants. *Foods*, *10*(1), 106.

Poudel, A., Gachumi, G., Wasan, K. M., Dallal, B. Z., El-Aneed, A., & Badea, I., (2019). Development and characterization of liposomal formulations containing phytosterols extracted from canola oil deodorizer distillate along with tocopherols as food additives. *Pharmaceutics*, *11*(4), 185.

Prado-Audelo, D., Luisa, M., Cortés, H., Caballero-Florán, I. H., González-Torres, M., Escutia-Guadarrama, L., Bernal-Chávez, S. A., et al., (2021). Therapeutic applications of terpenes on inflammatory diseases. *Front Pharmacol.*, 2114.

Prakash, D., & Kumar, N., (2011). Cost effective natural antioxidants. In: *Nutrients, Dietary Supplements, and Nutriceuticals* (pp. 163–187). Humana Press.

Prakash, D., Gupta, C., & Sharma, G., (2012). Importance of phytochemicals in nutraceuticals. *Journal of Chinese Medicine Research and Development*, *1*(3), 70–78.

Prakash, V., (2017). Terpenoids as source of anti-inflammatory compounds. *Asian J. Pharm. Clin. Res.*, *10*(3), 68–76.

Preedy, V. R., Zibadi, S., & Watson, R., (2013). *Polyphenols in Human Health and Disease*. Netherlands: Elsevier Science.

Qin, Y., Zhang, H., Liu, Q., Jiang, B., Chen, J., & Zhang, T., (2021). Sulforaphane attenuates oxidative stress and inflammation induced by fine particulate matter in human bronchial epithelial cells. *J. Funct. Foods*, *81*, 104460.

Radauer, C., Bublin, M., Wagner, S., Mari, A., & Breiteneder, H., (2008). Allergens are distributed into few protein families and possess a restricted number of biochemical functions. *J. Allergy Clin Immunol.*, *121*(4), 847–852.

Radauer, C., Nandy, A., Ferreira, F., Goodman, R. E., Larsen, J. N., Lidholm, J., Pomes, A., et al., (2014). Update of the WHO/IUIS allergen nomenclature database based on analysis of allergen sequences. *Allergy*, *69*(4), 413–419.

Radovich, T. J., (2011). *Biology and Classification of Vegetables* (p. 3). Wiley-Blackwell Publishing: USA.

Raghavan, S., (2006). *Handbook of Spices, Seasonings, and Flavorings*. United States: CRC Press.

Rane, B. R., Bharath, M. S., Patil, R. R., Keservani, R. K., & Jain, A. S., (2020b). Novel approaches in nutraceuticals; Chapter 11. In: Rajesh, K. K., Raj, K. K., & Anil, K. S., (eds.), *Enhancing the Therapeutic Efficacy of Herbal Formulations Through Novel Drug Delivery Systems* (pp. 241–266). IGI Global International Publisher, Pennsylvania, USA. ISBN13: 9781799844532.

Rane, B. R., Patil, A. S., Keservani, R. K., & Jain, A. S., (2020a). Novel approaches in herbal formulation; Chapter 2. In: Rajesh, K. K., Raj, K. K., & Anil, K. S., (eds.), *Enhancing the*

Therapeutic Efficacy of Herbal Formulations Through Novel Drug Delivery Systems (pp. 43–68). IGI Global International Publisher, Pennsylvania, USA. ISBN13: 9781799844532.

Rane, B. R., Tadavi, S. A., & Keservani, R. K., (2020c). Naturopathy; Chapter 12. In: Anil, K. S., Raj, K. K., & Surya, P. G., (eds.), *Herbal Product Development* (pp. 321–347). Apple Academic Press, CRC Press, Taylor & Francis Group. ISBN: 9781771888776.

Reboredo-Rodríguez, P., Varela-López, A., Forbes-Hernández, T. Y., Gasparrini, M., Afrin, S., Cianciosi, D., Zhang, J., et al., (2018). Phenolic compounds isolated from olive oil as nutraceutical tools for the prevention and management of cancer and cardiovascular diseases. *Int. J. Mol. Sci.*, *19*(8), 2305.

Rowland, I., (1999). Optimal nutrition: Fiber and phytochemicals. *Proc Nutr Soc.*, *58*(2), 415–419.

Sahin, I., Bilir, B., Ali, S., Sahin, K., & Kucuk, O., (2019). Soy isoflavones in integrative oncology: Increased efficacy and decreased toxicity of cancer therapy. *Integr Cancer Ther.*, *18*, 1534735419835310.

Salanță, L. A., (2020). *The Health Benefits of Foods: Current Knowledge and Further Development.* IntechOpen, London, U.K.

Sarabandi, K., Rafiee, Z., Khodaei, D., & Jafari, S. M., (2019). Encapsulation of food ingredients by nanoliposomes. In: *Lipid-Based Nanostructures for Food Encapsulation Purposes* (pp. 347–404). Academic Press.

Sarkar, D., Ankolekar, C., & Shetty, K., (2020). Beneficial lactic acid bacteria (LAB)-based biotransformation of plant and dairy substrates to enhance type 2 diabetes-relevant health benefits. In: *Functional Foods and Biotechnology* (pp. 345–360). CRC Press.

Sarkar, P., Dh, L. K., Dhumal, C., Panigrahi, S. S., & Choudhary, R., (2015). Traditional and ayurvedic foods of Indian origin. *J. Ethn. Foods*, *2*(3), 97–109.

Schiepers, O. J., De Groot, R. H., Van, B. M. P., Jolles, J., De Jong, A., Lütjohann, D., & Mensink, R. P., (2009). Consuming functional foods enriched with plant sterol or stanol esters for 85 weeks does not affect neurocognitive functioning or mood in statin-treated hypercholesterolemic individuals. *The Journal of Nutrition, 139*(7), 1368–1373.

Septembre-Malaterre, A., Remize, F., & Poucheret, P., (2018). Fruits and vegetables, as a source of nutritional compounds and phytochemicals: Changes in bioactive compounds during lactic fermentation. *Food Research International, 104*, 86–99.

Shah, B. N., (2009). *Textbook of Pharmacognosy and Phytochemistry*. India: Elsevier. A Division of Reed Elsevier India Pvt. Limited.

Shah, I., Shah, M. A., Nawaz, M. A., Pervez, S., Noreen, N., Vargas-De La, C. C., & Niaz, K., (2020). Analysis of other phenolics (capsaicin, gingerol, and alkylresorcinols). In: *Recent Advances in Natural Products Analysis* (pp. 255–271). Elsevier.

Shahali, Y., & Dadar, M., (2018). Plant food allergy: Influence of chemicals on plant allergens. *Food and Chemical Toxicology, 115*, 365–374.

Shahidi, F., & Ambigaipalan, P., (2015). Phenolics and polyphenolics in foods, beverages, and spices: Antioxidant activity and health effects: A review. *Journal of Functional Foods, 18*, 820–897.

Sharma, A. K., Keservani, R. K., & Gautam, S. P., (2020). *Herbal Product Development*. Apple Academic Press, CRC Press, Taylor & Francis Group. ISBN: 9781771888776.

Sharma, G., & Prakash, D., (2014). *Phytochemicals of Nutraceutical Importance*. United Kingdom: CABI.

Sharma, G., & Thakur, N., (2020). Curcumin–the healing herb: Properties and future prospective. *Asian J. Pharm. Clin. Res., 13*(2), 4–9.

Sharma, G., Prakash, D., Gupta, C., Prakash, D., & Sharma, G., (2014). Phytochemicals of nutraceutical importance: Do they defend against diseases. *Phytochemicals of Nutraceutical Importance*, 1.

Sharma, K., Kumar, V., Kaur, J., Tanwar, B., Goyal, A., Sharma, R., Gat, Y., & Kumar, A., (2021). Health effects, sources, utilization, and safety of tannins: A critical review. *Toxin Rev.*, *40*(4), 432–444.

Sharma, S. H., Thulasingam, S., & Nagarajan, S., (2017). Terpenoids as anti-colon cancer agents: A comprehensive review on its mechanistic perspectives. *Eur. J. Pharmacol.*, *795*, 169–178.

Sharma, V. P., Singh, S., Dhanjal, D. S., Singh, J., & Yadav, A. N., (2021). Potential strategies for control of agricultural occupational health hazards. In: *Current Trends in Microbial Biotechnology for Sustainable Agriculture* (pp. 387–402). Springer, Singapore.

Shukitt-Hale, B., Lau, F. C., Carey, A. N., Galli, R. L., Spangler, E. L., Ingram, D. K., & Joseph, J. A., (2008). Blueberry polyphenols attenuate kainic acid-induced decrements in cognition and alter inflammatory gene expression in rat hippocampus. *Nutr. Neurosci.*, *11*(4), 172–182.

Sikdar, S., Nandy, S., Mukherjee, A., Bhattacharyya, R., Pandey, D. K., & Dey, A., (2019). Phytoestrogens as anticancer therapeutics: A retrospective and future perspectives. *J. Biol. Act Prod Nat.*, *9*(3), 179–196.

Sikorska-Zimny, K., & Beneduce, L., (2021). The metabolism of glucosinolates by gut microbiota. *Nutrients*, *13*(8), 2750.

Singh, B., & Sharma, R. A., (2015). Plant terpenes: Defense responses, phylogenetic analysis, regulation, and clinical applications. *3 Biotech*, *5*(2), 129–151.

Singh, S., & Rawat, P. S., (2021). Nutraceuticals: An approach towards safe and effective medications. In: *Treating Endocrine and Metabolic Disorders with Herbal Medicines* (278–297). IGI Global.

Storniolo, C. E., Sacanella, I., Lamuela-Raventos, R. M., & Moreno, J. J., (2020). Bioactive compounds of Mediterranean cooked tomato sauce (sofrito) modulate intestinal epithelial cancer cell growth through oxidative stress/arachidonic acid cascade regulation. *ACS Omega*, *5*(28), 17071–17077.

Tanumihardjo, S. A., (2012). *Carotenoids and Human Health*. Springer Science & Business Media.

Taylor, C. M., Northstone, K., Wernimont, S. M., & Emmett, P. M., (2016). Macro-and micronutrient intakes in picky eaters: A cause for concern? *Am. J. Clin. Nutr.*, *104*(6), 1647–1656.

Thakur, A., & Sharma, R., (2018). Health promoting phytochemicals in vegetables: A mini review. *Int. J Food Ferment Technol.*, *8*(2), 107–117.

Tiwari, B. K., Brunton, N., & Brennan, C. S., (2015). *Handbook of Plant Food Phytochemicals*. Wiley-Blackwell.

Tomás-Barberán, F. A., & Robins, R. J., (1997). Phytochemistry of fruit and vegetables. In: *International Symposium of Phytochemistry of Fruit and Vegetables*. 1995: Murcia, Spain. Clarendon Press.

Van, B. S. G., & De Kok, T. M., (2018). Smart combinations of bioactive compounds in fruits and vegetables may guide new strategies for personalized prevention of chronic diseases. *Mol Nutr Food Res.*, *62*(1), 1700597.

Van, B., & Pajkovic, (2008). Multitargeted therapy of cancer by lycopene. *Cancer Lett.*, *269*(2), 339–351.

Verma, A. K., & Pratap, R., (2010). The biological potential of flavones. *Nat Prod Rep.*, *27*(11), 1571–1593.

Verma, V. C., (2020). Chapter 5, tomatoes: Rich source of lycopene. In: *Food Technology and Nutrition* (p. 75).

Wang, Y., Xiang, Y., Wang, R., Li, X., Wang, J., Yu, S., & Zhang, Y., (2022). Sulforaphane enhances Nrf2-mediated antioxidant responses of skeletal muscle induced by exhaustive exercise in HIIT mice. *Food Sci. Hum. Wellness, 11*(5), 1355–1361.

Watson, N., Brandel, J. P., Green, A., Hermann, P., Ladogana, A., Lindsay, T., Mackenzie, J., et al., (2021). The importance of ongoing international surveillance for Creutzfeldt–Jakob disease. *Nat Rev Neurol.*, *17*(6), 362–379.

Willcox, M., Bodeker, G., Rasoanaivo, P., & Addae-Kyereme, J., (2004). *Traditional Medicinal Plants and Malaria.* CRC press.

Wolters, M., Dejanovic, G. M., Asllanaj, E., Günther, K., Pohlabeln, H., Bramer, W. M., Ahrens, J., et al., (2020). Effects of phytoestrogen supplementation on intermediate cardiovascular disease risk factors among postmenopausal women: A meta-analysis of randomized controlled trials. *Menopause*, *27*(9), 1081–1092.

Xu, H., Linn, B. S., Zhang, Y., & Ren, J., (2019). A review on the antioxidative and prooxidative properties of luteolin. *React. Oxyg. Species, 7*, 136–147.

Yahia, E. M., García-Solís, P., & Celis, M. E. M., (2019). Contribution of fruits and vegetables to human nutrition and health. In: *Postharvest Physiology and Biochemistry of Fruits and Vegetables* (pp. 19–45). Woodhead Publishing.

Yamagata, K., (2019). Soy isoflavones inhibit endothelial cell dysfunction and prevent cardiovascular disease. *J Cardiovasc Pharmacol.*, *74*(3), 201–209.

Yanaka, A., (2017). Role of sulforaphane in protection of gastrointestinal tract against *H. pylori* and NSAID-induced oxidative stress. *Curr Pharm Des.*, *23*(27), 4066–4075.

Yerou, K. O., El Kadi, F. Z., Kanoun, K., Khelladi, H. M., Benzahia, H., & Bekhti, S., (2022). Preliminary phytochemical investigation and antioxidant potential of various extracts of dietary turnip (*Brassica Rapa* L.). *Food and Environment Safety Journal*, *21*(1).

Yeung, K. S., Gubili, J., & Cassileth, B., (2008). Evidence-based botanical research: Applications and challenges. *Hematol. Oncol. Cli. North Am.*, *22*(4), 661–670.

Zhang, Y. J., Gan, R. Y., Li, S., Zhou, Y., Li, A. N., Xu, D. P., & Li, H. B., (2015). Antioxidant phytochemicals for the prevention and treatment of chronic diseases. *Molecules*, *20*(12), 21138–21156.

Zia-Ul-Haq, M., (2021). Historical and introductory aspects of carotenoids. In: *Carotenoids: Structure and Function in the Human Body* (pp. 1–42). Springer, Cham.

CHAPTER 4

FLAVONOIDS IN VEGETABLES AS NUTRACEUTICALS

MAYUR S. BIDKAR, AKSHAY U. GHORPADE,
DINESH B. CHAUDHARI, SWATI G. TALELE, and ANIL G. JADHAV

Department of Pharmaceutics, Sandip Institute of Pharmaceutical Sciences, Nashik, Maharashtra, India

ABSTRACT

The usage of health-promoting foods has risen dramatically as an outcome of changing lifestyles, higher purchasing power, and better-informed customers. People are increasingly preferring natural and organic meals over manufactured foods. As a result, nutraceuticals are gaining popularity and attractive a well-organized component of people's diets. Vegetables, which are high in vitamins, minerals, and dietary fibers, contribute significantly to our nutritional needs and are regarded as "protecting food." Flavonoids provide the nutritional qualities to fruits and vegetables. Because of their importance in food organoleptic characteristics and human health, a better understanding of their structures and biological activities specifies their potential as therapeutic compounds and also for forecasting and managing food standards. Because of the heterogeneity of therapeutic activities in the human body, flavonoids are mostly utilized as "nutraceuticals."

4.1 INTRODUCTION

Due to greater health awareness, rising urbanization, and inadequate nutrition as an outcome of contemporary lifestyle choices, health-promoting

Plant Metabolites and Vegetables as Nutraceuticals.
Raj K. Keservani, Bui Thanh Tung, Rajesh K. Kesharwani, and Eknath D. Ahire (Eds.)

foods are obtaining great traction. Around 795 million people are undernourished in high society this day, this implies that every nine individuals does not obtain enough food to live a healthy and active life (FAO, 2015). Vegetables are regarded as a necessary component of a well-balanced nutritious diet. Vitamins (Singh et al., 2017a), carotenoids, phenolics, alkaloids, organosulfur compounds, and other antioxidant substances are primarily responsible for the good health benefits of vegetables. Vegetables include, nutritional supplements (Keservani et al., 2010a), nutraceuticals (Keservani et al., 2010b), phototherapeutic compounds, and other plant foods variety of phytochemicals that assist the body in maintaining its health and fight sickness (Rahal et al., 2014) Natural goods and health-promoting foods undergo obtain group observation lately, thanks to recent developments in medical and nutrition sciences. People began consuming more vegetables, and fruits in order to improve their quality of life.

Nutraceuticals (Keservani et al., 2017, 2020) are chemical compounds found in plants that can be considered food or a component of the diet and can aid with illness prevention and treatment in addition to providing nutrients for body development (Kalia & Muthu Kumar, 2015). They're also known as products that have been separated or purified from foods and are offered in therapeutic forms that aren't normally linked with food and have been shown to supply an anatomical edge or give protection against chronic illnesses. Dr. Stephen L. De Felice, creator and Chairperson of the Organization for Revolution in Medicine, New Chemise, USA, invented the phrase 'nutraceutical' in 1989 by merging the terms 'nutrition' and 'pharmaceutical. Nutraceuticals, designer foods, pharma foods, dietary supplements, and other terms are used to describe functional foods. In their interchangeable use by various persons at different times, there appears to be a thin separating line (Rajasekaran et al., 2008). Both locally and globally, practical nutriment and nutraceutical compounds provide a value-added development potential for the agri-food business. Nutraceuticals can be found in a heterogeneity of goods from (a) the catering industry; (b) the botanical and food supplement sector; (c) the pharmaceutical industry; and (d) newly integrated pharmaceutical/agribusiness/nutrition conglomerates. With the development of several nutraceuticals, the objective of obtaining almost suitable or Maximal nation of vitamins and fitness is becoming a greater difficulty (Gibson & Makrides, 2000). Although Most people of nutraceuticals now in use are well-known essential nutrients for the human body, many specifics such as dosage, drug-drug interactions, nutraceutical-drug interactions, and their impact on people with specific health issues are still unknown (Prabu, 2015).

4.2 FLAVONOIDS AS NUTRACEUTICAL

Stephen DE Felice created the word "nutraceutical" in 1979. It's described as "a food or component of a meal that has medicinal or health advantages, such as illness prevention and therapy." Nutraceuticals can Encompass The whole lot from remoted vitamins, nutritional supplements, and diets to genetically Changed "dressmaker" ingredients, natural gadgets, and processed meals which include cereals, soups, And drinks. Any Risk-free meal extract supplement with scientifically demonstrated Fitness advantages for both the remedy and prevention of ailment is referred to as a nutraceutical. Consumers are becoming more interested in nutraceuticals as epidemiological research shows that a certain diet or component of a diet is linked to a decreased risk of certain diseases.

Flavonoids (Keservani & Sharma, 2014; Keservani et al., 2016a, b) are the most active nutraceutical chemicals found in plants. They can operate as powerful antioxidants and metal chelators, as is characteristic of phenolic compounds. Anti-inflammatory, antiallergic, hepatoprotective, antithrombotic, antiviral, and anticarcinogenic properties have long been recognized, as mentioned in succeeding sections.

4.3 VEGETABLES AND HEALTH

Vegetables are considered protective foods because they include vital vitamins and minerals, dietary fiber, and phytochemicals such as carotenoids, glucosinolates (GSs), polyphenols, thiosulfides, ascorbate, and folate, among other things. Protein-rich veggies include peas, beans, cowpeas, and lablab beans, whereas carbohydrate-rich (Singh et al., 2017b) tuber crops include potato, tapioca, yams, Colocasia, and sweet potato. Calcium, phosphorus, and iron are deficient in cereals, but They're rich in peas, beans, and okra. Leafy greens like amaranth and spinach are high in cellulose and fiber, both of which are necessary for good digestion. Vegetable consumption is critical for maintaining good health and protecting the body from degenerative disorders (Paramaguru, 2012). Table 4.1 lists phenolic chemicals, terpenoids, sulfur compounds, pigments, and other natural antioxidants found in vegetables and plants that can be used to cure a variety of illnesses. Foods made from garlic, soybeans, cabbage, ginger, and umbelliferous vegetables are often utilized in cancer treatment (Prabu, 2015). Garlic powder preparations have been shown to reduce moderate hypertension in clinical trials. Potassium and magnesium are abundant in most vegetables and fruits, resulting in an alkaline pH that is beneficial to bones. It counteracts the acidic state that develops as an outcome of a meat-heavy diet.

TABLE 4.1 Vegetable Sources of Nutraceutical/Bioactive Compounds

Vegetables	Nutraceutical/Bioactive Compound
Allium vegetables (garlic, onion, leek).	Allyl sulfides
Cruciferous vegetables.	Indoles/glucosinolates, sulforaphane, isothiocyanate, thiols.
Solanaceous vegetables (tomato, chilly).	Lycopene
Umbelliferous vegetables (carrots, celery, parsley).	Umbelliferous vegetables (carrots, celery, parsley).
Compositae plants (artichoke).	Silymarin
Carrot, squash, broccoli, sweet potato, tomato, kale, cantaloupe, and pumpkin.	Vitamin A
Green leafy vegetables.	Vitamin E

4.4 PHYTOCHEMICALS IN VEGETABLES

Phytochemicals are plant-derived chemical molecules with health-promoting characteristics (Table 4.2). Phytonutrients or phytonutraceuticals are other terms for them. There are over 1,000 phytochemicals that have been identified. They are an important element of a plant's natural defense and healing mechanism, as well as pollinator attraction and seed dispersion. Phytochemicals are powerful antioxidants that can alter metabolic activity and help in carcinogen elimination. Rather than restricting oneself to fruits and vegetables with the highest antioxidant content, it is advisable to consume a range of them (Wargovich, 2000). Soy isoflavones mimic human estrogens and can aid with menopausal symptoms and osteoporosis. Indoles, which are contained in cabbages, activate enzymes that limit the effectiveness of estrogen, potentially lowering the chance of breast cancer. Some phytochemicals, such as saponins in beans, hinder cancer cells from duplicating by interfering with cell DNA replication. Capsaicin, a compound found in warm peppers, protects DNA from carcinogens. Some phytochemicals bond to cell walls physically, preventing pathogens from adhering to human cellular.

4.5 BASED ON COLOR

Lycopene, one of over 600 carotenoids, is the main pigment in reds and pinks. It has a molecular mass of 536 and is a tetraterpene hydrocarbon containing 40 carbon atoms and 56 hydrogen atoms. It's a naturally occurring antioxidant. Reduce your chances of developing cardiovascular disease (CVD) or prostate cancer (Giovanucci, 2002). Due to the fact lycopene is fats soluble, it needs to be fed on with at least some fat. Lycopene is more easily absorbed by the body when heated. Though tomatoes are the most frequent supply of lycopene, the seed aril of gac (*Momordica cochinchinensis*) fruits is the greatest source, with 70 times more lycopene than tomatoes (Ishida et al., 2004). Because lycopene is a carotenoid and soluble in oil, the bioavailability of lycopene from aril was high due to its oil content (Voung et al., 2005). Lycopene is also rich in red carrots (Surles et al., 2004).

Carrots, pumpkins, sweet potatoes, and cantaloupe all contain beta carotene, a common phytochemical that belongs to the carotenoid family.

It is a Vitamin A precursor. Maintain your heart's health, your vision's health, and your immune system's health. A group of Harvard researchers looked at the result of carotenoids on lung cancer. A total of 1,24,000 men and women took part in the 10-year study. The findings revealed that Those who ate a weight loss plan wealthy in lycopene and alpha-carotene had a 20–25% decreased risk of lung cancer, especially among current smokers (Willoughby, 2000). Carotene concentrations in dark orange carrot types range from 130 to 500 ppm (Simon et al., 2009).

1. **Greens:** Leafy greens inclusive of spinach, lettuce, kale, broccoli, cabbage, and others contain lutein, a potent antioxidant that can help prevent cataracts and macular degeneration, as well as increase immunological response (Moeller et al., 2000). Brinjal, Purple cabbage, purple carrot, and other blue and purple vegetables include the disease-fighting phytonutrients anthocyanin and polyphenols. Colored flavonoids are known as anthocyanins. They scavenge free radicals, especially singlet oxygen, slowing the aging process and lowering the risk of cancer, Alzheimer's disease (AD), and myocardial infarction (MI) (Rahal et al., 2014).

2. **Whites:** Garlic, onion, and leek may be colorless, but they're rich with phytochemicals like allicin, the most frequent phytochemical in this category. It aids in the reduction of cholesterol and blood pressure (Mousa & Mousa, 2007). Eating vegetables from the onion family reduces the incidence of heart attacks and helps to prevent the spread of stomach and colon cancer.

4.6 GLUCOSINOLATES (GSS)

This is a broad category of sulfur-containing chemicals found in all commercially important cruciferous plants. A sulfonated oxime, skeletal âthioglucose, and a variable side chain derived from methionine, tryptophan, or phenylalanine make up the molecule (Mithen et al., 2000). Intact GSs have little biological effect against cancer, but their breakdown products trigger mixed-function oxidases, which are involved in carcinogen detoxification. When plant tissue is injured, the natural enzyme myrosinase hydrolyzes GSs, releasing a variety of breakdown products, including isothiocynates, which are the main source of anticarcinogenic action.

TABLE 4.2 Antioxidant Capacity of Some Vegetables

Vegetables	Antioxidant Value (ORAC Units/100 g)
Garlic	1,940
Kale	1,770
Spinach	1,250
Brussels sprouts	980
Red bell pepper	710
Onion	450
Brinjal	390
Cauliflower	380
Cucumber	50

According to Chang et al. (2007), tiny Chinese cabbage, cabbage sprouts, and Chinese kale have the greatest amount of indole GSs among the locally farmed cruciferous vegetables in Taiwan. Sulforaphane, a GS produced from broccoli's glucoraphane, is anticarcinogenic. This is accomplished by inhibiting carcinogenic phase I enzymes while inducing detoxifying phase II enzymes (Piekarsha et al., 2013).

Beneforte' Broccoli, created at the Institute of Food Research in the United Kingdom, accumulates three times more glucoraphanin (Grn), a GS that is degraded to sulforaphane. Broccoli was crossed with wild *Brassica villosa* to create it. It has a 28-transcription factor allele from *B. villosa*. It has a softer, more agreeable flavor (Mithen, 2013).

4.7 POLYPHENOLICS

The most diverse collection of compounds found in plants. Anthcyanins, flavones, flavonols, isoflavonols, and terpenoids are examples of polyphenolics/flavonoids. The most significant flavonoid found in onions is quercetin (Crozier et al., 1997). Flavonoids have antibacterial, hepatoprotective, antiulcer, anti-inflammatory, antidiabetic, and cardioprotective properties, in addition to having dramatic effects on the central nervous system (CNS) (Tapas et al., 2008).

4.8 ANTIOXIDANTS IN VEGETABLES

Antioxidants are certainly taking place plant compounds that can prevent or reduce the oxidative damage produced by free radicals in the human body. Carotenoids, flavonoids, organosulfurs, GSs, Vitamin C, Vitamin E, and Selenium are all examples of antioxidants. The importance of vegetables as a source of antioxidants in disease prevention and anti-aging is widely understood. The condition of instability caused by Entities which include loose radicals is highly frequent in biological systems, and antioxidants play a critical role in managing free radicals and overcoming the instability (Sheokand et al., 2012).

A molecule or atom with as a minimum one unpaired electron in its outer shell is known as a free radical. It may either win or lose an argument.

You can either lose an electron and oxidize or acquire an electron and reduce. Free radicals are formed as a result of metabolism, cigarette smoke, pollution, ultraviolet light, and ionizing radiation, among other things. Free radical damage alters the functioning of cell proteins, causes DNA mutations, damages polyunsaturated lipids in cell membranes, and disrupts cell signaling. The USDA created a system to rate the antioxidant content material of vegetables. The ORAC scale stands for Oxygen Radical Absorbance. Table 4.2 lists the antioxidant capacity of several vegetables. According to Ou et al. (2002), the major sources of antioxidant activity against peroxyl radicals include green pepper, spinach, purple onion, broccoli, beet, and cauliflower. Lee and his colleagues (According to a study published in 2007, red cabbage exhibited the highest radical scavenging activity among cruciferous vegetables, followed by Chinese white cabbage, green cabbage, and mustard cabbage. Green leafy vegetables are high in beta carotene, nature's most powerful antioxidant that can help prevent heart disease and atherosclerosis (Subhasree et al., 2009).

4.9 STRATEGIES TO IMPROVE PHYTOCHEMICAL CONTENT IN VEGETABLES

When it comes to the number of phytochemicals available, there is a lot of variation amongst plants. People have chosen a variety of vegetable crops for food purposes from the dawn of time, whether consciously or unconsciously. It is feasible to produce novel vegetable varieties or incorporate favorable genes for nutraceuticals and bioactive chemicals into farmed varieties using traditional breeding (Kalia & Muthukumar, 2015). Advances in molecular biology and recombinant technology have paved the road for adopting marker-aided breeding and creating new vegetable crops to speed up the production of specific trait varieties.

Cucumber fruits have no mesocarp or placental coloration. The orange flesh of the Xishuangbanna gourd (*Cucumis sativus var. xishuangbannesis*) is high in â carotene (700 g/100 g). By introducing the carotene gene from the xishuangbanna gourd to cucumber, Simon & Navazio (1997) created three varieties: Early orange mass 400, early orange mass 402, and past due orange mass. Cucumber CC3 and XIS gourd SWCC8 were crossed by Bo et al. (2012). Selfing of hybrid progeny yielded 124 F7 RILs. The ore gene responsible for orange color was connected to 7 SSR markers on linkage organization three according to market analysis. SSR07706 provides general value for the creation of HQâC germplasm using MAS in cucumber, according to the association between marker and QâC.

Beijing Vegetable Research Center created an Orange Cauliflower Hybrid 'Jinyu 60' with a high carotene content (Ding & Jian, 2010). **It** was an orange mutant cauliflower with little uncompacted curds that was crossed with CMS92-60, which had white compact curds, to generate Jinyu60. The dry curd had 31.3 g/g of carotene, which became 10.8 instances higher than white cauliflower.

Just et al. (2009) used QAL (wild white) x B493 to do QTL analysis for carotenoid production in carrots (orange). Two key interacting loci, Y on LG-2 and Y (2) on LG-5, are reported to govern a lot of variances in carotenoid accumulation. They're linked to the gene's zeaxanthin epoxidase and carotene hydroxylase, which produce carotenoids.

In cauliflower, the purple (Pr) gene causes anthocyanin accumulation (Chiu et al., 2010). The Pr gene was extracted using a combination of Candidate gene evaluation and first-class mapping to determine the nature of the pr mutation in cauliflower. We used purple cvr. Graffiti and white cvr. Stovepipe from the store. For transformation research, Arabidopsis was employed. Pr was found to encode an R2R3 MYB transcription factor with tissue-specific expression.

Anthocyanin synthase genes in carrot were mapped by Yildiz et al. (2013). B9547, which was solid purple, B7262, which had purple phloem and orange xylem, and B493, which was pure orange, were the three USDA carrot genetic stocks utilized. Anthocyanin accumulation in carrot roots is controlled by a single dominant gene, P1. In a population segregating for purple root color, five structural genes-FLS1, F3H, LDOX2, PAL3, UFGT- and three regulatory genes-DcEFRI, DcMYB3, DcMYB5-were mapped.

Xu et al. (2014) investigated anthocyanin production in purple carrots at the molecular level and discovered six new structural genes involved in anthocyanin biosynthesis: PAL3/PAL4, CA4H1, 4CL1, CHS1, CHI1, and F3H1.

IARI released Pusa Kesar, Pusa Meghali, Pusa Rudhira, Pusa Rudhira, Pusa Vrishti, Pusa Asita, and Pusa Kulfi in carrot utilizing a standard breeding program with color of roots as one significant characteristic. The concentration of associated pigments Consisting of lycopene in red carrots, â carotene in orange carrots, lutein/xanthophylls in cream carrots, and anthocyanin in purple carrots shows a positive link with color intensity. In these cultivars, there was a wide variation in total carotenoids concentration as well as specific components like lycopene (28.5–1913.7 mg/100 g) and beta carotene (34.9–190.5 mg/100 g), showing room for improvement (Singh et al., 2015).

Soil type, mulching, irrigation, fertilization, and other methods may all alter the plant's water and nutrient supply, as well as the composition and quality of the plant. Altitude, soil pH, and salinity are some of the other environmental parameters to consider. Vegetable nutritional quality Is also prompted by means of production strategies, with variances between organic and conventional systems, as well as greenhouse and field grown plants (Goldman et al., 1999). Dumas et al. (2003) investigated the antioxidant content of tomatoes in relation to environmental conditions and farming approaches. The ideal temperature for the synthesis of lycopene and carotene was 22–250°C. Low soil water levels boosted lycopene synthesis but had no influence on â carotene production in red and pink big tomatoes.

Kuti & Konuru (2005) studied the Consequences of Genotype and cultivation surroundings on tomato lycopene content. The lycopene content of greenhouse-grown clusters and round tomatoes was higher than that of field-grown tomatoes. The lycopene concentration of cherry tomato varieties cultivated in the outdoors was greater than that of greenhouse fruits. Mitchell et al. (2007) looked at how Natural and conventional crop control strategies affected flavonoids in tomatoes. Under organic management, quercetin content increased by 79% and kaempferol content climbed by 97%. Plants with a lack of nitrogen collected more flavonoids than those with enough.

High nitrogen fertilization rates, according to Leskovar et al. (2009), decrease ascorbic acid and either decrease or enhance carotenoid concentration in vegetables. The synthesis and accumulation of health-promoting phytochemicals is reduced when mineral elements are abundant, whereas nutritional deficits have the reverse impact. Stress conditions generated by deficiency irrigation or excessive salinity in watermelon and tomato may increase carotenoid levels, such as lycopene, but at the cost of poorer output.

Mingchi et al. (2010) found that simulating drought stress with $CaCl_2$ boosted lycopene and vitamin C levels in tomatoes while lowering yields. The number of fruits remained nearly same, but the weight per fruit decreased.

Fortier et al. (2010) looked at how irrigation and nitrogen fertilizer affected broccoli polyphenolics. The absence of irrigation resulted in a considerable rise in the content of polyphenolics. A linear or curved drop in polyphenolics was seen when fertilizer N increased. In irrigated and non-irrigated plots, unfertilized plots had 16% and 86% more polyphenolics, respectively.

Phytochemical content in vegetables is affected by several post-harvest procedures. The phenolic compound in vegetables varies with preparation, storage, and cooking. Because ascorbic acid is heat sensitive, it may be destroyed during canning and prefreezing. Oxidation can also occur as an outcome of heat treatment. Because most frozen veggies must be cooked before being consumed, there may be extra nutritional loss; the amount of loss varies on the kind and time of heating. Ascorbic acid levels are greater in freshly selected veggies (Rickman et al., 2007).

Murica et al. (2009) looked on the results of industrial processing and refrigerated storage on vegetables. Around 25 different vegetables were utilized to test antioxidant activity (carrot, cauliflower, cucumber, broccoli, onion, and so on). Lipoperoxyl and hydroxyl radicals might be scavenged by all fresh veggies. Except for cucumber, broccoli, brussels sprouts, and leek, vegetables refrigerated in the refrigerator for seven days exhibited the same antioxidant activity. There was a greater decrease of antioxidant activity in canned veggies than in frozen vegetables. Total antioxidant activity was reduced by 19–48% after storage (8 months for frozen and 18 months for canned).

Tomato lycopene levels rose after thermal processing (Charles et al., 2014). Tomatoes were pulped and then boiled or fried for 2, 15, or 30 minutes. The lycopene concentration of the pulp increased from 24.2 to 32.9% when it was boiled or fried for 30 minutes.

The impact of several light treatments on onion quercetin content was investigated by Ko et al. (2015). At 25°C, whole and peeled onions were exposed to fluorescent, blue, red, and UV lights over three days, with exposure times of 8 hours and 16 hours. In peeled onions, all treatments increased the amount of quercetin blue mild produced the greatest quantity in whole onions, while fluorescent light produced the bottom. Onions interpret mild as a stress sign, which triggers flavonoid production via the polyphenol pathway.

4.10 REGULATIONS AND MARKET

Fast-shifting healthcare items (FMHG) is the designation given to nutraceuticals and dietary supplements sold in India (FMHG). The Meals Protection and Well-Known Act of 2006 changed into surpassed by way of the Meals Protection and Standards Authority of India (FSSAI). But it does not distinguish among purposeful foods, nutraceuticals, nutritional supplements, and so on. All goods are classified as food for specific dietary purposes.

Meal safety and requirements (food or fitness supplements, nutraceuticals, meals for unique nutritional uses, foods for Unique scientific functions, functional foods, and novel meals) rules, 2015, has now been drafted, which includes nutraceuticals.

It defines "nutraceuticals" as "a certainly happening chemical compound with a Physiological benefit or offering protection in opposition to persistent disorder, isolated and purified from meals or non-meals assets, and can be organized and marketed in the meals layout of granules, powder, tablet, tablet, liquid, or gel, and may be packed in sachet, Ampoule, bottle, and so on. And to be taken as measured unit quantities" (FSSAI, 2015).

Because of the consumer acceptance established in these regions, the United States and Japan are the most developed markets for nutraceuticals globally. India, China, and Brazil are growing countries with significant nutraceutical market potential. The nutraceutical industry's primary innovation centers have emerged as Germany and Israel. Ingredient firms drive the sector in Israel.

4.11 CONCLUSION

The changing food habits have led to increased incidence of many lifestyle diseases. The raised demands for health care have Dramatically increased the Value of hospital therapy. Consequently, people have attempted to obtain a higher first-class of life by Relying more on greens and culmination, taking nutritional supplements or nutraceuticals. Flavonoids contain a huge formation of therapeutic active compounds that are all over in plants, many of which have been used in established Eastern medication for Lots of years.

Flavonoids also constitute an inevitable component of the diet. In the present review, we have reviewed detailed structural features and biological properties of flavonoids. The synthetics and structural similarities of flavonoids with various biomolecules as well as their importance in plant-insect and plant-bacterial interchange Led them to an Attractive magnificence of phytoconstituents for therapeutic activity. Their general occurrence, broad spectrum Diversity, and herbal starting place cause them to be suitable chemical scaffolds for Novel biological agents. Of the many actions of flavonoids, antioxidant, and antiproliferative of extended effects.

KEYWORDS

- **antioxidant**
- **flavonoids**
- **glucosinolates**
- **lifestyle**
- **nutraceuticals**
- **phytochemicals**
- **therapeutic activity**

REFERENCES

Bo, K., Song, H., Shen, J., Qian, C., Staub, J. E., Simon, P. W., Lou, Q., & Chen, J., (2012). Inheritance and mapping of the ore gene controlling the quantity of â carotene in cucumber endocarp. *Mol. Breed., 30*, 335–344.

Chang, H. P., Tsai, I. H., Huang, S. Y., Dai, P. H., Chen, Y. C., & Chen, Y. H., (2007). Study of the contents of indole glucosinolates in fresh and pickled cruciferous vegetables cultivated in Taiwan. *Nutr. Sci. J., 32*(3), 102–107.

Charles, I. N., Jude, I. O., James, O. E., & Tochukwu, E. C., (2014). Effect of thermal processing on lycopene, beta- carotene and vitamin C content of tomato [Var.UC82B]. *J. Food Nutr. Sci., 2*(3), 87–92.

Chiu, L., Zhou, X., Burke, S., Wu, X., Prior, R. L., & Li, L., (2010). The purple cauliflower arises from activation of a MYB transcription factor. *Pl. Physiol., 154*, 1470–1480.

Crozier, A., Lean, M. E., McDonald, M. S., & Black, C., (1997). Quantitative analysis of the flavonoid content of commercial tomatoes, onions, lettuce and celery. *J. Agri. Food Chem., 45*, 590–595.

Ding, Y., & Jian, Y., (2010). An orange cauliflower hybrid 'Jinyu 60' with high â-Carotene. *Acta Hort., 856*, 261–264.

Dumas, Y., Dadomo, A., Lucca, G., & Grolier, P., (2003). Effects of environmental factors and agricultural techniques on antioxidant content of tomatoes. *J. Sci. Food Agric., 83*, 369–382.

FAO [Food and Agriculture Organization]. (2015). *The State of Food and Agriculture, Social Protection, and Agriculture: Breaking the Cycle of Rural Poverty*. Available: http://www.fao.org (accessed on 15 November 2023).

Fortier, E., Desjardins, Y., Tremblay, N., & Bélec, C., (2010). Influence of irrigation and nitrogen fertilization on broccoli polyphenolics concentration. *Acta Hort., 856*, 55–62.

FSSAI [Food Safety and Standards Authority of India], (2015). *Draft Regulations on Food or Health Suppliment, Nutraceuticals, Foods for Special Dietary Uses, Foods for Special Medical Purpose, Functional Foods, and Novel Food.* Available: http://www.fssai.gov.in (accessed on 15 November 2023).

Gibson, R. A., & Makrides, M., (2000). n-3 polyunsaturated fatty acid requirements of term infants. *Am. J. Clin. Nutr., 71*, 251–255.

Giovannucci, E., (2002). A review of epidemiologic studies of tomatoes, lycopene, and prostate cancer. *Exp. Biol. Med., 227*, 852–859.

Goldman, I. L., Kader, A. A., & Heintz, C., (1999). Influence of production, handling, and storage of phytonutrient content of foods. *Nutr. Rev., 57*, 546–552.

Ishida, B. K., Turner, C., Chapman, M. H., & Mc Keon, T. A., (2004). Fatty acid and carotenoid composition of Gac (*Momordica cochinchinensis*) fruit. *J. Agric. Food Chem., 52*(2), 274–279.

Just, B. J., Santos, C. A., Yandell, B. S., & Simon, P. W., (2009). Major QTL for carrot color are positionally associated with biosynthetic genes and interact epistatically in a domesticated x wild carrot cross. *Theor. Appl. Genet., 119*(7), 1159–1169.

Kalia, P., & Muthukumar, P., (2015). Â- carotene bio-fortification in cauliflower using marker assisted backcross breeding. In: Kalia, P., Mangal, M., & Srivastava, A., (eds.), *Winter School on Advances in Improvement of Vegetable Crops Using Biotechnological Approaches* (pp. 90–97). New Delhi.

Keservani, R. K., & Sharma, A. K., (2014). Flavonoids: Emerging trends and potential health benefits. *Journal of Chinese Pharmaceutical Sciences*, *23*(12), 815.

Keservani, R. K., Kesharwani, R. K., Sharma, A. K., Vyas, N., & Chadoker, A., (2010b). Nutritional supplements: An overview. *International Journal of Current Pharmaceutical Review and Research, 1*(1), 59–75.

Keservani, R. K., Kesharwani, R. K., Vyas, N., Jain, S., Raghuvanshi, R., & Sharma, A. K., (2010a). Nutraceutical and functional food as future food: A review. *Der Pharmacia Lettre, 2*(1), 106–116.

Keservani, R. K., Sharma, A. K., & Kesharwani, R. K., (2016a). Nutraceutical and functional foods for cardiovascular health. In: *Food Process Engineering* (pp. 257–278, 291–312). Apple Academic Press, CRC Press. ISBN: 9781771884020.

Keservani, R. K., Sharma, A. K., & Kesharwani, R. K., (2016b). Medicinal effect of nutraceutical fruits for the cognition and brain health. *Scientifica*, 10. Article ID 3109254. doi:10.1155/2016/3109254.

Keservani, R. K., Sharma, A. K., & Kesharwani, R. K., (2017). An overview and therapeutic applications of nutraceutical and functional foods. *Recent Advances in Drug Delivery Technology*, 160–201. ISBN:9781522507543.

Keservani, R. K., Sharma, A. K., & Kesharwani, R. K., (2020). *Nutraceuticals and Dietary Supplements: Applications in Health Improvement and Disease Management*. CRC Press. ISBN: 9781771888738.

Ko, E. Y., Nile, S. H., Sharma, K., Li, G. H., & Park, S. W., (2015). Effect of different exposed lights on quercetin and quercetin glucoside content in onion. *Saudi J. Biol. Sci., 22*, 398–403.

Kuti, J. O., & Konuru, H. B., (2005). Effects of genotype and cultivation environment on lycopene content in red ripe tomatoes. *J. Sci. Food. Agric., 85*, 2021–2026.

Lee, W. Y., Emmy, K. I., Abbe-Maleyki, M. J., & Amin, I., (2007). Antioxidant capacity and phenolic content of selected commercially available cruciferous vegetables. *Malay. J. Nutr., 13*(1), 71–80.

Leskovar, D. I., Crosby, K., & Jifon, J. L., (2009). Impact of agronomic practices on phytochemicals and quality of vegetable crops. *Acta Hort., 841*, 317–321.

Mingchi, L., Xiangli, L., Jing, H., & Lihong, G., (2010). Effect of simulated drought stress on plant growth, yield and fruit properties of tomato. *Acta Hort., 856*, 193–202.

Mitchell, A. E., Hong, Y., Koh, E., Barrett, D. M., Bryant, D. E., Denison, R. F., & Kafka, S., (2007). Ten-year comparison of the influence of organic and conventional crop management practices on the content of flavonoids in tomatoes. *J. Agric. Food Chem., 55*, 6154–6159.

Mithen, R. F., (2013). Development and commercialization of Beneforte broccoli and potential health benefits. *Acta Hort., 1005*, 67–70.

Mithen, R. F., Dekker, M., Verkerk, R., Rabot, S., & Johnson, I. T., (2000). The nutritional significance, biosynthesis, and bioavailability of glucosinolates in human foods. *J. Sci. Food Agric., 80*(7), 967–984.

Moeller, S. M., Jacques, P. F., & Blumberg, J. B., (2000). The potential role of dietary xanthophylls in cataract and age-related macular degeneration. *J. Am. Coll. Nutr., 19*, 522–527.

Mousa, A. S., & Mousa, S. A., (2007). Cellular effects of garlic supplements and antioxidant vitamins in lowering marginally high blood pressure in humans. *Nutr. Res., 27*, 119–123.

Murica, M. A., Jimenez, A. M., & Martinez-Tom, M., (2009). Vegetables antioxidant losses during industrial processing and refrigerated storage. *Food Res. Int., 42*, 1046–1052.

Ou, B., Huang, D., Hampsh, M., Flanegan, J. A., & Deemer, E. K., (2002). Analysis of antioxidant activities of common vegetables employing oxygen radical absorbance capacity and ferric reducing antioxidant power assays: A comparative study. *J. Agric. Food Chem., 50*, 3122–3128.

Paramaguru, P., Pugalendhi, L., & Velmurugan, M., (2012). Nutritional security through vegetables. In: Rana, M. K., (ed), *Modern Concepts of Vegetable Production* (pp. 1–20). Biotech Books, New Delhi.

Piekarsha, A., Szczyglowska, M., Kusznierewiez, B., & Kolodziejskei, D., (2013). The innovative exploitation of *Brassica* vegetables in the health quality food production chain. *Acta Hort., 1005*, 71–85.

Prabu, S. L., Suriyaprakash, T. N. K., Kumar, C. D., & Kumar, S. S., (2012). Nutraceuticals and their medicinal importance. *Int. J. Health Allied Sci., 1*(2), 47–53.

Rahal, A., Mahima, Verma, A. K., Kumar, A., Tiwari, R., Kapoor, S., Chakraborty, S., & Dhama, K., (2014). Phytonutrients and nutraceuticals in vegetables and their multi-dimensional medicinal and health benefits for humans and their companion animals: A review. *J. Biol. Sci., 14*(1), 1–19.

Rajasekaran, A., Sivagnanam, G., & Xavier, R., (2008). Nutraceuticals as therapeutic agents: A review. *Res. J. Pharm. Technol., 1*(4), 1–13.

Rickman, J. C., Barrett, D. M., & Bruhn, C. M., (2007). Review: Nutritional comparison of fresh, frozen, and canned fruits and vegetables. Part 1: Vitamin C and B and phenolic compounds. *J. Sci. Food Agric., 87*, 930–944.

Sheokand, S., Madan, S., & Kumari, B., (2012). Antioxidants in vegetables. In: Rana, M. K., (ed), *Modern Concepts of Vegetable Production* (pp. 21–43). Biotech Books, New Delhi.

Simon, P. W., & Navazio, J. P., (1997). Early orange mass 400, early orange mass 402 and late orange mass 404: High carotene cucumber germplasm. *HortScience, 32*(1), 144–145.

Simon, P., Pollak, L., Clevidence, B., Holden, J., & Haytowitz, B., (2009). Plant breeding for human nutritional quality. Pl. *Breed. Rev., 31*, 385–392.

Singh, P., Kesharwani, R. K., & Keservani, R. K., (2017a). Antioxidants and vitamins: Roles in cellular function and metabolism. In: *Sustained Energy for Enhanced Human Functions and Activity* (pp. 385–407). Academic Press. ISBN:9780128054130.

Singh, P., Kesharwani, R. K., & Keservani, R. K., (2017b). Protein, carbohydrates, and fats: Energy metabolism. In: *Sustained Energy for Enhanced Human Functions and Activity* (pp. 103–115). Academic Press. ISBN:9780128054130.

Singh, S., Kalia, P., & Sureja, A. K., (2015). Improving carotenoid content in vegetable crops for health security. In: Kalia, P., Mangal, M., & Srivastava, A., (eds), *Winter School on Advances in Improvement of Vegetable Crops Using Biotechnological Approaches* (pp. 111–126). New Delhi.

Subhasree, B., Baskar, R., Laxmi-Keerthana, R., Susan, L., & Rajasekaran, P., (2009). Evaluation of antioxidant potential in selected green leafy vegetables. *Food Chem., 115*(4), 1213–1220.

Surles, R. L., Weng, N., Simon, P. W., & Tanumihardjo, S. A., (2004). Carotenoid profiles and consumer sensory evaluation of specialty carrots of various colors. *J. Agric. Food Chem., 52*, 3417–3421.

Tapas, A. R., Sakarkar, D. M., & KAKDE, R. B., (2008). Flavonoids as nutraceuticals: A review. *Trop. J. Pharm. Res., 7*(3), 1089–1099.

Vuong, L. T., Franke, A. A., Custer, L. J., & Murphy, S. P., (2005). *Momordia cochinchinensis* fruit carotenoids reevaluated. *J. Food Comp. Anal., 19*, 664–668s.

Wargovich, M. J., (2000). Anticancer properties of fruits and vegetables. *HortScience, 35*, 573–575.

Willoughby, V., (2000). Tomatoes and carrots may reduce lung cancer risk. *Am. Veg. Grow., 48*, 9.

Xu, Z., Ying, H., Feng, W., Xiong, S., Guang-Long, W., & Ai-Sheng, X., (2014). Transcript profiling of structural genes involved in cyaniding-based anthocyanin biosynthesis between purple and non- purple carrot cultivar reveals distinct patterns. *BMC Plant Biol., 14*, 262.

Yildiz, M., David, K. W., Pablo, F. C., Massimo, I., Kazim, A., & Philip, W. S., (2013). Expression and mapping of anthocyanin biosynthesis genes in carrot. *Theor. Appl. Genet., 126*, 1689–1702.

CHAPTER 5

NUTRACEUTICAL BIOACTIVE COMPOUNDS IN FRUITS AND VEGETABLES

ANKUR SRIVASTAV,[1] AJAY KUMAR SHUKLA,[1] MANISH KUMAR,[2] SHAILENDRA KUMAR,[3] AND RAHUL KUMAR MISHRA[4]

[1]*Institute of Pharmacy, Dr. Ram Manohar Lohia Avadh University, Ayodhya, Uttar Pradesh, India*

[2]*Department of Pharmacy, Madhav University, Pindwara, Rajasthan, India*

[3]*Institute of Microbiology, Dr. Ram Manohar Lohia Avadh University, Ayodhya, Uttar Pradesh, India*

[4]*Department of Pharmaceutical Science and Technology, Madan Mohan Malaviya University of Technology, Gorakhpur, Uttar Pradesh, India*

ABSTRACT

A nutraceutical is a type of functional food or a component of food that offers therapeutic or health advantages, including the capacity to treat and prevent disease. It makes a large contribution to the diet of humans and is a significant source of nutraceuticals with physiological action. They have generated a lot of attention and are currently being explored as a potential cure and prevention approach for a number of disorders because they are safe, effective, and may even have nutritional benefits in addition to their medical benefits. Nutrient-rich fruits and vegetables have health benefits, including the ability to heal and prevent disease. The many health advantages

Plant Metabolites and Vegetables as Nutraceuticals.
Raj K. Keservani, Bui Thanh Tung, Rajesh K. Kesharwani, and Eknath D. Ahire (Eds.)

of phytochemicals, include ascorbic acid in oranges, lycopene in tomatoes, curcumin in turmeric, gingerol in ginger, organosulfur compounds in allium species, omega-3 fatty acids in seeds of cucurbitaceous vegetables, and others. The role and functions of bioactive substances present in fruits and vegetables are discussed in this chapter. These substances can be consumed in a number of food forms to treat a wide range of health issues as well as function as an antioxidant, anticancer, and antibacterial agent. Need to do more research work related to isolation of bioactive components from fruits and vegetables. It could pave the path for the creation of value-added products from the fruits and vegetables industry, making it more commercially viable and maintaining nutrition value of products from the fruits and vegetables industry. It also helps in the treatment and management of various critical upcoming diseases and increases our immunity.

5.1 INTRODUCTION

A nutraceutical is anything that is a food or a part of a food that has therapeutic or health benefits, including the prevention and treatment of ailments. The name "nutraceutical" was coined in 1989 by Dr. Stephen De Felice, chairman of the Foundation for Innovation in Medicine, by fusing the words "nutrition" and "pharmaceutical." a food item that has undergone extraction or purification; frequently sold in forms intended for therapeutic purposes unrelated to food. A nutraceutical has been demonstrated in research to provide physiological benefits or to offer protection against chronic illness (Gupta et al., 2013; Sarin et al., 2012). Vegetables are functional foods and nutraceuticals because they include minerals and nutrients that promote health. It is a significant source of physiologically active nutraceuticals and plays a significant role in the human diet. It has recently been investigated as a potential cure and prevention measure for a number of illnesses. Because they are safe, effective, and may even be nutritional in addition to their medicinal benefits, they have attracted a lot of interest. Since they are low in calories and high in vitamins, minerals, antioxidants, and phytochemicals, vegetables are one of the natural dietary supplements. Lycopene from tomatoes and carotenoids from carrots are two well-known phytonutrients. These substances serve as antioxidants, which lower disease risk factors and help prevent disease. It is now known that traditional vegetables have a variety of health benefits, and novel cuisines are being created employing additional nutraceuticals. It is required to precisely determine the plant chemicals essential to human nutrition in order to create cultivars with improved nutritional attributes utilizing both conventional and molecular breeding techniques (Rai et al., 2012). Vegetables high in nutrients have health advantages, including the potential to heal and prevent disease (Ramaa et al., 2006). The important substances flavonoids, carotenoids, anthocyanins, vitamins, and other polyphenolics are abundant in vegetables. By functioning as antioxidants, these substances assist in lowering disease risk factors and preventing illness. The various health benefits of phytochemicals, such as lycopene in tomatoes, curcumin in turmeric, gingerol in ginger, organosulfur compounds in allium species, omega-3 fatty acids in cucurbitaceous vegetable seeds, and others, are constantly being discovered by researchers (Sirtori et al., 2002; Ram et al., 2000; Holman et al., 1998).

The nutritional content of various vegetable crops has been enhanced by vegetable breeders, including broccoli with additional vitamins and potatoes with increased essential amino acids. Vegetables with specific nutraceuticals that are physiologically active have gained enormous consumer awareness. Value-added processed vegetables, dietary supplements, food supplements, and non-food supplements such as tablets, soft gels, capsules, etc., are among these goods (Whitman, 2001; Wildman, 2001). Currently, herbal medications are fashionable. All kinds of diseases are having a field day because to the modernized, competitive lifestyle and the continually demanding surroundings. There is an allopathic treatment available, but it is pricey for the majority of ailments. First off, just as with all other goods, the price of medicine is rising daily. Second, there are several adverse effects connected to allopathic medications. As a result, a growing number of people are deciding to adopt a new lifestyle that only involves using herbal goods. This enhances general health and aids in the prevention of several ailments. Modern and conventional pharmaceuticals can be successfully formulated using herbal or plant medicine.

In India's rural areas, where it is more prevalent, almost 80% of people receive their primary healthcare there. Despite little scientific evidence, medicinal plants are a rich source of compounds with numerous advantageous health benefits. These substances, also referred to as phytochemicals, can serve as (Dillard et al., 2000):

- Provides assistance to many biological processes;
- Enzymatic cofactors and inhibitors for diverse processes;
- Sequestrants and absorbents that bind to and take out impurities from the stomach;

- Ligands that act as agonists or antagonists on intracellular or cell surface receptors;
- Substances that is very reactive or harmful to remove with chemicals;
- Substances that increase the important nutrients' stability and/or absorption;
- Specific growth elements for beneficial gut flora.

Vegetables and fruits are heavily marketed for their health benefits. The 2010 Dietary Guidelines for Americans recommend that half of your plate consist of fruits and vegetables. Fruits and vegetables should make up half of the plate, according to Myplate.gov. Fruits and vegetables, among other plant foods, offer various amounts of nutrients and energy. Dietary fiber, which is linked to a lower risk of obesity and cardiovascular disease (CVD), is also found in fruits and vegetables. Fruits and vegetables are good sources of phytochemicals, which include anti-inflammatory, phytoestrogenic, and antioxidant effects in addition to vitamins and minerals. In this publication, we provide a summary of the current dietary guidelines for fruit and vegetable consumption. The classification of fruits and vegetables into groups based on similar chemical functions and structures is also examined. Fruits and vegetables have different nutritional profiles than vegetables (Slavin et al., 2012). Bioactive chemicals and certain physiological or metabolic responses indicate that they are safe for human consumption. Diabetes, heart disease, the common cold, arthritis, cancer, hypertension, dyslipidemia, inflammatory bowel disease (IBD), and depression can all be treated with the phytochemicals present in these foods. The health advantages of a diet high in fruits and vegetables are brought on by compounds like phenylpropanoids, isoprenoids, polyphenols, anthocynidins, flavonoids, terpenoids, carotenoids, phytoestrogens, and alkaloids, among others. It provides a rational explanation of how phytochemicals influence the expression or structure of genes to enhance human health. These have an impact on how some diseases begin, progress, and develop. Therefore, by demonstrating genetic information, nutrigenomics plays a crucial role in how nutraceuticals combat aging and numerous ailments. Melatonin (N-acetyl-5-methoxytryptamine), a biogenic amine generated by oxidative metabolism from kynuramine, is also a component of plant diets. Iriti et al. (2006) claim that kynuramine is an important antioxidant, a cyclooxygenase 2 inhibitor, and an enhancer of mitochondrial metabolism. The science of nutrigenomics examines the connections between dietary factors and cellular and genetic functions.

5.2 EMERGING TRENDS IN NUTRACEUTICALS

Herbal/natural products, dietary supplements, and functional foods are all examples of nutraceuticals. Herbal and natural items are among the categories that are expanding the quickest, followed by dietary supplements. The notion that plants bio-active chemicals have antioxidants and other health-promoting characteristics has come about as a result of science's associations between plant-based foods and good health. A high dietary intake of fiber, which may be obtained from fruits, vegetables, and whole grains, is significantly linked to a lower risk of developing chronic diseases like cancer and CVDs. Cancer growth is a protracted, intricate, and sequential process that, if left untreated, can lead to metastasis. Epidemiological research currently offers strong support for the theory that nutrition may affect the development of cancer. Numerous phytochemicals and some meals made from plants that include unidentified compounds have anti-mutagenic and anti-carcinogenic characteristics, which are also widely documented. Therefore, we cannot completely rule out the idea of employing these bioactive substances as chemopreventive medications in the future (Cherdshewasart et al., 2009). Isoflavonoids from soy products, like those from flaxseed, can lower total and LDL cholesterol while raising HDL cholesterol, which lowers the risk of CVD. It is also claimed that phytoestrogens can help prevent CVDs. Significant CVD risk factors such as diabetes, hypertension, hyperlipidemia, and obesity can all be mitigated by phytochemicals. Additionally, phytochemicals aid in lowering oxidative stress, which is linked to the development of atherosclerosis. The body's antioxidant defense system performs better thanks to nutritional supplements (Prakash et al., 2011).

5.3 CATEGORIES OF NUTRACEUTICALS

- Substances like vitamins, minerals, amino acids, fatty acids, polysaccharides, etc., that have been scientifically proven to be nutrient-dense.
- Herbal products, such as aloe vera, wheat grass, ginger, and garlic extracts, or concentrates.
- Substances that are derived from different sources and utilized to achieve particular objectives, such as meal replacements, weight loss supplements, and enhanced sports nutrition.

Nutrition due to their numerous positive effects, nutritional supplements have been used to treat and prevent a variety of diseases, including cancer, inflammation, hypertension, CVD, atherosclerosis, obesity, and diabetes. Numerous nutraceuticals, including silymarin, curcumin, vitamin E, docosahexaenoic acid, choline, and phosphatidylcholine, can be used to treat and prevent steatosis (Stellavato et al., 2018). Numerous nutraceuticals, including caffeine, curcumin, and gallic acid, have anti-aging and antioxidant characteristics (Bourbon et al., 2015; Zhang et al., 2019).

Fish oils rich in PUFA1 have qualities that improve brain function and lower the risk of coronary heart disease (CAD) (Liu et al., 2016). Numerous of its components, such as pomegranate, curcumin, and epigallocatechin gallate, are used to treat a variety of malignancies, including breast cancer, prostate cancer, and other types of cancer, as a result of the efficiency of nutraceuticals in the fight against cancer (Hu et al., 2012). Nutraceutical products that are high in nutrients are now widely accessible on a global basis. Because there aren't many tight regulations governing nutraceuticals, their market share has dramatically expanded (Nounou et al., 2018). On the other hand, pharmaceutical products are tightly monitored and frequently audited. Pharmaceutical products are subject to rigorous legal restrictions and may be prosecuted. Nutraceuticals have also been promoted as a potent, healthy alternative to drugs. Additionally, it has been asserted that these products have no negative side effects and can be used to prevent and treat a variety of health issues (Ekor et al., 2014). People have voiced concern about using medications due to their high cost and various negative side effects (Nounou et al., 2018). As a result, nutraceuticals now make up a larger percentage of the sector. Around 80% of individuals worldwide like nutraceuticals and dietary supplements (Ekor et al., 2014). Nutraceuticals are now the industry sector with the fastest growth, with a global market value of USD 117 billion forecast for 2017 (Swaroopa et al., 2017).

5.4 CLASSIFICATION OF NUTRACEUTICALS

Nutraceuticals can be divided into one of the following types based on their dietary sources:

1. **Dietary Fibers:** These are plant-based ingredients that are present in meals but not digested in the stomach or intestines, adding bulk to the intestinal contents. Fruits, barley, oats, lignin, cellulose, and pectin are a few examples. According to Das et al., a diet high in

these fibers is linked to a lower risk of colon cancer, gastrointestinal problems, obesity, diabetes, hypertension, and CVD (2012).

2. **Probiotics:** These live microbial feed supplements, which comprise lactobacilli and bifidobacteria, help maintain the proper balance of the host's gut flora when administered in sufficient proportions. They are associated to a lower incidence of allergies, asthma, cancer, ear infections, and urinary tract infections (UTI), according to study, when used less frequently (Lenoir-Wijnkoop et al., 2016).

3. **Prebiotics:** These are dietary components that help the host by carefully controlling the gut microbiota's makeup or metabolism. These are typically fructose-based oligosaccharides that are present naturally or are added and are indigestible to humans. Beans, tomatoes, alliums, chicory roots, bananas, and other produce are examples of this. These have been discovered to help treat certain tumors, as well as enhance lactose tolerance, detoxification, and dyslipidemia (Hord et al., 2008).

4. **Polyunsaturated Fatty Acids:** These fatty acids, which can be either omega 3 or omega 6 fatty acids like – linolenic acid, eicosapentaenoic acid, and docosahexaenoic acid, are present in foods like fatty fish, flaxseed, soybeans, and others (Das et al., 2012).

5. **Antioxidant Vitamins:** Vitamin E, vitamin C, and carotenoids are some vitamins that fight free radicals. The numerous fruits and vegetables that are rich in these vitamins have the ability to quench singlet oxygen and stop lipid peroxidation. Regular use of these helps to prevent certain diseases (Das et al., 2012).

6. **Polyphenols:** Plants manufacture these phytochemicals, such as flavonoids, anthocyanins, and phenolic acids, to defend themselves from the harm caused by reactive oxygen species (ROS) and photosynthetic stress. They have antioxidant and anti-inflammatory qualities and can be found in many foods, such as beans, tea, and soybeans (Das et al., 2012). Living plants manufacture these phytochemicals, including flavonoids, anthocyanins, and phenolic acids, to defend themselves against the harm that ROS and photosynthetic stress can do. They have antioxidant and anti-inflammatory qualities and can be found in a variety of foods, such as beans, tea, and soybeans (Das et al., 2012).

7. **Spices:** Exotic seasonings like spices are added to food to enhance the sensory appeal of cuisine. The bulk of components in spices are terpenes and other naturally occurring compounds that can be found in essential oils (EOs). Spices used in food have chemopreventive, anti-inflammatory, anti-mutagenic, and immune-modulating activities in small doses (Kochhar et al., 2008).

The quantity and kind of antioxidants present in fresh tissues of fruits and vegetables determine their antioxidant capacity. Nevertheless, each person's contribution to the body's overall antioxidant capacity differs greatly. Numerous studies have shown that phenols and flavonoids are more crucial for the antioxidant activity of fruits and vegetables than vitamin C, carotenoids, and other substances. According to past studies, the high vitamin C concentration of certain fruits only accounts for 10–20% of the total antioxidant potential. However, due to their low vitamin E content, fruits' potential to act as antioxidants is greatly exceeded by vitamin C antioxidants.

Due to their vibrant colors, distinctive flavors, and nutritional value, bananas, mangoes, pineapples, papayas, and guavas are some of the most well-liked tropical fruits in the world. Due to the rising demand in the USA and Europe, these fruits are shipped from subtropical and tropical regions. Growers are using the excellent nutritional value of exotic and tropical fruits as another strategy to boost market share because customers are eating more of them. As a source of carbohydrates, minerals, and dietary fiber, apples are among the most popular fruits in the world and are crucial to a person's diet. Apples are also rich in the physiologically active substances vitamin C, flavonoids, and phenol compounds, all of which are known to function as natural antioxidants. Consequently, apples are regarded as one of the main sources of phytochemicals in the diet of the average person (Palafox-Carlos et al., 2011).

Tocopherols and tocotrienols, two types of vitamin E, are also present in date seeds. This lipid-soluble vitamin cannot be synthesized by the body on its own. It functions as an antioxidant and dissipates free radicals. It protects against age-related diseases like cancer, cataracts, and arthritis as a result. Additionally, vitamin E helps to reduce prostaglandin formation. Fruit by-products are a rich source of bioactive ingredients, such as proteins (bioactive peptides), carotenoids (lycopene), polysaccharides (pectin), phytochemicals (flavonoids), and vitamins (-tocopherol). All of these nutrients are beneficial to the body and can be used in place of other ingredients to create a variety of food products with a high functional and nutraceutical potential. The bioactive compounds included in tomato, apple, date, and guava seeds are covered in this chapter. These chemicals can treat a range of cardiovascular and neurological problems in addition to acting as antioxidants, cancer-fighting, and antibacterial agents. The fruit industry's byproducts might be utilized to extract a variety of bioactive substances, enabling the development of commodities with added value that would increase the sector's financial viability and lessen its environmental impact (Allaqaband et al., 2022).

Bioactive natural substances have always been essential as cutting-edge medicines, regardless of where they came from. They significantly impact human health both directly and indirectly, and they also offer excellent therapeutic potential. All around the world, different fruit species are highly appreciated and devoured. The association between a high fruit diet and a low risk of chronic disease development is supported by epidemiologic evidence. For a variety of biological reasons, eating fruit may minimize or avoid chronic diseases. Fruits are excellent sources of fiber, vitamins, minerals, and other biologically active substances. They contain a lot of nutrients and energy as well. This review has also been broadened to cover the therapeutic properties of fruit crops' peels, leaves, and barks. Fru Among the most important activities covered in this review are their anti-diabetic, anti-cancer, anti-hypertensive, neuroprotective, anti-inflammatory, antioxidant, antimicrobial, antiviral, immune system stimulation, cell detoxification, cholesterol synthesis, anticonvulsant, and ability to lower blood pressure. Several phytochemicals are explored in this context with an emphasis on their structural features and connection to human illnesses (Marines Marli et al., 2018).

Food offers the vital nutrients needed for living, along with other bioactive compounds for promoting health and preventing disease. Dietary guidelines recommend eating plenty of plant-based foods at least 400 g of fruits and vegetables each day (Agudo et al., 2005). Eating fruits and vegetables can help avoid a number of diseases, such as hypertension, obesity, CAD, stroke risk, generalized cancer, eye difficulties, asthma, and osteoporosis (Boeing et al., 2012). A wide range of micronutrients and bioactive molecules that are not nutrients, such as vitamins, phytochemicals like carotenoids and (poly) phenolic compounds, minerals like potassium, calcium, and magnesium, and dietary fiber, are abundant in fruits and vegetables. In addition to although more than 5,000 distinct phytochemicals have reportedly been found in fruits, vegetables, and grains, a sizable portion of them is still unknown (Liu et al., 2013).

Secondary metabolites of plants are bioactive substances that have pharmacological and toxicological effects on both people and other animals. Plants generate compounds known as secondary metabolites that act as antioxidants and perform crucial functions in living cells, such as fending off free radicals and avoiding diseases brought on by oxidative stress (Kaur et al., 2001). Bioactive chemicals are produced in response to light; therefore, they normally build up in the leaves and skin of fruits and vegetables (Bernhoft et al., 2010). There are varied amounts of bioactive compounds in various fruits, vegetables, genetic characteristics, and environmental factors like light, ripeness, and postharvest treatments. The degree of fruit development during harvest may also have a major impact on the levels of bioactive compounds (Deepa et al., 2007; Vallejo et al., 2003).

The liquid form of several vitamins, minerals, and bioactive substances in fruits and vegetables increases their bioavailability in the human diet (Yildiz et al., 2010). It is advised that a varied plant-based diet be taken in order to reap the best health advantages from fruits and vegetables (Liu, 2013). It is commonly acknowledged that eating a diet high in fruits and vegetables lowers the incidence of oxidative stress-related illnesses like cancer and heart disease. Free radicals cause oxidative stress, which antioxidants can reduce or stop. A variety of antioxidant substances (phytochemicals) found in fruits and vegetables may assist to protect the biological system from oxidative damage and reduce the risk of chronic diseases (Liu et al., 2003). Nutraceutical bioactive compounds which are in fruit and vegetables are discussed briefly in subsections.

5.4.1 PHENOLICS

Due to their potential as antioxidants, high quantity in our diet, and potential significance in the prevention of numerous diseases linked to oxidative stress, including cancer and cardiovascular and neurological diseases, phenolic compounds have garnered a lot of attention (Scalbert et al., 2005). Fruits, beverages (fruit juice, tea, and coffee), vegetables, legumes, and grains are the main dietary sources of phenols. Up to 200–300 mg of phenolic compounds is found in fruits like apples, grapes, pears, cherries, and different berries for every 100 g of fresh weight; 100 mg are found in a glass of red wine, a cup of tea, or a cup of coffee. The average person consumes 1 g of phenolics per day through their food (Scalbert et al., 2005). Food can include intricate phenolic combinations. The amount of phenols in plants is significantly influenced by environmental factors as soil type, sunlight exposure, temperature, and rainfall. Additionally, phenol concentration can be impacted by genetic and technological factors, some of which can be controlled by improving growing practices and minimizing losses during industrial processing and home cooking. Ripeness also affects the ratio and amount of phenols; as fruit ripens, anthocyanin concentration increases while phenolic acid content decreases (Manach et al., 2004, 2005; Williamson et al., 2005).

According to Vasco et al. (2008), phenolic component concentrations can be divided into three categories: low (100 mg GAE/100 g), medium (100–500 mg GAE/100 g), and high (>500 mg GAE/100 g). The berries with high phenol content are strawberries and blackberries. These fruits contain a lot of phenols. Red raspberries, blueberries, and cherries are all considered to be great sources of phenols since they have a medium quantity of phenols. Numerous phenols have been identified that have at least one aromatic (phenolic) ring that contains hydroxyl groups. Large structural variants of them that belong to several molecular families, including benzoic acid derivatives, flavonoids, proanthocyanidins, lignans, lignins, and tannins, are created. Phenols come in two main forms: flavonoids and phenolic acids. The different classes of flavonoids include flavones, flavonols, flavanones, isoflavones, and anthocyanins (Scalbert et al., 2005; Manach et al., 2004).

5.4.2 FLAVONOIDS

The usage of flavonoids, which has persisted to the present, appears to be responsible for the majority of effective medical treatments from ancient times. Numerous biological characteristics of flavonoids include antioxidative, antibacterial, anticarcinogenic, and cardioprotective effects. Table 5.1 lists dietary sources of flavonoids by phenolic groups. Flavonol Quercetin is a common flavonoid of the flavonol class that is found in many foods, including apple, onion, and tea. It is also ingested virtually every day. According to D'Andrea et al. (2015), the average daily consumption of quercetin in the Western diet is between 0 and 30 mg.

Most 2015 epicatechin, gallocatechin, epicatechin gallate, epigallocatechin, gallocatechin gallate, and epigallocatechin gallate are among the most common catechins (flavan-3-ols) that belong to the flavanol group. Epicatechin gallate primarily affects how fruit tastes. As grapes ripen, catechins, the main tannin's ingredient, become less abundant (Gadkari et al., 2015). Hesperetin, a flavanone found in citrus fruits, has a brief biological life, poor water solubility, and a low bioavailability. These restrict its application for therapeutic purposes.

TABLE 5.1 Classification and Dietary Sources of Phenols

Phenolic Class	Bioactive Compound	Sources of Diet
Flavonols	Quercetin, Kaempferol	Tea, red wine, broccoli, cherry tomato, onion, and berries.
Flavones	Apigenin, Luteolin	Berries, celery wine, parsley, cereals.
Flavanones	Hesperetin, Naringenin, Catechins	Citrus fruits.
Flavanols	Cyanidin	Tea, red wine, cherries, apple wine, chocolate, beans, apricots, and berries.
Anthocyanidins	Daidzein	Berries, red wine, red cabbage wine, aubergine, and red wine.
Isoflavones	Genistein, Glycitein	Soy products and peas.
Phenolic Acids	Bioactive Compound	Sources of Diet
Hydroxybenzoic acid	Lactic acid Hydrogenated p-hydroxybenzoic acid, protocatechuic acid	Blackberry, raspberry, onion, strawberry, and berries.
Hydroxycinnamic acid	Coumaric acid, caffeine acid Sinapic acid, ferulic acid	Apple, pear, cherry, plum, kiwi, potato, and garlic.

Source *Bernhoft et al. (2010); Dimitrios et al. (2006); Shahidi et al. (1996); Clifford et al. (1999); Macheix et al. (1990); and Dao et al. (1992).*

Purposes citrus fruits should be consumed right away (Shen et al., 2016). Another natural flavanone called naringenin, which is abundant in citrus and grapefruits, enhances brain insulin signaling and cognitive processes (Ghofrani et al., 2015).

5.4.2.1 ANTHOCYANINS

Water-soluble plant pigments called anthocyanins are what give fruits their various hues, including red, blue, and purple (Rufino et al., 2010). They are mostly found as glycosides or acylglycosides of the corresponding aglycone anthocyanidins, and fresh plant materials rarely contain aglycone forms. Delphinidin, cyanidin, petunidin, pelargonidin, peonidin, and malvidin are the anthocyanidins found in higher plants most frequently (De Pascual-Teresa et al., 2008). They are abounding in berries including cranberries, red grapes, strawberries, raspberries, and cherries (Wu et al., 2006).

Recent years have seen increased interest in anthocyanidins due to potential health advantages (Galvano et al., 2004; Marko et al., 2004). According to epidemiological research, ingesting anthocyanins may lower the chance of developing cancer, diabetes, arthritis, and CVD (Prior et al., 2006). Anthocyanins are vulnerable to ascorbic acid, metal ions, UV radiation, oxidation, and temperature fluctuations (Santos-Buelga et al., 2003). Fruits and vegetables of the same variety may have different anthocyanin compositions depending on temperature, light type and intensity, postharvest treatments, processing, and storage in addition to genetic and agronomic considerations (De Pascual-Teresa et al., 2008).

Anthocyanin concentrations in red grapes, for example, can reach 250 mg/100 g, albeit they depend on the grape variety used, the type of vinification procedure employed, and most crucially, the length of time red wine is matured. However, young wines with a median value of 500 mg/L anthocyanins may be found (De Pascual-Teresa & Sanchez-Ballesta, 2008). The relative anthocyanin contents within raspberry cultivars have varied significantly (Borges et al., 2009; Beekwilder et al., 2005). In addition, many authors have noted significant seasonal variations in the anthocyanin content of raspberries (Koponen et al., 2007; Kassim et al., 2009).

Low and high temperatures, respectively, have long been regarded to favor and impede the production of anthocyanins in fruits and berries (Bobinaitis & Vikelis, 2013). The most common anthocyanin in fruits and vegetables, cyanidin, is what gives many fruits and flowers their vivid colors, including black currant, raspberry, strawberry, and others (red, orange, and blue). They are of great nutritional importance due to their high daily intake in the United States (180–215 mg/day), which is much higher than the intake of other flavonoids including quercetin, kaempferol, myricetin, apigenin, and luteolin (0–30 mg/day). It is also acknowledged as a strong antioxidant due to their peculiar chemical make-up and high level of reactivity caused by their absence of electrons (Galvano et al., 2004).

Isoflavone Daidzein, genistein, and glycitein are examples of the isoflavones class of phytoestrogens. They exhibit strong estrogenic activity. These compounds, which are members of the Leguminosae family, are typically found in soybeans and clover (Nemitz et al., 2016). Proanthocyanidins are oligomers or polymers of catechins (flavan-3-ol) joined by interflavan connections. They are also known as condensed tannins and the last of the flavonoids. When multiple catechin and epicatechin monomeric molecules combine, they are produced (Auger et al., 2004). The plant foods and beverages that include them include red wine, beans, and grains, especially broad bean, lentil, and chocolate. Fruits that contain them include plum, apple, strawberry, cherry, and other berries (Auger et al., 2004). Red wine, chocolate, cranberry juice, and four different kinds of apples were all investigated for their procyanidin concentration by Hammerstone et al. in 2000. In comparison to red wine and cranberry juice, they discovered that chocolate and apples had the highest procyanidin levels per serving (164.7 and 147.1 mg, respectively; 22.0 and 31.9 mg, respectively). The average amounts of procyanidin were found to be highest in the Red Delicious (207.7 mg/serving) and Granny Smith (183.3 mg/serving) varieties and lowest in the Golden Delicious (92.5 mg/serving) and McIntosh (105.0 mg/serving) varieties. Procyanidin content varied greatly between apple samples (12.3–252.4 mg/serving). According to Lee et al. (2003), an apple contains 9.35 mg of procyanidin on average per 100 grams of fresh fruit.

Fruits and vegetables' procyanidin contents can vary depending on their origin, maturity stage, postharvest storage, and preparation (Auger et al., 2004). Proanthocyanidin intake on average is thought to be 95 mg per day. Proanthocyanidins are more potent at neutralizing free radicals than resveratrol (RES) or ascorbic acid (Ou et al., 2014). Recent investigations have suggested that the procyanidins may have health benefits, especially for cardiovascular health, making them particularly interesting (Prior et al., 2001; Corder et al., 2006). They have antioxidant activity that may have health advantages and have been shown to prevent low-density lipoprotein (LDL) oxidation in tests (Da Silva Porto et al., 2003).

One of the possible explanations for the connections between fruit and vegetable consumption and a lower risk of CVD is the presence of procyanidin in fruits and vegetables (Prior et al., 2001). The unique astringency or bitterness of the fruits is also a result of proanthocyanidins (Scalbert et al., 2005).

5.4.2.2 *PHENOLIC ACIDS*

The other types of phenolics are the two families of hydroxybenzoic acids and hydroxycinnamic acids. Gallic, protocatechuic, and p-hydroxybenzoic acids are benzoic acid derivatives, whereas caffeic, coumaric, ferulic, and sinapic acids are derivatives of cinnamic acid. The hydroxybenzoic acid group content of edible plants is frequently fairly low, with the exception of some red fruits and onions, whereas the hydroxycinnamic acids are more common in fruits and vegetables (Clifford et al., 1999). Blueberry, kiwi, plum, cherry, and apple have the highest hydroxycinnamic acid contents (0.5–2 g/kg). Caffeine makes up the majority of hydroxycinnamic acid, the most abundant phenolic acid found in fruits. Although hydroxycinnamic acids are found in all fruit parts, the highest concentrations can be found in the mature fruit's outermost layers (Manach et al., 2005).

5.4.2.3 PHYTOESTROGENS

Phytoestrogens are examples of estrogenic compounds. They are chemicals that naturally occur and can be found in plants. The four major classes of phenolic compounds that plants produce that are hypothesized to function as phytoestrogens are isoflavonoids, stilbenes, lignans, and coumestans (Cornwell et al., 2004). Phytoestrogens have recently drawn attention because of their ability to defend against a range of diseases and conditions, such as cancer, CVD, osteoporosis, and menopausal symptoms (Yildiz et al., 2006; Adlercreutz et al., 2004; Cornwell et al., 2004; Webb et al., 2005). Numerous experimental studies have demonstrated that phytoestrogens may inhibit the growth of both hormone-dependent and hormone-independent cancer cells (Ganry et al., 2005; Peeters et al., 2003).

5.4.2.4 ISOFLAVONOID

Phytoestrogens have been discussed in relation to flavonoids because isoflavonoids are recognized as members of the flavonoid group. They structurally resemble mammalian estrogens. Genistein, daidzein, and glycitein, the three most important isoflavones, are widely present in soybeans and soy-based foods like miso and tofu. Because they consume a lot of isoflavones, Asians may have lower rates of breast and prostate cancer (Ganry et al., 2005; Adlercreutz et al., 2004).

5.4.2.5 COUMESTANS

Coumestans are frequently engaged in isoflavones because of their comparable structure and activity to isoflavones (Ganry et al., 2005). There are many different coumestans, but only a few of them primarily coumestrol 3 and 4-methoxycoumestrol—have demonstrated estrogenic action (Cornwell et al., 2004). The main food sources of coumestans are legumes, Brussel sprouts, spinach, clover, and soybean sprouts (Cornwell et al., 2004; Ganry et al., 2005).

5.4.2.6 LIGNANS

Lignans are secondary metabolites that are important sources of phytoestrogens in plants. The plant lignans that are most frequently found in food are lariciresinol, matairesinol, pinoresinol, and secoisolariciresinol. Some foods also include other lignans, such as sesamin, sesaminol, syringaresinol, sesamin, and sesamolin, which is found in rye, lemons, and sesame seeds (Smeds et al., 2007). However, they were also found in a variety of commonly consumed foods, including fruits, seeds, vegetables, and drinks like juices, beer, coffee, and wine. They are particularly abundant in flaxseed (335 mg/100 g) and sesame seeds (373 mg/100 g) (Milder et al., 2005; Smeds et al., 2007; Thompson et al., 2006), as well as in small amounts in cereal (Kuhnle et al., 2008). A number of factors, including genetic and environmental changes, sampling, storage, drying, and extraction methods, have an impact on the composition of food and the amount of lignan present (Smeds et al., 2012). Lignans are frequently found in the hull of cereal grains, seeds, and oilseeds (Smeds et al., 2012). Therefore, the lignans may be lost during seed dehulling. It is unclear why the lignan content of most foods' changes after processing and storage (Kuhnle et al., 2009; Thompson et al., 2006). However, some authors claim that processing doesn't significantly affect the lignan content of flaxseeds or sesame seeds (Hyvärinen et al., 2006a, b) or sesame seeds (Moazzami et al., 2007; Wu et al., 2007; Lee et al., 2010).

Touillaud et al. (2007) found that the daily consumption of lignans varied from 0 to 5 mg. Overall lignan consumption was comparable across men and women, according to Milder et al. (2005). However, compared to other groups, lignan intake was higher in vegetarians, non-smokers, people with higher socioeconomic status, and people who were older. Epidemiological and pharmaceutical studies have shown that lignans are helpful at lowering the risk of inflammation, atherosclerosis, cancer, stroke risk factors, and oxidative stress (Prasad et al., 2005; Saleem et al., 2005). Johnsen et al. (2010) compared the incidence of colon and rectal cancer and the plasma lignan content in more than 57,000 participants between the ages of 50 and 64. According to their findings, higher lignan concentrations are associated with a higher risk of rectal cancer in men and a decreased risk of colon cancer in women. According to Lemay et al. (2002), lignans have a protective effect against osteoporosis and menopausal symptoms.

The presence of stilbenes in food is uncommon (Balsano et al., 2009). Wine, peanuts, peanut butter, grapes, grape juice, and peanuts are the main dietary sources of stilbenes (Cassidy et al., 2000). In low concentrations, they are primarily found in the fruits' skin and flesh and are created in reaction to stress (Scalbert et al., 2000). One of these, RES, is produced by plants, particularly in grapevines, berries, pines, peanuts, and legumes (Burns et al., 2002). It is a bioactive element because of the many advantageous biological functions it serves. In addition to its chemopreventive, anti-inflammatory, antioxidant, anti-proliferative, pro-apoptotic, cardioprotective, and anticancer capabilities, RES has positive impacts on human health (Mason Gusman et al., 2001; Pervaiz et al., 2009). The wine is a naturally occurring polyphenolic compound that develops in response to UV rays, damage, stress, and fungus infection (Kasiotis et al., 2013). They are categorized as belonging to the phytoestrogen category because of their ability to interact with estrogen receptors (Fulda et al., 2010). They also belong to the phytoalexin category (Li et al., 2006). They are present in both the cis- and trans-configurations.

Most notably found in grapes, RES can be found in a variety of grape products, according to studies. White grape juice is said to contain lower concentrations than red grape juice. The levels of RES in various wine varieties vary greatly (Nikfardjam et al., 2000). The amount of RES in grapes varies depending on the plant variety. Burns et al. (2002) found values that varied greatly in their investigation of several grape types, ranging from 98 to 1.803 g/100 mL. The highest trans-RES concentration in red wines is around 8 mg/L, however, it varies depending on the grape variety. Levels in rose wines range from 1.38 to 2.93 mg/L, although levels in white wines are frequently low as a result of little interaction with the grape skins, which are the main source of RES, throughout the winemaking process (Hooper et al., 2006). Many foods naturally contain pterostilbenes, another stilbene with a structure similar to RES, especially blueberries (Fulda et al., 2010). The anti-cancer, anti-proliferative, pro-apoptotic, antioxidant, anti-inflammatory, anti-invasive, and antimetastatic actions of pterostilbenes have also been connected to a number of other good health effects (Pervaiz et al., 2009). Other stilbene kinds include trihydroxy stilbenes, 4,4-dihydroxy-trans-stilbene (DHS), isorhapontigenin, bridging stilbenes, piceatannol, and dihydro-RES. According to Anisimova et al. (2011); and Kasiotis et al. (2013), they possess cytotoxic and antioxidant qualities that prevent the spread of breast and prostate cancer cells.

5.4.2 CAROTENOIDS

There are thought to be over 700 natural carotenoids. Carotene, lycopene, lutein, zeaxanthin, flavoxanthin, canthaxanthin, capsanthin, capsorubin, and -cryptoxanthin are the most well-known carotenoids. Some of these are included in Table 5.2, along with their dietary sources. Carotenoids are chemical compounds with a variety of structures that are lipophilic and insoluble in water. The main structural element of carotenoids is their backbone, which is composed of eight isoprene units and conjugated polyene (Rivera et al., 2012). The basic structure is susceptible to a wide range of changes, including chain shortening, cyclization, hydrogenation, dehydrogenation, the addition of oxygen functions, and combinations of these (Rodriguez-Amaya et al., 2004). Yellow, orange, and red pigments, also referred to as carotenoids, are extensively present in nature. Plants, some bacteria, and microscopic fungi that make carotenoids can all produce them, but only food can allow humans and other animals to absorb them. Carotenoids, which are typically found in leaves alongside chlorophyll, give many fruits and vegetables their color. Different physiological effects of these chemicals on plants; they support photosynthesis and phototaxis (Krinsky et al., 2005). Different carotenoid compositions can be found in foods. For instance, minor carotenoids found in green vegetables include lutein, beta-carotene, violaxanthin, and neoxanthin, whereas major carotenoids include beta-carotene, beta- or beta-cryptoxanthin, zeaxanthin, and antheraxanthin. Compared to vegetables, fruits have a far more complex and varied carotenoids composition. While certain fruits, like pears, only contain trace amounts of carotenoids, others, like tomatoes, watermelons, and papayas, contain high amounts of carotenoids like lycopene (Turcsi et al., 2016).

TABLE 5.2 Carotenoids in Fruits and Vegetables

Carotenoids	Sources of Diet
Lutein	Spinach, kale, broccoli, and Brussel sprouts.
Zeaxanthin	Egg yolks, maize, and spinach.
β-Cryptoxanthin	Citrus fruits, avocado, papaya, and pepper.
Alpha-carotene	Carrots, pumpkin, and maize.
Beta-carotene	Carrots, spinach, and parsley.
Lycopene	Tomato and its products, watermelon, and guava

Carotenoids are found in the two main organelles of fruits and vegetables, chloroplasts, and chromoplasts. Fruits' chromoplasts, where they are paired with diverse substances including proteins, are where carotenoids accumulate (Liu et al., 2015). Chromoplasts have an endless capacity to synthesize carotenoids. Tomatoes may contain more lycopene if they are exposed to bright sunlight or if the temperature is raised during storage (Helyes et al., 2007). Orange tree leaves can produce photooxidative stress, which can increase the fruit's carotenoid concentration (Poiroux-Gonord et al., 2013). Almost no fruits have carotenoids in their chloroplasts (e.g., grapes). Foods like mangos and carambolas contain a lot of epoxide forms of carotenoids (Turcsi et al., 2016).

Carotenoids are found in chloroplasts of green leafy plants (such as spinach, cabbage, or lettuce), where they are almost entirely attached to the proteins that make up the photosynthetic apparatus (Esteban et al., 2015). Unlike other plant pigments like anthocyanins or betalains, carotenoids have a variety of functions in plants. When oxidative cleavages produce carotenoid derivatives in plants, they are sources of vital precursors for the production of bioactive chemicals. These substances function as signaling molecules (Ramel et al., 2012). Plant signals for fruit's color, scent, and flavor are likewise caused by chemicals created by carotenoid oxidation. Carotenoids are the pigments that give edible fruits most of their colors, and they also serve as the building blocks for many significant volatile flavor compounds, adding to the sensory appeal of these foods.

Evidently, the color of a fruit can signify its nutritional value or maturity (Schaefer et al., 2004; Walter et al., 2010). They have been linked to how plants respond to their environment (Walter et al., 2011). Carotenoids play a protective role against ROS because they are strong scavengers of other free radicals and very good physical and chemical quenchers of singlet oxygen (Stahl et al., 2003). Carotenoids are essential elements in mammals' diets because they are the precursors to vitamin A.

Lycopene, a lycoA carotenoid, reduces the risk of CVD and protects men against prostate cancer. Lycopene contains a total of 13 unsaturated double bonds, 11 of which are conjugated. Due to the considerable number of conjugated double bonds, it is recognized as one of the strongest antioxidants among carotenoids. Tomatoes and products manufactured from tomatoes are the main dietary sources of lycopene (Omoni et al., 2005). As a result, serum lycopene levels are seen to significantly rise after consuming tomato products and supplements (Agarwal et al., 2001).

Carotene, an essential component of the carotenoid family of molecules, is what gives fruits and vegetables their vivid colors of yellow, orange, and red (Kandlakunta et al., 2008). These carotenoids have antioxidant effects by removing oxygen radicals from the body and reducing oxidative stress.

5.4.2.8 VITAMINS

Because they naturally contain high levels of the antioxidant vitamins C (ascorbic acid) and E (tocopherols), fruits and vegetables are encouraged for consumption. Antioxidants have a variety of positive health effects, such as reducing DNA damage, preventing lipid oxidation, and sustaining immune system function (Gropper et al., 2012). Most plants and animals can make this water-soluble vitamin from d-glucose or d-galactose. Because humans lack the l-gulonolactone oxidase enzyme, they are unable to generate vitamin C (Szajdek et al., 2008). Fruits, especially those from tropical species and leafy greens, are rich in ascorbic acid. Excellent sources of vitamin C include citrus fruits (such as oranges and grapefruit), broccoli, leafy green vegetables, tomatoes, peppers, potatoes, cantaloupe (rock melon), and strawberries. Vitamin C inhibits the production of the cancer-causing nitroso-compounds that are created when nitrites and nitrates are used to preserve meat and animal products (Kaur et al., 2001). It strengthens immune system defenses, has antioxidant properties that prevent the oxidation process from beginning, and facilitates the absorption of nonheme iron (Szajdek et al., 2008).

Several factors, including species, genetics, maturation stage, harvesting time, ambient circumstances, postharvest treatment, storage conditions, and processing, affect the vitamin C concentration in fruits (Denardin et al., 2015). Fruits and vegetables create more ascorbic acid during their growing stages when they are exposed to sunlight.

While sunlight reduces ascorbic acid levels in fruits and vegetables during the course of the postharvest period (Lee et al., 2000), ascorbic acid loss is accelerated during storage at high temperatures. In a 9-month period of storage, potatoes lose 75–80% of their ascorbic acid concentration, according to Vicente et al. (2009). Different fruits and vegetables lose different amounts of vitamin C when they are in the same state. According to Kaur et al. (2001), ascorbic acid levels in carrots and broccoli showed a minor decline, however under the same storage circumstances, green beans showed significant losses. Between cultivars, the amount of vitamin C can also vary greatly. According to a study by Koh et al. (2012), vitamin C concentrations in 27 distinct spinach cultivars ranged from 24.57 to 62.87 mg/100 g of fresh weight. Cooking is another process that causes high ascorbic acid losses. Due to leaching and oxidation, vegetables that have been cooked may lose 40–80% of their vitamin C concentration (Vicente et al., 2009).

Vitamin E is one of the well-known bioactive compounds present in fruits and vegetables (tocopherol). Tocopherol was first isolated from green leafy vegetables in 1922, and ever since, it has been extensively studied (Peh et al., 2015). Vitamin E is composed of four tocotrienols and four tocopherol mimics (Halliwell et al., 2005). Vitamin E is well known for having potent antioxidant and anti-cancer properties. It could reduce the risk of heart disease and a number of cancers (Schwartz et al., 2008). Numerous studies have also demonstrated potential health benefits, including effects that may be anti-inflammatory, anti-hypertensive, anti-atherogenic, anti-allergic dermatitis, and hypolipidemic.

Fruits and vegetables are more stable since they have longer shelf-life (Saini et al., 2016). Only a low to moderate amount of vitamin E can be found in the majority of fruits and vegetables. But because there are so many foods made from plants in our diet, they are a good source of vitamin E (Chun et al., 2006). Almonds, avocados, pecans, peanuts, and olives are some fruits that are excellent sources of vitamin E. Broccoli and other green vegetables have higher vitamin E contents than other types of vegetables (Vicente et al., 2009). The amount of vitamin E in fruits and vegetables depends on a number of factors, including species; variety, maturity, weather, soil type, solar intensity, and the time of harvest (Bauernfeind et al., 1980). Processing and storage drastically alter the vitamin E's composition.

5.4.2.9 GLUCOSINOLATES (GSS)

The Brassicaceae plant family is noted for its sulfur-containing secondary metabolites known as GSs. The most well-known members of this family are cabbage, broccoli, cauliflower, and Brussels sprouts (Verkerk et al., 2009). The GSs perform defense roles in vegetable agriculture because certain of the breakdown products of GSs have been shown to be toxic or pathogen-repellent (Björkman et al., 2011). These elements have an impact on a plant's age, temperature, water stress, soil type, and genetic and environmental elements such GS concentration (Farnham et al., 2004). There are 200 known GSs, and the three main families are aliphatic, aromatic, and indolyl. These veggies' particular flavor and taste are influenced by GSs and the molecules that break down from these substances (Bernhoft et al., 2010). The GSs are hydrolyzed by the plant enzyme myrosinase, which creates biologically active compounds like indoles, thiocyanates, and isothiocyanates (Holst et al., 2004). Progoitrin, Sinigrin, and Gluconapin are a few of the GSs that form unstable isothiocyanates that could change further (Agerbirk et al., 2009). Due to the water solubility of GSs and the byproducts of their breakdown, their loss during cooking happens as a result of leaching into the cooking water (McNaughton et al., 2003). When cooked for 9 to 15 minutes, Brassica family vegetables lost 18 to 59% of their total GS content. Veggies (McNaughton et al., 2003).

Consuming foods from Brassica plants lowers a person's risk of developing some cancers. Studies employing cell cultures and *in vivo* animal models have shown that GS breakdown products such indoles and isothiocyanates have anticancer properties (Ciska et al., 2016). Consuming five to nine servings (2–6.5 cups) of fruits and vegetables each day is advised by the National Cancer Institute. Additionally, adults should consume at least 5 servings of vegetables from the Brassica family each week to prevent cancer, according to the findings of numerous prospective cohort studies (Giovannucci et al., 2003). By encouraging apoptosis and disrupting the cell cycle, sulforaphane, a GS isothiocyanate derivative found in broccoli, may inhibit the growth of tumors (Keum et al., 2004; Thornalley et al., 2002). Sulforaphane also guards against gastritis and stomach cancer brought on by Helicobacter pylori (Fahey et al., 2002). However, a few recent research (Latté et al., 2011; Wiesner et al., 2014) shown that, in some cases, GS breakdown products may promote the growth of cancer.

Boiling broccoli sprouts causes the sulfur-containing aliphatic GS to thermally breakdown, resulting in the formation of nitriles, claim Hanschen et al. (2012b). When processing takes longer, Fe^{2+} and a lower pH are present, which speeds up the production of nitriles (Hanschen et al., 2012a, b; Jones et al., 2010).

5.5 CONCLUSION

Plants produce a wide variety of secondary metabolites, some of which may be harmful to human health. GSs, carotenoids, phytoestrogens, phenolics, and phytoestrogens are some of these bioactive compounds. Fruits and vegetables contain a variety of bioactive compounds and micronutrients that do not constitute proper nutrition. Numerous researchers have proposed that eating a diet rich in fruits and vegetables may help people avoid a variety of health problems, such as osteoporosis, obesity, CAD, hypertension, and stroke risk. Most of the credit for those protective effects is attributed to bioactive molecules.

KEYWORDS

- **bioactive compound**
- **flavonoid**
- **fruits and vegetables**
- **functional food**
- **glucosinolates**
- **nutraceuticals**
- **phenol**
- **phytoestrogens**

REFERENCES

Adlercreutz, H., & Pietinen, P., (2003). Phyto-oestrogen database of foods and average intake in Finland. *Br. J. Nutr.*, *89*(S1), 31–38.

Adlercreutz, H., Heinonen, S. M., & Penalvo-Garcia, J., (2004). Phytoestrogens, cancer, and coronary heart disease. *BioFactors*, *22*(1–4), 229–236.

Agarwal, A., Shen, H., Agarwal, S., & Rao, A., (2001). Lycopene content of tomato products: Its stability, bioavailability and *in vivo* antioxidant properties. *J. Med. Food*, *4*(1), 9–15.

Agerbirk, N., De Vos, M., Kim, J. H., & Jander, G., (2009). Indole glucosinolate breakdown and its biological effects. *Phytochem. Rev.*, *8*(1), 101–120.

Agudo, A., (2005). *Measuring Intake of Fruit and Vegetables*. In: Joint FAO/WHO workshop on fruit and vegetables for health, Kobe, Japan. WHO/FAO, Kobe.

Allaqaband, S., Dar, A. H., Patel, U., Kumar, N., Nayik, G. A., Khan, S. A., Ansari, M. J., et al., (2022). Utilization of fruit seed-based bioactive compounds for formulating the nutraceuticals and functional food: A review. *Frontiers in Nutrition, 9*, 1–23.

Anisimova, N. Y., Kiselevsky, M. V., Sosnov, A. V., Sadovnikov, S. V., Stankov, I. N., & Gakh, A. A., (2011). Trans-, cis-, and dihydro-resveratrol: A comparative study. *Chem. Cent. J., 5*(1), 88.

Auger, C., Al-Awwadi, N., Bornet, A., Rouanet, J. M., Gasc, F., Cros, G., & Teissedre, P. L., (2004). Catechins and procyanidins in Mediterranean diets. *Food Res. Int.*, *37*(3), 233–245.

Balsano, C., & Alisi, A., (2009). Antioxidant effects of natural bioactive compounds. *Curr. Pharm. Des.*, *15*(26), 3063–3073.

Bauernfeind, J., (1980). Tocopherols in foods. In: Machlin, L. J., (ed.), *Vitamin E: A Comprehensive Treatise* (pp. 99–167). Marcel Dekker, New York.

Beekwilder, J., Jonker, H., Meesters, P., Hall, R. D., Van, D. M. I. M., & Ric De Vos, C., (2005). Antioxidants in raspberry: On-line analysis links antioxidant activity to a diversity of individual metabolites. *J. Agric. Food Chem.*, *53*(9), 3313–3320.

Bernhoft, A., (2010). A brief review. In: Bernhoft, A., (ed.), *Bioactive Compounds in Plants Benefits and Risks for Man and Animals* (pp. 11–17). The Norwegian Academy of Science and Letters, Oslo, Norway.
Björkman, M., Klingen, I., Birch, A. N., Bones, A. M., Bruce, T. J., Johansen, T. J., Meadow, R., et al., (2011). Phytochemicals of *Brassicaceae* in plant protection and human health–influences of climate, environment, and agronomic practice. *Phytochemistry*, *72*(7), 538–556.
Boca, R. M. C., Scalbert, A., Morand, C., Rémésy, C., & Jiménez, L., (2004). Polyphenols: Food sources and bioavailability. *Am. J. Clin. Nutr.*, *79*(5), 727–747.
Boeing, H., Bechthold, A., Bub, A., Ellinger, S., Haller, D., Kroke, A., Leschik-Bonnet, E., et al., (2012). Critical review: Vegetables and fruit in the prevention of chronic diseases. *Eur. J. Nutr.*, *51*(6), 637–663.
Borges, G., Degeneve, A., Mullen, W., & Crozier, A., (2009). Identification of flavonoid and phenolic antioxidants in black currants, blueberries, raspberries, red currants, and cranberries. *J. Agric. Food Chem.*, *58*(7), 3901–3909.
Bourbon, A. I., Pinheiro, A. C., Cerqueira, M. A., & Vicente, A. A., (2018). *In vitro* digestion of lactoferrin-glycomacropeptide nanohydrogels incorporating bioactive compounds: Effect of a chitosan coating. *Food Hydrocolloids*, *84*, 267–275.
Brasil, R. Md. S. M., Alves, R. E., De Brito, E. S., Pérez-Jiménez, J., Saura-Calixto, F., & Mancini-Filho, J., (2010). Bioactive compounds and antioxidant capacities of 18 non-traditional tropical fruits from Brazil. *Food Chem.*, *121*(4), 996–1002.
Burns, J., Yokota, T., Ashihara, H., Lean, M. E., & Crozier, A., (2002). Plant foods and herbal sources of resveratrol. *J. Agric. Food Chem.*, *50*(11), 3337–3340.
Cambridge, S. E., Şahin, S., Demir, C., Türkben, C., & Uylaşer, V., (2010). Phenolic content and antioxidant activity of raspberry and blackberry cultivars. *J. Food Sci.*, *75*(4), C328–C335.
Cassidy, A., Hanley, B., & Lamuela-Raventos, R. M., (2000). Isoflavones, lignans and stilbenes–origins, metabolism, and potential importance to human health. *J. Sci. Food Agric.*, *80*(7), 1044–1062.
Champaign, S. F., Naczk, M., & Griffiths, W., (1996). Food phenolics: Sources, chemistry, effects, applications. *Trends Food Sci. Technol.*, *7*(7), 243.
Cherdshewasart, W., Sutjit, W., Pulcharoen, K., & Chulasiri, M., (2009). The mutagenic and antimutagenic effects of the traditional phytoestrogen-rich herbs, *Pueraria mirifica* and *Pueraria lobata*. *Braz. J. Med. Biol. Res.*, *42*, 816–823.
Chun, J., Lee, J., Ye, L., Exler, J., & Eitenmiller, R. R., (2006). Tocopherol and tocotrienol contents of raw and processed fruits and vegetables in the United States diet. *Journal of Food Composition and Analysis*, *19*(2, 3), 196–204.
Ciska, E., Drabińska, N., Narwojsz, A., & Honke, J., (2016). Stability of glucosinolates and glucosinolate degradation products during storage of boiled white cabbage. *Food Chem.*, *203*, 340–347.
Clifford, M. N., (1999). Chlorogenic acids and other cinnamates–nature, occurrence, and dietary burden. *J. Sci. Food Agric.*, *79*, 362–372.
Clifford, M. N., (2000). Chlorogenic acids and other cinnamates–nature, occurrence, dietary burden, absorption, and metabolism. *J. Sci. Food Agric.*, *80*(7), 1033–1043.
Corder, R., Mullen, W., Khan, N., Marks, S., Wood, E., Carrier, M., & Crozier, A., (2006). Oenology: Red wine procyanidins and vascular health. *Nature*, *444*(7119), 566–566.

Cornwell, T., Cohick, W., & Raskin, I., (2004). Dietary phytoestrogens and health. *Phytochemistry*, *65*(8), 995–1016.

D'Andrea, G., (2015). Quercetin: A flavonol with multifaceted therapeutic applications. *Fitoterapia*.

Da Silva Porto, P. A. L., Laranjinha, J. A. N., & De Freitas, V. A. P., (2003). Antioxidant protection of low-density lipoprotein by procyanidins: Structure/activity relationships. *Biochem. Pharmacol.*, *66*(6), 947–954.

Da Silva, F. L., Escribano-Bailón, M. T., Alonso, J. J. P., Rivas-Gonzalo, J. C., & Santos-Buelga, C., (2007). Anthocyanin pigments in strawberry. *LWT-Food Sci. Technol.*, *40*(2), 374–382.

Dao, L., & Friedman, M., (1992). Chlorogenic acid content of fresh and processed potatoes determined by ultraviolet spectrophotometry. *J. Agric. Food Chem.*, *40*(11), 2152–2156.

Das, G. A., & Klein, K., (2014). Chapter 12 – fruits, vegetables, and nuts: Good sources of antioxidants. In: Klein, A. D., (ed.), *Ntioxidants in Food, Vitamins, and Supplements* (pp. 209–235). Elsevier.

Das, L., Bhaumik, E., Raychaudhuri, U., & Chakraborty, R., (2012). Role of nutraceuticals in human health. *J. Food Sci. Technol.*, *49*(2), 173–183.

De Pascual-Teresa, S., & Sanchez-Ballesta, M. T., (2008). Anthocyanins: From plant to health. *Phytochem. Rev.*, *7*(2), 281–299.

De Pascual-Teresa, S., Santos-Buelga, C., & Rivas-Gonzalo, J. C., (2000). Quantitative analysis of flavan-3-ols in Spanish foodstuffs and beverages. *J. Agric. Food Chem.*, *48*(11), 5331–5337.

Deepa, N., Kaur, C., George, B., Singh, B., & Kapoor, H., (2007). Antioxidant constituents in some sweet pepper (*Capsicum annuum* L.) genotypes during maturity. *LWT – Food Sci. Technol.*, *40*(1), 121–129.

Denardin, C. C., Hirsch, G. E., Da Rocha, R. F., Vizzotto, M., Henriques, A. T., Moreira, J. C. F., Guma, F. T. C. R., & Emanuelli, T., (2015). Antioxidant capacity and bioactive compounds of four Brazilian native fruits. *Journal of Food and Drug Analysis*, *23*(3), 387–398.

Dillard, C. J., & German, J. B., (2000). Phytochemicals: Nutraceuticals and human health. *J. Sci. Food Agric.*, *80*, 1744–1756.

Dimitrios, B., (2006). Sources of natural phenolic antioxidants. *Trends Food Sci. Technol.*, *17*(9), 505–512.

Ekor, M., (2014). The growing use of herbal medicines: Issues relating to adverse reactions and challenges in monitoring safety. *Front Pharmacol.*, *4*, 177.

Esteban, R., Barrutia, O., Artetxe, U., Fernández-Marín, B., Hernández, A., & García-Plazaola, J. I., (2015). Internal and external factors affecting photosynthetic pigment composition in plants: A meta-analytical approach. *New Phytol.*, *206*(1), 268–280.

Fahey, J. W., Haristoy, X., Dolan, P. M., Kensler, T. W., Scholtus, I., Stephenson, K. K., Talalay, P., & Lozniewski, A., (2002). Sulforaphane inhibits extracellular, intracellular, and antibiotic-resistant strains of helicobacter pylori and prevents benzo [a] pyrene-induced stomach tumors. *Proc Natl. Acad. Sci.*, *99*(11), 7610–7615.

Farnham, M., Wilson, P., Stephenson, K., & Fahey, J., (2004). Genetic and environmental effects on glucosinolate content and chemoprotective potency of broccoli. *Plant Breed*, *123*(1), 60–65.

Florkowski, W. J., Shewfelt, R. L., Brueckner, B., & Prussia, S. E., (1993). *Postharvest Handling: A System Approach* (pp. 58–93). Elsevier Inc.-Academic Press, San Diego.

Fulda, S., (2010). Resveratrol and derivatives for the prevention and treatment of cancer. *Drug Discov. Today*, *15*(17), 757–765.

Gadkari, P. V., & Balaraman, M., (2015). Catechins: Sources, extraction, and encapsulation: A review. *Food Bioprod. Process*, *93*, 122–138.

Galvano, F., La Fauci, L., Lazzarino, G., Fogliano, V., Ritieni, A., Ciappellano, S., Battistini, N. C., et al., (2004). Cyanidins: Metabolism and biological properties. *J. Nutr. Biochem.*, *15*(1), 2–11.

Ganry, O., (2005). Phytoestrogens and prostate cancer risk. *Prev Med.*, *41*(1), 1–6.

Ghofrani, S., Joghataei, M. T., Mohseni, S., Baluchnejadmojarad, T., Bagheri, M., Khamse, S., & Roghani, M., (2015). Naringenin improves learning and memory in an Alzheimer's disease rat model: Insights into the underlying mechanisms. *Eur. J. Pharmacol.*, *764*, 195–201.

Giovannucci, E., Rimm, E. B., Liu, Y., Stampfer, M. J., & Willett, W. C., (2003). A prospective study of cruciferous vegetables and prostate cancer. *Cancer Epidemiol Biomark Prev.*, *12*(12), 1403–1409.

Gupta, S. K., Yadav, S. K., & Patil, M. S. M., (2013). Nutraceutical-a bright scope and opportunity of Indian healthcare market. *International Journal of Research and Development in Pharmacy and Life Science*, *2*(4), 478–481.

Halliwell, B., Rafter, J., & Jenner, A., (2005). Health promotion by flavonoids, tocopherols, tocotrienols and other phenols: Direct or indirect effects? Antioxidant or not? *Am. J. Clin. Nutr.*, *81*(1), 268–276.

Hammerstone, J. F., Lazarus, S. A., & Schmitz, H. H., (2000). Procyanidin content and variation in some commonly consumed foods. *J. Nutr.*, *130*(8), 2086–2092.

Hanschen, F. S., Bauer, A., Mewis, I., Keil, C., Schreiner, M., Rohn, S., & Kroh, L. W., (2012). Thermally induced degradation of aliphatic glucosinolates: Identification of intermediary breakdown products and proposed degradation pathways. *J. Agric Food Chem.*, *60*(39), 9890–9899.

Hanschen, F. S., Platz, S., Mewis, I., Schreiner, M., Rohn, S., & Kroh, L. W., (2012). Thermally induced degradation of sulfur-containing aliphatic glucosinolates in broccoli sprouts (*Brassica oleracea* var. italica) and model systems. *J. Agric Food Chem.*, *60*(9), 2231–2241.

Helyes, L., Lugasi, A., & Pék, Z., (2007). Effect of natural light on surface temperature and lycopene content of vine ripened tomato fruit. *Canadian Journal of Plant Science*, *2007*, 927–929.

Holman, R. T., (1998). The slow discovery of the importance of omega 3 essential fatty acids in human health. *J. Nutr.*, *128*(2), 427S–433S.

Holst, B., & Williamson, G., (2004). A critical review of the bioavailability of glucosinolates and related compounds. *Nat. Prod Rep.*, *21*(3), 425–447.

Hooper, L., & Cassidy, A., (2006). A review of the health care potential of bioactive compounds. *J. Sci. Food Agric.*, *86*(12), 1805–1813.

Hord, N. G., (2008). Eukaryotic microbiotic crosstalk: Potential mechanisms for health benefits of prebiotics and probiotics. *Annu. Rev. Nutr.*, *28*, 215–231.

Hu, B., Ting, Y., Zeng, X., & Huang, Q., (2012). Cellular uptake and cytotoxicity of chitosan-caseinophosphopeptides nanocomplexes loaded with epigallocatechin gallate. *Carbohydr. Polym.*, *89*(2), 362–370.

Hyvärinen, H. K., Pihlava, J. M., Hiidenhovi, J. A., Hietaniemi, V., Korhonen, H. J., & Ryhänen, E. L., (2006). Effect of processing and storage on the stability of flaxseed lignan added to bakery products. *J. Agric. Food Chem.*, *54*(1), 48–53.

Hyvärinen, H. K., Pihlava, J. M., Hiidenhovi, J. A., Hietaniemi, V., Korhonen, H. J., & Ryhänen, E. L., (2006). Effect of processing and storage on the stability of flaxseed lignan added to dairy products. *J. Agric. Food Chem.*, *54*(23), 8788–8792.

Iriti, M., & Faoro, F., (2006). Grape phytochemicals: A bouquet of old and new nutraceuticals for human health. *Medical Hypotheses*, 833–838.

Johnsen, N. F., Olsen, A., Thomsen, B. L. R., Christensen, J., Egeberg, R., Knudsen, K. E. B., Loft, S., et al., (2010). Plasma enterolactone and risk of colon and rectal cancer in a case– cohort study of Danish men and women. *Cancer Causes Control*, *21*(1), 153–162.

Jones, R., Frisina, C., Winkler, S., Imsic, M., & Tomkins, R., (2010). Cooking method significantly effects glucosinolate content and sulforaphane production in broccoli florets. *Food Chem.*, *123*(2), 237–242.

Kalra, E. K., (2003). Nutraceutical- definition and introduction. *AAPS Pharm. Sci.*, *5*, 27, 28.

Kandlakunta, B., Rajendran, A., & Thingnganing, L., (2008). Carotene content of some common (cereals, pulses, vegetables, spices, and condiments) and unconventional sources of plant origin. *Food Chem.*, *106*(1), 85–89.

Kasiotis, K. M., Pratsinis, H., Kletsas, D., & Haroutounian, S. A., (2013). Resveratrol and related stilbenes: Their anti-aging and anti-angiogenic properties. *Food Chem Toxicol.*, *61*, 112–120.

Kassim, A., Poette, J., Paterson, A., Zait, D., McCallum, S., Woodhead, M., Smith, K., Hackett, C., & Graham, J., (2009). Environmental and seasonal influences on red raspberry anthocyanin antioxidant contents and identification of quantitative traits loci (QTL). *Mol. Nutr. Food Res.*, *53*(5), 625–634.

Kaur, C., & Kapoor, H. C., (2001). Antioxidants in fruits and vegetables–the millennium's health. *Int J. Food Sci. Technol.*, *36*(7), 703–725.

Keum, Y., Jeong, W. S., & Kong, A. T., (2004). Chemoprevention by isothiocyanates and their underlying molecular signaling mechanisms. *Mut. Res.*, *555*(1), 191–202.

Kochhar, K. P., (2008). Dietary spices in health and diseases: I. *Indian J. Physiol. Pharmacol.*, *52*, 106–122.

Koh, E., Charoenprasert, S., & Mitchell, A. E., (2012). Effect of organic and conventional cropping systems on ascorbic acid, vitamin C, flavonoids, nitrate, and oxalate in 27 varieties of spinach (*Spinacia oleracea* L.) *J. Agric. Food Chem.*, *60*(12), 3144–3150.

Koponen, J. M., Happonen, A. M., Mattila, P. H., & Törrönen, A. R., (2007). Contents of anthocyanins and ellagitannins in selected foods consumed in Finland. *J. Agric. Food Chem.*, *55*(4), 1612–1619.

Krinsky, N. I., & Johnson, E. J., (2005). Carotenoid actions and their relation to health and disease. *Mol. Asp Med.*, *26*(6), 459–516.

Kuhnle, G. G., Dell'Aquila, C., Aspinall, S. M., Runswick, S. A., Mulligan, A. A., & Bingham, S. A., (2008). Phytoestrogen content of beverages, nuts, seeds, and oils. *J. Agric. Food Chem.*, *56*(16), 7311–7315.

Kuhnle, G. G., Dell'Aquila, C., Aspinall, S. M., Runswick, S. A., Mulligan, A. A., & Bingham, S. A., (2009). Phytoestrogen content of cereals and cereal-based foods consumed in the UK. *Nutr. Cancer*, *61*(3), 302–309.

Latté, K. P., Appel, K. E., & Lampen, A., (2011). Health benefits and possible risks of broccoli – an overview. *Food Chem. Toxicol.*, *49*(12), 3287–3309.

Lee, K. W., Kim, Y. J., Kim, D. O., Lee, H. J., & Lee, C. Y., (2003). Major phenolics in apple and their contribution to the total antioxidant capacity. *J. Agric. Food Chem.*, *51*(22), 6516–6520.

Lee, S. K., & Kader, A. A., (2000). Preharvest and postharvest factors influencing vitamin C content of horticultural crops. *Postharvest Biol. Technol.*, *20*(3), 207–220.

Lee, S. W., Jeung, M. K., Park, M. H., Lee, S. Y., & Lee, J., (2010). Effects of roasting conditions of sesame seeds on the oxidative stability of pressed oil during thermal oxidation. *Food Chem.*, *118*(3), 681–685.

Lemay, A., Dodin, S., Kadri, N., Jacques, H., & Forest, J. C., (2002). Flax seed dietary supplement versus hormone replacement therapy in hypercholesterolemic menopausal women. *Obstet. Gynecol.*, *100*(3), 495–504.

Lenoir-Wijnkoop, L., Sanders, M. E., Cabana, M. D., Caglar, E., Corthier, G., Rayes, N., Sherman, P. M., & Timmerman, H. M., (2007). Probiotic and prebiotic influence beyond the intestinal tract. *Nutr. Rev.*, *65*, 469–489.

Li, Y., Cao, Z., & Zhu, H., (2006). Upregulation of endogenous antioxidants and phase 2 enzymes by the red wine polyphenol, resveratrol in cultured aortic smooth muscle cells leads to cytoprotection against oxidative and electrophilic stress. *Pharmacol. Res.*, *53*(1), 6–15.

Lin, J. Y., & Tang, C. Y., (2007). Determination of total phenolic and flavonoid contents in selected fruits and vegetables, as well as their stimulatory effects on mouse splenocyte proliferation. *Food Chem.*, *101*(1), 140–147.

Liu, F., Zhu, Z., Ma, C., et al., (2016). Fabrication of concentrated fish oil emulsions using dual-channel microfluidization: Impact of droplet concentration on physical properties and lipid oxidation. *J. Agric. Food Chem.*, *64*(50), 9532–9541.

Liu, L., Shao, Z., Zhang, M., & Wang, Q., (2015). Regulation of carotenoid metabolism in tomato. *Mol. Plant, 8*(1), 28–39.

Liu, R. H., (2003). Health benefits of fruit and vegetables are from additive and synergistic combinations of phytochemicals. *Am J. Clin. Nutr.*, *78*(3), 517–520.

Liu, R. H., (2013). Dietary bioactive compounds and their health implications. *J. Food Sci.*, *78*(s1), A18–A25.

Macheix, J. J., & Fleuriet, A., (1990). *Fruit Phenolics* (1st edn., p. 390). CRC Press.

Manach, C., Williamson, G., Morand, C., Scalbert, A., & Rémésy, C., (2005). Bioavailability and bioefficacy of polyphenols in humans. I. Review of 97 bioavailability studies. *Am. J. Clin. Nutr.*, *81*(1), 230–242.

Marines, M., Gniech, K., & Chakravarthi, M., (2018). Fruits as prospective reserves of bioactive compounds: A review. *Natural Products and Bioprospecting*, *8*, 335–346.

Marko, D., Puppel, N., Tjaden, Z., Jakobs, S., & Pahlke, G., (2004). The substitution pattern of anthocyanidins affects different cellular signaling cascades regulating cell proliferation. *Mol. Nutr. Food Res.*, *48*(4), 318–325.

Mason, G. J., Malonne, H., & Atassi, G., (2001). A reappraisal of the potential chemopreventive and chemotherapeutic properties of resveratrol. *Carcinogenesis*, *22*(8), 1111–1117.

McNaughton, S., & Marks, G., (2003). Development of a food composition database for the estimation of dietary intakes of glucosinolates, the biologically active constituents of cruciferous vegetables. *Br. J. Nutr.*, *90*(03), 687–697.

Milder, I. E., Arts, I. C., Putte, B., Venema, D. P., & Hollman, P. C., (2005). Lignan contents of Dutch plant foods: A database including lariciresinol, pinoresinol, secoisolariciresinol and matairesinol. *Br. J. Nutr.*, *93*(3), 393–402.

Milder, I. E., Feskens, E. J., Arts, I. C., De Mesquita, H. B. B., Hollman, P. C., & Kromhout, D., (2005). Intake of the plant lignans secoisolariciresinol, matairesinol, lariciresinol, and pinoresinol in Dutch men and women. *J. Nutr.*, *135*(5), 1202–1207.

Moazzami, A. A., Haese, S. L., & Kamal-Eldin, A., (2007). Lignan contents in sesame seeds and products. *Eur. J. Lipid Sci. Technol.*, *109*(10), 1022–1027.

Nemitz, M. C., Argenta, D. F., Koester, L. S., Bassani, V. L., Von, P. G. L., & Teixeira, H. F., (2016). The international scenario of patents concerning isoflavones. *Trends Food Sci. Technol*, *49*, 85–95.

Nikfardjam, M. P., Schmitt, K., Rühl, E., Patz, C., & Dietrich, H., (2000). Investigation of pure grape juices on their content of resveratrols. *Dtsch Lebensmitt Rundsch*, *96*(9), 319–324.

Nounou, M. I., Ko, Y., Helal, N. A., & Boltz, J. F., (2018). Adulteration and counterfeiting of online nutraceutical formulations in the United States: Time for intervention. *J. Diet Suppl.*, *15*(5), 789–804.

Omoni, A. O., & Aluko, R. E., (2005). The anti-carcinogenic and anti-atherogenic effects of lycopene: A review. *Trends Food Sci. Technol.*, *16*(8), 344–350.

Ou, K., & Gu, L., (2014). Absorption and metabolism of proanthocyanidins. *J. Funct. Foods, 7*, 43–53.

Palafox-Carlos, H. P., Ayala-Zavala, J. F., & Gonzalez-Aguilar, G. A., (2011). The role of dietary fiber in the bioaccessibility and bioavailability of fruit and vegetable antioxidants. *Journal of Food Science*, *76*(1), 1–10.

Pandey, M., Verma, R. K., & Saraf, S. A., (2010). Nutraceuticals: New era of medicine and health. *Asian J. Pharmaceuticals Clin Res.*, *3*, 11–15.

Peeters, P., Keinan-Boker, L., Van, D. S. Y., & Grobbee, D., (2003). Phytoestrogens and breast cancer risk. *Breast Cancer Res. Treat*, *77*(2), 171–183.

Peh, H. Y., Tan, W. D., Liao, W., & Wong, W. F., (2015). Vitamin E therapy beyond cancer: Tocopherol versus tocotrienol. *Pharmacol. Ther.*, *162*, 152–169.

Pervaiz, S., & Holme, A. L., (2009). Resveratrol: Its biologic targets and functional activity. *Antioxid Redox Signal*, *11*(11), 2851–2897.

Poiroux-Gonord, F., Santini, J., Fanciullino, A. L., Lopez-Lauri, F., Giannettini, J., Sallanon, H., Berti, L., & Urban, L., (2013). Metabolism in orange fruits is driven by photooxidative stress in the leaves. *Physiol. Plant*, *149*(2), 175–187.

Prakash, D., & Gupta, C., (2011). Role of phytoestrogens as nutraceuticals in human health: A review. biotechnology: *An Indian J.*, *5*, 1–8.

Prakash, D., Gupta, C., & Sharma, G., (2012). Importance of phytochemicals in nutraceuticals. *Journal of Chinese Medicine Research and Development*, *2012*, 70–78.

Prakash, D., Gupta, C., & Sharma, G., (2012). Importance of phytochemicals in nutraceuticals. *Journal of Chinese Medicine Research and Development*, 70–78.

Prakash, D., Upadhyay, G., Pushpangadan, P., & Gupta, C., (2011). Antioxidant and free radical scavenging activity activities of some fruits. *J. Complement Integr. Med., 8*, 1–19.

Prasad, K., (2005). Hypocholesterolemic and antiatherosclerotic effect of flax lignan complex isolated from flaxseed. *Atherosclerosis*, *179*(2), 269–275.

Prior, R. L., & Wu, X., (2006). Anthocyanins: Structural characteristics that result in unique metabolic patterns and biological activities. *Free Radic. Res.*, *40*(10), 1014–1028.

Prior, R. L., Lazarus, S. A., Cao, G., Muccitelli, H., & Hammerstone, J. F., (2001). Identification of procyanidins and anthocyanins in blueberries and cranberries (*Vaccinium* spp.) using high-performance liquid chromatography/mass spectrometry. *J. Agric. Food Chem*, *49*(3), 1270–1276.

Rai, S. K., Arora, N., Pandey, N., Meena, R. P., Shah, K., & Rai, S. P., (2012). Nutraceutical enriched vegetables: Molecular approaches for crop improvement. *International Journal of Pharma and Bio Sciences, 3*(2), 0975–6299.

Ram, S. M., Andrew, M. A., & Matthew, S. M., (2000). Isolation of curcumin from turmeric. *Journal of Chemical Education*, *77*(3), 359–360.
Ramaa, C. S., Shirode, A. R., Mundada, A. S., & Kadam, V. J., (2006). Nutraceuticals an emerging era in the treatment and prevention of cardiovascular diseases. *Curr. Pharm. Biotechnol.*, *7*, 15–23.
Ramel, F., Birtic, S., Ginies, C., Soubigou-Taconnat, L., Triantaphylidès, C., & Havaux, M., (2012). Carotenoid oxidation products are stress signals that mediate gene responses to singlet oxygen in plants. *Proc. Natl. Acad. Sci.*, *109*(14), 5535–5540.
Rivera, S., & Canela-Garayoa, R., (2012). Analytical tools for the analysis of carotenoids in diverse. *Journal of Chromatography*, *1224*, 1–10.
Rodriguez-Amaya, D. B., & Kimura, M., (2004). *Harvest Plus Handbook for Carotenoid Analysis, 106*, 256–271.
Rybarczyk-Plonska, A., Hagen, S. F., Borge, G. I. A., Bengtsson, G. B., Hansen, M. K., & Wold, A. B., (2016). Glucosinolates in broccoli (*Brassica oleracea* L. var. italica) as affected by postharvest temperature and radiation treatments. *Postharvest Biol. Technol*, *116*, 16–25.
Saini, R. K., & Keum, Y. S., (2016). Tocopherols and tocotrienols in plants and their products: A review on methods of extraction, chromatographic separation, and detection. *Food Res. Int., 82*, 59–70.
Saleem, M., Kim, H. J., Ali, M. S., & Lee, Y. S., (2005). An update on bioactive plant lignans. *Nat. Prod. Rep.*, *22*(6), 696–716.
Santini, A., & Novellino, E., (2018). Nutraceuticals - shedding light on the grey area between pharmaceuticals and food. *Expert. Rev. Clin. Pharmacol.*, *11*(6), 545–547.
Santos-Buelga, C., & Williamson, G., (2003). Methods in polyphenol analysis. *Royal Society of Chemistry*, *2003*.
Sarin, R. S. M., Manisha, S., Robin, S., & Sunil, K., (2012). Nutraceuticals a review. *International Research Journal of Pharmacy*, *3*(4), 95–99.
Scalbert, A., & Williamson, G., (2000). Dietary intake and bioavailability of polyphenols. *J. Nutr.*, *130*(8), 2073–2085.
Scalbert, A., Manach, C., Morand, C., Rémésy, C., & Jiménez, L., (2005). Dietary polyphenols and the prevention of diseases. *Crit. Rev. Food Sci. Nutr., 45*(4), 287–306.
Schaefer, H. M., Schaefer, V., & Levey, D. J., (2004). How plant–animal interactions signal new insights in communication. *Trends Ecol. Evol.*, *19*(11), 577–584.
Schwartz, H., Ollilainen, V., Piironen, V., & Lampi, A. M., (2008). Tocopherol, tocotrienol and plant sterol contents of vegetable oils and industrial fats. *J. Food Compos Anal.*, *21*(2), 152–161.
Shahidi, F., (1997). *Natural Antioxidants: Chemistry, Health Effects, and Applications* (p. 432). The American Oil Chemists Society, AOCS Press, Champaign, Illinois.
Shen, C., Chen, R., Qian, Z., Huang, C., Meng, X., Ma, T., Chen, Z., et al., (2016). A HPLC–MS/MS method for the quantitation of free, conjugated, and total HDND-7, a novel hesperetin derivative, in rat plasma and tissues: Application to the pharmacokinetic and tissue distribution study. *J. Pharm. Biomed. Anal.*, *118*, 149–160.
Sirtori, C. R., & Galli, C., (2002). Fatty acids and the omega 3. *Biomedecine and Pharmacotherapy*, *56*, 397–406.
Slavin, J. L., & Lloyd, B., (2012). Health benefits of fruits and vegetables. *American Society for Nutrition. Adv. Nutr., 3*, 506–516.

Smeds, A. I., Eklund, P. C., & Willför, S. M., (2012). Content, composition, and stereochemical characterization of lignans in berries and seeds. *Food Chem.*, *134*(4), 1991–1998.

Smeds, A. I., Eklund, P. C., Sjöholm, R. E., Willför, S. M., Nishibe, S., Deyama, T., & Holmbom, B. R., (2007). Quantification of a broad spectrum of lignans in cereals, oilseeds, and nuts. *J Agric Food Chem*, *55*(4), 1337–1346.

Stahl, W., & Sies, H., (2003). Antioxidant activity of carotenoids. *Mol. Asp. Med.*, *24*(6), 345–351.

Stellavato, A., Pirozzi, A. V. A., De Novellis, F., et al., (2018). *In vitro* assessment of nutraceutical compounds and novel nutraceutical formulations in a liver-steatosis-based model. *Lipids Health Dis.*, *17*, 24.

Swaroopa, G., & Srinath, D., (2017). Nutraceuticals and their health benefits. *Int. J. Pure App. Biosci.*, *5*(4), 1151–1155.

Szajdek, A., & Borowska, E., (2008). Bioactive compounds, and health-promoting properties of berry fruits: A review. *Plant Foods Hum. Nutr.*, *63*(4), 147–156.

Thompson, L. U., Boucher, B. A., Liu, Z., Cotterchio, M., & Kreiger, N., (2006). Phytoestrogen content of foods consumed in Canada, including isoflavones, lignans, and coumestan. *Nutr. Cancer*, *54*(2), 184–201.

Thornalley, P. J., (2002). Isothiocyanates: Mechanism of cancer chemopreventive action. *Anti-Cancer Drugs*, *13*(4), 331–338.

Touillaud, M. S., Thiébaut, A. C., Fournier, A., Niravong, M., Boutron-Ruault, M. C., & Clavel-Chapelon, F., (2007). Dietary lignan intake and postmenopausal breast cancer risk by estrogen and progesterone receptor status. *J. Natl. Cancer Inst.*, *99*(6), 475–486.

Turcsi, E., Nagy, V., & Deli, J., (2016). Study on the elution order of carotenoids on endcapped C 18 and C 30 reverse silica stationary phases. A review of the database. *J. Food Compos Anal.*, *47*, 101–112.

Vallejo, F., García-Viguera, C., & Tomás-Barberán, F. A., (2003). Changes in broccoli (*Brassica oleracea* L. var. italica) health-promoting compounds with inflorescence development. *J. Agric. Food Chem.*, *51*(13), 3776–3782.

Vasco, C., Ruales, J., & Kamal-Eldin, A., (2008). Total phenolic compounds and antioxidant capacities of major fruits from Ecuador. *Food Chem.*, *111*(4), 816–823.

Verkerk, R., Schreiner, M., Krumbein, A., Ciska, E., Holst, B., Rowland, I., De Schrijver, R., et al., (2009). Glucosinolates in Brassica vegetables: The influence of the food supply chain on intake, bioavailability, and human health. *Mol. Nutr. Food Res.*, *53*(S2), 219–219.

Vicente, A. R., Manganaris, G. A., Sozzi, G. O., & Crisosto, C. H., (2009). *Nutritional Quality of Fruits and Vegetables* (2nd edn., pp. 1–52). Elsevier Inc.-Academic Press.

Walter, M. H., & Strack, D., (2011). Carotenoids and their cleavage products: Biosynthesis and functions. *Nat. Prod. Rep.*, *28*(4), 663–692.

Walter, M. H., Floss, D. S., & Strack, D., (2010). Apocarotenoids: Hormones, mycorrhizal metabolites, and aroma volatiles. *Planta*, *232*, 1–17.

Webb, A. L., & McCullough, M. L. (2005). Dietary lignans: potential role in cancer prevention. *Nutrition and Cancer, 51*(2), 117–131.

Whitman, M., (2001). Understanding the perceived need for complementary and alternative nutraceuticals: Lifestyle issues. *Clin. J. Oncol. Nurs.*, *5*, 190–194.

Wiesner, M., Schreiner, M., & Glatt, H., (2014). High mutagenic activity of juice from pak choi (*Brassica rapa* ssp. chinensis) sprouts due to its content of 1-methoxy-3-indolylmethyl glucosinolate, and its enhancement by elicitation with methyl jasmonate. *Food Chem. Toxicol.*, *67*, 10–16.

Wildman, & Robert, E. C., (2001). *Handbook of Nutraceuticals and Functional Foods* (1st edn., p. 350). CRC Press.

Williamson, G., & Manach, C., (2005). Bioavailability and bioefficacy of polyphenols in humans. II. Review of 93 intervention studies. *Am. J. Clin. Nutr.*, *81*(1), 243–255.

Wu, W. H., (2007). The contents of lignans in commercial sesame oils of Taiwan and their changes during heating. *Food Chem.*, *104*(1), 341–344.

Wu, X., Beecher, G. R., Holden, J. M., Haytowitz, D. B., Gebhardt, S. E., & Prior, R. L., (2006). Concentrations of anthocyanins in common foods in the United States and estimation of normal consumption. *J. Agric. Food Chem.*, *54*(11), 4069–4075.

Yildiz, F., & Gültekin, E., (2006). Introduction to phytoestrogen. In: Yildiz, F., (ed.), *Phytoestrogen in Functional Foods* (pp. 3–18). CRC, Taylor and Francis Group, London/New York.

Yildiz, F., (2010). *Advances in Food Biochemistry* (pp. 409–456). CRC, Taylor and Francis Group, London/New York.

Zhang, X., Liu, J., Qian, C., Kan, J., & Jin, C. H., (2019). Effect of grafting method on the physical property and antioxidant potential of chitosan film functionalized with gallic acid. *Food Hydrocolloids*, *89*, 1–10.

PART III
Plant Metabolites and Vegetables in Prevention and Treatment of Diseases

CHAPTER 6

EFFECT OF PLANT AND VEGETABLE-BASED SECONDARY METABOLITES ON CANCER TREATMENT

ANKIT AWASTHI and YASHI SRIVASTAVA

Department of Applied Agriculture, Central University of Punjab, Ghudda, Bathinda, Punjab, India

ABSTRACT

Plants and vegetables produce a variety of organic compounds that are biologically active but directly do not contribute to growth and development known as secondary metabolites. The diverse class of secondary metabolites from plants and vegetables possess anti-cancer properties via various mechanisms of action. Interlinking between those mechanisms (antioxidant/prooxidants, apoptosis, necrosis, and autophagy) with the other metabolic regulations (miRNA regulation) are very critical but can be utilized for the development of an anticancer therapeutic drug with pathway specificity. The chapter focused on the involvement of vegetables and plants' metabolites in the different signaling pathways which helps in the prevention of cancer growth or carcinoma.

6.1 INTRODUCTION

Cancer or abnormal cell division in an uncontrolled way is a critical immunological cell response, that can create severe health problems in almost all countries. Chemotherapy and radiotherapy are currently used treatment for cancer patient, but it is not successful and reliable in many cases such as childhood cancer, blood cancer, etc. Furthermore, recent diversification

Plant Metabolites and Vegetables as Nutraceuticals.
Raj K. Keservani, Bui Thanh Tung, Rajesh K. Kesharwani, and Eknath D. Ahire (Eds.)

in cancer and coronavirus are negatively affecting the handling of cancer patient ultimately increasing the mortality rate in the entire world (Lee et al., 2020; Zaorsky et al., 2017). Natural secondary metabolites such as alkaloid, polyphenolic, organosulfur, betalains, capsaicinoids, diterpene, and triterpene compounds obtained from vegetables and plants are viewed as biologically, and toxicologically friendly as well as highly favorable in cancer therapy (Figure 6.1) (Seca et al., 2018; Herr & Buchler, 2010). Drug formation with the help of plants and vegetable-based metabolites shows noteworthy outcomes in cancer therapy and performs other therapeutic roles in patients (Abbaoui et al., 2018). Efficiency, bioactivity, and effective participation in the mechanism of those metabolites can be changed by certain factors including like physiological (age, individual variation, enterohepatic circulation, nutrition, intestinal flora, or sex differences), pathological, diet, and environment, that can be utilized for the significant targeting and distribution of vegetables and plant metabolite-based drugs (Gibson et al., 1986). Plant and vegetable-based bioactive compounds are classified into two basic categories in which carbohydrates (mono, di, oligo, and polysaccharides), lipids (saturated and unsaturated fats with glycerol backbone), and protein (amino acids) are considered primary metabolites whereas alkaloids, polyphenolic, organosulfur, betalains, capsaicinoids, diterpene, triterpene, etc., are included in secondary metabolites. The production of secondary metabolites can be achieved by chemical synthesis and field cultivation, that play a decisive role in plant interaction with the ecosystem which is necessary for the plant's well-being as well as providing health profits to the consumers (Lavecchia et al., 2013). Guerriero et al. (2018) surveyed engineering approaches in the metabolic system to increase the stability and activity of plant secondary metabolites.

Vegetables and plant metabolites used in anticancer drug discovery and emphasis on the searching for anticancer metabolites that not only prevent or work as a basic treatment for a cancer patient but also the presence of chemical metabolites or basic structures can help in the formulation of the drug that work as a permanent cure for various cancers. The most important steps for finding anticancer drugs are therapeutic targeting and main structural modifications that can perform a vital role in anticancer activity enhancement and in the formation of harmless anticancer drugs from plant sources (Seca et al., 2018). In the past several decades, researchers focus on the action of bioactive constituents originating from vegetables and plant sources. Mondal et al. (2022) evaluated the nano-formulations of metabolites that was sufficiently effective to inhibit the different cancers (breast, colorectal, gastric, liver,

FIGURE 6.1 Classification of some phytochemicals used as anticancer agents.

prostate, ovarian, oral, and skin) and its stages (initiation, promotion, and propagation). Most individuals in society prefer a mixture of treatments, like surgical treatment together with chemotherapy and radiotherapy. In addition to these other active treatments, such as gene therapy, immunotherapy, and thermal therapy kind of other techniques are also evolved as novel treatments for cancer. Vegetables and plants metabolites activities such as antioxidants and anticancer play a dynamic role in apoptosis, angiogenesis inhibition, cell cycle arrest, enzyme inhibition, and nuclear receptors modulation (Kamal et al., 2022; Liskova et al., 2021; Mokhtari et al., 2017). Metabolites found in plants exert a combined effect to reduce the chance of cancer via performing various therapeutic roles in the human body through various mechanisms including hormone regulation, carcinogenic pathways down-regulation, and inflammation attenuation. The chapter focused on the action mechanism of vegetables and plants' metabolites with the possible aspects of treatment and prevention approaches via target-specific signaling pathways through metabolites-based drugs in various cancer cell lines.

6.2 PLANT METABOLITES

6.2.1 PHYTOCHEMICALS

Phytochemicals are the naturally occurring chemical compounds obtained from green medicinal plants. Currently, phytochemicals gained high recognition as therapeutic and prevention agents (Mousavi et al., 2018). The various action mechanisms of phytochemicals such as antioxidant and pro-oxidant activities, apoptosis, necrosis, and autophagy induction, as well as miRNA regulation actively participate in the anticancer drug discovery and prevention of various cancers (Zhang et al., 2020). The action of plant-based phytochemicals has shown high selectivity toward cancer cells and inhibits cancer cell growth with fewer chances of toxicity (Cilwyn et al., 2021). Some anticancer drugs available in the market obtained from plants include vincristine, vinblastine, and podophyllotoxin and paclitaxel are usually used for treating cancer patients (Newman et al., 2011; Choudhari et al., 2020; Ashraf, 2020). Choudhari et al. (2020) reported that the understanding of molecular interaction with different signaling molecules, molecular targeting, and molecular docking were required for the successful preclinical stages of cancer treatment with minimum side effects. Thus, in recent trends of oncological research synergistic effects of phytochemicals and chemotherapeutics have been demonstrated due to their low intrinsic toxicity with a prominent effect on cancer cells acceptable for modern drug development (Choudhari et al., 2020; Pezzani et al., 2019). Wang et al. (2012) reported that the mechanism, formulation, drug-drug interaction, pharmacodynamic, metabolism, polymorphism, dosage regimen, and potential draggability of natural plant and vegetable phytochemicals were helpful in the curing process of various cancer like:

- Apigenin from parsley (cytotoxic activities – cell line from colon cancer (HCT 116), cell lines from breast cancer (MCF 7).
- Curcumin from Turmeric (colon and breast cancer, brain tumor, lung metastases).
- Crocetin from Saffron (inhibit nucleic acid synthesis, enhancement of the antioxidative system, induce apoptosis and obstructive growth factor signaling pathways).

- Epigallocatechin Gallate from Green Tea (Bcl-xl (Anti-apoptotic protein), hepatocyte growth factor inhibition and help in the treatment of various cancers).
- Genistein from lupine, soybeans, fava beans and coffee (Blocking uncontrolled cell growth, and inhibit tyrosine kinase).
- Vitamin E from plant oil (anti-tumor, anti-proliferative, constraining COX-2, and pro-apoptotic effects), etc.

Bruno et al. (2011) reported the active participation of apigenin on leptin-related pathways and found the initiation of cell apoptosis in the lung cell line (adenocarcinoma). Chen & Liu (2018) showed foremost phytochemicals and its effect on cell signaling transduction regulation counter to carcinogenesis through Akt, Nrf2, p53, CYP450, MAPK, 24 JAK/STAT, β-catenin/Wnt, NF-κB, and miRNAs related with cancer. According to the American Cancer Society cancer related to colorectal has a greater chance of occurrence in patients with inflammatory bowel disease (IBD). Mainly, the nuclear factor erythroid 2-related (Nrf2/KEAP1) pathway is responsible for oxidative stress, inflammation protection, and apoptosis in cancerous cells. Lee et al. (2021) reported that the phytochemicals have a significant anticancer effect on colorectal cancer through nrf2/KEAP1 signaling pathway (Table 6.1).

6.2.2 POLYPHENOLS

Polyphenols represent thousands of chemical compounds in various plant and vegetables (Lason & Le Skiewicz, 2013). Polyphenols as the chief class of secondary plant metabolites perform a chief role in the plant mechanism for defending against pathogens and UV rays (Hussain et al., 2022) besides inducing anticancer, anti-inflammatory, neuroprotective, and anti-SARS-CoV-2 activity in human (Fakhri et al., 2022). Two basic categories of polyphenols are flavonoids and nonflavonoids, that are further split into many subclasses by phenol units in structure, linkage, and presence of lignans, stilbenes, curcuminoids, phenolic acids, and flavonoids group (Singla et al., 2019).

The lignans are a small molecular weight diphenolic organic compound of the class phytoestrogen found mostly in fiber-rich plants and vegetables (Yoder et al., 2015; Lampe et al., 2006). Avila-Galvez et al. (2021) reported that the curcuminoids and other dietary polyphenols obtained from resveratrol (RES), turmeric extract, flaxseed extract, and red clover arrested the cell cycle by affecting the p53-wild type MCF-7 cells as well increased the

TABLE 6.1 List of Phytochemicals and Their Effect in *In-Vitro* and *In-Vivo* Studies

SL. No.	Phytochemical	Active Metabolites	Dietary Sources	Treatments *In Vitro* and *In Vivo*	Result/Benefits	References
1.	Lutein	3′-dehydro-lutein	Kale, spinach, parsley, and peas.	Outbred rats. Epithelial cells from the human colon. Adenocarcinoma cells from the human colon.	Reduction in the danger of colorectal neoplasms and colorectal cancer in women. Inhibit carcinoma cell growth rate. Decrease cellular proliferation and reduce the concentration of beta-catenin. Enhance apoptotic pathway, repairmen of damaged DNA and induces humoral immune response and perform quenching and scavenging of oxygen radicals.	Akuffo et al. (2015); Femia et al. (2008)
2.	Lycopene	Apo-10′-lycopene	Guava, grapefruit, tomato, papaya, and watermelon.	Colitis rat models. Outbred rats. NSIc rats/Fischer 344 rats cell HT-29.	Suppresses the carcinogenesis progression through DNA synthesis inhibition. Inhibits metastasis, angiogenesis, and cell invasion. Downregulates MMP-2, MMP-7, MMP-9, NF-kb, and AKT. Enzymes and pro-inflammatory mediators' reduction including TNF-a and COX-2. Cell cycle inhibitor protein (p21) upregulation and protein (PCNA and cyclin D1) expression suppression. Stimulate enzymatic activities of GSTs, GR, and GPX. Enhance apoptotic pathway and activation of MAPK signaling gene with the reduction of oxidative stress through scavenging oxygen-free radicals.	Lin et al. (2011); Boehm et al. (2016)

TABLE 6.1 *(Continued)*

SL. No.	Phytochemical	Active Metabolites	Dietary Sources	Treatments *In Vitro* and *In Vivo*	Result/Benefits	References
3.	β-Carotene	Falcarindiol 6-methoxymellein	Carrot	Rats treated with azoxymethane. CCD-33Co cells. HCT 116 cells, and HT-29 cells.	Neoplastic tumor formation inhibition. Inhibits the pleiotropic cytokines, and pathways including NF-κB, ERK/MAPK, and AKT/P13K. p53-dependent apoptosis pathway enhancement.	Shebaby et al. (2015); Pan et al. (2008)
4.	Quercetin	Isorhamnetin conjugates, methoxylated forms, and glucuronides of quercetin such as 3-(4-hydroxyphenyl)-propionic acid.	Cranberry, apple, onion, Parsley, tea, sage, etc.	Apc mice. Weanling rats (male). Cell lines (HCT116). HT-29 cell lines. Cancer cell line encyclopedia (CCLE)	Reduces small intestine tumor development, burden, multiplicity, and volume. Decreases the expression of inflammatory-related cytokine (TNF-α, IFN-γ). Signaling pathways such as AKT/PI3K and COX-2 activation inhibition. Inhibit MMP-2, MMP-9, and VEGF expression, proliferation, and the growth rate of the cancer cell, and induces apoptosis.	Pappas et al. (2009); Duthie et al. (2006)
5.	Ellagic acid	Peonidin-3-galactoside	Bilberry	Female Balb/c mice	Inflammation reduction and decreases development of tumor. Decreases cellular proliferation, and colonic ACFs formation. Topoisomerase I and II activities suppression. Inhibit NF-κB and induces cellular apoptosis.	Hodges et al. (2015); Lala et al. (2006)

TABLE 6.1 *(Continued)*

SL. No.	Phytochemical	Active Metabolites	Dietary Sources	Treatments *In Vitro* and *In Vivo*	Result/Benefits	References
6.	Allicin	Allyl methyl sulfide, sulfoxide, and sulfone (AMS, AMSO, $AMSO_2$).	Garlic	Cell line (HCT-116) of Mice-Xenograft nude. Cell lines of human gastric cancer (MGC-803, BGC-823, and SGC-7901).	Glycogen synthase kinase 3 (GSK-3) inhibition, stop NF-κB localization. Decreases cellular proliferation and growth of tumor cells. Apoptosis induction in the cancer cells (colon). Modulate EGFR (epithelial growth factor receptor). Nrf2 transcriptor activation. Cleaved of caspase 3 and increase the expression of p38.	Yi & Su (2013); Roy et al. (2008); Zhang et al. (2015)
7.	Allyl propyl disulfide	Quercetin 4-glucoside and 3,4-diglucoside	Onion	Cell lines (Caco-2 and SW620).	Reduces cellular proliferation, cancer cell migration, and tumor growth in the cancer cells (colorectal) as well as cellular apoptosis induction via G2/M phase arrest in the cell cycle.	Murayyan et al. (2017); He et al. (2014)
8.	Asparagusic acid	Asparagus polysaccharide dimethyl sulfide.	Asparagus	Cell lines (HCT-116 and Caco-2).	The cytotoxic effect is more than 5-FU against human colon cancer cells. Reduces cellular proliferation. Target Rho GTPase signaling pathway and Inhibit cell motility and cellular growth. Intrinsic apoptosis induction via toll-like receptor 4 with the BAX and Caspase 9 expressions enhancement.	Hamdi et al. (2017); Jaramillo-Carmona et al. (2018)

TABLE 6.1 *(Continued)*

SL. No.	Phytochemical	Active Metabolites	Dietary Sources	Treatments *In Vitro* and *In Vivo*	Result/Benefits	References
9.	Sulforaphane	Thiocyanates, isothiocyanates, epithionitrile nitrile	Broccoli, Brussels sprout, cabbage, and cauliflower	Squamous cell carcinoma	Adenomatous polyps risk reduction. Modulate miRNA and prevents cancer (colorectal). Apoptosis, cellular arrest, and detoxification of enzyme induction. Induce Nrf2 and show Cytoprotective effects.	Keum et al., 2006; Kim et al. (2016)
10	Curcumin	Curcumin glucuronide, Curcumin sulfate	Ginger	Cancer stem-like cells (CSC) Malignant cancer cell line (mesothelioma) including MM-F1, MM-B1, and H Meso-1.	Decreases tumor growth by $PPAR_{\Upsilon}$ pathway suppression. Generation of cellular apoptosis and stop proliferation. Upregulate expression (Caspase-3, cytochrome C, and BAX). Enhance the expression of p53, Bax, and bcl-2 ratio, caspase-9 activation, and PARP-1 cleavage.	Soltani et al. (2018); Ismail et al. (2019); Masuelli et al. (2017)
11.	p-Couramic acid	2-aminoadipate, N-methylpipecolate, vanillate, and piperidine.	Navy beans	FVB/N mice	Reduces oxidative stress. Induces anti-tumor activity and cellular apoptosis in the case of cancer (colorectal). Increase the phytochemicals, lipids, and amino acids amount in the stool.	He et al. (2011); Bennink (2002)
12.	Ferulic acid	α-ketoglutarate ϒ and β-tocopherol, ϒ-tocotrienol, tryptophan.	Rice bran	Mice APC. Caco-2 and HAT 29 cells.	Hinders cell cycle, cellular proliferation, and tumor growth. Induces cellular apoptosis, lipid peroxidation, and expression of nitric oxide synthase as well as activates the caspase-3 and NF-κB pathways. Increases the SCFAs production and decreases the β-catenin and COX-2 in the cancer cells.	Law et al. (2017); Ryan et al. (2011)

TABLE 6.1 *(Continued)*

SL. No.	Phytochemical	Active Metabolites	Dietary Sources	Treatments *In Vitro* and *In Vivo*	Result/Benefits	References
13.	Resveratrol	Trans derivatives of resveratrol including 4-O-glucuronide and sulfate, 3-O-glucuronide, and sulfate, dihydro resveratrol-glucuronide, and sulfate.	Wine, peanuts, and selected teas, grapes, cranberry, mulberry, pistachio, and peanut.	Carcinoma cell line from human colon SW480, and 620, B103, and HCT116.	Procaspases activation (3, 8, and FADD), resistance protein (breast cancer) activation (ABCG2), associated with drug resistance protein-2 (MRP2/ABCC2), and expressed P-glycoprotein transporter (MDR1/ABCB1).	Delmas et al. (2003); Rahman et al. (2012); Springer & Moco (2019); Teka et al. (2022)
14.	Diterpenoid	Eriocalyxin B (EriB)	Chinese herb *Isodon eriocalyx*	Pancreatic human cancer cells PANC-1, human pancreatic adenocarcinoma cell line (SW, 1990) CAPAN-1 and 2.	Activation of caspase 8, and 9, negatively regulate caspases 3,7 and PARP. Modulate various apoptotic-related factors, transcription factors, and enzymes like cyclooxygenase 2, matrix metalloproteinase 2, cytokines (ADP-ribose) polymerase, and protein kinases. Cytotoxicity potency against adenocarcinoma pancreatic cell lines. Activate the p53 pathway and custody of the G2/M phase of the cell cycle. Suppressed proteins and mRNAs of CDK1, CDK2, Cyclin A, and Cyclin B1.	Li et al. (2012); Mu et al. (2020); Zhou et al. (2016)

TABLE 6.1 *(Continued)*

SL. No.	Phytochemical	Active Metabolites	Dietary Sources	Treatments *In Vitro* and *In Vivo*	Result/Benefits	References
15.	β-Elemene	Monosubstituted amine derivative of β-elemene.	*Curcuma wenyujin*	Breast cancer cell lines MBA-MD-231, Bcap37 in humans. Induces apoptosis by modulating CDKs to arrest the cell cycle. Targets signaling pathways (Wnt/β-catenin, MAPK/ERK, notch, etc.), and regulates transcription factors (STAT3). Regulates main regulator molecules of tumor angiogenesis and metastasis.	Conservation of LC3-I-LC3-II. Adjust factors of inflammation like TNF-α, IL-6/10, IFN, and TGF-β. Downregulates the levels of iNOS and NO.	Guan et al. (2014); Xie et al. (2020); Zhai et al. (2018)
16.	Polycyclic polyprenylated acylphloroglucinols	Oblongifolin C	*Guttiferae* and related plant families such as *Garcinia yunnanensis* Hu.	Carcinoma cell lines of human breast (HeLa or MEF). Xenograft model (mouse).	Initiation of CASP3 and cleaved PARP. Act as a novel autophagic flux inhibitor. Autophagosomes increment and lessened degradation of SQSTM1/p62. Blocking the autophagosome-lysosome fusion and hinders lysosomal proteolytic activity by altering lysosomal acidification and downregulating the expression of lysosomal cathepsins.	Lao et al. (2014); Han et al. (2020)

TABLE 6.1 *(Continued)*

SL. No.	Phytochemical	Active Metabolites	Dietary Sources	Treatments *In Vitro* and *In Vivo*	Result/Benefits	References
17.	Apigenin	Genkwanin, acacetin	Parsley, celery, onions, chamomile, maize, rice, tea, wheat sprouts, peppers, plant-derived beverages, and some grasses.	Cancer cells (colorectal) SW480, HT-29, HCT-116 and LoVo.	Trigger NAG-1, p53, and p21. Modulate the expression of cytokines and other inflammatory molecules including cycloxygenase, lipoxygenase, heme oxygenase 1, inhibitor of nitric oxide synthase, prostaglandins, interleukins, tumor necrotic factor-alpha, interferon-gamma, X-linked inhibitor, and cellular FLICE like inhibitory protein.	Zhong et al. (2010); Ali et al. (2017)
18.	Anthocyanins	Phloroglucinolaldehyde, cyanidin-3-O-glucoside, hippuric acid.	Black rice (AEBR)	Breast cancer cell lines MCF-MDA-MB-231 and MDA-MB-453.	Reserve VEGF expression, MMP-9, MMP-2 suppression, and uPA expression.	Hui et al. (2010)
19.	Aspalathin	Rooibos O-glucosyl dihydrochalcone, phloridzin, and flavanoid.	Rooibos (South African legume).	Ovarian cancer cell Caov-3, resistant prostate cancer cells.	Reserve of Dox, reduction in p53 expression, and encourage AMPK and FOXO1. Inhibition of YAP signaling.	Lin et al. (2017)
20.	Baicalein	Oroxylin A	Tea, citrus fruits, dark chocolate, and herbs.	Cell lines (human hepatocellular carcinoma) SMMC-772 and Bel-7402.	ER stress increment and downregulate Bcl 2.	Wang et al. (2014)
21.	Berberine	Berberrubine, thalifendine, demethyleneberberine, and jatrorrhizine	Phellodendron, *Coptis chinensis*, *Hydrastis canadensis* (goldenseal), *Berberis aquifolium* (Oregon grape), *Berberis aristata* (Tree Turmeric), *Berberis vulgaris* (Barberry), and *Arcangelisia flava*	Cell lines (human glioma) U251 and U87 GBM.	AMPK/mTOR/ULK1 inhibition.	Peng et al. (2008); Wang et al. (2010); Yu et al. (2014); Guamán Ortiz et al. (2015); Wang et al. (2016a)

TABLE 6.1 *(Continued)*

SL. No.	Phytochemical	Active Metabolites	Dietary Sources	Treatments *In Vitro* and *In Vivo*	Result/Benefits	References
22.	Capsaicin	Capsaicinoids	Chili peppers	Carcinoma cell line (human nasopharyngeal) NPC-TW01.	PI3K/Akt/mTOR downstream and caspase-3 activity increment.	Lin et al. (2017b)
23.	Pentacyclic triterpenoid.	Celastrol	Celastraceae family, *Tripterygium wilfordii*	Cancer cell lines (Human prostate) LNCaP, 22Rv1, DU145 and PC-3.	miR-101 upstream	Guo et al. (2015)
24.	Cordyceps	Cordycepin	Cordyceps species	Cancer cell (human brain) SH-SY5Y and U-251.	Rise Bax/bcl-2 ratio, p53 expression, caspase 9 activation and cleavage of PARP-1.	Masuelli et al. (2017)
25.	Polyphenolic flavonoid	Epigallocatechin gallate (EGCG)	Tea	Cell line from vascular endothelial U-937.	Decrease TNF-α, and prevent VCAM1, LC3A, LC3B..	Yamagata et al. (2015)
26.	Quinazoline carboline alkaloid	Evodiamine	*Evodia rutaecarpa*	Cancer cell line (Gastric) SGC-7901.	Stimulates beclin-2, downregulates Bcl-2, Bax	Rasul et al. (2012); Hong et al. (2014)
27.	Fisetin	3,3′,4′,7-tetrahydroxy-flavone	Strawberry, apple, persimmon, grape, onion, and cucumber	Cancer cell lines (Prostate) PC3 and DU145	Suppressed Mtor and prevent Akt, stimulate AMPK	Suh et al. (2010)
28.	Isoflavones	Genistein	Tofu, fava beans, soybeans, kudzu, and lupin.	Cancer cell line (Ovarian) A2780.	Decreases Akt/mTOR phosphorylation.	Gossner et al. (2007)
29.	Beta-hydroxy ketone	Gingerol	Ginger	Cancer cell lines (human colon) SW480 and HCT116.	Prevention of JNK (C-Jun N-terminal kinase), ERK1-2, and P38 MAPK.	Shukla & Singh (2007); Baliga et al. (2011); Radhakrishnan et al. (2014)

TABLE 6.1 *(Continued)*

SL. No.	Phytochemical	Active Metabolites	Dietary Sources	Treatments *In Vitro* and *In Vivo*	Result/Benefits	References
30.	Tetracyclic triterpenoid	Ginsenoside F2	*Panax ginseng*	Cancer cell lines (Breast) MCF-7.	Raised up Atg-7, and cleaved PARP.	Mai et al. (2012); Tang & Eisenbrand (1992)
31.	Polyphenol	Hispolon	*Phellinus linteus, Phellinus igniarius, Inonotus hispidus, Phellinus merrillii, Phellinus lonicerinus*	Cancer cell lines (cervical) Hela and SiHa.	Lysosomal protease cathepsin S (CTSS) downregulated.	Chen et al. (2012); Sarfraz et al. (2020)
32.	Soy isoflavone daidzein.	3′-hydroxydaidzein	–	Melanoma cell line (Mouse) B16F1.	Decrease the α-MSH.	Kim et al. (2013)
33.	Toxicarioside O	–	*Antiaris toxicaria*	Cancer cell lines (human colorectal) HCT116 and SW480.	Prevention of the Akt/mTOR.	Huang et al. (2017)
34.	Polyacetylenic oxylipin	Falcarindiol	Carrot roots	Cancer cell lines (Human breast) MDA-MB-231, MDA-MB-468 and Her2.	FAD induces GRP78 expression.	Minto & Blacklock (2008); Jin et al. (2012); Lu et al. (2017)
35.	Pentacyclic triterpenoid	Oleanolic acid	Olives (*Olea europaea L.*)	Cancer cell line (human pancreatic) Panc-28.	Modify JNK and m-TOR pathway.	Pollier & Goossens (2012); Liu et al. (2014)
36.	Lignan	Honokiol	Magnolia genus – leaves, bark, and seed cones.	Cell lines (Human glioblastoma) LN229, GBM8401 and U373.	Decrease p-PI3K, p-Akt, and Ki67.	Cheng et al. (2016)

TABLE 6.1 *(Continued)*

SL. No.	Phytochemical	Active Metabolites	Dietary Sources	Treatments *In Vitro* and *In Vivo*	Result/Benefits	References
37.	Lignan	Magnolol	Magnolia bark	Cell lines (human glioblastoma) LN229, GBM8401 and U373.	Decrease of p-PI3K, p-Akt, and Ki67.	Cheng et al. (2016)
38.	Triterpenoid	Alisol B	*Rhizoma alismatis*	Cancer cell lines (Breast) MCF-7, SK-BR-3, and HeLa.	Initiation of AMPK/m-TOR/Ca^{2+}.	Law et al. (2010)

apoptosis and antiproliferation rate, consequently, decreased the chance of cancer (breast) in women. Liu et al. (2021) explained the lignans ingestion and its metabolite enterolactone concentration connection that effected on postmenopausal women and breast cancer survival. The plant-derived lignans, bioactive compounds (secoisolariciresinol, sesamin diglucoside, matairesinol, and enterodiol) are participating in the conversion of lignans (enterolactone and enterodiol) during the metabolic process through gut microflora show anticancer activity via antioxidative, anti-estrogenic, and pro-apoptotic mechanisms (Milder et al., 2005). Siriwarin & Weerapreeyakul (2016); and Khamphio et al. (2016) reported that the active compound in sesame, sesamol concentration of approximately (0.05 to 10 mM) within 48 hr. possibly induced the apoptosis in (SKLU-1) cancer cells (lungs) with a measured IC_{50} value of 2.7 ± 0.0014 mM and induced the arrest in the cell cycle at G0, G1-phase of cells having cancer (colorectal-HCT116) through an IC_{50} range of 2.58 ± 0.04 mM approximately.

Stilbenes are the natural polyphenolic organic compounds that originate from various plants including families like Fabaceae, Poaceae, Vitaceae, and Leguminosae. They are also derived from the *Yucca periculosas* (izote) bark and work as a possible chemo-preventive metabolite in tumor management (Ren et al., 2021; Jang et al., 1997). Stilbenes (resveratrol, piceatannol, and pterostilbene) have antimicrobial properties that are already utilized by agricultural industries and farmers against crop harming animals (bacteria, fungi, herbivores, etc.). Recently, researchers also noticed the nutraceutical (antitumor, antioxidant, neuroprotective, and cardioprotective) (Keservani et al., 2015a, b) effect of stilbenes by stopping the various phases in the colon and lung cancer cells HT-29, A-549, respectively (Hazafa et al., 2021). RES act as a lipophilic drug that is concentrated in lipophilic compartments and sensitive to obesity-related cancers (Ren et al., 2021).

Curcumin is an important phytochemical with various activities like anti-inflammatory, neuroprotective, and anti-SARS-CoV-2 effects with beneficial effects on cancer patients (Eskiler et al., 2019). Curcumin decreases the propagation of liver cancer cells through its antioxidant activity by enhancing apoptosis and signaling activations (caspase-8, HSP70, caspase-3, and MyD88, TLR4) (Li et al., 2014; Ren et al., 2018). Furthermore, curcumin stops the growth of cancer in the liver by diminishing the inflammatory factors (cyclooxygenase-2 (COX-2), prostaglandin E2 (PGE2), Interleukin-6, and 1b), inhibiting signaling pathways (TLR4, NF-kB), and colony-stimulating factors (granulocyte and granulocyte macrophage) reduction (Tian et al., 2021). Moreover, curcumin interferes with the pathways (EGFR, TLR4, and

MyD88), consequently increasing the cell cycle arrest in G2 and M phase and start declining the proliferation rate in other small lung cancer cells like NSCLC (non-small cell lung cancer cells) (Zhang et al., 2019). In addition, curcumin treatment decreased the capability of breast cancer cells through interferon-a/b inhibition and reduction in the TLR4 and IRF-3 expression (Eskiler et al., 2019). Quercetin a plant flavonoid is commonly dispersed in many plants and vegetables and synergistically displays beneficial biologically active properties with curcumin to encourage apoptosis in leukemia cells (K562) by disturbing pathways such as p53, transforming growth factor-a, and NF-κB (Yang et al., 2020). Quercetin obstructs the induction of NF-κB, p53, and arrest the cell cycle antagonistic towards the cancerous HeLa cervical cell line (Priyadarsini et al., 2010). Moreover, quercetin shows antitumor effects via programmed cell death and reduces the regulation of IL-6, STAT-3, and NF-kB in various cancers including breast (MCF-7) (Kas et al., 2010), lung (A549 and H460) (Youn et al., 2013; Mukherjee et al., 2015), prostate (PC3 and LNCaP) (Erdogan et al., 2018), and oral cancer cells (Zhang et al., 2017). Polyphenolic structure xanthohumol also displays anticancer properties by angiogenesis suppression, creation of inflammatory mediators, and breast cancer cell proliferation (Monteiro et al., 2008). In addition to this, xanthohumol reduces the CXCR4 expression and stops the invasion of cancerous cells (Wang et al., 2012). Furthermore, Fakhri et al. (2022) also stated that curcumin, apigenin, quercetin, and RES performed a dynamic role in the anticipation and treatment of cancer by interfering with the interconnected pathways such as TLR, NF-kB, NLRP, and help to fight against chemoresistance. George et al. (2021) reported that the immunomodulatory effect of polyphenolic compounds identifies and abolishes cancer cells by antiangiogenic activity as well as helps to find new therapeutic targets for enhancing chemotherapy efficacies via various pathways.

Largely distributed phenolic acid in plants and vegetables performs an important role in numerous metabolic functions like growth, reproduction, as well as protection of damaged plants from pathogens as well as performs an anticancer effect in the human body after consumption directly or through a drug. The Phenolic acids are classified according to phenolic cores in structure as hydroxybenzoic acids (ellagic acid, gallic acid, dihydroxybenzoic acid, etc.) and hydroxy-cinnamic acids (ferulic acid, prenylpropanoids, Caffeic acid, Rosmarinic acid, para-coumaric, sinapic acid, and 3,4-dimethoxycinnamic acid, etc.) (Isabel et al., 2009). Cateni et al. (2021) reported that phenolic acids are the main phenolic compound found in mushrooms in which hydroxybenzoic acid derivatives are bounded in the complex structures as hydrolysable tannins, lignins, organic acids, and sugars, whereas hydroxycinnamic acid derivatives bound with the cell wall components of mushrooms including lignins, cellulose, and linked by the organic acids such as quinic or tartaric acids through ester bonds. Furthermore, Kumar et al. (2018) evaluated the extracts obtained from different species of basella (alba and Rubra) for anti-proliferative and anti-angiogenic activity, flavones Hydroxy-benzoic, and hydroxycinnamic acids are identified in the extract of *B. alba* species is higher as compare to Rubra species with significant antioxidant activity and anti-proliferative activity (37.95–84.86%) as well as cytotoxicity on epidermoid carcinoma, hepatocellular carcinoma (HCC), and osteosarcoma cells show antiproliferative activity contrary to all the cell lines. In addition to this gallic acid action in the G2/M phase increases the kinase inhibitor proteins p27Kip1 and p21Cip1 by suppressing Skp2-mediated polyubiquitination, leading to the growth reduction in the breast cancer cells of humans (Hussain et all., 2022). Naturally, dihydroxy-cinnamic acid is available in vegetables, fruits, grains, and coffee (Kesharwani et al., 2016) it also has significant antioxidative, anti-inflammatory, and anti-proliferative in cells (Kanimozhi & Prasad, 2015). In addition to this, when verified against T47D a breast cancer line proliferation, it displays pro-oxidant properties and performs as a potential inhibitor (IC_{50} – 2.17 nM) as well as reducing the chances of DNA damage and inducing apoptotic cancer cell death (Kampa et al., 2004).

Flavonoids (Keservani & Sharma, 2014) are a complex and wide-ranging family of benzo-c-pyrone derivatives with thousands of identified compounds found in a free state and in their respective glycosides. They are also documented as pharmacological agents because of their properties against cancer, inflammation, mutagenic, and oxidative degradative properties (Kesarkar et al., 2009). Their activity displays association with the hydroxylation, conjugation, substitution, polymerization, and chemical nature of the compound (Kumar & Pandey, 2013). They are further divided into diverse subclasses including Flavanones, Anthocyanidins, Flavanols, Flavonols, isoflavones, and Flavones and the development of these compounds follows the phenylpropanoids pathway (Chinnabattigalla et al., 2021).

Khan et al. (2021) reported that flavones are participating in the metabolic regulations and control cell cycle progression, oxidative stress, angiogenesis, and metastasis through some molecular pathways that finally hindered the progression of various neurological diseases and cancers. Furthermore, Tiwari et al. (2021) reported that effective plant-based flavonoids such as chrysin, kaempferol, quercetin, luteolin, ECGC, apigenin, catechin, etc., use different flavonoid mediators including apigenin, quercetin, luteolin, fisetin, chrysin, epicatechin, EGCG, and Silibinin in apoptotic pathways for effectively arrest the cancer cell growth and tumors via intrinsic and extrinsic pathways. Zhang et al. (2021) conveyed that the protective autophagy induction is possibly attenuated by the anti-cancer properties of flavonoids, and it also enhances different cellular effects by the modulation of multifunctional autophagy. In addition to this, the flavonoids with autophagy reactivation properties might be more appropriate for the inhibiting mitigation, and initiation of secondary cancers developed due to the chemo and radiotherapy. Rajendran et al. (2021) conveyed the anticancer properties of secondary metabolite flavonoid myricetin with silicon molecular docking. Observations show that 73 μg/ml doses of myricetin were more effective against the growth of cancer cells by altering the capacity of the mitochondrial membrane in A549 cells of lung cancer and intensifying the P53 expression and relegating the expression of EGFR in cells. It promotes the aggregation of cells in the sub-G1 phase and decreases the cell fractions enter in the S and succeeding phase which is the indicator of apoptosis in cells. Plant polyphenols and its derivatives such as anthocyanidin, procyanidin, isoflavone, flavone, flavonol, flavanone, and lignans have anticancer properties with beneficial effects on breast cancer therapy. Polyphenols obtained from natural sources perform various anticancer properties via numerous mechanisms like enzymatic movement, proliferation inhibition, and apoptosis. The action of polyphenols depends upon the type of cancer, doses, cell lines, and application of modified structure of flavonoids that could increase the activeness of shielding in DNA as well as the efficacy of flavonoids (Sindhu et al., 2021).

Anthocyanidins are colored pigments soluble in water found in a diverse variety of plants, vegetables, fruits, and flowers. Most commonly they are classified according to the flavylium-B ring substituents which normally comprise cyanidin, procyanidin, pelargonidin, delphinidin, petunidin, and malvidin (Hussain et al., 2022). It possesses a significant amount of nutraceutical performances like antioxidant, anti-inflammatory, and antimicrobial (Keservani & Sharma, 2014; Keservani et al., 2016a, b; Khoo et al., 2017). Zhang et al. (2005) reported that Malvidin is the most efficient anthocyanidin to inhibit MCF-7 cell growth at approximately 74.7% through 200 mg/mL and shows an IC_{50} value of 32.2 Mm. Procyanidins (condensed flavanan-3-ols) and derivatives which are formed from epicatechins and (þ) catechin show health benefits like prevention of coronary thrombosis and heart diseases, inhibiting cell lines from breast cancer including MDA-MB 436, 231, 468, SKBR-3, MCF-7, and shows selective activity against cancer cell lines (human breast) (Hammerstone et al., 2000; Ramljak et al., 2005).

6.2.3 TERPENES AND TERPENOIDS

Terpenes are included in a diverse and largest class of plant secondary metabolites comprises of 5-carbons isoprene units (Tidgewell et al., 2010). Two main key players specifically limonene (colorless) and carnosine that is an acid are familiar as anticancer agents for breast carcinoma. Several Labiaceae plants, particularly rosemary leaves contain phenolic diterpene carnosic acid. It also shows effects on the mammary tumor, and the combination of curcumin will be helpful in the treatment of cancer form in the breast (Yesil-Celiktas et al., 2010; Einbond et al., 2012). The carnosol (a carnosic acid metabolite) shows antioxidant, and anticancer activity with IC_{50} values of 83 mM, and 25 mM at 24 hr. and 48 hr., respectively by targeting the G-2 phase in the cell cycle, induced apoptosis (ROS-dependent), and autophagy (BECN-1-independent) (Al-Dhaheri et al., 2014). Global sale of pharmaceuticals drugs that depend on terpene (antimalarial drug Artemisinin, and anticancer drugs) was nearly US $12 billion in 2002, recently global terpenes market was valued nearly at USD 610 million in 2020 beside it is expected to spread approximately USD 940.7 MM by the end of 2027 if growing at the rate of 6.4% CAGR (Shoker, 2020).

Terpenoids are widely distributed biologically active organic compounds derived from plant sources. They can be classified on the basis of cyclic structures as hemiterpenoids, monoterpenoids, tetraterpenoids, diterpenoids, triterpenoids, and polyterpenoids, etc. (Yang et al., 2010). The unique structure of terpenoids is accountable for various nutraceutical properties such as antioxidative effect, anti-inflammation activities, proliferation prevention, and apoptosis induction in cancer cells through complex molecular mechanisms (Palop et al., 2010; Keservani et al., 2017, 2020, 2020a, b,). Terpenoids also perform NF-kB pathway inhibition, which is helpful in the reduction of some gene expressions responsible for the production of compounds that are pro-inflammatory and able to induce processes like angiogenesis, cell proliferation, cellular survival, metastasis, and invasion of tumor cells (Palop et al., 2010; Yang et al., 2010). Bhattacharya et al. (2021) reported that terpenoids may be involved in the Bax protein (pro-apoptotic) upregulation and Bcl-2 protein (anti-apoptotic) downregulation, which eventually support the release and stimulation of cytochrome c, caspase, respectively in the cellular apoptosis process. Carveol a monoterpenoid extracted from celery shows anticancer properties counter to mammary cancer cell line human adenocarcinoma cells (MCF-7, IC_{50} – 1.2 mM) and carcinoma of the human mammary cells (MDA-MB-231) with IC 50–1.0 mM (Patel

& Thakkar, 2014). In addition to this thymol also possesses the cytotoxic capacity accompanying the potential to affect malignancies apoptotic gene expression in the various cells. It also shows the activity against the cancer cell line MCF-7 (IC 50–54 mg/ml (48 hr.), and 62 mg/ml (72 hr.)) with the induction of apoptotic conditions by disturbing the expression of p21 and 53 genes (Seresht et al., 2019). The Ginkgolides are counted in Diterpenoids comprising 20 carbon atom skeletons abundantly present in fruits of the ginkgo Biloba tree. It is found that ginkgolides change the appearance and movement of PBR (peripheral type benzodiazepine receptor) which is associated with the aggressive phenotype proliferation and expression in cancer cells (breast). It also inhibits the activity of exposed MDA-MB-231 without affecting the proliferation rate of non-aggressive cell lines (MCF-7) as well as can be helpful in tumor growth treatments (Papadopoulos et al., 2000). Chan (2007) reported that the treatment with ginkgolide induces apoptosis in cancer cells MCF-7 due to reactive oxygen species (ROS) generation, and intercellular elevation of nitric oxide (NO) content.

Triterpenoids are complex cyclic molecules with six isoprene units of carbon skeleton and bio-synthetically obtained from squalene (C_{10} hydrocarbon of acyclic). Triterpenoids are basically separated by approximately similar procedures, based on chromatography techniques (TLC, GLC, column), and identified through various experiments such as FT-IR, GLC-MS, NMR, melting point, and optical rotation (Cateni et al., 2021). Jesus et al. (2015) reported that the pentacyclic triterpenoids compound oleanolic acid and ursolic acid (isomer) are present in plant Kingdome species such as Olive, wield rose, honey mesquite, and *Wrightia tomentosa* with nutraceutical activities including anticancer, anti-inflammatory, hepatoprotective, and antioxidative properties. In contrast to this, oleanolic acid obtained from tomentosa shows repressive activity to counter the MDA-MB-231 and MCF-7 mammary cancer cell lines with IC_{50} (7.5 and 7 mg/mL) through the induction of the apoptotic pathway (Chakravarti et al., 2012). Kassi et al. (2009) also conveyed the availability of ursolic acid in plants such as basil, blueberries, cranberries, rosemary, etc., and their ability to encourage apoptosis, glucocorticoid receptors modulation as well as the activation of MCF-7 cells protein with IC_{50} (0.4 mM). Similarly, the Cucurbitaceae family contains triterpenoids known as cucurbitacins and has the ability to suppress the propagation of breast carcinoma cell line by inhibiting or through the induction of phosphor-STATE3 and apoptotic condition, respectively (Chen et al., 2005; Blaskovich et al., 2003). Wakimoto et al. (2008) reported that another triterpenoid from the Cucurbitaceae family known as Cucurbitacin

B shows induction of proliferation and anticancer effect against cell lines in cancer such as metastatic breast (231, 453), MCF-7, BT474, T47D, and ZR-75-1 due to the microtubules and F-actin disruption.

Lupeol is a triterpene having a pentacyclic structure that belongs to a family phytosterol (fagarsterol, clerodol, and lupenol) primarily obtained from green peppers, strawberries, cabbages, olives, mangos, and grapes. Balapure & Wadegaonkar (2014) reported the anti-tumor activity of lupeol to counter the cell line of breast cancer MCF-7 through the IC_{50} value of 50 Mm. Lupeol also arouses the effect in the RA4 DNA sequence (decoying effect) and increases gene expression levels in Era-negative mammary cancer cells (Lambertini et al., 2005). Moreover, the triterpene extract (betulinic acid) from birch trees shows a significant anticancer effect against cancer cells (breast), and ionic derivatives show higher cytotoxicity to counter MCF-7 (breast cancer cell line) (Cateni et al., 2021; Potze et al., 2014; Mertens-Talcott et al., 2013).

6.2.4 SULFUR-CONTAINING COMPOUNDS

Organosulfur is a subclass of sulfur-containing organic compounds obtained from innumerable animals and plant sources in the world. Basically, in plant sources, they mainly originated in Brassica and Eruca genera and allium species of vegetables. Goncharov et al. (2016) reported that those organic compounds consisting of sulfur are biologically active and help to constrain cancer (breast). Allicin is one of the sulfur-containing organic compounds from the Alliaceae family obtained from garlic (Block, 1985). Allicin shows a preventive effect in breast, colon, and uterine cancer cells accompanying many fibroblast lines via reducing the cellular glutathione (GSH) transiently level (Hirsch et al., 2000). Lee et al. (2015) reported that allicin inhibits the VCAM-1 protein expression developed by the TNF-alpha breast cancer cell line (MCF-7) in humans as well as decreases the migration and invasion rate of tumor cells. Naturally occurring phytochemicals have the capacity to block and converse the cancer process initiation. Broccoli and broccoli sprouts contain an isothiocyanate derivative name sulforaphane that possesses anti-cancer properties. Breast carcinogenesis is resulting due to regulation problems and difficulties in the normal mammary stem cell's resilience (Liu, Nguyen, & Colditz, 2015). Li et al. (2010) reported that sulforaphane is able to down-regulate the Wnt-β catenin pathway, which performs an important role in the development of resilience in the stem cells (breast cancer). Furthermore, Hussain et al. (2013) reported that sulforaphane cytotoxic potential is induced because of breast cancer cell apoptosis (time/concentration-dependent). Sulforaphane administration in a grouping with the anti-cancer drug "gemcitabine" proliferate MCF-7 breast cancer cells via time variant modulation in Bcl-2, and COX-2 expressions synergistically and induces the inhibitory effect evolution in MCF-7 cells. Consequently, the synergistic use of gemcitabine and sulforaphane boosts the effectiveness of gemcitabine as well as decreases the negative effects in normal cells. Moreover, the suppressing potential of sulforaphane in TNBC cells reduces the evolution of cancerous cells and retard the creation of tumor sphere (Castro et al., 2019). DADS (non-polar) are organosulfur compounds mostly taken out from garlic stems and additional plants of the family Alliaceae. Its non-polar structure maximizes the solubility in non-polar solvent and oils correspondingly accountable for the hard garlic aroma (Nakagawa et al., 2001). Several *in-vivo* and *in-vitro* models have been considered to evaluate the anticancer capacity of diallyl disulfide. In addition to this, Huang et al. (2015) examined the potential of DADS on TNBC cells line *in vitro* and found a reduction in viability, apoptotic elevation, and metastatic potential suppression. Furthermore, *in vivo* study also validated the anticancer outlooks of DADS in investigated xenograft mice (MDA-MB-231), and significantly reduces the physical size of tumor cells, together with the compact active b-catenin expression. Moreover, it was correspondingly reported that DADS pointedly prevents the expansion of cancer cells (human mammary) in estrogens receptors and is useful in breast cancer treatment (Huang et al., 2015). Linoleic acid stimulatory effects on cancer cells were mitigated by DADS although the harmonious upshot was detected when directed with cancer cell suppressor eicosapentaenoic acid (Nakagawa et al., 2001). S-allyl mercapto cysteine (SAMC) is also an organosulfur compound found in matured garlic and other allium species-related vegetables. It is observed that thioallyl significantly affects the hormones related to the various cancers cell and might work as a controlling factor in cancer treatment. In addition to this, Yi et al. (2019) observed that SAMC has efficiently deferred cell growth, and cell cycle arrest, made apoptotic circumstances in ER-positive and negative cell lines caused by the activation of caspase (3 and 9), downregulation of BCl-XL, and Bcl-2, and upregulation of p21, p53, and Bax.

6.2.5 BETALAINS

Betalains is a water-soluble heteroside compound that belongs to a yellow and red aromatic indole-derived pigment. It has been isolated from a diverse and heterogeneous order of flowering plants Caryophyllales and is also existing in bougainvillea and in a few higher fungi like Amanite (agarics species) (Baldassano et al., 2011). Betalains (tyrosine-based pigments) are closely related to the molecules of flavonoids, but they react destructively to the Erdmann test and are able to drift as anions even at very less pH – 2.4 (Piattelli, Minale, & Nicolaus, 1965). Mostly betalain pigments are divided on the basis of the color such as betaxanthins (yellow–orange) and betacyanin (reddish–violet) as well as used as a coloring agent. Betanin acts against oxidative stress and also reduces the evolution of numerous cancer cell lines including prostate, breast, colon, liver, and skin (Tesoriere et al., 2009; Lechner et al., 2010; Kapadia et al., 2013). In addition to this, weakly exhibit the expression of the epigenome-regulated (ER) gene in breast cancer cells (MCF-7) and encourage apoptosis in human myeloid leukemia cells (K562) (Madadi et al., 2020). Nowacki et al. (2015) reported that the betanin-enriched extracts reduce the progress of tumor cells which are p53-mutated and lack apoptosis induction.

6.2.6 CAROTENOIDS

Carotenoids are widely distributed yellow to red color pigments from the family of tetraterpenoids that are also created in the photosynthetic bacteria, algae, and plants (Tanaka, Shnimizu, & Moriwaki, 2012). Mainly, they are categorized on the basis of the oxygen molecules in their structure as carotenes (beta-carotene, lycopene) which are oxygen-lacking carotenoids, and oxygenated carotenoids xanthophylls (lutein, and zeaxanthin) (Hussain et al., 2022). It was demonstrated that lycopene decreases numerous types of cancers (lung, prostate, colon, stomach, and mammary) up to a high extent (IC_{50} – 1–2 Mm) (Levy et al., 1995; Giovannucci, 1999). Lutein belongs to a xanthophyll family of carotenoids and can be obtained from green leaves vegetables (kales, spinaches), and yellow carrots. Major carotenoids, like b-cryptoxanthin, b-carotene, and lycopene are already known for their anti-oxidant properties and are helpful in chemotherapy via quenchers of ROS, peroxyl radicals and making the enzymes (phase II) like GSH S-transferases, ligases (glutamate-cysteine), quinone oxidoreductase, and heme oxygenase-1 (HO-1) (Zare et al., 2021). Polyakov et al. (2001) reported that the carotenoids might propagate free radical induction-based reactions and serve as pro-oxidants. Furthermore, some other discoveries suggest that crocetin and b-carotene in heavy doses might excellently develop intracellular oxidative stresses by improving the generation of ROS in cancer cells; thus, in consequence, carotenoids can encourage cell cycle arrest and cancer cells apoptosis (Kim et al., 2014; Palozza et al., 2002, 2003). Moreover, the anti-angiogenic effects of b-carotene on male mice cells (mice-C57BL/6, cell-B16F-10) are helpful in cancer prevention and treatment (Guruvayoorappan & Kuttan, 2007; Sathasivam & Ki, 2018). The close relationship between inflammation and cancer creates high chances of cancer formation during chronic inflammation. Some studies established that phytochemicals such as astaxanthin, lycopene, and lutein show anti-inflammatory activity as well as put a significant effect on cancer cell inhibition (Gong et al., 2018; Rayburn, Ezell, & Zhang, 2009; Tanaka, Shnimizu, & Moriwaki, 2012).

6.2.7 CAPSAICINOIDS

Capsaicinoids are the phytochemicals obtained from chili peppers and are accountable for the pepper's spiciness (Tremblay, Arguin, & Panahi, 2016). Capsaicinoids generally comprise biologically active compounds such as dihydrocapsaicin, homocapsaicin, homodihydrocapsaicin, and nordihydrocapsaicin those poses anti-obesity, anticancer, anti-inflammatory, and antioxidant properties (Luo, Peng, & Li, 2011). Monsereenusorn et al. (1982) reported that capsaicin a form of capsaicinoid is the major principal component in chili pepper's pungency that acts as a vanilloid receptor agonist. It also stops hunger, decreases fullness, expenditure of energy, and oxidation of fat with the growth inhibitory effect against various cancers (Tremblay, Arguin, & Panahi, 2016). Thoennissen et al. (2010) demonstrated that capsaicin is related to cell-cycle custody and breast cancer cell apoptosis (ER-positive and ER-negative) as well as a standard moderator of some pathways (EGFR, and HER-2).

6.3 METABOLITES IN VEGETABLES

The diverse variety of primary (amino acids, carbs, fatty acids, and acids (organic)) and secondary metabolites (flavonoids, glucosinolates (GSs), sterols, phenolic acids, carotenoids, and alkaloids) are present in vegetables as well as participate in many functional mechanisms of plants and vegetables such as development and growth, photosynthesis, and respiration, hormone and synthesis of protein, color determination in greens, guard vegetation against herbivores and microbes, entice pollinators and seeds spreading animals, and behave like signaling molecules under stress conditions (Hounsome et al., 2008; Seiger, 1998; Crozier et al., 2008). Lignans are the di-phenolic complexes that help in the formation of an aquaphobic constituent (lignin) in the plant cell wall. Lignans perform numerous biological actions in humans (antioxidant, anti-estrogenic) and therefore decrease the chances of cancer growth (Adlercreutz & Mazur, 1997; Arts & Hollman, 2005). Furthermore, it was claimed that related to an entirely vegetarian and semi-vegetarian diet in many Asian nations, the western diet may modify hormone production, and metabolism at the cellular level and increases the chances of various cancers such as breast, colorectal, and prostate. Webb & McCullough (2005) reported that the higher lignan content in urinary excretion was associated with reduced cancer risk. GSs are sulfur-containing amino acid functional group including, sinigrin, progoitrin, and GSs (red and white cabbage, Brussels sprouts, and cauliflower), glucoraphanin (Grn), and glucoiberin (broccoli, radish (red), daikon), sinigrin, and gluconasturtiin (mustard, horseradish) (Fahey et al., 2001; Johnson, 2002). In addition to this, some breakdown components from GSs (glucotropaeolin, glucobrassin, and Grn) are connected with a certain type of cancer reduction (Das et al., 2000). Johnson (2002) reported the anticancer outcome of GSs by the enzyme activation, detoxification of carcinogens, enzyme inhibition, hormone metabolism modification (steroid), and guard against oxidative impairment in cells. Moreover, GSs obtained from cruciferous vegetables are converted into isothiocyanates through the action of plant myrosinase and gastrointestinal microflora and perform effective blocking of chemical carcinogenesis by the phase 2 proteins (GSH S-transferases, quinone oxidoreductase, glucuronosyltransferases, Epoxide hydrolases, dihydrodiol dehydrogenase, gamma-glutamyl cysteine synthetase, Heme oxygenase, Ferritin, Manganese superoxide dismutase (SOD), Leukotriene B4 dehydrogenase, etc.) induction (Talalay & Fahey, 2001). Indole-3-carbinol (13C) is formed by the breakdown of GS glucobrassicin and is mainly obtained from vegetables from the cruciferous family including broccoli, cabbage, brussels sprouts, collard greens, cauliflower, and kale, wherever diindolylmethane (DIM) is a breakdown derivative of 13 C formed in the stomach, both the derivatives 13 C and DIM show partial anticarcinogenic effects in the case of lung cancer which is facilitated via modulation of the tyrosine kinase/Akt signaling pathway and P13K receptors (Kim & Milner, 2005; Qian et al., 2011). Acharya et al. (2010) reported the unexpected effect of 13C and DIM against hormone-responsive cancers like ovarian, prostate, and breast cancer. DIM is more active than 13C in cell culture and shows anti-cancer properties via aryl hydrocarbon (Ah) receptor signaling transduction, Akt/Wnt/mTOR/NF-κB pathways, striking on cell cycle detention, modulation of CYP enzymes modulation, angiogenesis altering, metastasis, invasion, and epigenetic behavior of cancer cells (Bradlow & Zeligs, 2010; Banerjee et al., 2011). Lubet et al. (2011) observe that 13C acts as an AhR agonist in mammary cancer cells and DIM is not analogous to 13C in anti-cancer properties development.

Sulforaphane is an organosulfur complex found in cruciferous vegetables such as cabbages, broccoli, sprouts, and Brussels. Sulforaphane has revealed the initiation of stage II drug metabolism enzymes of xenobiotic transformation stage II drug metabolism enzymes induction including GSH S-transferase and quinine reductase with the enhancement in the tumor suppression proteins transcription (Hu et al., 2012). Sulforaphane downregulated the self-renewal pathway (Wnt/beta-catenin) in stem cells of breast cancer (); Akt, ERK signaling inhibition, and expression of D1 cyclin and COX-2 D1 protein, inducing apoptosis in SW620 cells in overall result it decreases the growth of cancer cell in a significant manner and can be utilized for the treatment and cure of cancers (Li et al., 2010; Shen et al., 2007; Rudolf et al., 2009).

6.4 MECHANISM OF METABOLITES

6.4.1 *ANTIOXIDANT AND PRO-OXIDANT ACTIVITY*

A molecule containing one or more unpaired electrons in its orbit (outer) with high reactivity and instability is known as free radical, unpaired electrons motivate them to catch electrons from other particles to formulate a stable molecule. Subsequently, the attacked molecule gains free space for electron and convert itself into a free radical chain reaction that starts and ends by damaging the healthy cell (Cilwyn et al., 2021). The free radicals comprise highly active oxygen and nitrogen molecules that are generated in the human body unsurprisingly due to various biological and chemical conditions in the human metabolic systems (Valko et al., 2007). Antioxidant or free radical scavenging properties of Phytochemicals such as ascorbate, polyphenols, tocopherols, and terpenoids perform a neutralization of reactive free radicals by donating electrons and quenching the initiator of the free radical mechanism resulting avoid harm to the healthy cells and generation of cancer (Rohman et al., 2006; Dimitrios, 2006). Antioxidant activity of phytochemicals diversely affects several interdependent pathways via inflowing the cell to initiate the pro-oxidant anticancer activity. The phytochemical growth reduces the excess of free radicals in cells and biomolecular changes (DNA, protein), conclusively avoiding cancer development (Cilwyn et al., 2021). Mainly ROS level imbalance and creation are the by-products generated through the mitochondrial electron transport due to the respiration (aerobic) and oxidation (metal-catalyzed) that is identified as oxidative stress, negative impact on intracellular biomolecules, genomic variability, and modifications in cell growth (Yedjou & Tchounwou, 2012). The intracellular ROS points are highly correlated with the cancer cells, most cancer cells are known by greater ROS levels that are essential to preserving fast proliferation rates due to the Warburg effect. The Reactive oxygen promotes tumor transformation by mutations in nuclear and mitochondrial DNA, which generate genomic unsteadiness shadowed by the sequential activation of various signaling pathways (Maciag et al., 2004; Ogrunc et al., 2014). Yuan et al. (2022) reported that the expression of oncogenes and suppressors may be affected by ROS via epigenetic modifications as well as able to regulate many proliferation-related genes including AKT/P13K and ERK/MAPK pathways. Furthermore, You et al. (1989); Herman & Singh (2004) reported that the organic sulfur complexes obtained from plants and vegetables might reduce the risk of cancer (gastric) by enhancing cellular antioxidative potential via induction of antioxidative enzymes. Xiao et al. (2004) reported that the growth inhibition capabilities of organosulfur complexes are directly associated with the number of sulfur atoms. Similarly, polyphenol's radical-scavenging capabilities are also interrelated to the hydroxyl group numbers (Kim et al., 2019). Cellular ROS level and ROS-intercede DNA damage have been playing a perilous role in carcinogenesis opening and progression, wherever revocable DNA damage might allow an interior restoration organization to stabilize the negative effect of ROS, but irretrievable damages are unable to utilize the appropriate functioning of the restoration system. As an effect, apoptosis suffers and develops a complication in cancer treatment (Aggarwal et al., 2019). Antioxidant activity of phytochemical significantly affects the carcinogenesis via various interdependent pathways and quenching of free radicals, as an outcome prevents the development of new free radicals by enzymes inclusion (dismutase, catalase, peroxidase, etc.), proteins (metal-binding-ceruloplasmin and ferritin, etc.), and minerals matter (Se, Cu, and Zn, etc.), prevent oxidative chain reactions and repair the biomolecular damage through antioxidative enzymes, DNA repair enzymes, and others (Mut-Salud et al., 2016; Cilwyn et al., 2021). Firstly, the phytochemical starts the pro-oxidant anticancer activity by entering the cells. Primarily, the phytochemicals accumulation reduces the excessive number of radicals. Afterward, the compact number of free radicals is non-much available to initiate the changes in cell biomolecules, especially DNA. Eventually, it nullifies the tremendously reactive free radicals in the cells and reduce the expansion of cancer (Cilwyn et al., 2021). Basically, free radicals from reactive nitrogen species, and oxygen species are considered one of the main factors for promoting cancer development in which hydrogen peroxide, nitric oxide, superoxide radicals, and singlet oxygen impart a recognizable potential impact on the cellular mechanism (Rahimi-Madiseh et al., 2016). These highly reactive free radicals disrupt the cellular membranes by damaging lipid bilayer and activate the sequence of enzymatic/non-enzymatic reactions in cells via interfering with the amino acids and DNA, which will make gene expressions unbalanced, mutate the gene, increases the chances of chromosomal abnormalities, and leading to failure in cell cycle with diminishing DNA capacity (George et al., 2021). Dizdaroglu et al. (2002) reported that the mechanism of highly reactive free radicals encouraged the inactivation of tumor suppressor genes, activation of oncogenes, replication errors, and malignancy (Figure 6.3).

The pro-oxidant activity of phytochemicals persuades oxidative stress either by hindering the cell antioxidant system or by the generation of free radicals (Rahal et al., 2014). Healthy functioning is essential for cell survival and can be determined by the redox level of cells. Several phytochemicals are able to target redox balance by sufficient creation of ROS that eventually leads to cell death (Kruk et al., 2019). González-Bartulos et al. (2015); and Wondrak (2009) reported that redox homeostasis and positive regulation (pro-oxidant-based) of nuclear-free radicals would aim precisely at cancer cells without negatively affecting the normal cells. Moreover, phytochemicals (quercetin, kaempferol, etc.) with transition metals such as iron and copper are more favorable for pro-oxidant therapies by inducing the steering redox reactions in the cancer cells and producing an excessive number of free radicals (hydroxyl radical) species through Fenton reaction (Rahal et al., 2014; Halliwell, 2008). The pro-oxidant action of phytochemicals might depend on diverse interdependent routes for its anticancer properties, primarily phytochemicals will enter into the cells to initiate an activity (pro-oxidant) and in liberal accumulation, cells start to produce an unwarranted number of free radicals. Afterward, a large number of free radical triggers DNA fragmentation and DNA damage (single- and double-strand DNA molecule disruptions, destruction to the deoxyribose sugar, purines loss in the DNA, DNA and protein cross-linking, and DNA repair system) via oxidative mechanism (Srinivas et al., 2019; Beckman & Ames, 1997). The fragmentation of DNA and damage lead to the custody of the cell cycle and efficiently reduce cancer cell growth (Nayak et al., 2021).

6.4.2 INDUCTION OF APOPTOSIS

Most plants and vegetables-originated phytochemicals like alkaloids, terpenoids, and phenolic acids have antitumor properties by adjusting the ROS signaling mechanism (Chirumbolo et al., 2018). In addition to the antioxidative capabilities of natural phytochemicals, several plants and vegetable components were identified and verified logically to make apoptosis through ROS creation. Miki et al. (2012) reported that RES induces caspase-based human colon cancer cell apoptosis by enhancing the ROS ranges intracellularly. Furthermore, quercetin also hinders the growth of cancer by developing apoptosis via COX-2-based ROS production. Moreover, several phytochemicals also show anticancer activity by apoptosis induction via upregulating expression of caspases such as hyperforin a phytochemical was reported to encourage caspase-dependent apoptosis in numerous leukemia cell lines by caspase-3, 8, and 9 upregulation (Hostanska et al., 2003). Mohan et al. (2012) reported a purified bioactive compound, pyranocycloartobiloxanthone-A plays an imperious part in apoptosis induction in cancer cells (breast) by Bcl-2 expression upregulating and downregulation of Bax expression, which results in cytochrome c release, and caspase cascade initiation. In addition to this, one of the phenolic phytochemical Scutellarin shows apoptosis dependent on caspase-3 promotion in human colon cancer via anti-apoptotic protein Bcl-2 suppression which further triggers the p53 pathway and leads to the Bax protein upregulation (Yang et al., 2017). Furthermore, gallic acid also takes a part in the induction of cancer cell apoptosis via upregulation of the p53, in this process of apoptosis, it alters the mitochondrial film potential, helps in the caspase-activator liberation, cytochrome c, and makes an apoptotic pathway intrinsically (Yang et al., 2018) (Figure 6.2). Moreover, capsaicin also induces p53-arbitrate apoptosis in numerous cancer cells have been well explained in previous studies (Clark & Lee, 2016; Garufi et al., 2016; Lee & Clark, 2016; Jin et al., 2014). The NF-kβ is extremely voiced in cancer cells due to its mechanism in anti-apoptotic and apoptotic genes regulation, phytochemicals are also aiming for nuclear factor kappa β to encourage cancer cell death (Kumar et al., 2016; Oh et al., 2012). In addition to this, phytochemicals, together with alkaloids and flavonoids, are also targeting NF-kB signaling pathway and induce apoptosis such as Corilagin (Gambari et al., 2012), Morusin (Lee et al., 2008), xanthohumol (Colgate et al., 2007), Magnolol (Tse et al., 2007), and ursolic acid (Manu et al., 2008) demonstrated the down-regulation of NF-kB in several cancer cells. Ashraf (2020) reported that phytochemicals are a vital basis of drugs for the prevention and handling of vinblastine and paclitaxel are developed from phytochemicals and several others are still in research. Mostly, these phytochemicals mark the apoptotic mechanism by intrusive with the caspase-based pathways.

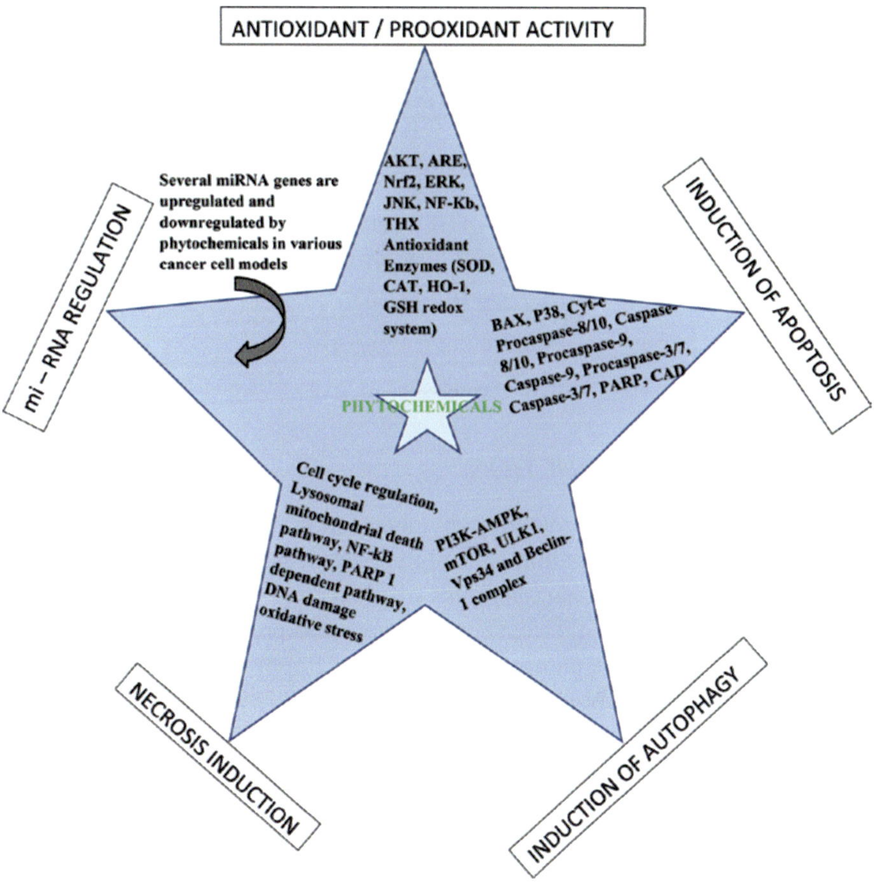

FIGURE 6.2 Signaling pathways and mechanisms involved in phytochemicals treatments.

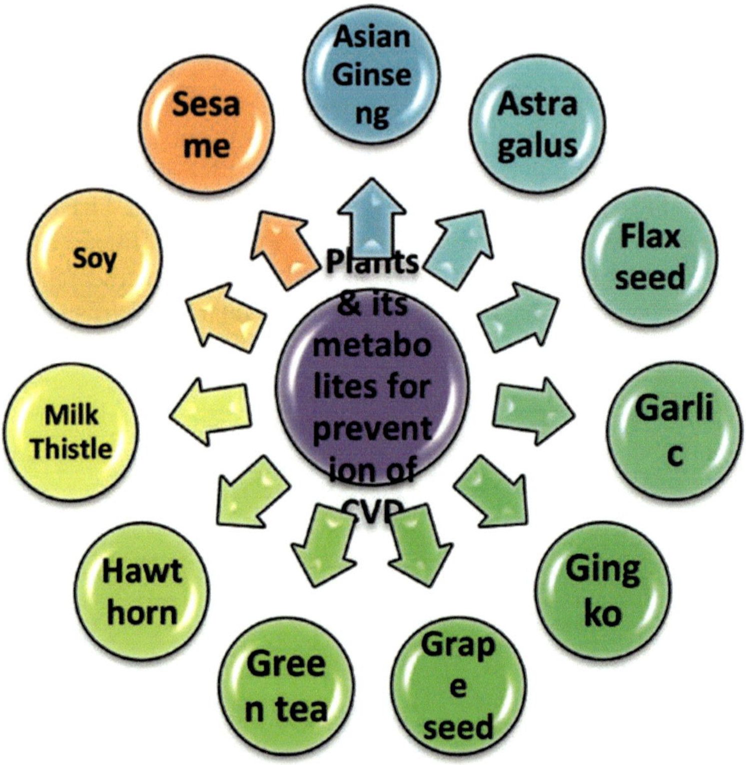

FIGURE 6.3 Antioxidant activity of phytochemicals in cancer treatment and prevention.

6.4.3 *NECROSIS INDUCTION*

Phytochemicals impart protection by interaction with various molecular mechanisms and pathways that control the cycles, survival, and death inside the human cells. These characteristics of plants and vegetables make them efficient in making drugs for cancer treatment and prevention. Necrosis is a non-apoptotic cell death pathway that plays an important part in inspecting cells (neoplastic) and abolishing tumor cells, necrosis induction mechanism is also used in some anticancer drugs like DNA-alkylating agent which is used for giving human cancers (Cho & Park, 2014). Phytochemical's function in signaling cascades is intricate in the generation of necrotic cell death will help us to create a novel drug for treating cancer. A cell takes the necrotic or apoptotic pathway based on its pathological, and physiological conditions. Lee et al. (2018) reported that Necrosis doesn't have its own molecular pathway and it also overlays various caspase-independent apoptotic pathways which conclude in disturbance of membrane integrity and organelle damage subsequent in the cellular content. In the induction of the necrotic pathway, cells go through high damage when cell content is released and creates a pro-inflammatory environment. Phytochemicals loaded with necrosis may exert a more effective tumor suppression property. Necroptosis is known as a programmed form of necrosis which mostly activated by stimuli such as the extrinsic apoptotic pathway. Necroptosis is basically activated by threonine and serine intermingling kinase receptor proteins 1 and 3 (RIP 1 and RIP 3) which induce necrotic cell death. Whereas RIP 1 and 3 can be ignited by various factors including signaling via receptors like tumor necrotic factors 1 and 2 (TNF-R1 and TNF-R2), toll-like receptors 3 and 4 (TLR3 and TLR4), PARP1 pathways induced during DNA damage, particularly when caspase-eight is either downregulated and hibernate by other regulators (De Giffoni De Carvalho et al., 2019). In addition to this, the progression of necrosis depends upon the caspase-eight and protein linked with Fas and death domain (FADD) that regulate RIP1 and RIP 3 stages. The initiation of necroptosis can be exemplified by the occurrence of TNF signaling pathway events and investigated intensively (Fulda, 2013). The binding actions of TNF to TNFR1 create the geometrical changes in the trimers of TNFR1- and lead to the multimeric protein complex I (membrane-bound, TRAF2) recruitment, cIAP1 (cellular inhibitor of apoptosis protein 1), TRAF5, cIAP2, TRADD (death associated domain – TNFR), and RIPK1 by TNFR1. In I-complex, cIAP1/2 polyubiquitinated the critical regulator of cell fate RIPK1 and induces the nuclear factor kappa B pathway, which induces cell survival by transactivation of cytoprotective genes (Gong et al., 2019; Vandenabeele et al., 2010). Furthermore, complex I proteins and post-translational modification can be changed due to the hasty internalization of TNFR-ligand-bound (Fulda, 2013). For instance, the development of deubiquitinated RIPK1 by deubiquitinase cylindromatosis (CYLD), afterward restricts the sustained instigation of NF-κB signaling and leads to the initiation of cell death pathways (Wright et al., 2007). Subsequently, a death-inducing signaling cytoplasmic compound (complex II) including death domain (TNFR-associated), protein kinase (Receptor-interacting), FAS-associated death domain protein, which is also recognized as "Ripoptosome" is formed and induces caspase-8 activation in the complex (Hitomi et al., 2008). In II-complex, caspase-8 activation led to cleavage and inactivation of both RIPK1 and RIPK3, resulting in the initiation of the proapoptotic caspase activation cascade and ultimately leading to apoptosis execution (Wu et al., 2012). However, phytochemicals inhibit the caspase-8, and develop pharmaceutical or genetic interventions in the RIP kinases cleavage restriction, as a result, the death pathway of the cell is diverted towards necrosis (Vandenabeele et al., 2010). After diversion towards necroptosis cell death mode RIPK1 is going through autophosphorylation, phosphorylated by the amino acid serine 161 (S161) at N-terminus, and activated. Activation results in the RIPK1 and RIPK3 interaction through the RIP homotypic interaction motif (RHIMs) and leads to the creation of the necrosome complex which acts as a main molecular signaling platform primarily including RIPK1 and RIPK2 in necroptosis (Li et al., 2012). Moreover, Wattanathamsan et al. (2019) demonstrated that downregulation of FADD and caspase 8 encourages RIP 3-dependent necrosis. Huang et al. (2013); and Lee et al. (2018) reported that the necrotic action mechanism can be activated by hypoxia and stress in the metabolic activity (oncogenic) by constructing Dlx-2 and transcription factors snail, which are apparently eliminated by enhanced anaerobic glycolysis and glucose uptake in cancer cells. Tumor progression in non-Hodgkins lymphoma is also correlated with the defects in necroptosis regulators (example: RIP3) and polymorphisms (Mishra et al., 2018). Lalaoui & Brumatti (2017) evidenced the increase in impaired necroptosis in cancer cells. Hence necroptosis targeting will be proven very promising in cancer therapy, and trials of necroptosis-inducing anticancer drugs (Lalaoui & Brumatti, 2017; Fulda, 2018). Phytochemicals show the induction of necrosis/necroptosis by aiming at several of the above molecular intermediaries and pathways. In addition to this, solamargine target, the death pathway (mitochondria lysosomal) in the lung, breast, and squamous cell carcinoma as well as leukemia cell lines with the induction of necrosis in melanoma and non-melanoma (skin) lines of cells (Al-Sinani et al., 2016). Similarly, Shikonin and Phenethyl isothiocyanate (necroptosis in apoptotic cell death), Lapachone (HCC SK-Hep1 necroptosis in human cells) (Diederich & Cerella, 2016).

6.4.4 INDUCTION OF AUTOPHAGY

Autophagy is created from the Greek word sense self-eating, this self-lysis style leads toward the cell level constituent's demotion (protein and organelles, etc.) to produce metabolic precursors and energy for cell existence (Glick et al., 2010). Basically, autophagosomes cover the cellular organelles and cytoplasmic proteins that are tainted by lysosomes via fusion in the autophagy mechanism. The cellular immune response toward stress regularly helps in the process of controlling quality and proceeds to cell death (apoptosis-independent). Several autophagy forms have been studied, together with microautophagy, chaperone-mediated autophagy, and macroautophagy (Sharma et al., 2014). Several phytochemicals, specifically phenols, flavones, alkaloids, and organic acids are known as regulators of autophagy, they may stop cytotoxicity and perform a cyto-protective role during cancer cells treatment by natural phytochemical products (Wang & Feng, 2014). In some conditions, both apoptosis and autophagy are parallelly necessary to create a condition of cell death (Mettlin, 1997). Furthermore, cancer cell characteristics such as enhancement in the proliferation rate of the cells, apoptotic pathways alteration, and cellular metabolism rearrangements, show an influence on autophagic induction (Linder & Kogel, 2019). Furthermost, chemotherapeutical drugs are autophagy promotors and are usually considered for a cryoprotective response (promote apoptotic cell death during inhibition) (Sharma et al., 2014). Moreover, RES creates molecular features of apoptosis, plus cytochrome c release and activates caspase from mitochondria, in addition to this, RES-treated cells displayed ultrastructural and morphological changes which indicate death by autophagocytic (Opipari et al., 2004). A soy-derived isoflavone 'Genistein' has induced both autophagy and apoptosis (Kueck et al., 2007). Gossner et al. (2007) also reported that genistein also encouraged the autophagy mechanism in cancer cells (ovarian), representing the active presence and positioning of LC3-II to autophagosomes. The utmost copious Polyphenol present in *Camellia Sinensis* (green tea) is EGCG is able to induce autophagy via highly prompting ROS in humanoid mesothelioma cell death (Satoh et al., 2013). In addition to this, Epicatechins also show modulation in the ER and autophagy which induces cell death through the apoptotic pathways (Zhang et al., 2020). In conditions of stress, it can encourage cell death, known as autophagy-dependent cell death (ADCD). Fairly, a lot of reports have an indication of moving toward the use of natural sources for cancer treatment and their capability to induce autophagy.

6.4.5 MIRNA REGULATION

Plant and vegetable-derived bioactive compounds, often called phytochemicals can release therapeutic potential abilities via directly acting on the cell system or decreasing the side effects of antitumor drugs and progression via regulating micro-RNAs (mi-RNAs) expression, mi-RNA can regulate protein-coding expression by translation restraining or mRNA (messenger ribonucleic acid) degradation. Chemotherapy is considered a great relief in cancer treatment with the combination of chemotherapy drugs, that not only arranges the side effects of drugs but also worsens the intensity of tumor induction after some period of relief. Some Phytochemicals from natural plant and vegetable sources showed anticancer potential with nearly negligible cytotoxicity to the normal cell physiology (Rao et al., 2007). Deng et al. (2008) reported that the irregularity in the transcription process, mutations, epigenetic, miRNA biosynthesis variation, and the incidence of different DNA numbers encourage the dysregulation of miRNA in human cancer. Moreover, researchers uncovered the capability of phytochemicals in the variation of miRNA expression, and the mode of operation in expressing mi RNA with an exception to the transcriptional directive (Cilwyn et al., 2021). Green tea constituents especially epigallocatechin-3-gallate (EGCG) is well known most efficient catechin in the treatment of cancer (Mukherjee et al., 2015). In addition to this, Wang et al. (2011) reported that EGCG incites the conclusiveness of factor-1a (hypoxia-induction) to the promoter section of miR-210, ultimately misleading the cells to halt the proliferation of cells and favoring autonomous growth as noted in human non-small cancer cell line like H1299, H460, and A549. In synchronous to this, Yamada et al. (2016) had a notable discovery about EGCG upregulating miRNA and energizing laminin-based receptors to hinder B16 melanoma cancer cells growth. Furthermore, ECGC also shows the downregulation in the expression of miR-25 with the apoptosis of MCF-7 cells via the intensification of various caspases and PARP (Zan et al., 2019). Kumar et al. (2017) reported the biological outcomes of RES in cancer (prostate) where it was able to create a down-regulation in the miR-221 expression, while in the lung cancer case high outpouring stayed testified in the miR-200c expression (Bai et al., 2014). In addition to this, RES dysregulates the two miRNAs primarily miR-542-3P (decrement) and miR-122-5P (increment) collectively to endorse apoptosis cell death in estrogen-responsive breast cancer cells, similarly, the triple-negative breast cancer cells only show a rise in miR-122-5p while treating with RES (Venkatadri et al., 2016). The Osteosarcoma violent cancer cells (U20S and MG63) were introduced to RES, and as a result, impressive effects are perceived such as a reduction in the expression of miR-139-5P and it has a tendency to facilitate the osteosarcoma via joining at the 3'UTR area of NOTCH1 (Xiao et al., 2020). Coker-Gurkan et al. (2019) reported that curcumin shows anti-proliferation activity in T47D cells via downregulating miR-96, miR-183 and miR-182 together with inhibition of NF-kB signaling. Furthermore, curcumin leads to the inhibition of SW480 cells proliferation in colon cancer via elevating miR-130a which leads to the pause in the Wnt/b-catenin pathway (Dou et al., 2017). Comparable to additional polyphenols, quercetin found in broccoli and onions was glaringly studied for its anticancer performance. In particular, after the action of quercetin on tumor cells (pancreatic) it revealed a plethora of 105 miRNA at which a large number of miRNAs were down-regulated and upregulated, in which let-7c is the most expressed and connected miRNA with the anti-proliferation mechanism in pancreatic cancer (Nwaeburu et al., 2016). Furthermore, Zhao et al. (2019) also studied the impact of quercetin using oral cancer cells (*in vitro* – SCC-9 and HSC-6). Cells with the application of quercetin show advanced miR-16 expression which further mediates the inhibition of the HOXA10 level. Jointly, the triggered miR-16 countenance behaves like a tumor suppressor in cancer (oral) by disturbing HOXA10, which is positively correlated with the consequence of cancer cell proliferation in the neck and head squamous cell carcinoma (Guo et al., 2018; Zhao et al., 2019). A derivative (isothiocyanate) of cruciferous plants (broccoli, cauliflower, etc.) Sulforaphane is recognized for its specificity to target SFN with the active anticancer properties in the human body. EGFR signaling is connected with EMT (epithelial-mesenchymal transition) of cancer in the lungs and when treated with lung cancer (non-small) cells (H1299, 95D, and 95C) it revealed clampdown of miR-616, which modulates the Glycogen synthase kinase-3β/β-catenin signaling path usually concerned in cancer intrusiveness (Wang et al., 2017; Zhang & Tang, 2007) (Figure 6.2).

6.5 FACTORS AFFECTING THE METABOLITES IN PLANTS AND VEGETABLES

Uddin et al. (2019) reported that the environmental factors and association of plants with abiotic stress have a significant effect on plant development and biosynthesis of secondary metabolites. Plant metabolites are highly affected by genomic, ecological factors, transference, and conditions of storage, in addition to these growth reasons such as temperature, air moisture content, soil type, damage by the insects and microorganisms, application of fertilizers, stress encouraged by UV radiation, pesticides, and heavy metallic element modify the metabolite composition of plants (Hounsome et al., 2008). Ramakrishna & Ravishankar (2011) reported that the climatic change drastically influences the water availability, salinity, and several other soil conditions which overall affect the productivity of plants and the bioavailability of phytochemicals which is further controlled by the utilization of genetic tools, cell culture, and *in vitro* plant production for the production of chemicals and pharmaceuticals medicines. Furthermore, Appolloni et al. (2021) elaborate on the effect of indoor LED light on dedicated metabolite content in microgreens, flowers (edible), plants used for medicinal purposes, and herbs (Sharma et al., 2020; Rane et al., 2020a–c) (aromatic) by the use of the light concoction. In the study, it is found that the addition of distant red or green to the lightning blend improved the synthesis of focused metabolites in the treatment of light with a photoperiod of (12–16 hr.). Moreover, Neugart et al. (2017) reported the effect of genetic, environmental, and agronomic factors on the secondary plant metabolite of brassicaceous vegetables.

6.6 CONCLUSION

A large number of plants and vegetables-based secondary metabolites are associated with the reduction in various types of cancer and studies also include the role and use of metabolites in nano-delivery systems for the treatment of different cancer cell lines. In addition, the therapeutic potential of orally administrative phytochemicals is narrowed because of their instability and oral bioavailability in the gastrointestinal tract. Solubility and cleavage by gastric enzymes highly affect the absorption of phytochemicals and activity in cell signaling for the treatment of cancer. For the specific treatment in a particular cancer cell line basic understanding of the cancer mechanisms such as Autophagy, Necrosis, Apoptosis, Antioxidant, Pro-Oxidant, and mi-RNA regulations perform significant importance in the evaluation of phytochemicals activity. Phytochemicals including polyphenols, betalains, carotenoids, capsaicinoids, terpenes, terpenoids, nitrogen-containing alkaloids, and sulfur-containing compounds actively participate in the reduction of ROS and oxidative stress by several metabolic pathways which are effective against the growth of tumor cells and provide safety to the normal cells from cancer growth. Phytochemicals not only affect the normal cells it also reduces cancer cell activity by taking part in various pathways related to cancer.

KEYWORDS

- **antioxidant**
- **glucosinolates**
- **metabolites**
- **NF-kB pathway**
- **organosulfur**
- **phytochemicals**
- **polyphenols**
- **terpenes**

REFERENCES

Abbaoui, B., Lucas, C. R., Riedl, K. M., Clinton, S. K., & Mortazavi, A., (2018). Cruciferous vegetables, isothiocyanates, and bladder cancer prevention. *Molecular Nutrition & Food Research, 62*(18), 1800079.

Acharya, A., Das, I., Singh, S., & Saha, T., (2010). Chemopreventive properties of indole-3-carbinol, diindolylmethane and other constituents of cardamom against carcinogenesis. *Recent Patents on Food, Nutrition & Agriculture, 2*(2), 166–177.

Adlercreutz, H., & Mazur, W., (1997). Phyto-oestrogens and Western diseases. *Annals of Medicine, 29*(2), 95–120.

Akuffo, K. O., Nolan, J., Stack, J., Moran, R., Feeney, J., Kenny, R. A., & Beatty, S., (2015). Prevalence of age-related macular degeneration in the Republic of Ireland. *British Journal of Ophthalmology, 99*(8), 1037–1044.

Al Dhaheri, Y., Attoub, S., Ramadan, G., Arafat, K., Bajbouj, K., Karuvantevida, N., & Iratni, R., (2014). Carnosol induces ROS-mediated beclin1-independent autophagy and apoptosis in triple negative breast cancer. *PloS One, 9*(10), e109630.

Appolloni, E., Pennisi, G., Zauli, I., Carotti, L., Paucek, I., Quaini, S., & Gianquinto, G., (2022). Beyond vegetables: Effects of indoor LED light on specialized metabolite biosynthesis in medicinal and aromatic plants, edible flowers, and microgreens. *Journal of the Science of Food and Agriculture, 102*(2), 472–487.

Arts, I. C., & Hollman, P. C., (2005). Polyphenols and disease risk in epidemiologic studies. *The American Journal of Clinical Nutrition, 81*(1), 317S–325S.

Ashraf, M. A., (2020). Phytochemicals as potential anticancer drugs: Time to ponder nature's bounty. *BioMed Research International, 2020*, 8602879. https://doi.org/10.1155/2020/8602879.

Ávila-Gálvez, M. Á., González-Sarrías, A., Martínez-Díaz, F., Abellán, B., Martínez-Torrano, A. J., Fernández-López, A. J., & Espín, J. C., (2021). Disposition of dietary polyphenols in

breast cancer patients' tumors, and their associated anticancer activity: The particular case of curcumin. *Molecular Nutrition & Food Research, 65*(12), 2100163.
Balapure, S., & Wadegaonkar, V., (2014). Determination of antitrypsin activity of lupeol by Folin lowery method. *The International Journal of Science and Technoledge, 2*(4), 1.
Banerjee, S., Kong, D., Wang, Z., Bao, B., Hillman, G. G., & Sarkar, F. H., (2011). Attenuation of multi-targeted proliferation-linked signaling by 3, 3′-diindolylmethane (DIM): From bench to clinic. *Mutation Research/Reviews in Mutation Research, 728*(1, 2), 47–66.
Bayat, M. R., Homayouni, T. S., Baluch, N., Morgatskaya, E., Kumar, S., Das, B., & Yeger, H., (2017). Combination therapy in combating cancer. *Oncotarget, 8*, 38022–38043.
Beckman, K., & Ames, B., (1997). Oxidative DNA decay. *J. Biol. Chem., 272*, 19633–19636.
Bennink, M. R., (2002). Consumption of black beans and navy beans (*Phaseolus vulgaris*) reduced azoxymethane-induced colon cancer in rats. *Nutrition and Cancer, 44*(1), 60–65.
Bhattacharya, T., Dutta, S., Akter, R., Rahman, M. H., Karthika, C., Nagaswarupa, H. P., & Bungau, S., (2021). Role of phytonutrients in nutrigenetics and nutrigenomics perspective in curing breast cancer. *Biomolecules, 11*(8), 1176.
Blaskovich, M. A., Sun, J., Cantor, A., Turkson, J., Jove, R., & Sebti, S. M., (2003). Discovery of JSI-124 (cucurbitacin I), a selective Janus kinase/signal transducer and activator of transcription 3 signaling pathway inhibitor with potent antitumor activity against human and murine cancer cells in mice. *Cancer Research, 63*(6), 1270–1279.
Block, E., (1985). The chemistry of garlic and onions. *Scientific American, 252*(3), 114–121.
Boehm, F., Edge, R., Truscott, T. G., & Witt, C., (2016). A dramatic effect of oxygen on protection of human cells against γ-radiation by lycopene. *FEBS Letters, 590*(8), 1086–1093.
Bradlow, H. L., & Zeligs, M. A., (2010). Diindolylmethane (DIM) spontaneously forms from indole-3-carbinol (I3C) during cell culture experiments. *In Vivo, 24*(4), 387–391.
Bruno, A., Siena, L., Gerbino, S., Ferraro, M., Chanez, P., Giammanco, M., Gjomarkaj, M., Pace, E. Apigenin affects leptin/leptin receptor pathway and induces cell apoptosis in lung adenocarcinoma cell line. *Eur. J. Cancer, 2011, 47*(13), 2042–2051.
Castro, N. P., Rangel, M. C., Merchant, A. S., MacKinnon, G., Cuttitta, F., Salomon, D. S., & Kim, Y. S., (2019). Sulforaphane suppresses the growth of triple-negative breast cancer stem-like cells *in vitro* and *in vivo* sulforaphane and breast cancer stem-like cells. *Cancer Prevention Research, 12*(3), 147–158.
Cateni, F., Gargano, M. L., Procida, G., Venturella, G., Cirlincione, F., & Ferraro, V., (2021). Mycochemicals in wild and cultivated mushrooms: Nutrition and health. *Phytochemistry Reviews*, 1–45.
Chakravarti, B., Maurya, R., Siddiqui, J. A., Bid, H. K., Rajendran, S. M., Yadav, P. P., & Konwar, R., (2012). *In vitro* anti-breast cancer activity of ethanolic extract of *Wrightia tomentosa*: Role of pro-apoptotic effects of oleanolic acid and urosolic acid. *Journal of Ethnopharmacology, 142*(1), 72–79.
Chan, W. H., (2007). The signaling cascades of ginkgolide B-induced apoptosis in MCF-7 breast cancer cells. *International Journal of Molecular Sciences, 8*(11), 1177–1195.
Chen, J. C., Chiu, M. H., Nie, R. L., Cordell, G. A., & Qiu, S. X., (2005). Cucurbitacins and cucurbitane glycosides: Structures and biological activities. *Natural Product Reports, 22*(3), 386–399.
Chinnabattigalla, S., Dakoju, R. K., & Gedu, S., (2021). Recent advances on the synthesis of flavans, isoflavans, and neoflavans. *Journal of Heterocyclic Chemistry, 58*(2), 415–441.

Chirumbolo, S., Bjørklund, G., Lysiuk, R., Vella, A., Lenchyk, L., & Upyr, T., (2018). Targeting cancer with phytochemicals via their fine tuning of the cell survival signaling pathways. *International Journal of Molecular Sciences, 19*(11), 3568.

Cho, Y. S., & Park, S. Y., (2014). Harnessing of programmed necrosis for fighting against cancers. *Biomolecules & Therapeutics, 22*(3), 167.

Choudhari, A. S., Mandave, P. C., Deshpande, M., Ranjekar, P., & Prakash, O., (2020). Phytochemicals in cancer treatment: From preclinical studies to clinical practice. *Frontiers in Pharmacology*, 1614.

Cilwyn, B., Vijayarathna, S., Jegathambigai, R. N., Sreeramanan, S., Chen, Y., & Sasidharan, S., (2021). The role of phytochemicals in cancer prevention and cure. In: *Bioactive Natural Products for Pharmaceutical Applications* (pp. 127–150). Springer, Cham.

Clark, R., & Lee, S. H., (2016). Anticancer properties of capsaicin against human cancer. *Anticancer Res., 36*, 837–843.

Coker-Gurkan, A., Bulut, D., Genc, R., Arisan, E. D., Obakan-Yerlikaya, P., & Palavan-Unsal, N., (2019). Curcumin prevented human autocrine growth hormone (GH) signaling mediated NF-κB activation and miR-183-96-182 cluster stimulated epithelial mesenchymal transition in T47D breast cancer cells. *Molecular Biology Reports, 46*(1), 355–369.

Colgate, E. C., Miranda, C. L., Stevens, J. F., Bray, T. M., & Ho, E., (2007). Xanthohumol, a prenylflavonoid derived from hops induces apoptosis and inhibits NF-kappaB activation in prostate epithelial cells. *Cancer Letters, 246*(1, 2), 201–209.

Crozier, A., Clifford, M. N., & Ashihara, H., (2008). *Plant Secondary Metabolites: Occurrence, Structure and Role in the Human Diet*. John Wiley & Sons.

Das, S., Tyagi, A. K., & Kaur, H., (2000). Cancer modulation by glucosinolates: A review. *Current Science*, 1665–1671.

De Giffoni De, C. J. T., Da Silva, B. D., Leite, D. F., De Araújo, L. C. A., De Toledo, E. P. P., Antunes, K. A., & Dos Santos, E. L., (2019). Medicinal plants from Brazilian Cerrado: Antioxidant and anticancer potential and protection against chemotherapy toxicity. *Oxidative Medicine and Cellular Longevity, 2019*.

Deng, S., Calin, G. A., Croce, C. M., Coukos, G., & Zhang, L., (2008). Mechanisms of microRNA deregulation in human cancer. *Cell Cycle, 7*(17), 2643–2646.

Dimitrios, B., (2006). *Trend in Food Sci. Technol., 17*(9), 505–512.

Dizdaroglu, M., Jaruga, P., Birincioglu, M., & Rodriguez, H., (2002). Free radical-induced damage to DNA: Mechanisms and measurement. *Free Radical Biology and Medicine, 32*(11), 1102–1115.

Dou, H., Shen, R., Tao, J., Huang, L., Shi, H., Chen, H., & Wang, T., (2017). Curcumin suppresses the colon cancer proliferation by inhibiting Wnt/β-catenin pathways via miR-130a. *Frontiers in Pharmacology, 8*, 877.

Duthie, S. J., Jenkinson, A. M., Crozier, A., Mullen, W., Pirie, L., Kyle, J., & Duthie, G. G., (2006). The effects of cranberry juice consumption on antioxidant status and biomarkers relating to heart disease and cancer in healthy human volunteers. *European Journal of Nutrition, 45*(2), 113–122.

Einbond, L. S., Wu, H. A., Kashiwazaki, R., He, K., Roller, M., Su, T., & Goldsberry, S., (2012). Carnosic acid inhibits the growth of ER-negative human breast cancer cells and synergizes with curcumin. *Fitoterapia, 83*(7), 1160–1168.

Erdogan, S., Turkekul, K., Dibirdik, I., Doganlar, O., Doganlar, Z. B., Bilir, A., & Oktem, G., (2018). Midkine downregulation increases the efficacy of quercetin on prostate cancer stem cell survival and migration through PI3K/AKT and MAPK/ERK pathway. *Biomedicine & Pharmacotherapy, 107*, 793–805.

Eskiler, G. G., Özkan, A. D., Kaleli, S., & Bilir, C., (2019). Inhibition of TLR4/TRIF/IRF3 signaling pathway by curcumin in breast cancer cells. *J. Pharm. Pharm. Sci., 22*, 281–291. doi: 10.18433/jpps30493.

Fahey, J. W., Zalcmann, A. T., & Talalay, P., (2001). The chemical diversity and distribution of glucosinolates and isothiocyanates among plants. *Phytochemistry, 56*(1), 5–51.

Fakhri, S., Moradi, S. Z., Yarmohammadi, A., Narimani, F., Wallace, C. E., & Bishayee, A., (2022). Modulation of TLR/NF-κB/NLRP signaling by bioactive phytocompounds: A promising strategy to augment cancer chemotherapy and immunotherapy. *Frontiers in Oncology, 12*.

Femia, A. P., Tarquini, E., Salvadori, M., Ferri, S., Giannini, A., Dolara, P., & Caderni, G., (2008). K-ras mutations and mucin profile in preneoplastic lesions and colon tumors induced in rats by 1, 2-dimethylhydrazine. *International Journal of Cancer, 122*(1), 117–123.

Fulda, S., (2013). The mechanism of necroptosis in normal and cancer cells. *Cancer Biology & Therapy, 14*(11), 999–1004.

Fulda, S., (2018). Repurposing anticancer drugs for targeting necroptosis. *Cell Cycle, 17*(7), 829–832.

Gambari, R., Borgatti, M., Lampronti, I., Fabbri, E., Brognara, E., Bianchi, N., & Chui, C. H., (2012). Corilagin is a potent inhibitor of NF-kappaB activity and downregulates TNF-alpha induced expression of IL-8 gene in cystic fibrosis IB3-1 cells. *International Immunopharmacology, 13*(3), 308–315.

Garufi, A., Pistritto, G., Cirone, M., & D'Orazi, G., (2016). Reactivation of mutant p53 by capsaicin, the major constituent of peppers. *Journal of Experimental & Clinical Cancer Research, 35*(1), 1–9.

George, B. P., Chandran, R., & Abrahamse, H., (2021). Role of phytochemicals in cancer chemoprevention: Insights. *Antioxidants (Basel, Switzerland), 10*(9), 1455. https://doi.org/10.3390/antiox10091455.

Gibson, G. G., & Skett, P., (1986). Factors affecting drug metabolism: Internal factors. In: *Introduction to Drug Metabolism* (pp. 113–146). Springer, Boston, MA.

Giovannucci, E., (1999). Tomatoes, tomato-based products, lycopene, and cancer: Review of the epidemiologic literature. *Journal of the National Cancer Institute, 91*(4), 317–331.

Goncharov, N., Orekhov, A. N., Voitenko, N., Ukolov, A., Jenkins, R., & Avdonin, P., (2016). Organosulfur compounds as nutraceuticals. *Nutraceuticals*.

Gong, X., Smith, J. R., Swanson, H. M., & Rubin, L. P., (2018). Carotenoid lutein selectively inhibits breast cancer cell growth and potentiates the effect of chemotherapeutic agents through ROS-mediated mechanisms. *Molecules, 23*(4), 905.

Gong, Y., Fan, Z., Luo, G., Yang, C., Huang, Q., Fan, K., & Liu, C., (2019). The role of necroptosis in cancer biology and therapy. *Molecular Cancer, 18*(1), 1–17.

Gonzalez-Bartulos, M., Aceves-Luquero, C., Qualai, J., Cusso, O., Martínez, M. A., Fernandez De, M. S., & Massaguer, A., (2015). Pro-oxidant activity of amine-pyridine-based iron complexes efficiently kills cancer and cancer stem-like cells. *PLoS One, 10*(9), e0137800.

Guerriero, G., Berni, R., Muñoz-Sanchez, J. A., Apone, F., Abdel-Salam, E. M., Qahtan, A. A., & Faisal, M., (2018). Production of plant secondary metabolites: Examples, tips and suggestions for biotechnologists. *Genes, 9*(6), 309.

Guo, L. M., Ding, G. F., Xu, W., Ge, H., Jiang, Y., Chen, X. J., & Lu, Y., (2018). MiR-135a-5p represses proliferation of HNSCC by targeting HOXA10. *Cancer Biol. Ther., 19*, 973–983.

Gupta, C., & Prakash, D., (2014). Phytonutrients as therapeutic agents. *Journal of Complementary and Integrative Medicine, 11*(3), 151–169.

Guruvayoorappan, C., & Kuttan, G., (2007). β-carotene inhibits tumor-specific angiogenesis by altering the cytokine profile and inhibits the nuclear translocation of transcription factors in B16F-10 melanoma cells. *Integrative Cancer Therapies, 6*(3), 258–270.

Halliwell, B., (2008). Are polyphenols antioxidants or pro-oxidants? What do we learn from cell culture and *in vivo* studies? *Archives of Biochemistry and Biophysics, 476*(2), 107–112.

Hamdi, A., Jaramillo-Carmona, S., Beji, R. S., Tej, R., Zaoui, S., Rodríguez-Arcos, R., & Guillén-Bejarano, R., (2017). The phytochemical and bioactivity profiles of wild *Asparagus albus* L. plant. *Food Research International, 99*, 720–729.

Hammerstone, J. F., Lazarus, S. A., & Schmitz, H. H., (2000). Procyanidin content and variation in some commonly consumed foods. *The Journal of Nutrition, 130*(8), 2086S–2092S.

Hazafa, A., Iqbal, M. O., Javaid, U., Tareen, M. B. K., Amna, D., Ramzan, A., & Naeem, M., (2021). Inhibitory effect of polyphenols (phenolic acids, lignans, and stilbenes) on cancer by regulating signal transduction pathways: A review. *Clinical and Translational Oncology*, 1–14.

He, S., Simpson, B. K., Sun, H., Ngadi, M. O., Ma, Y., & Huang, T., (2018). Phaseolus vulgaris lectins: A systematic review of characteristics and health implications. *Critical Reviews in Food Science and Nutrition, 58*(1), 70–83.

He, Y., Jin, H., Gong, W., Zhang, C., & Zhou, A., (2014). Effect of onion flavonoids on colorectal cancer with hyperlipidemia: An *in vivo* study. *Onco. Targets Ther., 7,* 101–110.

Herman-Antosiewicz, A., & Singh, S. V., (2004). Signal transduction pathways leading to cell cycle arrest and apoptosis induction in cancer cells by *Allium* vegetable-derived organosulfur compounds: A review. *Mutation Research/Fundamental and Molecular Mechanisms of Mutagenesis, 555*(1, 2), 121–131.

Herr, I., & Büchler, M. W., (2010). Dietary constituents of broccoli and other cruciferous vegetables: Implications for prevention and therapy of cancer. *Cancer Treatment Reviews, 36*(5), 377–383.

Hirsch, K., Danilenko, M., Giat, J., Miron, T., Rabinkov, A., Wilchek, M., & Sharoni, Y., (2000). Effect of purified allicin, the major ingredient of freshly crushed garlic, on cancer cell proliferation. *Nutrition and Cancer, 38*(2), 245–254.

Hitomi, J., Christofferson, D. E., Ng, A., Yao, J., Degterev, A., Xavier, R. J., & Yuan, J., (2008). Identification of a molecular signaling network that regulates a cellular necrotic cell death pathway. *Cell, 135*(7), 1311–1323.

Hodges, R. E., & Minich, D. M., (2015). Modulation of metabolic detoxification pathways using foods and food-derived components: A scientific review with clinical application. *Journal of Nutrition and Metabolism, 2015.*

Hostanska, K., Reichling, J., Bommer, S., Weber, M., & Saller, R., (2003). Hyperforin a constituent of St John's wort (*Hypericum perforatum* L.) extract induces apoptosis by triggering activation of caspases and with hypericin synergistically exerts cytotoxicity towards human malignant cell lines. *Eur. J. Pharm. Biopharm., 56*, 121–132.

Hounsome, N., Hounsome, B., Tomos, D., & Edwards-Jones, G., (2008). Plant metabolites and nutritional quality of vegetables. *Journal of Food Science, 73*(4), R48–R65.

Huang, C. Y., Kuo, W. T., Huang, Y. C., Lee, T. C., & Yu, L. C. H., (2013). Resistance to hypoxia-induced necroptosis is conferred by glycolytic pyruvate scavenging of mitochondrial superoxide in colorectal cancer cells. *Cell Death & Disease, 4*(5), e622–e622.

Huang, J., Yang, B., Xiang, T., Peng, W., Qiu, Z., Wan, J., & Ren, G., (2015). Diallyl disulfide inhibits growth and metastatic potential of human triple-negative breast cancer cells through inactivation of the β-catenin signaling pathway. *Molecular Nutrition & Food Research, 59*(6), 1063–1075.

Hussain, A., Bourguet-Kondracki, M. L., Hussain, F., Rauf, A., Ibrahim, M., Khalid, M., & Rengasamy, K. R., (2022). The potential role of dietary plant ingredients against mammary cancer: A comprehensive review. *Critical Reviews in Food Science and Nutrition, 62*(10), 2580–2605.

Hussain, A., Bourguet-Kondracki, M. L., Hussain, F., Rauf, A., Ibrahim, M., Khalid, M., & Rengasamy, K. R., (2022). The potential role of dietary plant ingredients against mammary cancer: A comprehensive review. *Critical Reviews in Food Science and Nutrition, 62*(10), 2580–2605.

Hussain, A., Mohsin, J., Prabhu, S. A., Begum, S., Nusri, Q. E. A., Harish, G., ... & Sharma, C. (2013). Sulforaphane inhibits growth of human breast cancer cells and augments the therapeutic index of the chemotherapeutic drug, gemcitabine. *Asian Pacific Journal of Cancer Prevention, 14*(10), 5855.

Isabel, C. F. R. F., Lillian, B., & Rui, M. A., (2009). Antioxidants in wild mushrooms. *Current Medicinal Chemistry, 16*, 1543–1560.

Ismail, N. I., Othman, I., Abas, F., Lajis, N. H., & Naidu, R., (2019). Mechanism of apoptosis induced by curcumin in colorectal cancer. *International Journal of Molecular Sciences, 20*(10), 2454.

Jalili-Nik, M., Soltani, A., Moussavi, S., Ghayour-Mobarhan, M., Ferns, G. A., Hassanian, S. M., & Avan, A., (2018). Current status and future prospective of curcumin as a potential therapeutic agent in the treatment of colorectal cancer. *Journal of Cellular Physiology, 233*(9), 6337–6345.

Jaramillo-Carmona, S., Guillén-Bejarano, R., Jiménez-Araujo, A., Rodríguez-Arcos, R., & López, S., (2018). *In vitro* toxicity of asparagus saponins in distinct multidrug-resistant colon cancer cells. *Chemistry & Biodiversity, 15*(11), e1800282.

Jesus, J. A., Lago, J. H. G., Laurenti, M. D., Yamamoto, E. S., & Passero, L. F. D., (2015). Antimicrobial activity of oleanolic and ursolic acids: An update. *Evidence-Based Complementary and Alternative Medicine, 2015*.

Jin, J., Lin, G., Huang, H., Xu, D., Yu, H., Ma, X., & Jiang, H., (2014). Capsaicin mediates cell cycle arrest and apoptosis in human colon cancer cells via stabilizing and activating p53. *International Journal of Biological Sciences, 10*(3), 285.

Johnson, I., (2002). Glucosinolates in the human diet. Bioavailability and implications for health. *Phytochemistry Reviews, 1*(2), 183–188.

Kamal, N., Ilowefah, M. A., Hilles, A. R., Anua, N. A., Awin, T., Alshwyeh, H. A., & Mediani, A., (2022). Genesis and mechanism of some cancer types and an overview on the role of diet and nutrition in cancer prevention. *Molecules, 27*(6), 1794.

Kampa, M., Alexaki, V. I., Notas, G., Nifli, A. P., Nistikaki, A., Hatzoglou, A., & Castanas, E., (2004). Antiproliferative and apoptotic effects of selective phenolic acids on T47D human breast cancer cells: Potential mechanisms of action. *Breast Cancer Research, 6*(2), 1–12.

Kanimozhi, G., & Prasad, N. R., (2015). Anticancer effect of caffeic acid on human cervical cancer cells. In: *Coffee in Health and Disease Prevention* (pp. 655–661). Academic Press.

Kas, R., Sevinc, E., Topal, U., & Acar, H. Y., (2010). A universal method for the preparation of magnetic and luminescent hybrid nanoparticles. *The Journal of Physical Chemistry C, 114*(17), 7758–7766.

Kassi, E., Sourlingas, T. G., Spiliotaki, M., Papoutsi, Z., Pratsinis, H., Aligiannis, N., & Moutsatsou, P., (2009). Ursolic acid triggers apoptosis and Bcl-2 downregulation in MCF-7 breast cancer cells. *Cancer Investigation, 27*(7), 723–733.

Kesarkar, S., Bhandage, A., Deshmukh, S., Shevkar, K., & Abhyankar, M., (2009). Flavonoids: An overview. *Journal of Pharmacy Research, 2*(6), 1148–1154.

Keservani, R. K., & Sharma, A. K., (2014). Flavonoids: Emerging trends and potential health benefits. *Journal of Chinese Pharmaceutical Sciences, 23*(12), 815.

Keservani, R. K., Kesharwani, R. K., Sharma, A. K., & Ahmad, M. F., (2015a). Stress, Parkinson's, Alzheimer's disease: Role of dietary supplements; Chapter 14. In: Debasis, B., Harry, G. P., & Anand, S., (eds), *Nutraceutical and Functional Foods in Human Life and Disease Prevention* (pp. 241–253). Boca Raton, Fl. CRC, Press, Taylor, and Francis. ISBN: 9781482237214.

Keservani, R. K., Kesharwani, R. K., Sharma, A. K., Vyas, N., & Chadoker, A., (2010b). Nutritional supplements: An overview. *International Journal of Current Pharmaceutical Review and Research, 1*(1), 59–75.

Keservani, R. K., Kesharwani, R. K., Vyas, N., Jain, S., Raghuvanshi, R., & Sharma, A. K., (2010a). Nutraceutical and functional food as future food: A review. *Der Pharmacia Lettre, 2*(1), 106–116.

Keservani, R. K., Sharma, A. K., & Kesharwani, R. K., (2016a). Nutraceutical and functional foods for cardiovascular health. In: *Food Process Engineering* (pp. 257–278, 291–312). Apple Academic Press, CRC Press. ISBN: 9781771884020.

Keservani, R. K., Sharma, A. K., & Kesharwani, R. K., (2016b). Medicinal effect of nutraceutical fruits for the cognition and brain health. *Scientifica*, 10. Article ID 3109254. doi:10.1155/2016/3109254.

Keservani, R. K., Sharma, A. K., & Kesharwani, R. K., (2017). An overview and therapeutic applications of nutraceutical and functional foods. *Recent Advances in Drug Delivery Technology*, 160–201. ISBN:9781522507543.

Keservani, R. K., Sharma, A. K., & Kesharwani, R. K., (2020). *Nutraceuticals and Dietary Supplements: Applications in Health Improvement and Disease Management*. CRC Press. ISBN: 9781771888738.

Keservani, R. K., Singh, S., Singh, V., Kesharwani, R. K., & Sharma, A. K., (2015b). Nutraceuticals and functional foods in the prevention of mental disorder; Chapter 15b. In: Debasis, B., Harry, G. P., & Anand, S., (eds.), *Nutraceutical and Functional Foods in Human Life and Disease Prevention* (pp. 255–268). Boca Raton, Fl. CRC Press, Taylor, and Francis. ISBN: 9781482237214.

Kesharwani, R. K., Singh, P., & Keservani, R. K., (2016). Green coffee bean extract and chlorogenic acids: Chemistry and novel antioxidant benefits; Chapter 1. In: Bagchi, D., Moriyama, H., & Swaroop, A., (eds.), *Green Coffee Bean Extract in Human Health* (pp. 1–17). Boca Raton, Taylor and Francis, USA. ISBN: 9781498716376.

Keum, Y. S., Yu, S., Chang, P. P. J., Yuan, X., Kim, J. H., Xu, C., & Kong, A. N. T., (2006). Mechanism of action of sulforaphane: Inhibition of p38 mitogen-activated protein kinase isoforms contributing to the induction of antioxidant response element–mediated heme oxygenase-1 in human hepatoma HepG2 cells. *Cancer Research, 66*(17), 8804–8813.

Khamphio, M., Barusrux, S., & Weerapreeyakul, N., (2016). Sesamol induces mitochondrial apoptosis pathway in HCT116 human colon cancer cells via pro-oxidant effect. *Life Sciences, 158*, 46–56.

Khan, A. U., Dagur, H. S., Khan, M., Malik, N., Alam, M., & Mushtaque, M., (2021). Therapeutic role of flavonoids and flavones in cancer prevention: Current trends and future perspectives. *European Journal of Medicinal Chemistry Reports, 3*, 100010.

Khoo, H. E., Azlan, A., Tang, S. T., & Lim, S. M., (2017). Anthocyanidins and anthocyanins: Colored pigments as food, pharmaceutical ingredients, and the potential health benefits. *Food & Nutrition Research, 61*(1), 1361779.

Kim, D. H., Kim, M. J., Kim, D. W., Kim, G. Y., Kim, J. K., Gebru, Y. A., & Kim, M. K., (2019). Changes of phytochemical components (urushiols, polyphenols, gallotannins) and antioxidant capacity during *Fomitella fraxinea*–mediated fermentation of *Toxicodendron vernicifluum* bark. *Molecules, 24*(4), 683.

Kim, J. K., Gallaher, D. D., Chen, C., Gallaher, C. M., Yao, D., & Trudo, S. P., (2016). Phenethyl isothiocyanate and indole-3-carbinol from cruciferous vegetables, but not furanocoumarins from apiaceous vegetables, reduced PhIP-induced DNA adducts in Wistar rats. *Molecular Nutrition & Food Research, 60*(9), 1956–1966.

Kim, S. H., Lee, J. M., Kim, S. C., Park, C. B., & Lee, P. C., (2014). Proposed cytotoxic mechanisms of the saffron carotenoids crocin and crocetin on cancer cell lines. *Biochemistry and Cell Biology, 92*(2), 105–111.

Kim, Y. S., & Milner, J. A., (2005). Targets for indole-3-carbinol in cancer prevention. *J. Nutr. Biochem., 16*(2), 65–73. [PubMed: 15681163].

Kruk, J., Aboul-Enein, H. Y., Kładna, A., & Bowser, J. E., (2019). Oxidative stress in biological systems and its relation with pathophysiological functions: The effect of physical activity on cellular redox homeostasis. *Free Radical Research, 53*(5), 497–521.

Kumar, A., Rimando, A. M., & Levenson, A. S., (2017). Resveratrol and pterostilbene as a microRNA-mediated chemopreventive and therapeutic strategy in prostate cancer. *Annals of the New York Academy of Sciences, 1403*(1), 15–26.

Kumar, M., Kaur, V., Kumar, S., & Kaur, S., (2016). Phytoconstituents as apoptosis inducing agents: Strategy to combat cancer. *Cytotechnology, 68*(4), 531–563.

Kumar, S., & Pandey, A. K., (2013). Chemistry and biological activities of flavonoids: An overview. *The Scientific World Journal, 2013.*

Lala, G., Malik, M., Zhao, C., He, J., Kwon, Y., Giusti, M. M., & Magnuson, B. A., (2006). Anthocyanin-rich extracts inhibit multiple biomarkers of colon cancer in rats. *Nutrition and Cancer, 54*(1), 84–93.

Lalaoui, N., & Brumatti, G., (2017). Relevance of necroptosis in cancer. *Immunology and Cell Biology, 95*(2), 137–145.

Lambertini, E., Lampronti, I., Penolazzi, L., Khan, M. T. H., Ather, A., Giorgi, G., & Piva, R., (2005). Expression of estrogen receptor α gene in breast cancer cells treated with transcription factor decoy is modulated by Bangladeshi natural plant extracts. *Oncology Research Featuring Preclinical and Clinical Cancer Therapeutics, 15*(2), 69–79.

Lampe, J. W., Atkinson, C., & Hullar, M. A., (2006). Assessing exposure to lignans and their metabolites in humans. *Journal of AOAC International, 89*(4), 1174–1181.

Lasoń, W., & Leśkiewicz, M., (2013). Effect of plant polyphenols on seizures–animal studies. *Journal of Epileptology, 21*(2).

Lavecchia, T., Rea, G., Antonacci, A., & Giardi, M. T., (2013). Healthy and adverse effects of plant-derived functional metabolites: The need of revealing their content and bioactivity in a complex food matrix. *Critical Reviews in Food Science and Nutrition, 53*(2), 198–213. https://doi.org/10.1080/10408398.2010.520829.

Law, B. M., Waye, M. M., So, W. K., & Chair, S. Y., (2017). Hypotheses on the potential of rice bran intake to prevent gastrointestinal cancer through the modulation of oxidative stress. *International Journal of Molecular Sciences, 18*(7), 1352.

Lee, C. G., Lee, H. W., Kim, B. O., Rhee, D. K., & Pyo, S., (2015). Allicin inhibits invasion and migration of breast cancer cells through the suppression of VCAM-1: Regulation of association between p65 and ER-α. *Journal of Functional Foods, 15*, 172–185.

Lee, D. Y., Song, M. Y., & Kim, E. H., (2021). Role of oxidative stress and Nrf2/keap1 signaling in colorectal cancer: Mechanisms and therapeutic perspectives with phytochemicals. *Antioxidants, 10*(5), 743.

Lee, J. C., Won, S. J., Chao, C. L., Wu, F. L., Liu, H. S., Ling, P., & Su, C. L., (2008). Morusin induces apoptosis and suppresses NF-κB activity in human colorectal cancer HT-29 cells. *Biochemical and Biophysical Research Communications, 372*(1), 236–242.

Lee, L. Y., Cazier, J. B., Angelis, V., Arnold, R., Bisht, V., Campton, N. A., & Middleton, G., (2020). COVID-19 mortality in patients with cancer on chemotherapy or other anticancer treatments: A prospective cohort study. *The Lancet, 395*(10241), 1919–1926.

Lee, S. H., & Clark, R., (2016). Anti-tumorigenic effects of capsaicin in colon cancer. *J. Food Chem. Nanotechnol., 2*(4), 162–167.

Levy, Joseph, et al. (1995). "Lycopene is a more potent inhibitor of human cancer cell proliferation than either α-carotene or β-carotene." 257–266.

Li, J., McQuade, T., Siemer, A. B., Napetschnig, J., Moriwaki, K., Hsiao, Y. S., & Wu, H., (2012). The RIP1/RIP3 necrosome forms a functional amyloid signaling complex required for programmed necrosis. *Cell, 150*(2), 339–350.

Li, P. M., Li, Y. L., Liu, B., Wang, W. J., Wang, Y. Z., & Li, Z., (2014). Curcumin inhibits MHCC97H liver cancer cells by activating ROS/TLR-4/. *Asian Pac. J. Cancer Prev., 15*(5), 2329–2334. doi: 10.7314/ APJCP.2014.15.5.2329 caspase signaling pathway.

Li, Y., Zhang, T., Korkaya, H., Liu, S., Lee, H. F., Newman, B., & Sun, D., (2010). Sulforaphane, a dietary component of broccoli/broccoli sprouts, inhibits breast cancer stem cells sulforaphane inhibits breast cancer stem cells. *Clinical Cancer Research, 16*(9), 2580–2590.

Lin, M. C., Wang, F. Y., Kuo, Y. H., & Tang, F. Y., (2011). Cancer chemopreventive effects of lycopene: Suppression of MMP-7 expression and cell invasion in human colon cancer cells. *Journal of Agricultural and Food Chemistry, 59*(20), 11304–11318.

Liskova, A., Samec, M., Koklesova, L., Brockmueller, A., Zhai, K., Abdellatif, B., & Kubatka, P., (2021). Flavonoids as an effective sensitizer for anti-cancer therapy: Insights into multi-faceted mechanisms and applicability towards individualized patient profiles. *EPMA Journal, 12*(2), 155–176.

Liu, Y., Nguyen, N., & Colditz, G. A., (2015). Links between alcohol consumption and breast cancer: A look at the evidence. *Women's Health, 11*(1), 65–77.

Liu, Z., Fei, Y. J., Cao, X. H., Xu, D., Tang, W. J., Yang, K., Xu, W. X., & Tang, J. H., (2021). Lignans intake and enterolactone concentration and prognosis of breast cancer: A systematic review and meta-analysis. *Journal of Cancer, 12*(9), 2787–2796. https://doi.org/10.7150/jca.55477.

Lubet, R. A., Heckman, B. M., De Flora, S. L., Steele, V. E., Crowell, J. A., Juliana, M. M., & Grubbs, C. J., (2011). Effects of 5, 6-benzoflavone, indole-3-carbinol (I3C) and diindolylmethane (DIM) on chemically-induced mammary carcinogenesis: Is DIM a substitute for I3C? *Oncology Reports, 26*(3), 731–736.

Luo, X. J., Peng, J., & Li, Y. J., (2011). Recent advances in the study on capsaicinoids and capsinoids. *European Journal of Pharmacology, 650*(1), 1–7.

Maciag, A., Sithanandam, G., & Anderson, L. M., (2004). Mutant K-ras V12 increases COX-2, peroxides, and DNA damage in lung cells. *Carcinogenesis, 25*(11), 2231–2237.

Madadi, E., Mazloum-Ravasan, S., Yu, J. S., Ha, J. W., Hamishehkar, H., & Kim, K. H., (2020). Therapeutic application of betalains: A review. *Plants, 9*(9), 1219.

Manu, K. A., & Kuttan, G., (2008). Ursolic acid induces apoptosis by activating p53 and caspase-3 gene expressions and suppressing NF-κB mediated activation of bcl-2 in B16F-10 melanoma cells. *International Immunopharmacology, 8*(7), 974–981.

Masuelli, L., Benvenuto, M., Di Stefano, E., Mattera, R., Fantini, M., De Feudis, G., & Bei, R., (2017). Curcumin blocks autophagy and activates apoptosis of malignant mesothelioma cell lines and increases the survival of mice intraperitoneally transplanted with a malignant mesothelioma cell line. *Oncotarget, 8*(21), 34405.

Mertens-Talcott, S. U., Noratto, G. D., Li, X., Angel-Morales, G., Bertoldi, M. C., & Safe, S., (2013). Betulinic acid decreases ER-negative breast cancer cell growth *in vitro* and *in vivo*: Role of Sp transcription factors and microRNA-27a: ZBTB10. *Molecular Carcinogenesis, 52*(8), 591–602. doi: 10.1002/mc.21893.

Mertens-Talcott, S. U., Noratto, G. D., Li, X., Angel-Morales, G., Bertoldi, M. C., & Safe, S., (2013). Betulinic acid decreases ER-negative breast cancer cell growth *in vitro* and *in vivo*: Role of Sp transcription factors and microRNA-27a: ZBTB10. *Molecular Carcinogenesis, 52*(8), 591–602.

Miki, H., Uehara, N., Kimura, A., Sasaki, T., Yuri, T., Yoshizawa, K., & Tsubura, A., (2012). Resveratrol induces apoptosis via ROS-triggered autophagy in human colon cancer cells. *International Journal of Oncology, 40*(4), 1020–1028.

Milder, I. E., Arts, I. C., Van De, P. B., Venema, D. P., & Hollman, P. C., (2005). Lignan contents of Dutch plant foods: A database including lariciresinol, pinoresinol, secoisolariciresinol and matairesinol. *British Journal of Nutrition, 93*(3), 393–402.

Mishra, A. P., Salehi, B., Sharifi-Rad, M., Pezzani, R., Kobarfard, F., Sharifi-Rad, J., & Nigam, M., (2018). Programmed cell death, from a cancer perspective: An overview. *Molecular Diagnosis & Therapy, 22*(3), 281–295.

Mohan, S., Abdelwahab, S. I., Kamalidehghan, B., Syam, S., May, K. S., Harmal, N. S. M., & Zajmi, A., (2012). Involvement of NF-κB and Bcl2/Bax signaling pathways in the apoptosis of MCF7 cells induced by a xanthone compound Pyranocycloartobiloxanthone A. *Phytomedicine, 19*(11), 1007–1015.

Mondal, A., Banerjee, S., Bose, S., Mazumder, S., Haber, R. A., Farzaei, M. H., & Bishayee, A., (2022). Garlic constituents for cancer prevention and therapy: From phytochemistry to novel formulations. *Pharmacological Research, 175*, 105837.

Monsereenusorn, Y., Kongsamut, S., Pezalla, P. D., & Yaksh, T. L., (1982). Capsaicin—A literature survey. *CRC Critical Reviews in Toxicology, 10*(4), 321–339.

Monteiro, R., Calhau, C., Silva, A. O. E., Pinheiro-Silva, S., Guerreiro, S., Gärtner, F., & Soares, R., (2008). Xanthohumol inhibits inflammatory factor production and angiogenesis in breast cancer xenografts. *Journal of Cellular Biochemistry, 104*(5), 1699–1707.

Mousavi, L., Salleh, R. M., & Murugaiyah, V., (2018). Phytochemical and bioactive compounds identification of *Ocimum tenuiflorum* leaves of methanol extract and its fraction with an anti-diabetic potential. *Int. J. Food Prop., 21*, 2390–2399.

Mukherjee, A., & Khuda-Bukhsh, A. R., (2015). Quercetin down-regulates IL-6/STAT-3 signals to induce mitochondrial-mediated apoptosis in a nonsmall-cell lung-cancer cell line, A549. *Journal of Pharmacopuncture, 18*(1), 19.

Mukherjee, S., Ghosh, S., Das, D. K., Chakraborty, P., Choudhury, S., Gupta, P., & Chattopadhyay, S., (2015). Gold-conjugated green tea nanoparticles for enhanced anti-tumor activities and hepatoprotection—Synthesis, characterization, and *in vitro* evaluation. *The Journal of Nutritional Biochemistry, 26*(11), 1283–1297.

Murayyan, A. I., Manohar, C. M., Hayward, G., & Neethirajan, S., (2017). Antiproliferative activity of Ontario grown onions against colorectal adenocarcinoma cells. *Food Research International, 96*, 12–18.

Mut-Salud, N., Álvarez, P. J., Garrido, J. M., Carrasco, E., Aránega, A., & Rodríguez-Serrano, F., (2016). Antioxidant intake and antitumor therapy: Toward nutritional recommendations for optimal results. *Oxidative Medicine and Cellular Longevity, 2016*.

Nakagawa, H., Tsuta, K., Kiuchi, K., Senzaki, H., Tanaka, K., Hioki, K., & Tsubura, A., (2001). Growth inhibitory effects of diallyl disulfide on human breast cancer cell lines. *Carcinogenesis, 22*(6), 891–897.

Nayak, A. K., Hasnain, M. S., Dhara, A. K., & Pal, D., (2021). Plant polysaccharides in pharmaceutical applications. In: *Bioactive Natural Products for Pharmaceutical Applications* (pp. 93–125). Springer, Cham.

Neugart, S., Baldermann, S., Hanschen, F. S., Klopsch, R., Wiesner-Reinhold, M., & Schreiner, M., (2018). The intrinsic quality of brassicaceous vegetables: How secondary plant metabolites are affected by genetic, environmental, and agronomic factors. *Scientia Horticulturae, 233*, 460–478.

Newman, D. J., Cragg, G. M., & Kingston, D. G., (2011). *Anticancer Agents from Natural Products.* CRC Press Taylor & Francis, Boca Ratan.

Nwaeburu, C. C., Bauer, N., Zhao, Z., Abukiwan, A., Gladkich, J., Benner, A., & Herr, I., (2016). Up-regulation of microRNA let-7c by quercetin inhibits pancreatic cancer progression by activation of Numbl. *Oncotarget, 7*(36), 58367.

Ogrunc, M., Di Micco, R., Liontos, M., Bombardelli, L., Mione, M., Fumagalli, M., & D'Adda Di, F. F., (2014). Oncogene-induced reactive oxygen species fuel hyperproliferation and DNA damage response activation. *Cell Death & Differentiation, 21*(6), 998–1012.

Oh, J. H., Ban, J. O., Cho, M. C., Jo, M., Jung, J. K., Ahn, B., & Hong, J. T., (2012). 4-O-methylhonokiol inhibits colon tumor growth via p21-mediated suppression of NF-κB activity. *The Journal of Nutritional Biochemistry, 23*(7), 706–715.

Palop, J. J., Mucke, L., & Roberson, E. D., (2010). Quantifying biomarkers of cognitive dysfunction and neuronal network hyperexcitability in mouse models of Alzheimer's disease: Depletion of calcium-dependent proteins and inhibitory hippocampal remodeling. In: *Alzheimer's Disease and Frontotemporal Dementia* (pp. 245–262). Humana Press, Totowa, NJ.

Palozza, P., Serini, S., Maggiano, N., Angelini, M., Boninsegna, A., Di Nicuolo, F., & Calviello, G., (2002). Induction of cell cycle arrest and apoptosis in human colon adenocarcinoma cell lines by β-carotene through down-regulation of cyclin A and Bcl-2 family proteins. *Carcinogenesis, 23*(1), 11–18.

Palozza, P., Serini, S., Torsello, A., Di Nicuolo, F., Piccioni, E., Ubaldi, V., & Calviello, G., (2003). β-Carotene regulates NF-κB DNA-binding activity by a redox mechanism in human leukemia and colon adenocarcinoma cells. *The Journal of Nutrition, 133*(2), 381–388.

Pan, M. H., & Ho, C. T., (2008). Chemopreventive effects of natural dietary compounds on cancer development. *Chemical Society Reviews, 37*(11), 2558–2574.

Papadopoulos, V., Kapsis, A., Li, H., Amri, H., Hardwick, M., Culty, M., & Drieu, K., (2000). Drug-induced inhibition of the peripheral-type benzodiazepine receptor expression and cell proliferation in human breast cancer cells. *Anticancer Research, 20*(5A), 2835–2847.

Pappas, E., & Schaich, K. M., (2009). Phytochemicals of cranberries and cranberry products: Characterization, potential health effects, and processing stability. *Critical Reviews in Food Science and Nutrition, 49*(9), 741–781.

Patel, P. B., & Thakkar, V. R., (2014). L-carvone induces p53, caspase 3 mediated apoptosis and inhibits the migration of breast cancer cell lines. *Nutrition and Cancer, 66*(3), 453–462.

Pezzani, R., Salehi, B., Vitalini, S., Iriti, M., Zuñiga, F. A., Sharifi-Rad, J., & Martins, N., (2019). Synergistic effects of plant derivatives and conventional chemotherapeutic agents: An update on the cancer perspective. *Medicina, 55*(4), 110.

Polyakov, N. E., Leshina, T. V., Konovalova, T. A., & Kispert, L. D., (2001). Carotenoids as scavengers of free radicals in a Fenton reaction: Antioxidants or pro-oxidants? *Free Radical Biology and Medicine, 31*(3), 398–404.

Priyadarsini, R. V., Murugan, R. S., Maitreyi, S., Ramalingam, K., Karunagaran, D., & Nagini, S., (2010). The flavonoid quercetin induces cell cycle arrest and mitochondria-mediated apoptosis in human cervical cancer (HeLa) cells through p53 induction and NF-κB inhibition. *European Journal of Pharmacology, 649*(1–3), 84–91.

Qian, X., Melkamu, T., Upadhyaya, P., & Kassie, F., (2011). Indole-3-carbinol inhibited tobacco smoke carcinogen-induced lung adenocarcinoma in A/J mice when administered during the post-initiation or progression phase of lung tumorigenesis. *Cancer Letters, 311*(1), 57–65.

Rahal, A., Kumar, A., Singh, V., Yadav, B., Tiwari, R., Chakraborty, S., & Dhama, K., (2014). Oxidative stress, prooxidants, and antioxidants: The interplay. *BioMed Research International, 2014.*

Rahimi-Madiseh, M., Malekpour-Tehrani, A., Bahmani, M., & Rafieian-Kopaei, M., (2016). The research and development on the antioxidants in prevention of diabetic complications. Asian *Pacific Journal of Tropical Medicine, 9*(9), 825–831.

Rajendran, P., Maheshwari, U., Muthukrishnan, A., Muthuswamy, R., Anand, K., Ravindran, B., & Chung, W. J., (2021). Myricetin: Versatile plant-based flavonoid for cancer treatment by inducing cell cycle arrest and ROS–reliant mitochondria-facilitated apoptosis in A549 lung cancer cells and in silico prediction. *Molecular and Cellular Biochemistry, 476*(1), 57–68.

Ramljak, D., Romanczyk, L. J., Metheny-Barlow, L. J., Thompson, N., Knezevic, V., Galperin, M., & Dickson, R. B., (2005). Pentameric procyanidin from Theobroma cacao selectively inhibits growth of human breast cancer cells. *Molecular Cancer Therapeutics, 4*(4), 537–546.

Rane, B. R., Bharath, M. S., Patil, R. R., Keservani, R. K., & Jain, A. S., (2020b). novel approaches in nutraceuticals; Chapter 11. In: Rajesh, K. K., Raj, K. K., & Anil, K. S., (eds.), *Enhancing the Therapeutic Efficacy of Herbal Formulations through Novel Drug Delivery Systems* (pp. 241–266). IGI Global International Publisher, Pennsylvania, USA. ISBN13: 9781799844532.

Rane, B. R., Patil, A. S., Keservani, R. K., & Jain, A. S., (2020a). Novel approaches in herbal formulation; Chapter 2. In: Rajesh, K. K., Raj, K. K., & Anil, K. S., (eds.), *Enhancing the*

Therapeutic Efficacy of Herbal Formulations Through Novel Drug Delivery Systems (pp. 43–68). IGI Global International Publisher, Pennsylvania, US. ISBN13: 9781799844532.

Rane, B. R., Tadavi, S. A., & Keservani, R. K., (2020c). Naturopathy; Chapter 12. In: Anil, K. S., Raj, K. K., & Surya, P. G., (eds.), *Herbal Product Development* (pp. 321–347). Apple Academic Press, CRC Press, Taylor & Francis Group. ISBN: 9781771888776.

Rao, Y. K., Geethangili, M., Fang, S. H., & Tzeng, Y. M., (2007). Antioxidant and cytotoxic activities of naturally occurring phenolic and related compounds: A comparative study. *Food and Chemical Toxicology, 45*(9), 1770–1776.

Rayburn, E. R., Ezell, S. J., & Zhang, R., (2009). Anti-inflammatory agents for cancer therapy. *Molecular and Cellular Pharmacology, 1*(1), 29.

Ren, B., Kwah, M. X. Y., Liu, C., Ma, Z., Shanmugam, M. K., Ding, L., & Goh, B. C., (2021). Resveratrol for cancer therapy: Challenges and future perspectives. *Cancer Letters, 515*, 63–72.

Ren, B., Kwah, M. X., Liu, C., Ma, Z., Shanmugam, M. K., Ding, L., Xiang, X., et al., (2021). Resveratrol for cancer therapy: Challenges and future perspectives. *Cancer Letters, 515*, 63–72. https://doi.org/10.1016/j.canlet.2021.05.001.

Ren, B., Luo, S., Tian, X., Jiang, Z., Zou, G., Xu, F., et al., (2018). Curcumin inhibits liver cancer by inhibiting DAMP molecule HSP70 and TLR4 signaling. *Oncol. Rep., 40*(2), 895–901. doi: 10.3892/or.2018.6485.

Rohman, A., Riyanto, S., & Utari, D., (2006). Antioxidant activities, total phenolic and flavonoid contents of ethyl acetate extract of Mengkudu (*Morinda citrifolia*, L) fruit and its fractions. *Indonesian Journal of Pharmacy*, 136–142.

Roy, N., Nazeem, P. A., Babu, T. D., Abida, P. S., Narayanankutty, A., Valsalan, R., & Raghavamenon, A. C., (2018). EGFR gene regulation in colorectal cancer cells by garlic phytocompounds with special emphasis on S-allyl-L-cysteine sulfoxide. *Interdisciplinary Sciences: Computational Life Sciences, 10*(4), 686–693.

Rudolf, E., Andělová, H., & Červinka, M., (2009). Activation of several concurrent proapoptic pathways by sulforaphane in human colon cancer cells SW620. *Food and Chemical Toxicology, 47*(9), 2366–2373.

Ryan, E. P., Heuberger, A. L., Weir, T. L., Barnett, B., Broeckling, C. D., & Prenni, J. E., (2011). Rice bran fermented with *Saccharomyces boulardii* generates novel metabolite profiles with bioactivity. *Journal of Agricultural and Food Chemistry, 59*(5), 1862–1870.

Sathasivam, R., & Ki, J. S., (2018). A review of the biological activities of microalgal carotenoids and their potential use in healthcare and cosmetic industries. *Marine Drugs, 16*(1), 26.

Satoh, M., Takemura, Y., Hamada, H., Sekido, Y., & Kubota, S., (2013). EGCG induces human mesothelioma cell death by inducing reactive oxygen species and autophagy. *Cancer Cell International, 13*(1), 1–8.

Seca, A. M., & Pinto, D. C., (2018). Plant secondary metabolites as anticancer agents: Successes in clinical trials and therapeutic application. *International Journal of Molecular Sciences, 19*(1), 263.

Seiger, D. S., (1998). *Plant Secondary Metabolism* (p. 592). Dordrecht, The Netherlands: Kluwer Academic Publishers.

Seresht, H. R., Albadry, B. J., Al-mosawi, A. K. M., Gholami, O., & Cheshomi, H., (2019). The cytotoxic effects of thymol as the major component of *Trachyspermum ammi* on breast cancer (MCF-7) cells. *Pharmaceutical Chemistry Journal, 53*(2), 101–107.

Sharma, A. K., Keservani, R. K., & Gautam, S. P., (2020). *Herbal Product Development*. Apple Academic Press, CRC Press, Taylor & Francis Group. ISBN: 9781771888776.
Shebaby, W. N., Bodman-Smith, K. B., Mansour, A., Mroueh, M., Taleb, R. I., El-Sibai, M., & Daher, C. F., (2015). Daucus carota pentane-based fractions suppress proliferation and induce apoptosis in human colon adenocarcinoma HT-29 cells by inhibiting the MAPK and PI3K pathways. *Journal of Medicinal Food, 18*(7), 745–752.
Shen, G., Khor, T. O., Hu, R., Yu, S., Nair, S., Ho, C. T., & Kong, A. N. T., (2007). Chemoprevention of familial adenomatous polyposis by natural dietary compounds sulforaphane and dibenzoylmethane alone and in combination in Apc Min/+ mouse. *Cancer Research, 67*(20), 9937–9944.
Shoker, R. M., (2020). A review article: The importance of the major groups of plants secondary metabolism phenols, alkaloids, and terpenes. *International Journal for Research in Applied Sciences and Biotechnology, 7*(5), 354–358.
Sindhu, R. K., Verma, R., Salgotra, T., Rahman, M. H., Shah, M., Akter, R., & Al-Kuraishy, H. M., (2021). Impacting the remedial potential of nano delivery-based flavonoids for breast cancer treatment. *Molecules, 26*(17), 5163.
Singla, R. K., Dubey, A. K., Garg, A., Sharma, R. K., Fiorino, M., Ameen, S. M., & Al-Hiary, M., (2019). Natural polyphenols: Chemical classification, definition of classes, subcategories, and structures. *Journal of AOAC International, 102*(5), 1397–1400.
Siriwarin, B., & Weerapreeyakul, N., (2016). Sesamol induced apoptotic effect in lung adenocarcinoma cells through both intrinsic and extrinsic pathways. *Chemico-Biological Interactions, 254*, 109–116.
Srinivas, U. S., Tan, B. W., Vellayappan, B. A., & Jeyasekharan, A. D., (2019). ROS and the DNA damage response in cancer. *Redox Biology, 25*, 101084.
Talalay, P., & Fahey, J. W., (2001). Phytochemicals from cruciferous plants protect against cancer by modulating carcinogen metabolism. *The Journal of Nutrition, 131*(11), 3027S–3033S.
Tanaka, T., Shnimizu, M., & Moriwaki, H., (2012). Cancer chemoprevention by caroteno. *Molecules, 17*(3), 3202–3242.
Thoennissen, N. H., O'kelly, J., Lu, D., Iwanski, G. B., La, D. T., Abbassi, S., & Koeffler, H. P., (2010). Capsaicin causes cell-cycle arrest and apoptosis in ER-positive and-negative breast cancer cells by modulating the EGFR/HER-2 pathway. *Oncogene, 29*(2), 285–296.
Tian, S., Liao, L., Zhou, Q., Huang, X., Zheng, P., Guo, Y., et al., (2021). Curcumin inhibits the growth of liver cancer by impairing myeloid−derived suppressor cells in murine tumor *Tissues. Oncol Lett., 21*(4), 1–10.
Tidgewell, K., Clark, B. R., & Gerwick, W. H., (2010). 2.06-The natural products chemistry of cyanobacteria. In: Liu, H. W., & Mander, L., (eds.), *Comprehensive Natural Products*.
Tremblay, A., Arguin, H., & Panahi, S., (2016). Capsaicinoids: A spicy solution to the management of obesity? *International Journal of Obesity, 40*(8), 1198–1204.
Tse, A. K. W., Wan, C. K., Zhu, G. Y., Shen, X. L., Cheung, H. Y., Yang, M., & Fong, W. F., (2007). Magnolol suppresses NF-κB activation and NF-κB regulated gene expression through inhibition of IkappaB kinase activation. *Molecular Immunology, 44*(10), 2647–2658.
Valko, M., Leibfritz, D., Moncol, J., Cronin, M. T., Mazur, M., & Telser, J., (2007). Free radicals and antioxidants in normal physiological functions and human disease. *The International Journal of Biochemistry & Cell Biology, 39*(1), 44–84.

Vandenabeele, P., Galluzzi, L., Vanden, B. T., & Kroemer, G., (2010). Molecular mechanisms of necroptosis: An ordered cellular explosion. *Nature Reviews Molecular Cell Biology, 11*(10), 700–714.

Venkatadri, R., Muni, T., Iyer, A. K. V., Yakisich, J. S., & Azad, N., (2016). Role of apoptosis-related miRNAs in resveratrol-induced breast cancer cell death. *Cell Death & Disease, 7*(2), e2104-e2104.

Wakimoto, N., Yin, D., O'Kelly, J., Haritunians, T., Karlan, B., Said, J., & Koeffler, H. P., (2008). Cucurbitacin B has a potent antiproliferative effect on breast cancer cells *in vitro* and *in vivo*. *Cancer Science, 99*(9), 1793–1797.

Wang, D. X., Zou, Y. J., Zhuang, X. B., Chen, S. X., Lin, Y., Li, W. L., & Lin, Z. Q., (2017). Sulforaphane suppresses EMT and metastasis in human lung cancer through miR-616–5p-mediated GSK3β/β-catenin signaling pathways. *Acta Pharmacologica Sinica, 38*(2), 241–251.

Wang, H., Bian, S., & Yang, C. S., (2011). Green tea polyphenol EGCG suppresses lung cancer cell growth through upregulating miR-210 expression caused by stabilizing HIF-1α. *Carcinogenesis, 32*(12), 1881–1889.

Wang, H., Oo Khor, T., Shu, L., Su, Z. Y., Fuentes, F., Lee, J. H., & Tony, K. A. N., (2012). Plants vs. cancer: A review on natural phytochemicals in preventing and treating cancers and their druggability. *Anti-Cancer Agents in Medicinal Chemistry (Formerly Current Medicinal Chemistry-Anti-Cancer Agents), 12*(10), 1281–1305.

Wang, Y., Chen, Y., Wang, J., Chen, J., Aggarwal, B. B., Pang, X., & Liu, M., (2012). Xanthohumol, a prenylated chalcone derived from hops, suppresses cancer cell invasion through inhibiting the expression of CXCR4 chemokine receptor. *Current Molecular Medicine, 12*(2), 153–162.

Wattanathamsan, O., Hayakawa, Y., & Pongrakhananon, V., (2019). Molecular mechanisms of natural compounds in cell death induction and sensitization to chemotherapeutic drugs in lung cancer. *Phytotherapy Research, 33*(10), 2531–2547.

Webb, A. L., & McCullough, M. L., (2005). Dietary lignans: Potential role in cancer prevention. *Nutrition and Cancer, 51*(2), 117–131.

Wondrak, G. T., (2009). Redox-directed cancer therapeutics: Molecular mechanisms and opportunities. *Antioxidants & Redox Signaling, 11*(12), 3013–3069.

Wright, A., Reiley, W. W., Chang, M., Jin, W., Lee, A. J., Zhang, M., & Sun, S. C., (2007). Regulation of early wave of germ cell apoptosis and spermatogenesis by deubiquitinating enzyme CYLD. *Developmental Cell, 13*(5), 705–716.

Wu, W., Liu, P., & Li, J., (2012). Necroptosis: An emerging form of programmed cell death. *Critical Reviews in Oncology/Hematology, 82*(3), 249–258.

Xiao, D., Choi, S., Johnson, D. E., Vogel, V. G., Johnson, C. S., Trump, D. L., & Singh, S. V., (2004). Diallyl trisulfide-induced apoptosis in human prostate cancer cells involves c-Jun N-terminal kinase and extracellular-signal regulated kinase-mediated phosphorylation of Bcl-2. *Oncogene, 23*(33), 5594–5606.

Xiao, X., Zhang, Y., Pan, W., & Chen, F., (2020). miR-139-mediated NOTCH1 regulation is crucial for the inhibition of osteosarcoma progression caused by resveratrol. *Life Sciences, 242*, 117215.

Yamada, S., Tsukamoto, S., Huang, Y., Makio, A., Kumazoe, M., Yamashita, S., & Tachibana, H., (2016). Epigallocatechin-3-O-gallate up-regulates microRNA-let-7b expression by activating 67-kDa laminin receptor signaling in melanoma cells. *Scientific Reports, 6*(1), 1–8.

Yang, C., Xie, X., Tang, H., Dong, X., Zhang, X., & Huang, F., (2018). Transcriptome analysis reveals GA induced apoptosis in HCT116 human colon cancer cells through calcium and p53 signal pathways. *RSC Advances, 8*(22), 12449–12458.

Yang, H., & Ping, D. Q., (2010). Targeting apoptosis pathway with natural terpenoids: Implications for treatment of breast and prostate cancer. *Current Drug Targets, 11*(6), 733–744.

Yang, L., Hu, Z., Zhu, J., Liang, Q., Zhou, H., Li, J., & Fei, B., (2020). Systematic elucidation of the mechanism of quercetin against gastric cancer via network pharmacology approach. *BioMed Research International, 2020.*

Yang, N., Zhao, Y., Wang, Z., Liu, Y., & Zhang, Y., (2017). Scutellarin suppresses growth and causes apoptosis of human colorectal cancer cells by regulating the p53 pathway. *Molecular Medicine Reports, 15*(2), 929–935.

Yedjou, C. G., & Tchounwou, P. B., (2012). *In vitro* assessment of oxidative stress and apoptotic mechanisms of garlic extract in the treatment of acute promyelocytic leukemia. *Journal of Cancer Science & Therapy, 2012*(Suppl 3), 006.

Yesil-Celiktas, O., Sevimli, C., Bedir, E., & Vardar-Sukan, F., (2010). Inhibitory effects of rosemary extracts, carnosic acid and rosmarinic acid on the growth of various human cancer cell lines. *Plant Foods for Human Nutrition, 65*(2), 158–163.

Yi, L. V., Kwok-Fai, S. O., Nai-Kei, W. O. N. G., & Jia, X. I. A. O., (2019). Anti-cancer activities of S-allylmercaptocysteine from aged garlic. *Chinese Journal of Natural Medicines, 17*(1), 43–49.

Yi, L., & Su, Q., (2013). Molecular mechanisms for the anti-cancer effects of diallyl disulfide. *Food and Chemical Toxicology, 57*, 362–370.

Yoder, S. C., Lancaster, S. M., Hullar, M. A., & Lampe, J. W., (2015). Gut microbial metabolism of plant lignans: Influence on human health. In: *Diet-Microbe Interactions in the Gut* (pp. 103–117). Academic Press.

You, W. C., Blot, W. J., Chang, Y. S., Ershow, A., Yang, Z. T., An, Q., & Wang, T. G., (1989). Allium vegetables and reduced risk of stomach cancer. *JNCI: Journal of the National Cancer Institute, 81*(2), 162–164.

Youn, H., Jeong, J. C., Jeong, Y. S., Kim, E. J., & Um, S. J., (2013). Quercetin potentiates apoptosis by inhibiting nuclear factor-kappaB signaling in H460 lung cancer cells. *Biological and Pharmaceutical Bulletin, 36*(6), 944–951.

Yuan, M., Zhang, G., Bai, W., Han, X., Li, C., & Bian, S., (2022). The role of bioactive compounds in natural products extracted from plants in cancer treatment and their mechanisms related to anticancer effects. *Oxidative Medicine and Cellular Longevity, 2022.*

Zan, L., Chen, Q., Zhang, L., & Li, X., (2019). Epigallocatechin gallate (EGCG) suppresses growth and tumorigenicity in breast cancer cells by downregulation of miR-25. *Bioengineered, 10*(1), 374–382.

Zaorsky, N. G., Churilla, T., Egleston, B., Fisher, S., Ridge, J., Horwitz, E., & Meyer, J., (2017). Causes of death among cancer patients. *Ann. Oncol., 28*, 400–407.

Zare, M., Norouzi, R. Z., Assadpour, E., & Jafari, S. M., (2021). Improving the cancer prevention/treatment role of carotenoids through various nano-delivery systems. *Critical Reviews in Food Science and Nutrition, 61*(3), 522–534.

Zhang, B., (2015). MicroRNA: A new target for improving plant tolerance to abiotic stress. *Journal of Experimental Botany, 66*(7), 1749–1761.

Zhang, L., Chinnathambi, A., Alharbi, S. A., Veeraraghavan, V. P., Mohan, S. K., & Zhang, G., (2020). Punicalagin promotes the apoptosis in human cervical cancer (ME-180) cells through mitochondrial pathway and by inhibiting the NF-kB signaling pathway. *Saudi J. Biol. Sci., 27*, 1100–1106.

Zhang, L., Tao, X., Fu, Q., Ge, C., Li, R., Li, Z., et al., (2019). Curcumin inhibits cell proliferation and migration in NSCLC through a synergistic effect on the TLR4/MyD88 and EGFR pathways. *Oncol. Rep., 42*(5), 1843–1855. doi: 10.3892/or.2019.7278.

Zhang, S., Cao, M., & Fang, F., (2020). The role of epigallocatechin-3-gallate in autophagy and endoplasmic reticulum stress (ERS)-induced apoptosis of human diseases. *Medical Science Monitor: International Medical Journal of Experimental and Clinical Research, 26*, e924558–1.

Zhang, W., Yin, G., Dai, J., Sun, Y. U., Hoffman, R. M., Yang, Z., & Fan, Y., (2017). Chemoprevention by quercetin of oral squamous cell carcinoma by suppression of the NF-κB signaling pathway in DMBA-treated hamsters. *Anticancer Research, 37*(8), 4041–4049.

Zhang, X., Zhu, Y., Duan, W., Feng, C., & He, X., (2015). Allicin induces apoptosis of the MGC-803 human gastric carcinoma cell line through the p38 mitogen-activated protein kinase/caspase-3 signaling pathway. *Molecular Medicine Reports, 11*(4), 2755–2760.

Zhang, Y., & Tang, L., (2007). Discovery and development of sulforaphane as a cancer chemopreventive phytochemical 1. *Acta Pharmacologica Sinica, 28*(9), 1343–1354.

Zhang, Y., Vareed, S. K., & Nair, M. G., (2005). Human tumor cell growth inhibition by nontoxic anthocyanidins, the pigments in fruits and vegetables. *Life Sciences, 76*(13), 1465–1472.

Zhang, Z., Shi, J., Nice, E. C., Huang, C., & Shi, Z., (2021). The multifaceted role of flavonoids in cancer therapy: Leveraging autophagy with a double-edged sword. *Antioxidants, 10*(7), 1138.

Zhao, J., Fang, Z., Zha, Z., Sun, Q., Wang, H., Sun, M., & Qiao, B., (2019). Quercetin inhibits cell viability, migration, and invasion by regulating miR-16/HOXA10 axis in oral cancer. *European Journal of Pharmacology, 847*, 11–18.

CHAPTER 7

PLANT METABOLITES AND VEGETABLES FOR GOUT PREVENTION AND TREATMENT

BUI THANH TUNG,[1] NGUYEN NHU SON,[1] NGUYEN BAO KIM,[1] and RAJ K. KESERVANI[2]

[1]*VNU University of Medicine and Pharmacy, Vietnam National University, Hanoi, Vietnam*

[2]*Faculty of B. Pharmacy, CSM Group of Institutions, Prayagraj, Uttar Pradesh, India*

ABSTRACT

Gout is the most common form of inflammatory arthropathy globally and is caused by the deposition of monosodium urate (MSU) crystals. People with persistent hyperuricemia can lead to chronic gouty arthritis, tophi, irreversible structural joint damage, and even disability. Some commonly approved drugs such as allopurinol have been associated with serious dose-dependent side effects. Therefore, there are needs to develop safe and useful new gout treatments from medicinal plants. Traditional medicines have long been known to contain various biologically active compounds with anti-hyperuricemia effects including colchicine, berberine, resveratrol (RES), epigallocatechin-3-gallate (EGCG), morin, quercetin, caffeic acid, etc. The authors aim to provide a concise overview of the pathogenesis and primary therapeutic targets associated with medicinal plants in the management of gouty arthritis. Additionally, the chapter presents various potential herbal plants and isolated phytochemicals that have been explored and utilized in research to prevent and alleviate symptoms associated with this condition.

Plant Metabolites and Vegetables as Nutraceuticals.
Raj K. Keservani, Bui Thanh Tung, Rajesh K. Kesharwani, and Eknath D. Ahire (Eds.)

7.1 INTRODUCTION

Gout is the most common form of inflammatory arthropathy globally (Dehlin et al., 2020). According to the Global Health Data Exchange (GHDx) and the World Health Organization (WHO) database, there were an estimated 7.44 million cases of gout around the world in 2017 (incidence, 0.097%), with 41.22 million prevalent cases (0.54%), causing 1.28 million disability-adjusted life years (DALYs) (0.051% of all DALYs). Gout incidence, prevalence, and health damage significantly magnified throughout the last 25 years, particularly in high SDI (socio-demographic index) countries and are much higher in men than in women. Gout's burden will increase linearly with aging until the age of 64 years, and gout mortality is expected to escalate by 55% in 2060 (Mattiuzzi & Lippi, 2020).

A high concentration of serum urate is the most crucial risk factor for gout's development. Several factors that affect hyperuricemia levels are also risk factors for incident gout, including chronic kidney disease, metabolic syndrome, and some medications such as diuretics and ciclosporin (Dalbeth et al., 2021). Gout has also long been associated alcohol and excessive dietary intake. The prevalence heightened proportional to red meat, seafood, and fructose consumption and beer intake, whereas high purine content vegetables and moderate wine consumption show no effect. Rates of gout are also increased with raised body-mass index but decreased with weight loss (Richette & Bardin, 2010). The relative effect of genetic and environmental risk factors for gout is controversial (Dalbeth et al., 2021).

The progression of hyperuricemia and gout can be divided into four primary pathophysiological stages: hyperuricemia development, monosodium urate (MSU) crystals deposition, gout flares first clinical presentation due to an acute inflammatory response initiated by deposited crystals, and advanced disease clinical presentation which characterized by tophi (Dalbeth & Stamp, 2014).

Gout is caused by the deposition of MSU crystals in articular and non-articular structures. Precipitated MSU crystals act as pro-inflammatory stimuli which trigger, amplify, and sustain an intense immuno-inflammatory response (Richette & Bardin, 2010). After released from preformed deposits in the joints, MSU crystals will be phagocytosed by monocytes as particles, thus initiating a typical inflammatory response through the release of pro-inflammatory mediators such as interleukin 1β, tumor necrosis factor α, and interleukin 8 (Pétrilli & Martinon, 2007). Mechanisms by which urate crystals activate cells in the joint cavity have just been partly explained (Lioté

& Ea, 2007; So, 2008). MSU crystals might activate monocytes through the toll-like receptor (TLR) pathway and the inflammasome (Dalbeth et al., 2021).

The typical first presentation of gout is usually an intensely painful, rapidly developed acute inflammatory monoarticular arthritis (gout flare) affecting a lower limb joint, marked by swelling and redness. The gout flare in most cases, without treatment, is self-limiting over a period of 7–14 days. After that, a pain-free asymptomatic period (intercritical gout) will occur before another gout flare happens (Dalbeth et al., 2021). Over time, people with persistent hyperuricemia might develop chronic gouty arthritis (persistent joint inflammation), irreversible structural joint damage and even disability with massively deformed joints. Tophi (i.e., subcutaneous nodules contain MSU crystals in a matrix structure comprised of lipids, proteins, and mucopolysaccharides) may also form at the joint space (Schlesinger & Thiele, 2010). The commonly affected regions include the midtarsal joints, knees, ankles, wrists, elbows, fingers, and especially the first metatarsophalangeal joint. Urate crystals can also be precipitated throughout the whole body (e.g., skin, vertebrae, soft tissues) and mimicking other disease states (Ning & Keenan, 2010). Microscopy of joint fluid can also be used in diagnosis but not much often, mostly in equivocal cases.

The differential diagnosis for acute monoarticular joint swelling is pseudogout, infection, and trauma. Pseudogout, or calcium pyrophosphate (CPP) deposition disease, can be just like gout in term of clinical appearance and might also respond to nonsteroidal anti-inflammatory drugs (NSAIDs). Detecting CPP crystals on joint fluid analysis and normal serum uric acid (UA) levels can help to differentiate pseudogout from gout (Hainer et al., 2014).

7.1.1 PREVENTION

Serum urate-lowering therapy ought to be initiated to prevent recurrence in individuals with a history of gout and any of the following: at least two flares per year (one per year in individuals with chronic kidney disease stage 2 or higher), tophi, or a history of kidney stones (Janssens et al., 2008).

Serum urate levels should be reduced to a target lower than 5 to 6 mg per dL (297 to 357 μmol per L), depending on the crystal and tophaceous burden (Janssens et al., 2008). Normal serum urate levels cannot exclude the diagnosis of gout. They should be observed regularly for assessment of preventive therapy in individuals with recurrent gout and a history of high urate levels (Burns & Wortmann, 2012). Urate-lowering therapy should be maintained for three to six months following a flare if no ongoing symptoms detected. Therapy should be carried on indefinitely if there are signs of ongoing symptoms (e.g., one or more tophi on examination) (Janssens et al., 2008).

7.1.2 TREATMENT

For prompt and definitive resolution of symptoms, treatment for acute gout should begin within 24 hours of symptoms. Oral corticosteroids, intravenous corticosteroids, NSAIDs, and colchicine are all equally effective in acute gout flares treatment (Janssens et al., 2008). NSAIDs are now considered the first-line treatment. Indomethacin (Indocin) was previously the preferred choice; however, there is no substantiation it is more effective than any other NSAID. Intramuscular ketorolac appears to be equally effective (Khanna et al., 2012). Any oral NSAID can be used at maximum dosage and continued for one to two days after remission of symptoms (Hainer et al., 2014).

Corticosteroids are a proper alternative for patients who are intolerant to NSAIDs or colchicine (Burns & Wortmann, 2012). Diabetic patients may be given corticosteroids for short use with proper monitoring for hyperglycemia. When gout only affects a single joint, intra-articular corticosteroid injections might be preferable to systemic corticosteroids due to fewer side effects (Zhang et al., 2006). Discontinued corticosteroid therapy in acute gout usually lead to rebound flares. To lessen the possibility of rebound flares, prophylaxis is recommended, and a tapering course of corticosteroids is initiated within 10 to 14 days of resolution of symptoms (Hainer et al., 2014).

Another treatment option for acute gout is colchicine. Generic colchicine, which has been used for many years, failed to endure formal review by the U.S. Food and Drug Administration (FDA) for this indication till 2009, once branded colchicine (Colcrys) was approved. However, Colcrys is pricey, and generic colchicine is not any longer accessible. Additionally, colchicine does not possess analgesic properties and may be less effective in treating acute flares when used 72 to 96 hours after symptom onset. Common side effects include nausea, vomiting, and diarrhea (Khanna et al., 2012; Zhang et al., 2006). Colchicine should be cautiously used in hepatic or renal impairment patients (Burns & Wortmann, 2012).

Xanthine oxidase (XOD) is the rate-limiting enzyme takes part in UA production. XOD regulates the oxidation process of xanthine to UA in livers, which was considered to be an promising therapeutic target for treatment of gout and hyperuricemia (Mandal & Mount, 2015). In fact, several XOD inhibitors including allopurinol and febuxostat have been developed as anti-hyperuricemia and anti-gout medicine commonly used clinically (Chen et al., 2016). However, both febuxostat and allopurinol have been connected with serious adverse effects, while their use was demonstrated to be not effective for acute gouty arthritis (Hilmi et al., 2012; Sattui & Gaffo, 2016). On the other hand, NOD-like receptor 3 (NLRP3) which constituted by NLRP3, apoptosis-associated speck-like protein (ASC) and caspase-1 could be activated after recognizing MSU, and then lead to the inflammatory cascade (Hoffman et al., 2010). Since NLRP3 has been seen as an important recognition receptor involved in the pathogenesis of acute gouty arthritis (Kingsbury et al., 2011), a few NLRP3 inhibitors for gout therapy have entered the phase of clinical research, such as bucillamine (Zhou et al., 2018). Recently, hyperuricemia and gout treatment from natural products such as medicinal plants, phytochemicals, and botanical herb extracts have attracted a great deal of attention due to their nontoxic properties as an alternative to synthetic chemicals (Jhang et al., 2018; Li et al., 2015; Mehmood et al., 2019). Among them, many have shown significant XOD and NLRP3 inflammasome inhibitory activities. In this review, we aim to focus on phytochemicals and medicinal plants that have been demonstrated to exert therapeutic effect on gout mainly through one or both important mechanisms mentioned above.

7.2 MEDICINAL Plants Using for Treatment of Gouty Arthritis

7.2.1 GLORIOSA superba L. (Colchicaceae)

Gloriosa superba L. (Colchicaceae) also known as Glory lily, is a high-value medicinal plant indigenous to Africa, India, and Southeast Asia (Pandey et al., 2021). This herbal is useful in curing inflammation, bleeding piles, ulcers, skin diseases, leprosy, and snakebites but the most well-known use is against rheumatism and gout (Jana & Shekhawat, 2011). As a result of its over-exploitation as well as problems faced during field cultivation, Gloriosa superba L. was on the verge of extinction and therefore, has been cited as a threatened species in the Red Data Book (2001) by the International Union for Conservation of Nature and Natural Resources (IUCN) (Khandel et al.,

2012; Sivakumar & Krishnamurthy, 2000). The therapeutic value of the Glory lily is particularly owing to the alkaloids presence in all parts of the plant, especially the colchicines (colchicine and gloriosine) (Sivakumar et al., 2004). Although the use of colchicine has an associated risk of toxicity, it still gained the US FDA approval in 1992 for its extraordinary effect in the treatment of gout, by preventing microtubule assembly thereby disrupts inflammasome activation, microtubule-based inflammatory cell chemotaxis, generation of leukotrienes and cytokines and phagocytosis (Dalbeth et al., 2014). The seeds of G. superba make up a potential commercial source of colchicine and colchicoside as it contains up to 0.9% colchicine and 0.8% colchicoside (Roberge et al., 1996). John et al. reported that hydroalcoholic extract of aerial part of G. superba possesses a moderate anti-inflammatory effect and is safe up to 2,000 mg/kg body weight when tested on Wistar rat model weighing 150–200 g (John et al., 2009).

7.2.2 ASTER glehni

Originated from Ulleung Island, Korea, A. glehni has long been used as food and traditional medicine by Korean. A. Glehni has been reported to possess beneficial effects on diabetes mellitus, hypercholesterolemia, cardiovascular disease (CVD), sedative effects, and insomnia (Lee et al., 2013). The hypouricemic activity of ethanol extract from Asterglehni leaves was evaluated in an experimental model of hyperuricemia in mice induced by potassium oxonate (PO). The results showed that A. glehni extract had both in vitro and in vivo inhibitory effects against XOD, thus remarkably reduced serum urate concentration in hyperuricemic rats. A. glehni extract, therefore, might be a promising therapeutic medicinal plant for treatment of gout (Park et al., 2018). In addition, A. glehni leaves have also been identified to contain a significant level of flavonoids (Keservani & Sharma, 2014) which are mainly responsible for the strong XOD inhibitory activity. The detected 3,5-dicaffeoylquinic acid methyl ester and 4,5-dicaffeoylquinic acid methyl ester from isolated Asterglehni leaves in ethyl acetate soluble fraction showed the most potent inhibitory activity against XOD (Son et al., 2015). XOD generates reactive oxygen species (ROS) by transferring electrons to oxygen from hypoxanthine and xanthine to produce H2O2, which shows that antioxidants could be potential in treating hyperuricemia by inhibiting XOD activity. Oxidative stress in uremia is the result of higher ROS production and XOD is one of the main factors to create excessive

ROS (Duni et al., 2017). A. glehni extract contains both XOD inhibitory activity and antioxidant activity. These findings indicated that antioxidant activity might contribute to the mechanism of action of A. glehni extract in hyperuricemic rodents. Moreover, in comparison with Trolox, A. glehni extract also showed moderate antioxidant activity. Therefore, A. glehni can be considered promising to ameliorate diseases like hyperuricemia and gout and worth to be further researched, either alone or combined with other gout treatments (Park et al., 2018).

7.2.3 SIEGESBECKIA orientalis L.

Siegesbeckia orientalis, belongs to the Asteraceae family, is widely spread in the South-East Asia region and especially in Vietnam (Wang et al., 2011). The aerial part of this medicinal plant has been used for therapeutic purposes in many inflammatory disorders, such as neurasthenia, impetigo, insomnia, furuncle, wound, and burn (El-Tantawy, 2019). The crude ethanol extract (CEE) and its butanolic fraction (BuOH fr.) were prepared from the aerial parts of S. Orientalis collected in the mountainous region of Vietnam. Antihyperuricemic activities of the CEE and BuOH fr. were then tested in vivo with the help of the oxonate-induced hyperuricemia rats' model through determination of serum urate levels and XOD inhibitory effects in rat liver. The results suggested that CEE of Siegesbeckia orientalis showed anti-hyperuricemic activity and the most powerful portion of the extract was the BuOH fraction. On the other hand, phenolic compounds were identified as significant components of the BuOH fraction and therefore are thought to be preeminently associated with the anti-gout and anti-inflammatory effects of S. orientalis (Ling & Bochu, 2014). The major constituents in this fraction comprise of caffeic acid analogs: chlorogenic acid, caffeic acid, 3-CQA (3-caffeoylquinic acid), 5-CQA (5-caffeoylquinic acid), diCQAs (3,4-dicaffeoylquinic acid and 3,5-dicaffeoylquinic acid. The BuOH fraction also contains various flavonones constituents such as rutin, quercitrin, kaempferol-3-O-rutinoside, and kaempferol-3-O-rhamnoside (Nguyen et al., 2017). Further in vivo studies on this fraction witnessed a 31.4% reduction of serum urate levels and 32.7% inhibition of XOD. Multiple studies have indicated that caffeic acid analogs possess hyperuricemia and anti-inflammatory activities (Masuda et al., 2014), which, in S. Orientalis, have been suggested to be mainly linked to XOD inhibitory activity of the phenolic components. The less recognized flavonones is also reported to exert hyperuricemia and anti-inflammatory effects as a dual inhibitor of XOD and NLRP3 (Zhou et al., 2018). In short, these findings supported the use of this herb in the treatment of gout and other inflammatory diseases although further research are still needed (Nguyen et al., 2017).

7.2.4 *PLANTAGO asiatica L.*

It is a perennial herbal plant commonly used as a component in dietary supplements, and the dried ripe seeds of P. asiatica, Plantaginis semen, are one of the most common folk herbal medicines in China as well as other Asian countries for treatment of gout and gouty nephropathy, with recovery rate up to 86.7% and symptoms were significantly improved (El-Tantawy, 2019; Zhao et al., 2021). It is reported that P. asiatica possesses XOD inhibitory activity in vitro and in vivo and could therefore be used to develop a more powerful XOD inhibitor against hyperuricemia. The XOD inhibitory activities of Plantaginis Semen and four representative component compounds, acteoside, isoacteoside, 1H-indolo-3-carbaldehyde and myristic acid, were analyzed through electron transfer signal blocking activities (ETSBAS), which is based on the electron transfer signal of XOD enzymatic reaction. The inhibitory activities were measured using an electrochemical biosensing method. The effects were significantly noticed after the addition of P. asiatica extract, acteoside, and 1H-indolo-3-carbaldehyde as compared with control. The IC50 values of the extract and acteoside were 89.14 and 7.55 mg/mL, respectively (Zeng et al., 2018). The results suggested that P. asiatisca blocked the electron transfer activities of the enzymatic reactions (Shan et al., 2009). On the other hand, Zhao et al. illustrated that Plantaginis Semen polysaccharides could ameliorate the inflammatory response and regulating the (NOD)-like receptor protein 3 (NLRP3) protein in gouty nephropathy rats (Zhao et al., 2021). It is also reported that Plantaginis Semen could attenuate PO-induced hyperuricemia through lipid metabolism disorder regulation. Additionally, Plantaginis Semen might exert anti-hyperuricemia effect through URAT1 and PI3K/Akt pathway, but the mechanism needs to be further studied (Yang et al., 2021). In short, P. asiatica is a strongly effective anti-gout candidate and need deeper studies both in clinical and preclinical.

7.2.5 *GNAPHALIUM affine D. Don*

Gnaphalium affine, belonging to Asteraceae, is an annual herbaceous plant which is widely distributed in diverse regions of China. Locally named Qing

Ming vegetable, itis reported to be used as a folk medicine due to the anti-inflammatory, antitussive, gout, and expectorant activities (Li et al., 2013; Meng et al., 2021). It is reported that the primary mechanism of Gnaphalium affine in the treatment of gout could be attributed to its inhibitory effect on XOD (Lin et al., 2014). Previous research showed that G. affine extracts ingredients were rich of flavonoids, caffeoylquinates, phenolic acids and adenine derivatives which appeared to have potent XOD inhibitory activity in vitro and also related to anti-inflammatory activity (Zhang et al., 2018). Among the chemical constituents from Gnaphalium affine, luteolin have been consensually demonstrated to possess significant uricosuric effects in hyperuricemia mice, predominantly through decreasing renal mURAT1 and also inhibiting XOD activity, with IC50 value of 2.42 ± 0.13 μmol/L as compared with control allopurinol (IC50 value of 1.62 ± 0.16 μmol/L) (Lin et al., 2018; Zhang et al., 2018). In a PO-induced hyperuricemia mice model, the tested G. affine extract was shown to possess potent uricosuric effect in hyperuricemia mice through effect renal mGLUT9 (glucose transporter 9) and mURAT1 (urate transporter 1) primarily and inhibiting XO activity to a certain extent, which are attributable to the improvement of UA excretion and protect hyperuricemia-induced renal dysfunction. Meanwhile, G. Affine also showed significant anti-inflammatory activity, by potently attenuated paw swelling in MSU crystal-induced acute gouty arthritis mice, which in part seemingly due to the downregulation of the production of IL-1β and TNF-α (Zhang et al., 2017). To conclude, Gnaphalium affine extracts can be considered a promising candidate for gout and anti-hyperuricemia treatment and future studies are needed to evaluate the safety and potential clinical utility.

7.2.6 SELAGINELLA moellendorffii Hieron.

Selaginella moellendorffii, in the family of Selaginellaceae, is a traditional Chinese herb with extremely rich resources. It has long been applied as a folk medicine to treat bleeding and chronic inflammation, such as arthritis, hepatitis, gonorrhea, and mastitis (Cao et al., 2010). The potency of aqueous Selaginella moellendorffii extract on hyperuricemia and gouty arthritis was examined. The extract was shown to be efficient in reducing hyperuricemia in a dose-dependent manner, and the levels of blood urea nitrogen and creatinine in high dose group were remarkably decreased as compared with hyperuricemic control group (Zhao et al., 2017). Many previous studies on

Selaginella moellendorffii revealed the presence of phytochemical groups such as bioflavonoids (da Silva Almeida et al., 2013), lignans (Wang et al., 2011), and phenols (Chao et al., 1987). Containing flavonoids and phenols ingredients, the medicinal plant is an ideal candidate for the treatment of hyperuricemia and gouty arthritis, on account of the antioxidant, anti-inflammatory, XOD inhibitory and NLRP3 inflammasome inhibitory activities (Lim et al., 2018). Apigenin glycosides, the primary phytochemicals of Selaginella moellendorffii aqueous extract, might be able to quantitatively transformed into apigenin in vivo. This is important because the flavone aglycone apigenin exhibited the most potent effect on XOD in vitro. The extract was reported to be effective in reducing hyperuricemia and inflammation caused by mono-sodium urate crystals. The mechanism is believed to be chiefly due to the suppression of XOD activity, the production of pro-inflammatory cytokines and the neutrophil accumulation, decreased lipid peroxidation, and the increase of antioxidant status. In conclusion, Selaginella moellendorffii has great potential for the treatment of hyperuricemia and gout arthritis since they possess both anti-hyperuricemic and anti-inflammatory properties (Zhao et al., 2017).

7.2.7 ARISTOLOCHIA bracteolata Retz.

Aristolochia is an important genus belongs to Aristolochiaceae family and can be found across tropical Asia, Africa, and South America. Aristolochia bracteolata is also known as "worm killer" in English because of supposed anthelmintic activity and trypanocidal effect (El-Tantawy, 2019). It is commonly used in traditional medicine as a gastric stimulant and in the treatment of cancer, dysentery, lung inflammation, and snakebites. The plant leaves are known to bear antioxidant activity through free radical scavenging effect (Shirwaikar et al., 2003). Also, the excessively bitter plant leaves are known to possess anti-inflammatory and anthelmintic activity (Negi et al., 2003). The hypouricemic effect of crude extract of A. bracteolate leaves (extracted using ether and chloroform) was analyzed by studying levels of UA in serum and in urine of arthritis induced rats. Results indicate that administration of A. bracteolate presented a substantial change in UA concentration, augmented by PO administration in rats. The reduction in UA levels was roughly the same as allopurinol. A. bracteolate extracts are able to increase the amount of urate excreted (Li et al., 2017). In addition, it could attenuate hyperuricemia in arthritic mouse model probably due to its antioxidant effects due to the presence of phenolic ingredients, such as phenolic diterpenes, acids, and different flavonoids. Aristolochia bracteolata needs more particular investigations to discover exactly anti-hyperuricemia benefits and its active compounds.

7.2.8 *CINNAMOMUM cassia*

Cinnamomum cassia (C. Cassia of Lauraceae family) bark is commonly used to treat dyspepsia, gastritis, blood circulation, and inflammatory diseases (Sung et al., 2010). It is indigenous to China and is considered as one of the 50 fundamental herbs in Chinese folk medicine (Kawatra & Rajagopalan, 2015; Sung et al., 2010). Of the several components that C. Cassia possesses, cinnamaldehyde was found to be the main active constituent (Ho et al., 2013) with significant anti-inflammatory properties (Bandara et al., 2012). C. cassia inhibited XOD activity in vitro and exerted a protective effect on MSU-induced mouse peritonitis by reducing IL-1β production and neutrophils infiltration through inhibition of NLRP3 inflammasome activation due to the suppression of ASC oligomerization (Shin et al., 2017; Zhao et al., 2006). Methanol extracts of the twig of Cinnamomum cassia is also shown inhibitory effect on XOD at the corresponding IC50 values of 1.06 μg/ml while the IC50 value of allopurinol is also 1.06 μg/ml under the assay condition (Kong et al., 2000). In addition, the mixture of Chrysanthemum indicum Linne flower and Cinnamomum cassia (L.) J. Presl bark extracts, which have been used as the main ingredients in several prescriptions for treatment of hyperuricemia and gout in traditional medicine, is demonstrated in vitro, in vivo as well as in randomized clinical trial to reduce serum UA levels and promoted the excretion of UA in urine without side effects or toxicity (Lee et al., 2017, 2018; Park et al., 2020). In short, the bioactive compounds of this plant and its use in combination with Chrysanthemum indicum are necessary to exact research about its anti-gout mechanism.

7.3 BIOACTIVE COMPOUNDS FROM THE MEDICINAL PLANT FOR TREATMENT OF GOUT

7.3.1 COLCHICINE

Colchicine is a tricyclic, lipid soluble alkaloid isolated from plants of the Lily family Colchicum autumnale (autumn crocus) (Angelidis et al., 2018). Colchicine is commonly used for treatment of gout, familial Mediterranean fever (FMF) and other diseases including pericarditis, Behcet's disease (BD), coronary artery disease as well as other inflammatory and fibrotic conditions (Slobodnick et al., 2018). There are multiple mechanisms through which colchicine exerts its anti-inflammatory effects (Andreu & Timasheff, 1982). The most studied mechanism is the capacity of colchicine to bind to free tubulin dimers which, once incorporated into microtubules, will block the assembly and polymerization of microtubules (Dalbeth et al., 2014). This mechanism of action is dose-dependent and is pointed out in vitro to be directly responsible for colchicine's effects on cell migration, cytokine release and intracellular trafficking. It also plays an important role in disrupting inflammatory cell activities (Cronstein et al., 1995). Colchicine showed inhibitory effect on neutrophil adhesion, extravasation, and recruitment by altering neutrophil L-selectin expression and endothelial cell E-selectin distribution, blocking the release of the chemotactic agent leukotriene B4 (Paschke et al., 2013), as well as altering the deformability of neutrophil (Chia et al., 2008). Colchicine also alters leucocyte-mediated inflammatory activities, such as inhibiting leucocyte superoxide production and release of various cytokines and pyrogens (Li et al., 1996; Martinon et al., 2006). Recent studies have revealed that at high concentrations, colchicine can suppress both MSU and CPP crystal-induced activation of the Nod-Like Receptor Protein 3 (NLRP3) inflammasome, which lead to the suppression of the caspase-1 activation and, subsequently, the release of IL-1β and IL-18 (Centola et al., 2000). For the NLRP3 inflammasome is known to be expressed largely in myeloid lineage cells (Tufan et al., 2007), these findings are matched with the clinical awareness that colchicine is more useful in treating ailments associated with innate immune system (associated with neutrophils and monocytes/ macrophages) than those of adaptive immune system. In vitro evidence suggests that suppression of inflammasome activity happens only with doses much higher than therapeutic levels of colchicine. However, for leucocytes accumulate colchicine plus intracellular neutrophil colchicine levels have been proved to be 16 times higher than plasma concentrations, clinical doses of colchicine might still enough for inflammasome suppression (Slobodnick et al., 2018).

7.3.2 BERBERINE (BBR)

Berberine (BBR) is a crucial quaternary benzylisoquinoline alkaloid that can be extracted from many important medicinal plants, such as Argemone mexicana L. (Papaveraceae), Berberis kansuensis C. K. Schneid. (Berberidaceae), Coptis chinensis Franch. (Ranunculaceae), Coscinium fenestratum

(Goetgh.) Colebr. (Menispermaceae), and Phellodendron amurense Rupr. (Rutaceae) (Xu et al., 2021). Multiple studies have reported that BBR has evident hypoglycemic, hypolipidemic, hepatoprotective, anti-obesity, and anti-inflammatory activities (Tillhon et al., 2012; Wang et al., 2017). Numerous studies have provided evidence that BBR can significantly reduce fasting serum urate levels which is beneficial for the treatment of hyperuricemia or gout (Wang et al., 2018; Wu et al., 2010). BBR can also dilate blood vessels, regulate blood flow as well as ameliorate kidney function, leading to increased excretion of UA (Wu et al., 2010). Zhao et al. Reported treatment with BBR (100 mg/kg/d) for 8 weeks could markedly reduce the activity of XOD enzyme in the serum and kidney tissues of rats which in turn leads to the reduction of UA production (Xinmei Xu et al., 2021). In another study, BBR also reported to inhibit XOD potently, with IC50 value of 4.86 ± 1.12 μg/ml (Chen et al., 2016). Moreover, Wu et al. found that the reduction of serum UA level by BBR was closely related to the decrease in LDL-C level (Wu et al., 2010). LDL-C, by reducing the expression of the human urate transporter (hUAT) gene in the renal tubules, can suppress renal excretion of UA and increase serum UA concentrations (Chen et al., 2009). Liu et al. demonstrated that BBR could alleviate MSU crystal-induced inflammation – a key factor in the development of gouty arthritis by downregulating the expression of NLRP3 inflammasome and IL-1β in human monocytic THP-1 cells (Liu et al., 2016). Wu et al. found that the oral administration of BBR and oryzanol for 2 months considerably reduced serum UA levels in patients with hyperuricemia with an effective rate of 83% (Wang et al., 2018). Recently, Wang et al. reported that the combination of BBR with benzbromarone in the treatment of gout could remarkably decrease serum UA levels and improve metabolic disorders (Xinmei Xu et al., 2021). Furthermore, BBR combined with amlodipine in the treatment of elderly patients with gout and hypertension has a substantial clinical effect and could effectively reduce blood pressure and UA levels (Cai, 2017; Ganlin, 2013).

7.3.3 RESVERATROL (RES)

Resveratrol (RES) is a phytoalexin polyphenolic stilbene collected from the root of Veratrum grandiflorum and Polygonum cupsidatum and can be found in numerous nutritional foods such as grapes, blueberry, cranberry, peanut, purple grape, and juice (Tian & Liu, 2020). This phytochemical possesses various healing properties, including anti-glycation, antioxidant, anti-inflammation, anti-neurodegeneration, and anticancer (Galiniak et al., 2019). RES has been reported to suppress inflammation in both in vivo and in vitro experiments and to lower serum UA levels due to the suppression of the NLRP3 inflammasome activation which in turn leads to the production of IL-1β (Chen et al., 2016; Elmali et al., 2005; Misawa et al., 2015). In addition, Chen et al. found that RES, as SIRT1 activator, not only prevented MSU-induced gouty arthritis but also inhibited the increased level of serum UA in a dose and time-dependent manner (Chen et al., 2016; Yang et al., 2019). Furthermore, RES is demonstrated to be capable of promoting MSU-induced autophagy to inhibit gouty inflammation (Chen et al., 2016; Shi et al., 2012; Wu et al., 2016) and to be a significant XOD inhibitor with IC50 value of 7.14 μM as compared to standard drug allopurinol with IC50 of 8.95 μM (Wu et al., 2011).

7.3.4 QUERCETIN

Dietary quercetin (3,3,4,5,7-pentahydroxyflavone) is a major flavonoid that is usually found in apples, cherries, berries, onions, asparagus, and red leaf lettuce (Costa et al., 2016; Manach et al., 2005). Quercetin, in animal models and in vitro has been proved to have various medical benefits including anti-inflammatory, anticancer, antiviral, lipid peroxidation, platelet aggregation, capillary permeability (Li et al., 2016). Quercetin is thought to promote lowering UA levels due to potent XOD inhibitory activity (Nagao et al., 1999). Mice administered quercetin suppressed MSU crystals-induced leukocyte recruitment, cytokine levels, oxidative stress, and inflammasome component gene expression in the knee joints (Ruiz-Miyazawa et al., 2017). Moreover, rats treated with allopurinol (lowering UA as a positive control), quercetin, and rutin showed significantly ameliorated NLRP3 inflammasome expression, pro-inflammatory cytokine secretion, and lipid accumulation (Hu et al., 2012). Quercetin from Nitraria retusa extract exhibited significant XOD inhibitory activities (IC50 = 6.0 μM), close to reference drug allopurinol (IC50 = 2.4 μM) (Salem et al., 2011). Wang et al. in addition demonstrated that quercetin reduced blood UA levels through regulating UA transport, including organic anion transport 1 (OAT1), OAT3 and electrogenic urate transporter (UAT) (Wang et al., 2012). A randomized-controlled clinical trial in healthy male adults showed that supplementation with quercetin at 500 mg/day can progressively reduce plasma UA levels approximately 8% with high significance (P value of 0·008) without changing BMI (Shi & Williamson,

2016). Lately, a phase IV clinical trial has been projected in the University of Pavia on nutraceutical containing quercetin as an active component to assess its hypouricemic effect in subjects with UA level between ≥ 6 mg/dL and < 7 mg/dL after three months of therapy (clinicaltrial.gov registry number: NCT04161872) (NCT04161872, 2019).

7.3.5 MORIN

Morin belongs to the group of flavonols, a class of flavonoids contains 3-hydroxyflavon backbone and can be extracted from the branches of almond (Psidium guajava), white mulberry (Morus alba L.), Osage orange (Maclura pomifera), fig (Chlorophora tinctoria), mill (Prunus dulcis) and old fustic (Maclura tinctoria). Morin owns many types of pharmacological effects, from antioxidant, anti-inflammatory, XOD inhibitor property, prevention of low-density lipoprotein oxidation to anticancer activity (Sinha et al., 2016). Treatment with morin significantly down-regulated the mRNA level of NLRP3 component and attenuated ankle swelling and the levels of pro-inflammatory cytokines with an increased antioxidant status in the joint space of MSU crystal-induced rats (Dhanasekar & Rasool, 2016). Morin possesses protective effect on lipopolysaccharide (LPS)-induced acute lung injury in Balb/c mice. Also, morin suppressed IL-1β, IL-18, TNF-α, and IL-6 secretion and decreased myeloperoxidase activity and inflammatory cell infiltration by suppressing gout-inducing NLRP3 inflammasome in lungs (Tianzhu et al., 2014). In vitro, morin treatment significantly decreased the levels of inflammatory cytokines (IL-1β, IL-6, TNF-α, MCP-1 and VEGF), inflammatory mediators (NO and PEG2), and lysosomal enzymes (acid phosphatase, β-galactosidase, cathepsin D and N-acetyl glucosamindase) in MSU-crystals stimulated macrophage cells (Dhanasekar et al., 2015). Moreover, it is also demonstrated that morin hypouricemic activity was probably due to the inhibitory effect of morin on XOD (IC50 = 44 μMin compared with allopurinol, IC50 = 5.43 μM) (Fernandes et al., 2002; Yu et al., 2006).

7.3.6 EPIGALLOCATECHIN-3-GALLATE (EGCG)

Epigallocatechin-3-gallate (EGCG), a major bioactive polyphenol in green tea, belongs to the flavanols class and is well-known to possess antioxidant, anti-inflammatory activities, and free radical-scavenging ability (Rahman et al., 2006; Singh et al., 2011). EGCG inhibited MSU crystals-induced neutrophil infiltration and IL-1β secretion through the downregulation of neutrophil cytosolic factor 1 and NLRP3 protein expression. In addition, treatment with EGCG suppressed TXNIP/NLRP3 conjugation in MSU crystals-challenged THP-1 monocytes (Jhang et al., 2016). EGCG also inhibited melanoma growth by modulating NLRP1 and TXNIP (thioredoxin interacting protein) expression (Ellis et al., 2011). Lee et al. found that administration of EGCG attenuated the activation of the NLRP3 inflammasome in macrophages via the blockade of mitochondrial DNA synthesis, contributing to the prevention of gouty inflammation (Lee et al., 2019). EGCG was also evaluated as XO inhibitor, with a determined IC50 of 7.4 μM by Lin et al. as compared with allopurinol (IC50 value of 0.68 μM). These authors suggested that EGCG might act as "suicide substrate," similarly to allopurinol (Lin et al., 2000). Recently, Zhu et al. reported that EGCG exerted inhibitory effects on hepatic XOD at the dose of 50 mg/kg. Moreover, this compound showed a significant dose-dependent effect on lowering serum UA levels and in the regulation of GLUT9 and URAT1 mRNA expression levels in hyperuricemic mice (Zhu et al., 2018).

7.3.7 PROCYANIDIN

Procyanidins are members of the flavanol class of polyphenols and are composed of oligomers or polymers of (+)-catechin and/or (-)-epicatechin units. They are mainly found in apples, cocoa, and grapes (Gu et al., 2004; Hammerstone et al., 2000). Procyanidins significantly reduced the degree of pain and suppressed ankle swelling in MSU crystals-challenged CD-1 mice. Procyanidins also suppressed MSU crystals-induced increasing expression of IL-1β, Cathepsin B and NLRP3 in mice gout groups accompanied by the downregulation of prostaglandin E2 (PGE2), and decreased ROS levels (XOD inhibitory activity) observed in Raw.7 murine macrophage cells (Liu et al., 2017; Qiao et al., 2020). Maffei et al. found that procyanidin extracted from Vitis vinifera L. exerted inhibitory effect on XOD with IC50 value of 4.24 mmol/L (Maffei Facino et al., 1994). Additionally, procyanidin B2 suppressed the activation of NLRP3 inflammasome via the suppression of the AP-1 pathway in human umbilical vein endothelial cells (Yang et al., 2014), and it is also reported to decrease the cytoplasmic pool of NLRP3 and pro-IL-1β, thereby inhibiting subsequent caspase-1 activity and IL-1β secretion (Martinez-Micaelo et al., 2015).

7.4 CONCLUSION

Currently, the application of herbs in the therapy of gouty arthritis is a potential and effective method with fewer adverse effects than treatment with chemical synthetic drugs such as allopurinol or febuxostat. The source of herbs, as well as their phytochemicals, is immensely prolific. Further studies are needed to evaluate the safety and potential clinical utility, in order to develop an optimal natural-originated treatment for gout.

KEYWORDS

- **active compounds**
- **epigallocatechin-3-gallate**
- **gout**
- **gouty arthritis**
- **medicinal plants**
- **quercetin**

REFERENCES

Andreu, J. M., & Timasheff, S. N., (1982). Tubulin bound to colchicine forms polymers different from microtubules. Proc. Natl. Acad. Sci. U S A, 79(22), 6753–6756. https://doi.org/10.1073/pnas.79.22.6753.

Angelidis, C., Kotsialou, Z., Kossyvakis, C., Vrettou, A. R., Zacharoulis, A., Kolokathis, F., Kekeris, V., & Giannopoulos, G., (2018). Colchicine pharmacokinetics and mechanism of action. Curr. Pharm. Des., 24(6), 659–663. https://doi.org/10.2174/1381612824666180123110042.

Bandara, T., Uluwaduge, I., & Jansz, E. R., (2012). Bioactivity of cinnamon with special emphasis on diabetes mellitus: A review. Int. J. Food Sci. Nutr., 63(3), 380–386. https://doi.org/10.3109/09637486.2011.627849.

Burns, C. M., & Wortmann, R. L., (2012). Latest evidence on gout management: What the clinician needs to know. Ther. Adv. Chronic. Dis., 3(6), 271–286. https://doi.org/10.1177/2040622312462056.

Cai, S. J. C. R., (2017). Clinical Effect and Safety of Combination Medicine for Elderly Patients with Hypertension and Gout., 26(11), 84–89.

Cao, Y., Tan, N. H., Chen, J. J., Zeng, G. Z., Ma, Y. B., Wu, Y. P., Yan, H., et al., (2010). Bioactive flavones and biflavones from Selaginella moellendorffii hieron. Fitoterapia, 81(4), 253–258. https://doi.org/10.1016/j.fitote.2009.09.007.

Centola, M., Wood, G., Frucht, D. M., Galon, J., Aringer, M., Farrell, C., Kingma, D. W., et al., (2000). The gene for familial Mediterranean fever, MEFV, is expressed in early leukocyte development and is regulated in response to inflammatory mediators. Blood, 95(10), 3223–3231.

Chao, L. R., Seguin, E., Tillequin, F., & Koch, M., (1987). New alkaloid glycosides from Selaginella doederleinii. Journal of Natural Product, 50(3), 422–426.

Chen, C., Lü, J. M., & Yao, Q., (2016). Hyperuricemia-related diseases and xanthine oxidoreductase (XOR) inhibitors: An overview. Med. Sci. Monit., 22, 2501–2512. https://doi.org/10.12659/msm.899852.

Chen, H., Zheng, S., Wang, Y., Zhu, H., Liu, Q., Xue, Y., Qiu, J., Zou, H., & Zhu, X., (2016). The effect of resveratrol on the recurrent attacks of gouty arthritis. Clin. Rheumatol., 35(5), 1189–1195. https://doi.org/10.1007/s10067-014-2836-3.

Chen, W. J., Wu, Y., Zhao, X., Liu, S., Song, F. R., Liu, Z. Y., & Liu, Z. Q., (2016). Screening the anti-gout traditional herbs from TCM using an in vitro method. Chinese Chemical Letters, 27(11), 1701–1707. https://doi.org/https://doi.org/10.1016/j.cclet.2016.05.022.

Chen, Y., Miao, Z., & Yan, S., (2009). Effect of Various Lipids on the Expression of Human Urate Transporter Gene in HK-2 Cells, 25(1), 22–24.

Chia, E. W., Grainger, R., & Harper, J. L., (2008). Colchicine suppresses neutrophil superoxide production in a murine model of gouty arthritis: A rationale for use of low-dose colchicine. Br J. Pharmacol, 153(6), 1288–1295. https://doi.org/10.1038/bjp.2008.20.

Costa, L. G., Garrick, J. M., Roquè, P. J., & Pellacani, C., (2016). Mechanisms of neuroprotection by quercetin: Counteracting oxidative stress and more. Oxid. Med. Cell Longev., 2016, 2986796. https://doi.org/10.1155/2016/2986796.

Cronstein, B. N., Molad, Y., Reibman, J., Balakhane, E., Levin, R. I., & Weissmann, G., (1995). Colchicine alters the quantitative and qualitative display of selectins on endothelial cells and neutrophils. J. Clin. Invest., 96(2), 994–1002. https://doi.org/10.1172/jci118147.

Da Silva, A. J. R. G., De Sá, P. G. S., De Oliveira, M. L. A. R., De Siqueira, F. J. A., De Oliveira, V. R., & Barbosa, F. J. M., (2013). Phytochemistry of the genus Selaginella (Selaginellaceae). Journal of Medical Plant Research, 7(25), 1858–1868.

Dalbeth, N., & Stamp, L., (2014). Hyperuricaemia and gout: Time for a new staging system? Ann. Rheum Dis., 73(9), 1598–1600. https://doi.org/10.1136/annrheumdis-2014-205304.

Dalbeth, N., Gosling, A. L., Gaffo, A., & Abhishek, A., (2021). Gout. Lancet, 397(10287), 1843–1855. https://doi.org/10.1016/s0140-6736(21)00569-9.

Dalbeth, N., Lauterio, T. J., & Wolfe, H. R., (2014). Mechanism of action of colchicine in the treatment of gout. Clin. Ther., 36(10), 1465–1479. https://doi.org/10.1016/j.clinthera.2014.07.017.

Dehlin, M., Jacobsson, L., & Roddy, E., (2020). Global epidemiology of gout: Prevalence, incidence, treatment patterns and risk factors. Nat Rev Rheumatol, 16(7), 380–390. https://doi.org/10.1038/s41584-020-0441-1.

Dhanasekar, C., & Rasool, M., (2016). Morin, a dietary bioflavonol suppresses monosodium urate crystal-induced inflammation in an animal model of acute gouty arthritis with reference to NLRP3 inflammasome, hypo-xanthine phospho-ribosyl transferase, and inflammatory mediators. Eur. J. Pharmacol., 786, 116–127. https://doi.org/10.1016/j.ejphar.2016.06.005.

Dhanasekar, C., Kalaiselvan, S., & Rasool, M., (2015). Morin, a bioflavonoid suppresses monosodium urate crystal-induced inflammatory immune response in RAW 264.7 macrophages through the inhibition of inflammatory mediators, intracellular ROS levels and NF-κB activation. PLoS One, 10(12), e0145093. https://doi.org/10.1371/journal.pone.0145093.

Duni, A., Liakopoulos, V., Rapsomanikis, K. P., & Dounousi, E., (2017). Chronic kidney disease and disproportionally increased cardiovascular damage: Does oxidative stress explain the burden? Oxid. Med. Cell Longev., 2017, 9036450. https://doi.org/10.1155/2017/9036450.

Ellis, L. Z., Liu, W., Luo, Y., Okamoto, M., Qu, D., Dunn, J. H., & Fujita, M., (2011). Green tea polyphenol epigallocatechin-3-gallate suppresses melanoma growth by inhibiting inflammasome and IL-1β secretion. Biochem. Biophys. Res. Commun., 414(3), 551–556. https://doi.org/10.1016/j.bbrc.2011.09.115.

Elmali, N., Esenkaya, I., Harma, A., Ertem, K., Turkoz, Y., & Mizrak, B., (2005). Effect of resveratrol in experimental osteoarthritis in rabbits. Inflamm. Res., 54(4), 158–162. https://doi.org/10.1007/s00011-004-1341-6.

El-Tantawy, W., (2019). Natural products for the management of hyperuricaemia and gout: A review. Archives of Physiology and Biochemistry, 127, 1–12. https://doi.org/10.1080/13813455.2019.1610779.

Fernandes, E., Carvalho, F., Silva, A. M., Santos, C. M., Pinto, D. C., Cavaleiro, J. A., & Bastos, M. L., (2002). 2-styrylchromones as novel inhibitors of xanthine oxidase. A structure-activity study. J. Enzyme Inhib. Med. Chem., 17(1), 45–48. https://doi.org/10.1080/14756360290019944.

Galiniak, S., Aebisher, D., & Bartusik-Aebisher, D., (2019). Health benefits of resveratrol administration. Acta Biochim. Pol., 66(1), 13–21. https://doi.org/10.18388/abp.2018_2749.

Gu, L., Kelm, M. A., Hammerstone, J. F., Beecher, G., Holden, J., Haytowitz, D., Gebhardt, S., & Prior, R. L., (2004). Concentrations of proanthocyanidins in common foods and estimations of normal consumption. J. Nutr., 134(3), 613–617. https://doi.org/10.1093/jn/134.3.613.

Hainer, B. L., Matheson, E., & Wilkes, R. T., (2014). Diagnosis, treatment, and prevention of gout. Am. Fam. Physician, 90(12), 831–836.

Hammerstone, J. F., Lazarus, S. A., & Schmitz, H. H., (2000). Procyanidin content and variation in some commonly consumed foods. J. Nutr., 130(8S Suppl), 2086s–2092s. https://doi.org/10.1093/jn/130.8.2086S.

Hilmi, B. A., Asmahan, M. I., & Rosman, A., (2012). Use of newly available febuxostat in a case of chronic tophaceous gout contraindicated to allopurinol and probenecid. Med. J. Malaysia, 67(1), 125–126.

Ho, S. C., Chang, K. S., & Chang, P. W., (2013). Inhibition of neuroinflammation by cinnamon and its main components. Food Chem., 138(4), 2275–2282. https://doi.org/10.1016/j.foodchem.2012.12.020.

Hoffman, H. M., Scott, P., Mueller, J. L., Misaghi, A., Stevens, S., Yancopoulos, G. D., Murphy, A., et al., (2010). Role of the leucine-rich repeat domain of cryopyrin/NALP3 in monosodium urate crystal-induced inflammation in mice. Arthritis Rheum., 62(7), 2170–2179. https://doi.org/10.1002/art.27456.

Hu, Q. H., Zhang, X., Pan, Y., Li, Y. C., & Kong, L. D., (2012). Allopurinol, quercetin and rutin ameliorate renal NLRP3 inflammasome activation and lipid accumulation in fructose-fed rats. Biochem. Pharmacol., 84(1), 113–125. https://doi.org/10.1016/j.bcp.2012.03.005.

Huang, G. L. (2013). Effect of amlodipine and berberine in mild and midrange hypertension patients complicated with hyperuricemia arthrolithiasis. China Pharm, 22(5), 32–33.

Jana, S., & Shekhawat, G. S., (2011). Critical review on medicinally potent plant species: Gloriosa superba. Fitoterapia, 82(3), 293–301. https://doi.org/10.1016/j.fitote.2010.11.008.

Janssens, H. J., Janssen, M., Van De, L. E. H., Van, R. P. L., & Van, W. C., (2008). Use of oral prednisolone or naproxen for the treatment of gout arthritis: A double-blind, randomized equivalence trial. Lancet, 371(9627), 1854–1860. https://doi.org/10.1016/s0140-6736(08)60799-0.

Jhang, J. J., Lin, J. H., & Yen, G. C., (2018). Beneficial properties of phytochemicals on NLRP3 inflammasome-mediated gout and complication. J. Agric. Food Chem., 66(4), 765–772. https://doi.org/10.1021/acs.jafc.7b05113.

Jhang, J. J., Lu, C. C., & Yen, G. C., (2016). Epigallocatechin gallate inhibits urate crystals-induced peritoneal inflammation in C57BL/6 mice. Mol. Nutr. Food Res., 60(10), 2297–2303. https://doi.org/10.1002/mnfr.201600106.

John, J. C., Fernandes, J., Nandgude, T., Niphade, S. R., Savla, A., & Deshmukh, P. T., (2009). Analgesic and anti-inflammatory activities of the hydroalcoholic extract from Gloriosa superba Linn. International Journal of green Pharmacy, 3, 215–219.

Kawatra, P., & Rajagopalan, R., (2015). Cinnamon: Mystic powers of a minute ingredient. Pharmacognosy Res., 7(Suppl 1), S1–6. https://doi.org/10.4103/0974-8490.157990.

Keservani, R. K., & Sharma, A. K., (2014). Flavonoids: Emerging trends and potential health benefits. Journal of Chinese Pharmaceutical Sciences, 23(12), 815.

Khandel, A., Ganguly, S., Bajaj, A., & Khan, S., (2012). New records, ethno pharmacological applications & indigenous uses of Gloriosa superba L. (Glory lily) practices by tribes of pachmarhi biosphere reserve, Madhya Pradesh, Central India. Journal of Nature of Science, 10(5), 23–48.

Khanna, D., Khanna, P. P., Fitzgerald, J. D., Singh, M. K., Bae, S., Neogi, T., Pillinger, M. H., et al., (2012). 2012 American College of Rheumatology guidelines for management of gout. Part 2: Therapy and anti-inflammatory prophylaxis of acute gouty arthritis. Arthritis Care Res. (Hoboken), 64(10), 1447–1461. https://doi.org/10.1002/acr.21773.

Kingsbury, S. R., Conaghan, P. G., & McDermott, M. F., (2011). The role of the NLRP3 inflammasome in gout. J. Inflamm. Res., 4, 39–49. https://doi.org/10.2147/jir.S11330.

Kong, L. D., Cai, Y., Huang, W. W., Cheng, C. H., & Tan, R. X., (2000). Inhibition of xanthine oxidase by some Chinese medicinal plants used to treat gout. J. Ethnopharmacol., 73(1, 2), 199–207. https://doi.org/10.1016/s0378-8741(00)00305-6.

Lee, H. E., Yang, G., Park, Y. B., Kang, H. C., Cho, Y. Y., Lee, H. S., & Lee, J. Y., (2019). Epigallocatechin-3-gallate prevents acute gout by suppressing NLRP3 inflammasome activation and mitochondrial DNA synthesis. Molecules, 24(11). https://doi.org/10.3390/molecules24112138.

Lee, H. M., Yang, G., Ahn, T. G., Kim, M. D., Nugroho, A., Park, H. J., Lee, K. T., et al., (2013). Antiadipogenic effects of Aster glehni extract: In vivo and in vitro effects. Evid. Based Complement Alternat Med., 2013, 859624. https://doi.org/10.1155/2013/859624.

Lee, Y. S., Kim, S. H., Yuk, H. J., & Kim, D. S., (2018). DKB114, a mixture of Chrysanthemum indicum Linne flower and Cinnamomum cassia (L.) J. Presl bark extracts, improves hyperuricemia through inhibition of xanthine oxidase activity and increasing urine excretion. Nutrients, 10(10). https://doi.org/10.3390/nu10101381.

Lee, Y. S., Son, E., Kim, S. H., Lee, Y. M., Kim, O. S., & Kim, D. S., (2017). Synergistic uric acid-lowering effects of the combination of Chrysanthemum indicum Linne flower and Cinnamomum cassia (L.) J. Persl bark extracts. Evid. Based Complement Alternat Med., 2017, 9764843. https://doi.org/10.1155/2017/9764843.

Li, J., Huang, D., Chen, W., Xi, Z., Chen, C., Huang, G., & Sun, L., (2013). Two new phenolic glycosides from Gnaphalium affine D. Don and their anti-complementary activity. Molecules, 18(7), 7751–7760. https://doi.org/10.3390/molecules18077751.

Li, S., Tang, Y., Liu, C., Li, J., Guo, L., & Zhang, Y., (2015). Development of a method to screen and isolate potential xanthine oxidase inhibitors from Panax japlcus var via ultrafiltration liquid chromatography combined with counter-current chromatography. Talanta, 134, 665–673. https://doi.org/10.1016/j.talanta.2014.12.005.

Li, Y. P., Wu, S., Ran, A., Xu, D. Y., Wei, J. M., & Zhao, Z. L., (2017). Aristolochia bracteolate Retz. attenuates hyperuricemia in a metabolic arthritis rat model. Afr. J. Tradit. Complement Altern. Med., 14(4), 180–187. https://doi.org/10.21010/ajtcam.v14i4.21.

Li, Y., Yao, J., Han, C., Yang, J., Chaudhry, M. T., Wang, S., Liu, H., & Yin, Y., (2016). Quercetin, inflammation, and immunity. Nutrients, 8(3), 167. https://doi.org/10.3390/nu8030167.

Li, Z., Davis, G. S., Mohr, C., Nain, M., & Gemsa, D., (1996). Inhibition of LPS-induced tumor necrosis factor-alpha production by colchicine and other microtubule disrupting drugs. Immunobiology, 195(4, 5), 624–639. https://doi.org/10.1016/s0171-2985(96)80027-1.

Lim, H., Min, D. S., Park, H., & Kim, H. P., (2018). Flavonoids interfere with NLRP3 inflammasome activation. Toxicology and Applied Pharmacology, 355, 93–102. https://www.unboundmedicine.com/medline/citation/29960001/Flavonoids_interfere_with_NLRP3_inflammasome_activation_; https://linkinghub.elsevier.com/retrieve/pii/S0041-008X(18)30288-6 (accessed on 15 November 2023).

Lin, J. K., Chen, P. C., Ho, C. T., & Lin-Shiau, S. Y., (2000). Inhibition of xanthine oxidase and suppression of intracellular reactive oxygen species in HL-60 cells by theaflavin-3,3'-digallate,(−)-epigallocatechin-3-gallate, and propyl gallate. J. Agric. Food Chem., 48(7), 2736–2743.

Lin, W. Q., Xie, J. X., Wu, X. M., Yang, L., & Wang, H. D., (2014). Inhibition of xanthine oxidase activity by gnaphalium affine extract. Chin Med Sci J, 29(4), 225–230. https://doi.org/10.1016/s1001-9294(14)60075-4.

Lin, Y., Liu, P. G., Liang, W. Q., Hu, Y. J., Xu, P., Zhou, J., Pu, J. B., & Zhang, H. J., (2018). Luteolin-4'-O-glucoside and its aglycone, two major flavones of Gnaphalium affine D. Don, resist hyperuricemia and acute gouty arthritis activity in animal models. Phytomedicine, 41, 54–61. https://doi.org/10.1016/j.phymed.2018.02.002.

Ling, X., & Bochu, W., (2014). A review of phytotherapy of gout: Perspective of new pharmacological treatments. Pharmazie, 69(4), 243–256.

Lioté, F., & Ea, H. K., (2007). Recent developments in crystal-induced inflammation pathogenesis and management. Curr. Rheumatol. Rep., 9(3), 243–250. https://doi.org/10.1007/s11926-007-0039-5.

Liu, H. J., Pan, X. X., Liu, B. Q., Gui, X., Hu, L., Jiang, C. Y., Han, Y., et al., (2017). Grape seed-derived procyanidins alleviate gout pain via NLRP3 inflammasome suppression. J Neuroinflammation, 14(1), 74. https://doi.org/10.1186/s12974-017-0849-y.

Liu, Y. F., Wen, C. Y., Chen, Z., Wang, Y., Huang, Y., & Tu, S. H., (2016). Effects of berberine on NLRP3 and IL-1β expressions in monocytic THP-1 cells with monosodium urate crystals-induced inflammation. Biomed. Res. Int., 2016, 2503703. https://doi.org/10.1155/2016/2503703.

Maffei, F. R., Carini, M., Aldini, G., Bombardelli, E., Morazzoni, P., & Morelli, R., (1994). Free radicals scavenging action and anti-enzyme activities of procyanidines from Vitis vinifera. A mechanism for their capillary protective action. Arzneimittelforschung, 44(5), 592–601.

Manach, C., Williamson, G., Morand, C., Scalbert, A., & Rémésy, C., (2005). Bioavailability and bioefficacy of polyphenols in humans. I. Review of 97 bioavailability studies. Am. J. Clin. Nutr., 81(1 Suppl), 230s–242s. https://doi.org/10.1093/ajcn/81.1.230S.

Mandal, A. K., & Mount, D. B., (2015). The molecular physiology of uric acid homeostasis. Annu. Rev. Physiol., 77, 323–345. https://doi.org/10.1146/annurev-physiol-021113-170343.

Martinez-Micaelo, N., González-Abuín, N., Pinent, M., Ardévol, A., & Blay, M., (2015). Procyanidin B2 inhibits inflammasome-mediated IL-1β production in lipopolysaccharide-stimulated macrophages. Mol. Nutr. Food Res., 59(2), 262–269. https://doi.org/10.1002/mnfr.201400370.

Martinon, F., Pétrilli, V., Mayor, A., Tardivel, A., & Tschopp, J., (2006). Gout-associated uric acid crystals activate the NALP3 inflammasome. Nature, 440(7081), 237–241. https://doi.org/10.1038/nature04516.

Masuda, T., Shingai, Y., Takahashi, C., Inai, M., Miura, Y., Honda, S., & Masuda, A., (2014). Identification of a potent xanthine oxidase inhibitor from oxidation of caffeic acid. Free Radic Biol. Med., 69, 300–307. https://doi.org/10.1016/j.freeradbiomed.2014.01.016.

Mattiuzzi, C., & Lippi, G., (2020). Recent updates on worldwide gout epidemiology. Clin. Rheumatol., 39(4), 1061–1063. https://doi.org/10.1007/s10067-019-04868-9.

Mehmood, A., Ishaq, M., Zhao, L., Safdar, B., Rehman, A. U., Munir, M., Raza, A., et al., (2019). Natural compounds with xanthine oxidase inhibitory activity: A review. Chem Biol. Drug Des., 93(4), 387–418. https://doi.org/10.1111/cbdd.13437.

Meng, X. W., He, C. X., Chen, X., Yang, X. S., & Liu, C., (2021). The extract of Gnaphalium affine D. Don protects against H(2)O(2)-induced apoptosis by targeting PI3K/AKT/GSK-3β signaling pathway in cardiomyocytes. J. Ethnopharmacol., 268, 113579. https://doi.org/10.1016/j.jep.2020.113579.

Misawa, T., Saitoh, T., Kozaki, T., Park, S., Takahama, M., & Akira, S., (2015). Resveratrol inhibits the acetylated α-tubulin-mediated assembly of the NLRP3-inflammasome. Int. Immunol., 27(9), 425–434. https://doi.org/10.1093/intimm/dxv018.

Nagao, A., Seki, M., & Kobayashi, H., (1999). Inhibition of xanthine oxidase by flavonoids. Biosci. Biotechnol. Biochem., 63(10), 1787–1790. https://doi.org/10.1271/bbb.63.1787.

NCT04161872, C. G., (2019). Nutraceutical on Hyperuricemia. https://clinicaltrials.gov/ct2/show/NCT04161872?term=Quercetin&cond=Hyperuricemia&draw=04161872&rank=04161871 (accessed on 15 November 2023).

Negi, P. S., Anandharamakrishnan, C., & Jayaprakasha, G. K., (2003). Antibacterial activity of Aristolochia bracteata root extracts. J. Med. Food, 6(4), 401–403. https://doi.org/10.1089/109662003772519994.

Nguyen, T. D., Thuong, P. T., Hwang, I. H., Hoang, T. K., Nguyen, M. K., Nguyen, H. A., & Na, M., (2017). Anti-hyperuricemic, anti-inflammatory and analgesic effects of Siegesbeckia orientalis L. resulting from the fraction with high phenolic content. BMC Complement Altern. Med., 17(1), 191. https://doi.org/10.1186/s12906-017-1698-z.

Ning, T. C., & Keenan, R. T., (2010). Unusual clinical presentations of gout. Curr. Opin. Rheumatol., 22(2), 181–187. https://doi.org/10.1097/BOR.0b013e3283361ac0.

Pandey, D. K., Kaur, P., Kumar, V., Banik, R. M., Malik, T., & Dey, A., (2021). Screening the elite chemotypes of Gloriosa superba L. in India for the production of anticancer colchicine: Simultaneous microwave-assisted extraction and HPTLC studies. BMC Plant Biol., 21(1), 77. https://doi.org/10.1186/s12870-021-02843-8.

Park, J. E., Yeom, Z., Park, K. T., Han, E. H., Yu, H. J., Kang, H. S., & Lim, Y. H., (2018). Hypouricemic effect of ethanol extract of Aster glehni leaves in potassium oxonate-induced hyperuricemic rats. Clin. Nutr. Res., 7(2), 126–135. https://doi.org/10.7762/cnr.2018.7.2.126.

Park, Y. H., Kim, D. H., Lee, J. S., Jeong, H. I., Lee, K. W., & Kang, T. H., (2020). A 12-week, multicenter, randomized, double-blind, placebo-controlled clinical trial for evaluation of the efficacy and safety of DKB114 on reduction of uric acid in serum. Nutrients, 12(12). https://doi.org/10.3390/nu12123794.

Paschke, S., Weidner, A. F., Paust, T., Marti, O., Beil, M., & Ben-Chetrit, E., (2013). Technical advance: Inhibition of neutrophil chemotaxis by colchicine is modulated through viscoelastic properties of subcellular compartments. J. Leukoc. Biol., 94(5), 1091–1096. https://doi.org/10.1189/jlb.1012510.

Pétrilli, V., & Martinon, F., (2007). The inflammasome, autoinflammatory diseases, and gout. Joint Bone Spine, 74(6), 571–576. https://doi.org/10.1016/j.jbspin.2007.04.004.

Qiao, C. Y., Li, Y., Shang, Y., Jiang, M., Liu, J., Zhan, Z. Y., Ye, H., et al., (2020). Management of gout-associated MSU crystals-induced NLRP3 inflammasome activation by procyanidin B2: Targeting IL-1β and Cathepsin B in macrophages. Inflammopharmacology, 28(6), 1481–1493. https://doi.org/10.1007/s10787-020-00758-8.

Rahman, I., Biswas, S. K., & Kirkham, P. A., (2006). Regulation of inflammation and redox signaling by dietary polyphenols. Biochem. Pharmacol., 72(11), 1439–1452. https://doi.org/10.1016/j.bcp.2006.07.004.

Richette, P., & Bardin, T., (2010). Gout. Lancet, 375(9711), 318–328. https://doi.org/10.1016/s0140-6736(09)60883-7.

Roberge, C. J., Gaudry, M., Gilbert, C., Malawista, S. E., De Médicis, R., Lussier, A., Poubelle, P. E., & Naccache, P. H., (1996). Paradoxical effects of colchicine on the activation of human neutrophilis by chemotactic factors and inflammatory microcrystal. J. Leukoc. Biol, 59(6), 864–871. https://doi.org/10.1002/jlb.59.6.864.

Ruiz-Miyazawa, K. W., Staurengo-Ferrari, L., Mizokami, S. S., Domiciano, T. P., Vicentini, F., Camilios-Neto, D., Pavanelli, W. R., et al., (2017). Quercetin inhibits gout arthritis in mice: Induction of an opioid-dependent regulation of inflammasome. Inflammopharmacology. https://doi.org/10.1007/s10787-017-0356-x.

Salem, J., Chevalot, I., Harscoat-Schiavo, C., Paris, C., Fick, M., & Humeau, C., (2011). Biological activities of flavonoids from Nitraria retusa (Forssk.) Asch. and their acylated derivatives. Food Chemistry, 124, 486–494. https://doi.org/10.1016/j.foodchem.2010.06.059.

Sattui, S. E., & Gaffo, A. L., (2016). Treatment of hyperuricemia in gout: Current therapeutic options, latest developments, and clinical implications. Ther. Adv. Musculoskelet Dis., 8(4), 145–159. https://doi.org/10.1177/1759720x16646703.

Schlesinger, N., & Thiele, R. G., (2010). The pathogenesis of bone erosions in gouty arthritis. Ann. Rheum Dis., 69(11), 1907–1912. https://doi.org/10.1136/ard.2010.128454.

Shan, D., Wang, Y. N., Xue, H. G., & Cosnier, S., (2009). Xanthine oxidase/laponite nanoparticles immobilized on glassy carbon electrode: Direct electron transfer and multielectrocatalysis. Biosensors & Bioelectronics, 24, 3556–3561. https://doi.org/10.1016/j.bios.2009.05.009.

Shi, C. S., Shenderov, K., Huang, N. N., Kabat, J., Abu-Asab, M., Fitzgerald, K. A., Sher, A., & Kehrl, J. H., (2012). Activation of autophagy by inflammatory signals limits IL-1β

production by targeting ubiquitinated inflammasomes for destruction. Nat. Immunol., 13(3), 255–263. https://doi.org/10.1038/ni.2215.
Shi, Y., & Williamson, G., (2016). Quercetin lowers plasma uric acid in pre-hyperuricaemic males: A randomized, double-blinded, placebo-controlled, cross-over trial. Br. J. Nutr., 115(5), 800–806. https://doi.org/10.1017/s0007114515005310.
Shin, W. Y., Shim, D. W., Kim, M. K., Sun, X., Koppula, S., Yu, S. H., Kim, H. B., et al., (2017). Protective effects of Cinnamomum cassia (Lamaceae) against gout and septic responses via attenuation of inflammasome activation in experimental models. J. Ethnopharmacol., 205, 173–177. https://doi.org/10.1016/j.jep.2017.03.043.
Shirwaikar, A., Somashekar, A. P., Udupa, A. L., Udupa, S. L., & Somashekar, S., (2003). Wound healing studies of Aristolochia bracteolata Lam. with supportive action of antioxidant enzymes. Phytomedicine, 10(6, 7), 558–562. https://doi.org/10.1078/094471103322331548.
Singh, B. N., Shankar, S., & Srivastava, R. K., (2011). Green tea catechin, epigallocatechin-3-gallate (EGCG): Mechanisms, perspectives and clinical applications. Biochem. Pharmacol., 82(12), 1807–1821. https://doi.org/10.1016/j.bcp.2011.07.093.
Sinha, K., Ghosh, J., & Sil, P. C., (2016). Morin and its role in chronic diseases. Adv. Exp. Med. Biol., 928, 453–471. https://doi.org/10.1007/978-3-319-41334-1_19.
Sivakumar, G., & Krishnamurthy, K., (2000). Micropropagation of Gloriosa superba L. – an endangered species of Asia and Africa. Current Science, 78, 30–32.
Sivakumar, G., Krishnamurthy, K., & Hahn, E. J., (2004). Enhanced in vitro production of colchicine in Gloriosa superba L. – An emerging industrial medicinal crop in South India. Journal of Horticultural Science and Biotechnology, 79, 602–605. https://doi.org/10.1080/14620316.2004.11511813.
Slobodnick, A., Shah, B., Krasnokutsky, S., & Pillinger, M. H., (2018). Update on colchicine, 2017. Rheumatology (Oxford), 57(suppl_1), i4–i11. https://doi.org/10.1093/rheumatology/kex453.
So, A., (2008). Developments in the scientific and clinical understanding of gout. Arthritis Res. Ther., 10(5), 221. https://doi.org/10.1186/ar2509.
Son, M., Jin, C., Lee, Y., Lee, J., & Kim, H., (2015). Characterization of caffeoyl glucoside derivatives and hypouricemic activity of the ethyl acetate fraction from Aster glehni. Bulletin of the Korean Chemical Society, 36, 503–512. https://doi.org/10.1002/bkcs.10095.
Sung, Y. Y., Yoon, T., Jang, J., Park, S. J., Jeong, G. H., & Kim, H., (2010). Inhibitory effects of Cinnamomum cassia extract on atopic dermatitis-like skin lesions induced by mite antigen in NC/Nga mice. J. Ethnopharmacol., 133, 621–628. https://doi.org/10.1016/j.jep.2010.10.043.
Tian, B., & Liu, J., (2020). Resveratrol: A review of plant sources, synthesis, stability, modification, and food application. J. Sci. Food Agric., 100(4), 1392–1404. https://doi.org/10.1002/jsfa.10152.
Tianzhu, Z., Shihai, Y., & Juan, D., (2014). The effects of morin on lipopolysaccharide-induced acute lung injury by suppressing the lung NLRP3 inflammasome. Inflammation, 37(6), 1976–1983. https://doi.org/10.1007/s10753-014-9930-1.
Tillhon, M., Guamán, O. L. M., Lombardi, P., & Scovassi, A. I., (2012). Berberine: New perspectives for old remedies. Biochem. Pharmacol., 84(10), 1260–1267. https://doi.org/10.1016/j.bcp.2012.07.018.
Tufan, A., Babaoglu, M. O., Akdogan, A., Yasar, U., Calguneri, M., Kalyoncu, U., Karadag, O., et al., (2007). Association of drug transporter gene ABCB1 (MDR1) 3435C to T

polymorphism with colchicine response in familial Mediterranean fever. J. Rheumatol., 34(7), 1540–1544.
Wang, C., Pan, Y., Zhang, Q. Y., Wang, F. M., & Kong, L. D., (2012). Quercetin and allopurinol ameliorate kidney injury in STZ-treated rats with regulation of renal NLRP3 inflammasome activation and lipid accumulation. PLoS One., 7(6), e38285. https://doi.org/10.1371/journal.pone.0038285.
Wang, J. P., Zhou, Y. M., Ye, Y. J., Shang, X. M., Cai, Y. L., Xiong, C. M., Wu, Y. X., & Xu, H. X., (2011). Topical anti-inflammatory and analgesic activity of kirenol isolated from Siegesbeckia orientalis. J. Ethnopharmacol., 137(3), 1089–1094. https://doi.org/10.1016/j.jep.2011.07.016.
Wang, K., Feng, X., Chai, L., Cao, S., & Qiu, F., (2017). The metabolism of berberine and its contribution to the pharmacological effects. Drug Metab. Rev., 49(2), 139–157. https://doi.org/10.1080/03602532.2017.1306544.
Wang, Y., Yang, R., Shan, L., & Yuan, J. (2018). Efficacy and safety of berberine combined with benzbromarone in the treatment of hyperuricemia with gout. Chin J Mod Drug Appl, 12(7), 3–6.
Wu, J., Li, X., Zhu, G., Zhang, Y., He, M., & Zhang, J., (2016). The role of resveratrol-induced mitophagy/autophagy in peritoneal mesothelial cells inflammatory injury via NLRP3 inflammasome activation triggered by mitochondrial ROS. Exp. Cell Res., 341(1), 42–53. https://doi.org/10.1016/j.yexcr.2016.01.014.
Wu, L. G., Chen, M., & Wu, Y. (2010). Effect of berberine on hyperuricemia. Chin J Integr Tradit West Med, 30(6), 659–662.
Wu, N., Kong, Y., Fu, Y., Zu, Y., Yang, Z., Yang, M., Peng, X., & Efferth, T., (2011). In vitro antioxidant properties, DNA damage protective activity, and xanthine oxidase inhibitory effect of cajaninstilbene acid, a stilbene compound derived from pigeon pea [Cajanus cajan (L.) Millsp.] leaves. J. Agric. Food Chem., 59(1), 437–443. https://doi.org/10.1021/jf103970b.
Xu, X., Yi, H., Wu, J., Kuang, T., Zhang, J., Li, Q., Du, H., et al., (2021). Therapeutic effect of berberine on metabolic diseases: Both pharmacological data and clinical evidence. Biomed Pharmacother., 133, 110984. https://doi.org/10.1016/j.biopha.2020.110984.
Yang, F., Shi, W., Wang, L., Qin, N., Wang, C., Guo, Y., Xu, G., et al., (2021). Lipidomics study of the therapeutic mechanism of plantaginis semen in potassium oxonate-induced hyperuricemia rat. BMC Complementary Medicine and Therapies, 21. https://doi.org/10.1186/s12906-021-03350-x.
Yang, H., Xiao, L., Yuan, Y., Luo, X., Jiang, M., Ni, J., & Wang, N., (2014). Procyanidin B2 inhibits NLRP3 inflammasome activation in human vascular endothelial cells. Biochem. Pharmacol., 92(4), 599–606. https://doi.org/10.1016/j.bcp.2014.10.001.
Yang, Q. B., He, Y. L., Zhong, X. W., Xie, W. G., & Zhou, J. G., (2019). Resveratrol ameliorates gouty inflammation via upregulation of sirtuin 1 to promote autophagy in gout patients. Inflammopharmacology, 27(1), 47–56. https://doi.org/10.1007/s10787-018-00555-4.
Yu, Z., Fong, W. P., & Cheng, C. H., (2006). The dual actions of morin (3,5,7,2,'4'-pentahydroxyflavone) as a hypouricemic agent: Uricosuric effect and xanthine oxidase inhibitory activity. J. Pharmacol. Exp. Ther., 316(1), 169–175. https://doi.org/10.1124/jpet.105.092684.
Zeng, J. X., Wang, J., Zhang, S. W., Zhu, J. X., Li, M., Huang, W. H., Wan, J. Y., et al., (2018). Antigout effects of Plantago asiatica: Xanthine oxidase inhibitory activities assessed by

electrochemical biosensing method. Evid. Based Complement Alternat Med., 2018, 1364617. https://doi.org/10.1155/2018/1364617.

Zhang, H. J., Li, L. N., Zhou, J., Yang, Q. Q., Liu, P. G., Xu, P., Liang, W. Q., et al., (2017). Effects of Gnaphalium affine D. Don on hyperuricemia and acute gouty arthritis. J. Ethnopharmacol., 203, 304–311. https://doi.org/10.1016/j.jep.2017.03.057.

Zhang, W., Doherty, M., Bardin, T., Pascual, E., Barskova, V., Conaghan, P., Gerster, J., et al., (2006). EULAR evidence-based recommendations for gout. Part II: Management. Report of a task force of the EULAR standing committee for international clinical studies including therapeutics (ESCISIT). Ann. Rheum. Dis, 65(10), 1312–1324. https://doi.org/10.1136/ard.2006.055269.

Zhang, W., Wu, C. Z., & Fan, S. Y., (2018). Chemical constituents from Gnaphalium affine and their xanthine oxidase inhibitory activity. Chin. J. Nat. Med., 16(5), 347–353. https://doi.org/10.1016/s1875-5364(18)30066-9.

Zhao, H., Xu, J., Wang, R., Tang, W., Kong, L., Wang, W., Wang, L., et al., (2021). Plantaginis semen polysaccharides ameliorate renal damage through regulating NLRP3 inflammasome in gouty nephropathy rats. Food Funct., 12(6), 2543–2553. https://doi.org/10.1039/d0fo03143g.

Zhao, P., Chen, K. L., Zhang, G. L., Deng, G. R., & Li, J., (2017). Pharmacological basis for use of Selaginella moellendorffii in gouty arthritis: Antihyperuricemic, anti-inflammatory, and xanthine oxidase inhibition. Evid. Based Complement Alternat Med., 2017, 2103254. https://doi.org/10.1155/2017/2103254.

Zhao, X., Zhu, J. X., Mo, S. F., Pan, Y., & Kong, L. D., (2006). Effects of cassia oil on serum and hepatic uric acid levels in oxonate-induced mice and xanthine dehydrogenase and xanthine oxidase activities in mouse liver. J. Ethnopharmacol., 103(3), 357–365. https://doi.org/10.1016/j.jep.2005.08.040.

Zhou, M., Li, S., Song, L., Hu, Q., & Liu, W., (2018). 4-(2-(4-chlorophenyl)-1-((4-chlorophenyl)amino)ethyl)benzene-1, 3-diol is a potential agent for gout therapy as a dual inhibitor of XOD and NLRP3. Phytomedicine, 42, 9–17. https://doi.org/10.1016/j.phymed.2018.03.007.

Zhu, C., Xu, Y., Liu, Z. H., Wan, X. C., Li, D. X., & Ling-Ling Tai, (2018). The anti-hyperuricemic effect of epigallocatechin-3-gallate (EGCG) on hyperuricemic mice. Journal of Biomedicine & Pharmacotherapy, 97, 168–173.

CHAPTER 8

PLANT METABOLITES AND VEGETABLES FOR HEART DISEASE PREVENTION AND TREATMENT

ASHWINI V. SHELKE,[1,2] SWATI G. TALELE,[1] TUSHAR LOKHANDE,[2] and ANIL G. JADHAV[1]

[1]*Sandip Institute of Pharmaceutical Sciences, Mahirawani, Nashik, Maharashtra, India*

[2]*Mahatma Gandhi Vidyamandir's, Pharmacy College, Panchavati, Nashik, Maharashtra, India*

ABSTRACT

Cardiovascular disease (CVD) has surpassed cancer as the top cause of mortality worldwide and consider as significant public health problem. The recent increase in popularity of natural products and alternative medicine has renewed interest in traditional heart disease treatments. The anti-oxidative, antihypercholesterolemic, anti-ischemic, and inhibitory effects of platelet aggregation in different plants and vegetables may lower the risk of heart-related diseases. According to evidence from prospective group studies, consuming a regular plant-based diet such as green vegetables is linked to reduce risk of stroke and coronary artery disease and. Dietary variables are crucial in the diagnosis of CVD and may influence CVD risk to a great extent, but they have received less attention. Certain dietary habits have been linked to better cardiovascular health in epidemiological research. Vitamin K and nitrates are rich in green leafy vegetables, which can help enhance vascular function and lower blood pressure. Nitrates are a chemical that aids in the opening of blood arteries, allowing blood rich in oxygen to reach our

Plant Metabolites and Vegetables as Nutraceuticals.
Raj K. Keservani, Bui Thanh Tung, Rajesh K. Kesharwani, and Eknath D. Ahire (Eds.)

heart. Vegetable eating has been proven to be inversely associated to the risk of CVD in epidemiological studies. A higher diet of leafy greens has been linked to a lower risk of heart disease in studies. Multiple beneficial plant metabolites such as mono- and polyunsaturated fatty acids, n3 fatty acids, antioxidant vitamins, minerals, phytochemicals, fiber, and plant protein are thought to be involved in the protective effects of these foods. Here plant-based metabolites and foods for Heart Disease Prevention and Treatment were discovered by a comprehensive search of web databases using relevant terms. Epidemiologic studies on plant metabolites and vegetables in connection to heart disease prevention and treatment are summarized in succeeding sections.

8.1 INTRODUCTION

In recent years, there has been a surge in interest in using plant-based meals to prevent cardiovascular disease (CVD) such as heart attacks, myocardial infarction (MI), coronary heart disease (CAD), cerebrovascular disease (stroke), heart failure (HF), hypertension, peripheral artery disease (PAD), congenital heart disease and rheumatic heart disease, which has resulted in a slew of epidemiologic and clinical studies (Hu et al., 2002). Traditional Mediterranean and Asian diets include plant-based foods such as fruits and vegetables, nuts, natural vegetable oils, and whole grains (Kushi et al., 1995). In these areas, such diets have been linked to lower rates of heart disease and mortality. Such research is critical for expanding our knowledge of the diet-heart disease relationship beyond what we already know about specific nutrients. Plant-based diets are often thought to have to be low in fat. However, substantial data from epidemiologic and clinical studies suggests that the types of fat consumed in the diet are more relevant than the total amount of fat consumed in predicting the risk of coronary artery disease (Hu et al., 2003). When it comes to suggesting diets to lower the risk of heart disease and treating individuals who have already developed illnesses, the composition of carbohydrate is critical (Mann et al., 2007). Nature is clearly an essential source for discovering novel medications that lead to disease treatment. Aspirin comes from the *Salix alba* L. tree, digoxin (cardiac glycoside) comes from *Digitalis purpurea*, ephedrine comes from *Ephedra sinica*, lovastatin comes from *Monascus purpureus* L., taxol comes from *Taxus brevifolia*, reserpine comes from *Rauvolfia serpentina*, and so on (Harvey et al., 2000). In 1985, the World Health Organization (WHO) estimated that nearly 65% of the world's population relied mostly on traditional medicines produced from plants. The WHO's Traditional Medicine Centers identified 122 substances that are regularly utilized in the countries where the Centers are located. Surprisingly, the 122 chemicals are derived from only 94 plant species and are employed for identical ethnomedical therapies in each of the host countries (Farnsworth et al., 1985). Papaverine from *Papaver somniferum*, which is used to make the hypertension medication verapamil, is an example of such a molecule (Fabricant et al., 2001). Traditional herbal (Sharma et al., 2020; Rane et al., 2020a–c) and plant-derived extracts are becoming more popular as scientific study demonstrates their value in illness prevention and treatment (Frishman et al., 2009). Despite the fact that numerous and chemically diverse secondary metabolites have been purified from plant bioactives and have been refined for exerting a biological effect, they remain untested for therapeutic use. Natural-derived molecules will, nonetheless, be a major source of novel medications, according to newly published scientific evidence, technical developments, and research trends (Davison et al., 2019). Despite advances in combinatorial chemical synthesis and the manufacturing of modern synthetic pharmaceuticals, this has been a driving force for the increasing popularity of traditional herbal and plant-derived therapies among researchers. Another explanation for the renewed interest in medicinal plant products is that rural populations in developing nations have turned to traditional herbal and plant-derived treatments in their efforts to control diseases in the face of limited socioeconomic resources. This is due to a number of variables, including the fact that plant-based medicines are a less expensive option with less side effects (Frishman et al., 2009). Herbal and plant therapies are not only cost-effective, but they also contain thousands of bioactive ingredients with established medicinal properties (Pan et al., 2013). The goal of this study is to summarize the existing information on the efficacy and safety of the most regularly used herbal drugs in cardiovascular care. The present information on the efficacy and safety of herbal drugs for the treatment of cardiovascular illnesses was reported through an unsystematic, narrative collection of published data. The National Center for Complementary Medicine of the National Institutes of Health (NCCMNIH) selected for information on herbal resources used in the treatment of any CVD (Bethesda, 2010). Near about 42 herbal sources are reported and identifies as possible indications in report of NCCMNIH. In this chapter, we selected plant metabolites and vegetables with a possible indication for the treatment of one or more of the following cardiovascular conditions: PAD, hypertension, dyslipidemia, coronary artery disease, thromboembolic disorders, and HF. We used somewhat data reported from PubMed Central, the Cochrane Database of Systematic Reviews, and the U.S. National Center for Complementary and Integrative Health website (Rosa et al., 2017). Figure 8.1 represents various types of plants and its metabolites which can be used for prevention and treatment of heart disease.

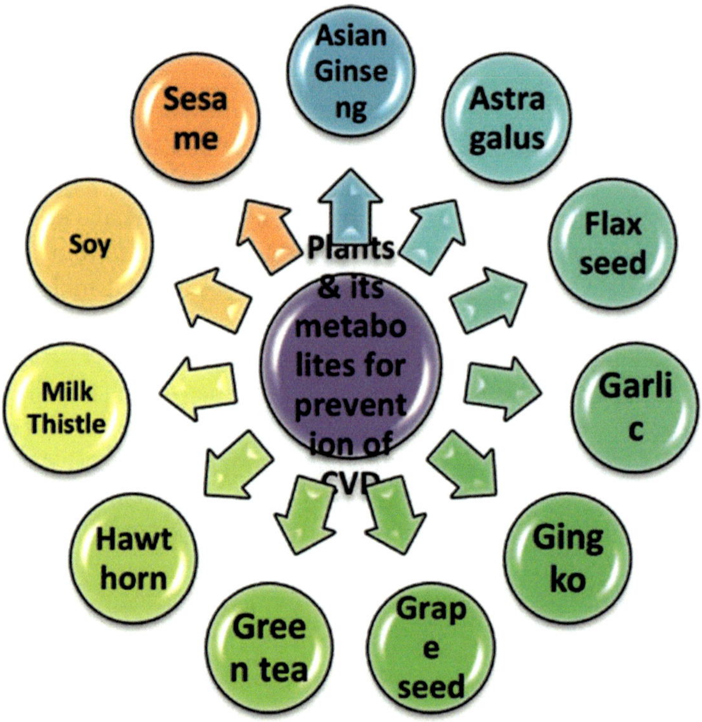

FIGURE 8.1 Various types of plants and its metabolites which can be used for prevention and treatment of heart disease.

8.2 PLANT METABOLITES USED IN THE PREVENTION AND TREATMENT OF HEART DISEASE

8.2.1 ASIAN GINSENG (PANAX GINSENG)

Asian ginseng is said to improve blood pressure regulation and lower blood glucose and cholesterol levels, among its many other medical benefits. It also helps the body's overall health and immune system, it is a type of antioxidant (Emboden, 1974). According to preclinical data from *in vitro* and *in vivo* studies, Asian ginseng contains saponins with a partial agonistic effect on steroidal receptors, known as ginsenosides, which may have a beneficial impact on the cardiovascular system (Kim, 2012). When used for a short period of time and at the prescribed doses, Asian ginseng appears to be usually safe. But for long-term use potential side effects such as Increased blood pressure, nausea, diarrhea, headache, sleeplessness, and allergic responses have been reported (Izzo, 2009).

8.2.2 ASTRAGALUS (*ASTRAGALUS MEMBRANACEUS*)

Astragalus is a tonic that boosts the immune system and is commonly used to treat the common cold and upper respiratory infections. It's also been suggested that it could have an influence on HF and CAD. Scientific data from *in vivo* and *in-vitro* research suggests that astragalus may have a favorable effect on cardiomyocytes due to its antioxidant and anti-inflammatory properties (Huang et al., 2016; Zhang et al., 2007; Yang et al., 2013). In some controlled studies including patients with ischemic heart disease and congestive HF, astragalus has been shown to have a beneficial symptomatic effect on angina, dyspnea, and chest pain as well as an enhancement in electrocardiographic measures and cardiac output (Luo et al., 1995; Li et al., 1995; Lei et al., 1994).

8.2.3 FLAXSEED OIL (*LINUM USITATISSIMUM*)

Because of its high soluble fiber content, flaxseed is widely used as a laxative. Symptomatic relief in arthritis, osteoporosis, and menopause, as well as cancer prevention are all possible health advantages related with flaxseed and flaxseed oil use (Goyal et al., 2014). Flaxseed is high in alpha-linolenic acid, a type of important n–3 polyunsaturated fatty acid that may help with hyperlipidemia, coronary artery disease. Hence along with abovementioned effects lowering blood pressure and cholesterol levels, slowing the progression of atherosclerosis, and preventing arrhythmias are all said to have favorable benefits of flax seed on cardiovascular health (Fleming et al., 2014). Flaxseed also includes lignans, which function as antioxidants as well as soluble and insoluble fiber, which have been shown to lower blood cholesterol levels by boosting bile acid excretion and inhibiting synthesis of endogenous cholesterol (Anderson et al., 2009). Flaxseed has been proven in animal experiments to slow the growth of atherosclerotic plaques and promote their reversal (Francis et al., 2013). Flaxseed's anti-inflammatory qualities, which include the potential to reduce platelet and endothelial reactivity and proliferation, could help explain the antiatherogenic impact (Haliga et al., 2013).

8.2.4 GARLIC (ALLIUM SATIVUM)

Garlic has long been known as having anti-inflammatory and anti-cancer qualities in a variety of medical ailments, including heart disease, cancer, and infections. Garlic has been shown to lower blood pressure, decreasing cholesterol and triglyceride levels in the blood and inhibit platelet aggregation (Rahman et al., 2006; Breithaupt-Grögler et al., 1997). Garlic's direct influence on various biological pathways involved in the etiology of hypertension, such as oxidative stress and endothelial NO and hydrogen sulfide generation, could mediate decreased peripheral vascular resistance, vasodilation, and smooth muscle cell relaxation (Varshney et al., 2016; Ried et al., 2014). Garlic has also been demonstrated to lower cholesterol absorption and endogenous production (Shouk et al., 2014; Gebhardt et al., 1996). Despite the fact that most research lack a sufficient assessment of the safety profile garlic supplementation appears to be generally safe for people because it is a natural plant-derived part.

8.2.5 GINKGO (GINKGO BILOBA)

Very recently, gingko has been studied in regard to CAD. Ginkgolides, one of the most active components of ginkgo leaf extract include, Antioxidant bilobalides and flavonoids, platelet inhibition, and anti-inflammatory action (Zhou et al., 2004). Furthermore, some clinical trials have revealed a possible symptomatic improvement in patients with PAD.

8.2.6 GRAPE (VITIS VINIFERA) SEEDS

Many cardiovascular disorders, including atherosclerosis, hypertension, hypercholesterolemia, chronic venous insufficiency is thought to benefit from grapeseed extract (GSE). Grapes are recognized for having a high concentration of antioxidant chemicals. Resveratrol (RES), for instance, is concentrated in the skin of the grape, whereas oligomeric proanthocyanidin complexes, vitamin E, flavonoids, and linoleic acid are concentrated in the seeds (Bertelli et al., 2009; Shafiee et al., 2003). Furthermore, grape seeds have been shown to impair platelet function and platelet-dependent inflammatory reactions (Vitseva et al., 2005; Albers et al., 2004). Grape Seed Extract has recently been discovered to lower systolic and diastolic blood pressure in people who are pre-hypertensive (Park et al., 2016). Grape Seed Extract has also been demonstrated to lower LDL cholesterol and protect it from oxidation (Preuss et al., 2000; Vigna et al., 2003).

8.2.7 GREEN TEA (CAMELLIA SINENSIS)

Green tea has been suggested to reduce levels of cholesterol, and prevention of cardiovascular events is among the proposed cardiovascular outcomes (Stangl et al., 2006). When taken in high dosages (at least 3 cups per day) and over time, large observational studies have indicated that green tea may reduce cardiovascular morbidity and mortality (Zhang et al., 2015). The effectiveness of tea for the primary prevention of CVD was evaluated in a comprehensive Cochrane study. Green tea had a statistically significant effect on lowering total cholesterol, LDL cholesterol, SBP, and DBP, according to the researchers (Hartley et al., 2013).

8.2.8 HAWTHORN (CRATAEGUS)

Hawthorn has long been used to treat digestive, cardiovascular, and kidney problems. Hawthorn has been shown to have beneficial effects on a variety of cardiovascular disorders, including hypertension, hyperlipidemia, arrhythmia, and congestive HF (Chang et al., 2005). Hawthorn has been demonstrated to have antioxidant effects in biological studies. Furthermore, hawthorn appeared to have a favorable inotropic impact on cardiomyocytes through modulating the sodium-potassium (Na/K)-ATPase and boosting calcium transport (Rodriguez et al., 2008). It may also have antiarrhythmic properties via a mechanism similar to that of class III antiarrhythmics (Müller et al., 1999). Hawthorn has been shown to have a vasodilating impact in both coronary and peripheral vessels, and it may directly influence NO levels in the endothelium and inhibit angiotensin-converting enzyme (Brixius et al., 2006). Hawthorn is also said to have anti-atherosclerotic properties, lowering endothelial barrier dysfunction (Bubik et al., 2012). Furthermore, hawthorn has been shown to reduce cholesterol synthesis in the liver and lipid absorption in the gut (Zhang et al., 2002; Wang et al., 2011). A positive effect of hawthorn on hypertension, especially on lowering diastolic blood pressure, was reported in few random clinical trials (Asher et al., 2012; Walker et al., 2006).

8.2.9 MILK THISTLE (SILYBUM MARIANUM)

Milk thistle has been suggested as a natural treatment for liver and gallbladder problems. A possible effect on CVD has also been suggested recently. Milk thistle's recognized active component, silymarin, contains a group of flavonoids that may be responsible for the plant's potent antioxidant properties (Surai, 2015). Recently, it was suggested that silymarin could protect tissues, particularly the heart, from ischemia reperfusion injury by altering preconditioning pathways.

8.2.10 SOY (GLYCINE MAX)

Soy has been claimed to help with a variety of medical ailments, including menopausal symptoms, osteoporosis, cognitive problems, and cancer (Xiao, 2008). It is thought to lower cholesterol and blood pressure, therefore lowering cardiovascular risk (Carroll, 1991). Soybeans, which are the seeds of the soy plant, are high in protein. Animal research revealed that soy protein decreased cholesterol levels and avoided atherosclerosis when compared to animal proteins (Carroll, 1982). Because of its high level of polyunsaturated fats (Singh et al., 2017b), fiber, vitamins (Singh et al., 2017a), and minerals, and low content of saturated fat, the American Heart Association has concluded that soy dietary products may be advantageous to cardiovascular health (EFSA J., 2012).

8.2.11 SESAME

Sesame metabolites have been shown to have substantial antioxidant, anti-atherogenic, anti-thrombotic, and anti-hypertensive activity; consequently, eating sesame whole grains or purified bioactive components on a daily basis should provide effective protection against CVDs. Although fat-soluble sesame components were assumed to help the cardiovascular system, several research showed that defatted sesame seed extracts (DSSE) had preventive effects as well (Visavadiya, 2009).

8.3 VEGETABLES PREVENTION AND TREATMENT OF HEART DISEASE

Including greens in our diet can colorfully color your plate and healthfully nourish your heart (Figure 8.2). Consuming leafy green and cruciferous veggies lowers your chance of heart disease and possibly death. Antioxidation, anti-inflammation, anti-platelet effects, regulating blood pressure, blood glucose, and lipid profile, attenuating myocardial damage, and modulating relevant enzyme activities, gene expression, and signaling pathways, as well as some other biomarkers linked to CVDs, may all be involved in the cardioprotective effects of vegetables. These nutrient-dense green vegetables are high in a number of elements that work together to combat CVD risk factors like cholesterol, blood pressure, and inflammation. Figure 8.3 represents various vegetables for healthy heart.

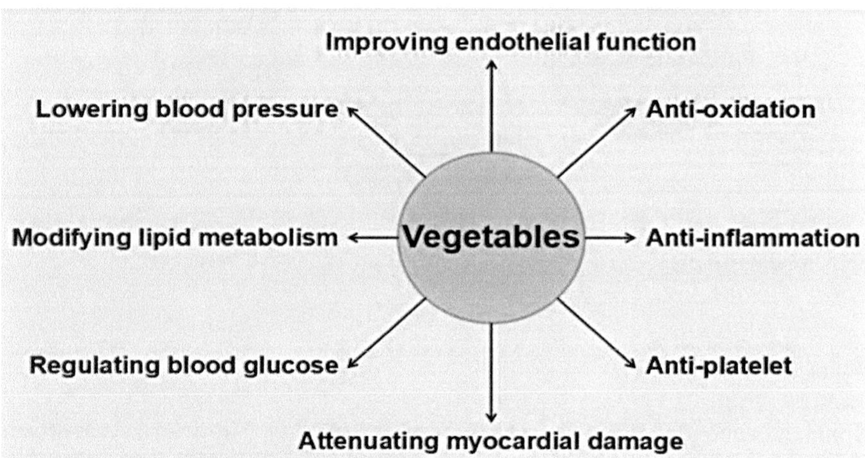

FIGURE 8.2 Health benefits of vegetables.

8.3.1 SWISS CHARD

Potassium and magnesium are abundant in this dark green, leafy vegetable. These minerals aid in the regulation of blood pressure. Swiss chard also contains fiber, vitamin A, and the antioxidants lutein and zeaxanthin, all of which are good for your heart (Hung et al., 2004).

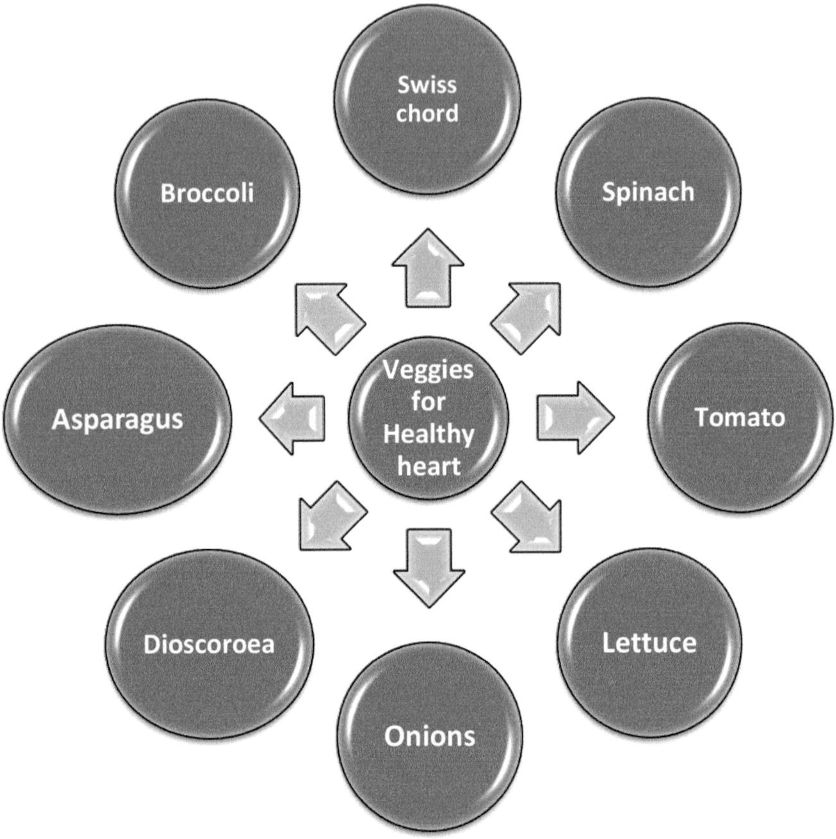

FIGURE 8.3 Various vegetables for heart disease prevention and treatment.

8.3.2 SPINACH

Spinach is one of the best sources of nitrate in the diet. The arterial waveform is suggestive of arterial stiffness, as well as central and peripheral blood pressure, can be affected by short-term consumption of spinach, a vegetable high in dietary nitric oxide (Jovanovski et al., 2015). Dietary nitrate, probably via the nitrate-nitrite NO pathway as an alternative to the oxygen-dependent L-arginine-endothelial NO synthase-NO pathway, is thought to contribute to anaerobic NO generation. The anaerobic process, which provides an alternative source of NO, may be particularly relevant in the pathophysiology of CVD, which is characterized by a reduction in blood and oxygen flow and a loss in endothelial NO-derived activities (Machha et al., 2011, 2012).

8.3.3 TOMATO

Vegetables like tomatoes are common in most dietary patterns around the world, and their health benefits have been demonstrated in longitudinal epidemiologic research. Self-reported high intakes of tomato and tomato products, as well as dietary lycopene (a carotenoid molecule), are linked to a lower risk of CVD. While epidemiological research suggests a link between tomato products and/or lycopene and a reduced risk of CVD (Sesso et al., 2003; Jacques et al., 2013). Lycopene is one of the most powerful antioxidants and the most abundant carotenoid in human plasma, and it's thought to be one of the active molecules behind tomato's health advantages (Mein et al., 2008).

8.3.4 LETTUCE

Due to its high antioxidant content, lettuce diet boosts total cholesterol end-products excretion and enhances antioxidant status (vitamins C, E, and carotenoids). In certain studies, lettuce has been shown to have a positive impact on lipid metabolism and tissue oxidation. As a result, eating lettuce on a daily basis should help to prevent from cardiovascular illnesses (Nicolle et al., 2004).

8.3.5 ONIONS

Onions are a widely used vegetable that contains bioactive components such as phytochemicals. Onion extracts had strong anti-atherogenic properties that were linked to a number of bioactivities. In a study, it was discovered that onion (Allium cepa L.) extracts, as well as the bioactive components quercetin and catechin, increased paraoxonase-1 activity and radical scavenging activity, preventing LDL oxidation and lipid peroxidation. It was discovered that onion extract reduced atherosclerotic lesions, increased endogenous aortic hydrogen sulfide (H_2S) synthesis, and decreased plasma and aortic adrenomedullin concentration (Li et al., 2011). Onions were a prospective candidate for anti-atherogenic therapy because of all of these properties.

8.3.6 DIOSCOREA

Dioscorea zingiberensis is a type of tuberous root vegetable. It is grown in China for the manufacturing of diosgenin, a key pharmaceutical intermediary in the creation of steroid hormones. As part of a treatment for cardiovascular illness, the rhizomes generate steroidal saponins (TSS). The antioxidant and anti-apoptotic properties of diosgenin isolated from Dioscorea, as well as cyclic guanosine monophosphate modifying actions, protected the heart against cardiotoxicity caused by doxorubicin (Li et al., 2010).

8.3.7 ASPARAGUS (ASPARAGUS OFFICINALIS)

Asparagus is a perennial flowering plant species in the Asparagus genus. Its young shoots are eaten as a vegetable in the spring. Asparagus is a nutritious and healthy vegetable that also contains useful components. Chronic renal failure is caused by high blood pressure, which aggravates renal function and structure while also increasing the risk of CVD. Asparagus reduces hypertension in the kidney by inhibiting angiotensin-converting enzyme activity (Sanae et al., 2013).

8.3.8 Broccoli

Intake of broccoli sprouts high in glucosinolate (GS) and glucoraphanin (Grn) reduced oxidative stress and inflammation in the kidneys and cardiovascular system, according to certain studies (Lingyun et al., 2004).

8.4 CONCLUSIONS

According to this chapter, intake of cruciferous and leafy green vegetables, may provide significant cardiovascular health advantages. Increased vegetable consumption, particularly leafy green and cruciferous vegetables, may provide the best cardiovascular health advantages. When these dietary modifications are combined with other recommended lifestyle adjustments, the health advantages will be maximized. Consuming a diet rich in vegetables, fruits, and whole grains; low-fat dairy products, chicken, fish, legumes, non-tropical vegetable oils, and nuts; and reducing saturated and trans fats, sweets, sugar-sweetened beverages, and red meats are all examples of lifestyle adjustments. Other lifestyle modifications include increasing physical exercise, reducing cigarette smoking and alcohol consumption, and eating enough calories to maintain a healthy body weight. Large, well-designed epidemiological studies evaluating the cardiovascular health benefits of various vegetable varieties are needed. To establish the causal effects of the individual vegetables identified to be most beneficial for cardiovascular health, large long-term randomized controlled trials are required.

KEYWORDS

- **asparagus**
- **cardiovascular disease**
- **heart disease**
- **milk thistle**
- **phytochemicals**
- **plant metabolites**
- **vegetables**

REFERENCES

Albers, A. R., Varghese, S., Vitseva, O., et al., (2004). The anti-inflammatory effects of purple grape juice consumption in subjects with stable coronary artery disease. *Arterioscler. Thromb. Vasc. Biol., 24*, 179, 180.

Anderson, J. W., Baird, P., Davis, R. H. Jr., et al., (2009). Health benefits of dietary fiber. *Nutr. Rev., 67*, 188–205.

Asher, G. N., Viera, A. J., Weaver, M. A., et al., (2012). Effect of hawthorn standardized extract on flow mediated dilation in prehypertensive and mildly hypertensive adults: A randomized, controlled cross-over trial. *BMC Complement Altern. Med., 12*, 26.

Bertelli, A. A., & Das, D. K., (2009). Grapes, wines, resveratrol, and heart health. *J Cardiovasc Pharmacol., 54*, 468–476.

Bethesda, M. D., (2010). *Herbs at a Glance: A Quick Guide to Herbal Supplements*. U.S. Department of Health and Human Services, National Institutes of Health, National Center for Complementary and Alternative Medicine.

Breithaupt-Grögler, K., Ling, M., Boudoulas, H., et al., (1997). Protective effect of chronic garlic intake on elastic properties of aorta in the elderly. *Circulation, 96*, 2649–2655.

Brixius, K., Willms, S., Napp, A., et al., (2006). Crataegus special extract WS 1442 induces an endothelium dependent, NO-mediated vasorelaxation via eNOS-phosphorylation at serine 1177. *Cardiovasc Drugs Ther., 20*, 177–184.

Bubik, M. F., Willer, E. A., Bihari, P., et al., (2012). A novel approach to prevent endothelial hyperpermeability: The Crataegus extract WS 1442 targets the cAMP/Rap1 pathway. *J. Mol. Cell Cardiol., 52*, 196–205.

Carroll, K. K., (1982). Hypercholesterolemia and atherosclerosis: Effects of dietary protein. *Fed. Proc., 41*, 2792–2796.

Carroll, K. K., (1991). Review of clinical studies on cholesterol-lowering response to soy protein. *J. Am. Diet. Assoc., 91*, 820–827.

Chang, W. T., Dao, J., & Shao, Z. H., (2005). Hawthorn: Potential roles in cardiovascular disease. *Am. J. Chin. Med., 33*, 1–10.

Davison, E. K., & Brimble, M. A., (2019). Natural product derived privileged scaffolds in drug discovery. *Curr. Opin. Chem. Biol., 52*, 1–8.

EFSA Panel on Dietetic Products, Nutrition and Allergies (NDA). (2012). Scientific Opinion on the substantiation of a health claim related to isolated soy protein and reduction of blood LDL-cholesterol concentrations pursuant to Article 14 of Regulation (EC) No 1924/2006. *EFSA Journal, 10*(2), 2555.

Emboden, W. A., (1974). *Bizarre Plants* (p. 167). New York: MacMillam Pub. Co. Inc.

Fabricant, D. S., & Farnsworth, N. R., (2001). The value of plants used in traditional medicine for drug discovery. *Environ. Health Perspect., 109* (Suppl 1), 69–75.

Farnsworth, N. R., Akerele, O., Bingel, A. S., Soejarto, D. D., & Guo, Z., (1985). *Medicinal Plants in Therapy, 63*(6), 965–981.

Fleming, J. A., & Kris-Etherton, P. M., (2014). The evidence for a-linolenic acid and cardiovascular disease benefits: Comparisons with eicosapentaenoic acid and docosahexaenoic acid. *Adv. Nutr., 5*, 863S–876S.

Francis, A. A., Deniset, J. F., Austria, J. A., et al., (2013). Effects of dietary flaxseed on atherosclerotic plaque regression. *Am. J. Physiol. Heart Circ. Physiol., 304*, H1743–1751.

Frishman, W. H., Beravol, P., & Carosella, C., (2009). Alternative and complementary medicine for preventing and treating cardiovascular disease. *Dis. Mon., 55* (3), 121–192.

Gebhardt, R., & Beck, H., (1996). Differential inhibitory effects of garlic-derived organosulfur compounds on cholesterol biosynthesis in primary rat heaptocyte cultures. *Lipids*, *31*, 1269–1276.

Goyal, A., Sharma, V., Upadhyay, N., et al., (2014). Flax and flaxseed oil: An ancient medicine & modern functional food. *J. Food Sci. Technol., 51*, 1633–1653.

Haliga, R. E., Iancu, R. I., Butcovan, D., et al., (2013). Flaxseed prevents leukocyte and platelet adhesion to endothelial cells in experimental atherosclerosis by reducing sVCAM-1 and vWF. *Sci. World J.,* 303950.

Hartley, L., Flowers, N., Holmes, J., et al., (2013). Green and black tea for the primary prevention of cardiovascular disease. *Cochrane Database Syst. Rev., 6*, CD009934.

Harvey, A., (2000). Strategies for discovering drugs from previously unexplored natural products. *Drug Discovery Today, 5*(7), 294–300.

Hu, F. B., & Willett, W. C., (2002). Optimal diets for prevention of coronary heart disease. *JAMA., 288*, 2569–2578.

Hu, F. B., (2003). Plant-based foods and prevention of cardiovascular disease: An overview. *Am J Clin Nutr., 78*(suppl), 544S–551S.

Huang, H., Lai, S., Wan, Q., et al., (2016). Astragaloside IV protects cardiomyocytes from anoxia/reoxygenation injury by upregulating the expression of Hes1 protein. *Can J. Physiol. Pharmacol., 94*, 542–553.

Hung, H. C., Joshipura, K. J., Jiang, R., Hu, F. B., Hunter, D., Smith-Warner, S. A., Colditz, G. A., et al., (2004). Fruit and vegetable intake and risk of major chronic disease. *J. Natl. Cancer Inst., 96*, 1577–1584.

Izzo, A. A., & Ernst, E., (2009). Interactions between herbal medicines and prescribed drugs: An updated systematic review. *Drugs, 69*, 1777–1798.

Jacques, P. F., Lyass, A., Massaro, J. M., Vasan, R. S., & D'Agostino, R. B., (2013). Relationship of lycopene intake and consumption of tomato products to incident CVD. *Br. J. Nutr., 110*(3), 545–551.

Jovanovski, E., et al., (2015). Effect of spinach, a high dietary nitrate source, on arterial stiffness and related hemodynamic measures: A randomized, controlled trial in healthy adults. *Clin. Nutr. Res., 4*, 160–167.

Kim, J. H., (2012). Cardiovascular diseases and panax ginseng: A review on molecular mechanisms and medical applications. *J. Ginseng Res., 36*, 16–26.

Kushi, L. H., Lenart, E. B., & Willett, W. C., (1995). Health implications of Mediterranean diets in light of contemporary knowledge, I: Plant foods and dairy products. *Am. J. Clin. Nutr., 61*(suppl), 1407S–1415S.

Lei, Z. Y., Qin, H., & Liao, J. Z., (1994). Action of astragalus membranaceus on left ventricular function of angina pectoris. *Zhongguo Zhong Xi Yi Jie He Za Zhi., 14*(4), 195, 199–202.

Li, H., Huang, W., Wen, Y., Gong, G., Zhao, Q., & Yu, G., (2010). Anti-thrombotic activity and chemical characterization of steroidal saponins from *Dioscorea zingiberensis* C.H. Wright. *Fitoterapia, 81*(8), 1147–1156.

Li, S. Q., Yuan, R. X., & Gao, H., (1995). Clinical observation on the treatment of ischemic heart disease with *Astragalus membranaceus*. *Zhongguo Zhong Xi Yi Jie He Za Zhi, 15*, 77–80.

Li, W., Tang, C., Jin, H., & Du, J., (2011). Effects of onion extract on endogenous vascular H_2S and adrenomedulin in rat atherosclerosis. *Curr. Pharm. Biotechnol., 12*, 1427–1439.

Lingyun, W., et al., (2004). Dietary approach to attenuate oxidative stress, hypertension, and inflammation in the cardiovascular system. *PNAS, 4*, 101–183.

Luo, H. M., Dai, R. H., & Li, Y., (1995). *Nuclear Cardiology Study on Effective Ingredients of Astragalus Membranaceus in Treating Heart Failure, 15*, 707–709.

Machha, A., & Schechter, A. N., (2011). Dietary nitrite and nitrate: A review of potential mechanisms of cardiovascular benefits. *Eur. J. Nutr., 50*, 293–303.

Machha, A., & Schechter, A. N., (2012). Inorganic nitrate: A major player in the cardiovascular health benefits of vegetables? *Nutr. Rev., 70*, 367–372.

Mann, J., (2007). Dietary carbohydrate: Relationship to cardiovascular disease and disorders of carbohydrate metabolism. *European Journal of Clinical Nutrition., 61*(Suppl 1), S100–S111.

Mein, J. R., Lian, F., & Wang, X. D., (2008). Biological activity of lycopene metabolites: Implications for cancer prevention. *Nutr. Rev., 66*(12), 667–683.

Müller, A., Linke, W., & Klaus, W., (1999). Crataegus extract blocks potassium currents in guinea pig ventricular cardiac myocytes. *Planta Med., 65*, 335–339.

Nicolle, C., et al., (2004). Health effect of vegetable-based diet: Lettuce consumption improves cholesterol metabolism and antioxidant status in the rat. *Clinical Nutrition, 23*, 605–614.

Pan, S. Y., Zhou, S. F., Gao, S. H., Yu, Z. L., Zhang, S. F., Tang, M. K., et al., (2013). New perspectives on how to discover drugs from herbal medicines: CAM's outstanding contribution to modern therapeutics. *Evid. Based Complement. Alternat. Med.* Article ID. 627375.

Park, E., Edirisinghe, I., Choy, Y. Y., et al., (2016). Effects of grape seed extract beverage on blood pressure and metabolic indices in individuals with prehypertension: A randomized, double-blinded, two-arm, parallel, placebo-controlled trial. *Br. J. Nutr., 115*, 226–238.

Preuss, H. G., Wallerstedt, D., Talpur, N., et al., (2000). Effects of niacin-bound chromium and grape seed proanthocyanidin extract on the lipid profile of hypercholesterolemic subjects: A pilot study. *J. Med., 31*, 227–246.

Rahman, K., & Lowe, G. M., (2006). Garlic and cardiovascular disease: A critical review. *J. Nutr., 136*, 736S–40S.

Rane, B. R., Bharath, M. S., Patil, R. R., Keservani, R. K., & Jain, A. S., (2020b). Novel approaches in nutraceuticals; Chapter 11. In: Rajesh, K. K., Raj, K. K., & Anil, K. S., (eds.), *Enhancing the Therapeutic Efficacy of Herbal Formulations Through Novel Drug Delivery Systems* (pp. 241–266). IGI Global International Publisher, Pennsylvania, USA. ISBN13: 9781799844532.

Rane, B. R., Patil, A. S., Keservani, R. K., & Jain, A. S., (2020a). Novel approaches in herbal formulation; Chapter 2. In: Rajesh, K. K., Raj, K. K., & Anil, K. S., (eds.), *Enhancing the Therapeutic Efficacy of Herbal Formulations Through Novel Drug Delivery Systems* (pp. 43–68). IGI Global International Publisher, Pennsylvania, USA. ISBN13: 9781799844532.

Rane, B. R., Tadavi, S. A., & Keservani, R. K., (2020c). Naturopathy; Chapter 12. In: Anil, K. S., Raj, K. K., & Surya, P. G., (eds.), *Herbal Product Development* (pp. 321–347). Apple Academic Press, CRC Press, Taylor & Francis Group. ISBN: 9781771888776.

Ried, K., & Fakler, P., (2014). Potential of garlic (*Allium sativum*) in lowering high blood pressure: Mechanisms of action and clinical relevance. *Integr. Blood Press Control, 7*, 71–82.

Rodriguez, M. E., Poindexter, B. J., Bick, R. J., et al., (2008). A comparison of the effects of commercially available hawthorn preparations on calcium transients of isolated cardiomyocytes. *J. Med. Food, 1*, 680–686.

Rosa, L., et al., (2017). Herbal medications in cardiovascular medicine. *Herbal Medications in Cardiology, 69*(9), 1188–1199.

Sanae, M., & Yasuo, A. (2013). Green asparagus (Asparagus officinalis) prevented hypertension by an inhibitory effect on angiotensin-converting enzyme activity in the kidney of spontaneously hypertensive rats. *Journal of Agricultural and Food Chemistry, 61*(23), 5520–5525.

Sesso, H. D., Liu, S., Gaziano, J. M., & Buring, J. E., (2003). Dietary lycopene, tomato-based food products and cardiovascular disease in women, *J. Nutr., 133*(7), 2336–2341.

Shafiee, M., Carbonneau, M. A., Urban, N., et al., (2003). Grape and grape seed extract capacities at protecting LDL against oxidation generated by Cu2þ, AAPH or SIN-1 and at decreasing superoxide THP-1 cell production. A comparison to other extracts or compounds. *Free Radic. Res., 37*, 573–584.

Sharma, A. K., Keservani, R. K., & Gautam, S. P., (2020). *Herbal Product Development*. Apple Academic Press, CRC Press, Taylor & Francis Group. ISBN: 9781771888776.

Shouk, R., Abdou, A., Shetty, K., et al., (2014). Mechanisms underlying the antihypertensive effects of garlic bioactives. *Nutr. Res., 34*, 106–115.

Singh, P., Kesharwani, R. K., & Keservani, R. K., (2017a). Antioxidants and vitamins: Roles in cellular function and metabolism. In: *Sustained Energy for Enhanced Human Functions and Activity* (pp. 385–407). Academic Press. ISBN:9780128054130.

Singh, P., Kesharwani, R. K., & Keservani, R. K., (2017b). Protein, carbohydrates, and fats: Energy metabolism. In: *Sustained Energy for Enhanced Human Functions and Activity* (pp. 103–115). Academic Press. ISBN:9780128054130.

Stangl, V., Lorenz, M., & Stangl, K., (2006). The role of tea and tea flavonoids in cardiovascular health. *Mol. Nutr. Food Res., 50*, 218–228.

Surai, P. F., (2015). Silymarin as a natural antioxidant: An overview of the current evidence and perspectives. *Antioxidants (Basel), 4*, 204–247.

Varshney, R., & Budoff, M. J., (2016). Garlic and heart disease. *J. Nutr., 146*, 416S–421S.

Vigna, G. B., Costantini, F., Aldini, G., et al., (2003). Effect of a standardized grape seed extract on low-density lipoprotein susceptibility to oxidation in heavy smokers. *Metabolism, 52*, 1250–1257.

Visavadiya, N. P., Soni, B., & Dalwadi, N., (2009). Free radical scavenging and antiatherogenic activities of *Sesamum indicum* seed extracts in chemical and biological model systems. *Food Chem. Toxicol., 47*, 2507–2515.

Vitseva, O., Varghese, S., Chakrabarti, S., et al., (2005). Grape seed and skin extracts inhibit platelet function and release of reactive oxygen intermediates. *J. Cardiovasc. Pharmacol., 46*, 445–451.

Walker, A. F., Marakis, G., Simpson, E., et al., (2006). Hypotensive effects of hawthorn for patients with diabetes taking prescription drugs: A randomized controlled trial. *Br. J. Gen. Pract., 56*, 437–443.

Wang, T., An, Y., Zhao, C., et al., (2011). Regulation effects of *Crataegus pinnatifida* leaf on glucose and lipids metabolism. *J. Agric. Food Chem., 59*, 4987–4994.

Xiao, C. W., (2008). Health effects of soy protein and isoflavones in humans. *J. Nutr., 138*, 1244S–1249S.

Yang, J., Wang, H. X., Zhang, Y. J., et al., (2013). Astragaloside IV attenuates inflammatory cytokines by inhibiting TLR4/NF-IB signaling pathway in isoproterenol-induced myocardial hypertrophy. *J. Ethnopharmacol., 150*, 1062–1070.

Zhang, B. Q., Hu, S. J., Qiu, L. H., et al., (2007). Effects of astragalus membranaceus and its main components on the acute phase endothelial dysfunction induced by homocysteine. *Vascul. Pharmacol., 46*, 278–285.

Zhang, C., Qin, Y. Y., Wei, X., et al., (2015). Tea consumption and risk of cardiovascular outcomes and total mortality: A systematic review and meta-analysis of prospective observational studies. *Eur. J. Epidemiol., 30*, 103–113.

Zhang, Z., Ho, W. K. K., Huang, Y., et al., (2002). Hawthorn fruit is hypolipidemic in rabbits fed a high cholesterol diet. *J. Nutr., 132*, 5–10.

Zhou, W., Chai, H., Lin, P. H., et al., (2004). Clinical use and molecular mechanisms of action of extract of Ginkgo biloba leaves in cardiovascular disease. *Cardiovasc Drug Rev., 22*, 309–319.

CHAPTER 9

PLANT METABOLITES AND VEGETABLES FOR LIVER DISEASE PREVENTION AND TREATMENT

DARCIA D'MELLO,[1] S. O. SADASHIV,[2] S. RENUKA JYOTHI,[3] and SHARANGOUDA J. PATIL[4]

[1]*Department of Biotechnology, School of Life Sciences, Garden City University, Bangalore, Karnataka, India*

[2]*Department of Studies in Food Technology, Shivagangothri, Davangere University, Davanagere, Karnataka, India*

[3]*Department of Life Science, School of Sciences, Jain University, Bangalore, Karnataka, India*

[4]*Department of Zoology, NMKRV College for Women (Autonomous), Bangalore, Karnataka, India*

ABSTRACT

In the human body liver is the largest organ which plays a prominent role in the regulation of metabolism, detoxifying toxic substances, and synthesizing useful biomolecules. However, multiple conditions such as non-alcoholic fatty liver disease (NAFLD), long-term alcohol intake, hepatitis, and chronic level of drug administration can affect its regenerative capacity which may lead to hepatic scaring, cirrhosis, and increase in tissue lipid peroxidation causing chronic liver diseases. As stated by the World Health Organization (WHO), around 1.4 million deaths occur annually due to chronic liver diseases, while liver cirrhosis and drug-induced liver injury is the 9th leading issue for the death in many of developing countries. In the recent advances in

Plant Metabolites and Vegetables as Nutraceuticals.
Raj K. Keservani, Bui Thanh Tung, Rajesh K. Kesharwani, and Eknath D. Ahire (Eds.)

therapeutic and innovative development of medicines are not enough, there is a still a need for exploring the full potential of novel therapeutic agents to treat liver diseases. Even though, the current synthetic drugs are strong radical scavengers, they are also carcinogenic in nature and causes server liver damage. Since, plants serves as a huge literature by means of repository of bioactive principles such as phenols, flavonoids, alkaloids, carotenoids, etc., the consuming nutraceutical foods which contain such bioactive molecules in large proportion seems to be significant role in prevention and treatment of liver diseases. Thus, in this chapter we intend to provide an overview of the current progress in research on plant derived metabolites that serve as potential advanced therapeutic molecules for the liver treatment and their other consequences.

9.1 INTRODUCTION

Being the second largest glandular organ in the human body, the liver plays a vital role in the maintenance, performance, and regulation of homeostasis of the body. It performs various functions such as detoxifying and excreting many of the endogenous as well as exogenous compounds. It is also seen to be associated with the metabolism of carbohydrates along with proteins and lipid metabolism to convert into as complex sugars into glycogen, which is used in times hypoglycemia. Apart from this, the liver serves as the chief site for the production of bile juices which plays a crucial role in the process of digestion and also participates in function so f innate immunity. Furthermore, the liver has the capacity of regeneration and reversible in earlier form completely in severe and non-iterative situations (Gao et al., 2008; Cornelius, 1980). Various chemical substances such as ethanol, acetaminophen, chemotherapeutic agents, carbon tetrachloride (CCl_4), and thioacetamide, which may be present in the blood, are detoxified by the liver (Singh et al., 1998). However, very high concentrations of certain chemicals known as hepatotoxins or certain pathogens such as hepatotoxic viruses and parasites like amoebas or giardia also major cause for liver damage (Bhawna et al., 2009). Approximately, 50% of all acute liver failures (ALFs) and 5% of total admitted patients exhibited liver associated hepatotoxicity due to drug induced conditions (Dey et al., 2013).

As stated by the World Health Organization (WHO), around 1.4 million deaths occur annually due to chronic liver diseases, while liver cirrhosis and drug-induced liver injury is the 9th leading cause of death in Western and developing countries (Asrani et al., 2008). Fatty liver diseases (FLDs), long term alcohol abuse, viral hepatitis, and chronic use of medications are some of the conditions that can affect the regenerative capacity of the liver (Williams, 2006). The liver damage is often associated with cellular necrosis, increase tissue lipid peroxidation and elevation in the glutathione (GSH) levels. In addition, the serum levels of many biochemical markers' transaminases, alkaline phosphatase, bilirubin, triglycerides, and cholesterol are elevated in liver disease. Thus, in such situation, the liver eventually loses its vitality, which lead to an imbalance in the normal metabolic phenomenon, leading to many other fatal conditions (Schiff et al., 1993). Despite the considerable amount of recent research on liver diseases, there is still a lot of hindrance in finding an effective hepatoprotective drugs than the currently available medications. Additionally, the easy availability, the less toxic nature, and the easy acceptability of plant-based hepatoprotective agents have increased their use in alternative therapy (Mohamed et al., 2010).

9.2 LIVER DISEASE

Liver diseases are one of the major causes for morbidity or mortality in humans and animals all over the globe. In statistics report on liver disease and around 2 million deaths were accounted per year, while more than 1 million deaths are due to cirrhosis or viral hepatitis and hepatocellular carcinoma (HCC) (Mokdad et al., 2014). Multiple genetic and environmental problems are major factors for the etiopathogenesis of liver diseases. Generally, the various liver diseases can be subdivided as hepatocellular, cholestatic or mixed. In hepatocellular diseases such as alcoholic liver disease (ALD), viral hepatitis, non-alcoholic fatty liver and toxic hepatitis, features of liver injury, inflammation, and necrosis predominate. While on the other hand, in cholestatic diseases which include primary biliary cirrhosis, gall stone, and malignant obstruction, features of inhibition of bile flow predominate (Ghany et al., 2014). However, a mixed pattern includes a combination of both hepatocellular and cholestatic injuries. Chronic liver disease also includes hereditary diseases such as hemochromatosis, alpha-1-antitrypsin deficiency, Wilson's disease, non-alcoholic steatohepatitis (NASH), and a group of liver diseases with no identifiable cause like cryptogenic liver disease (Uhl et al., 2014).

9.2.1 SYMPTOMS OF LIVER DISEASE

Being one of the key organs for the metabolism of xenobiotics and detoxification, the liver is continuously exposed to a variety of chemicals agent that can lead to various liver diseases. Some of the major symptoms that are associated with liver diseases are jaundice, fatigue, right upper quadrant pain, abdominal distention, and intestinal bleeding (Sivakrishnan et al., 2019). When liver cells get affected by any incidence, subsequently repair mechanisms are systematically replaced by scar tissue with clinical manifestations of appetite loss, hepatomegaly, splenomegaly, ascites, edema, diarrhea, and anemia (Balachandran et al., 2003). Hepatotoxicity due to drugs such as paracetamol, isoniazid, sulfonamides, pesticides, insecticides, fungicides, alcohol, CCl_4, and carcinogens drugs appears to be the most common contributing factor for liver diseases. Hepatotoxicity can be present in the form of necrosis, fatty infiltration, centrilobular necrosis, vascular lesions, and cholestasis with or without portal inflammation (Ferrell et al., 2010).

9.3 TYPES AND CAUSES OF LIVER DISEASES

9.3.1 FATTY LIVER DISEASE (FLD)

It is most common and chronic diseases in global scenario, the FLD can further be classified as non-alcoholic and alcoholic in nature (NAFLD and AFLD) (Mitra et al., 2020). Approximately 2 billion people are facing obesity and more than 400 million facing diabetes issue; these two factors became a risk for NAFLD and HCC (Zhou et al., 2020). While on the other hand, 75 million individuals are diagnosed with AFLD due to the excessive consumption of alcohol (Asrani et al., 2019). The starting stages that are associated with fat accumulation in hepatocytes that is characterized by excess triglyceride deposition which leads to the development of fatty liver (steatosis). In liver accumulation of lipid biomolecule due to increasing oxidative stress level and liver inflammation serves as the fundamental pathological characteristics of both NAFLD and AFLD, however, they both have different inducers (Xiao et al., 2014).

Besides this, other studies revealed that the disorders of several important signaling pathways such as adenosine 5′-monophosphate-activated protein kinase (AMPK), sirtuin 1 (SIRT1), nuclear factor erythroid 2-related factor 2 (Nrf2), toll-like receptor 4 (TLR4), and nuclear factor kappa-B (NF-κB) also plays a vital role in the progression of FLD (Yan et al., 2020; Li et al., 2021).

9.3.1.1 NON-ALCOHOLIC FATTY LIVER DISEASE (NAFLD)

Non-alcoholic fatty liver disease (NAFLD) is generally caused due to unhealthy dietary habits such as intake of high-fat diet, over intake of fructose, and animal protein (Rietman et al., 2018). Generally, NAFLD is associated with elements of the metabolic syndrome, hypertension, insulin resistance, glucose intolerance, clinical constellation of obesity and hyperlipidemia, which encompasses a broad spectrum of liver disorders from the simple hepatic steatosis to the more ominous condition known as NASH (Farrell et al., 2006). The oxidative stress serves as the primary underlying mechanism that leads to the development and progression of NAFLD (Polimeni et al., 2015; Sumida et al., 2013). During an oxidative stress, the imbalance between oxidants and antioxidant can potentially induces cell and tissue damage (Sies, 1997). The presence of excessive lipids in the liver during a hepatic steatosis, normally induces an immune response recruiting the immune cells such as macrophages and T-cells. Thereby, inducing the development of insulin resistance in liver may be due to increase in the reactive oxygen species (ROS). Thus, the generated ROS induces necrosis, apoptosis, and inflammatory response of hepatocytes, which eventually leads to fibrogenesis, which serves as a fundamental characteristic of the progression from NAFL to NASH (Sánchez-Valle et al., 2012). The liver also houses a type of macrophages called Kupffer cells, hepatic stellate cells, endothelial cells are highly sensitive against oxidative stress and their related molecules. With the response of oxidative stress, the Kupffer cells and the tumor necrosis factor α (TNF-α) which generate an inflammatory response or even help to initiation of apoptosis (Li et al., 2015; Sánchez-Valle et al., 2012).

9.3.1.2 ALCOHOLIC FATTY LIVER DISEASE (AFLD)

On the other hand, excessive consumption of alcohol is the major trigger of ALD. Generally, on ingestion of the ethanol, the organic compound gets oxidized with acetaldehyde under the influence of alcohol dehydrogenases enzyme (ADHs). It is seen that the presence of acetaldehyde is detrimental to liver health, as it is carcinogenic is nature, thereby inducing ALD. In addition, steatosis in hepatocytes is closely associated with oxidative stress, inflammation, gut dysbiosis, autophagy, and apoptosis (Gao et al., 2016). Thus, if alcohol consumption is continued, fatty liver may progress to hepatitis, inflammation, fibrosis, which may eventually lead to liver cirrhosis (Szabo et al., 2010; Tilg et al., 2011).

9.3.2 *VIRAL HEPATITIS*

The chronic hepatitis is a blood-borne infection, with hepatitis B and C virus serving as major risk factors for the development of HCCs, chronic liver diseases, and cirrhosis (Levrero, 2006). Thereby, inducing chronic inflammation, liver cell proliferation and assist the cytokines in the development of fibrosis. It also serves as a chief reason for majority of the liver transplantation in adults. Approximately, more than 400 million people worldwide are chronically infected by the hepatitis virus (Lee, 1997). Among the 5 known hepatitis viruses, the hepatitis B, the hepatitis C, and the hepatitis D virus can cause severity in infection to consider it as chronic hepatitis. While on the other hand, the hepatitis A and E, can also cause infection in acute level and self-limited disease (Lai et al., 2003; Worm et al., 2002).

9.3.3 *GENETIC LIVER DISEASE*

9.3.3.1 *ALPHA-1 ANTITRYPSIN DEFICIENCY*

Deficiency of alpha-1 antitrypsin is an autosomal recessive disease that mainly affects the functions of the lung and liver. The protein of alpha-1 antitrypsin encoded with the gene SERPINA1 which is located on the long arm of chromosome 14 (14q31–32.3) and serves as a serine protease inhibitor that is predominantly produced in the liver. However, the deficiency of which is shown to manifest benign neonatal hepatitis syndrome while liver fibrosis may eventually develop in adults with the progression to cirrhosis and HCC (Köhnlein et al., 2008).

9.3.3.2 *HEREDITARY HEMOCHROMATOSIS*

Hemochromatosis is autosomal recessive disorder characterized by normal iron-driven erythropoiesis and toxic accumulation of iron in parenchymal cells of liver, cardiac, and endocrine glands. It can be caused due to a mutation involving the high iron Fe (HFE) gene which helps in regulation of iron absorption in the intestine, or mutations of any of the proteins that limits the iron entry in the blood. If not treated, the overload of iron (ferritin and hemosiderin) particularly in the parenchymal cells, may eventually generate highly ROS that can cause organ fibrosis (Bacon et al., 1999; Pietrangelo, 2010).

9.3.3.3 WILSON DISEASE

It is a genetically inherited disorder characterized as autosomal recessive allele due to the pathological accumulation of inorganic element copper (Bandmann et al., 2015). It is caused due to mutations in ATP7B, a gene that encodes a trans-membrane copper-transporting ATPase, dysfunction of which can lead to impairing the copper homeostasis and generate a copper overload in the organs like liver, brain, etc. This disease is predominant at the stage of adolescence and also found in women. Some of the prominent symptoms that is associated with these individuals may be that of liver diseases which is often associated with anemia due to breakdown of red blood cells and mental confusion (Członkowska et al., 2018) (Figure 9.1).

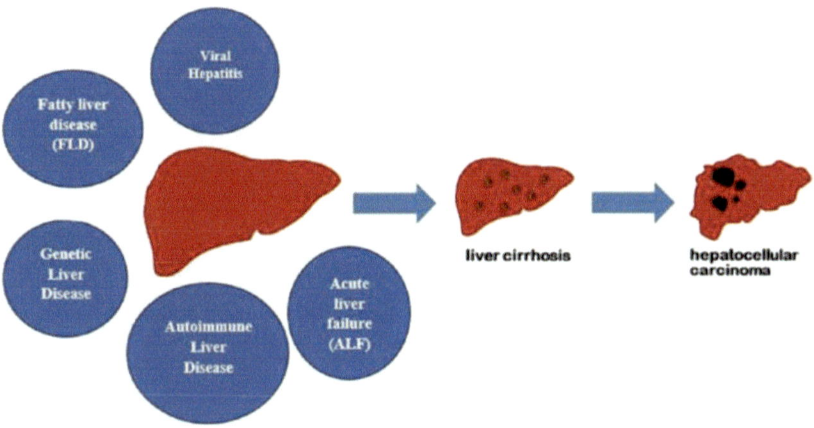

FIGURE 9.1 Diagrammatic representation of liver diseases.

9.3.4 AUTOIMMUNE LIVER DISEASE

9.3.4.1 AUTOIMMUNE HEPATITIS (AIH)

It is progressive disorder of liver, exhibiting inflammation an affecting mainly in female individuals, characterized by increasing level of enzyme transaminases, immune protein like immunoglobulin G (IgG) and presence of issue specific autoantibodies. According to the autoantibody profile, AIH can be of 2 types: anti-nuclear (ANA) and anti-smooth muscle (SMA) as type 1 positive antibodies whereas anti-liver kidney microsomal (anti-LKM-1) as type 2 positive antibody. Even though, AIH responds against immunosuppressive treatment significantly, however, if left untreated, it is progressive in liver to rapid action to become cirrhosis and leads to liver failure (Mieli-Vergani et al., 2018).

9.3.4.2 PRIMARY BILIARY CIRRHOSIS (PBC)

It is an autoimmune disorder of liver characterized site-specific antibody as a serum anti-mitochondrial protein. Thereby, inducing series of destruction in the bile ducts of intrahepatic site result into as chronic cholestasis, threshold of inflammation and fibrosis which lead to later liver cirrhosis and as an end liver failure. PBC is seen to be predominant in middle-aged women (Bowlus et al., 2014).

9.3.4.3 PRIMARY SCLEROSING CHOLANGITIS (PSC)

It is a chronic liver disease characterized as cholestatic condition by decreasing the size in hepatic bile ducts in of intra and extra cellular level due to inflammation in liver and fibrosis action. PSC may eventually progress to hepatic cirrhosis and subsequent liver failure and is accompanied by an increased risk of cholangiocarcinoma. In majority of cases, it occurs in associated disorder with other organs like inflammatory bowel disease (IBD), which in turn precedes the progressive stage of PSC (Lee et al., 1995).

9.3.5 POLYCYSTIC LIVER DISEASE (PLD)

It is a rare genetically inherited disorder characterized as Mendelian traits by the development of multiple hepatic cysts. It is caused due to the malformation of embryonic ductal plate and the development within the liver intrahepatic biliary tree (Raynaud et al., 2011). Thus, even though it is present at an early age, it still remains asymptomatic until cyst growth is initiated during adulthood. As a result, in the liver numerous cysts are developed in the parenchyma cell surfaces. Furthermore, in advanced stages, the patients diagnosed with PLD may have massively enlarged liver and have shown symptoms of abdominal pain, abdominal distension, and polycystic kidneys it leads to renal failure (Cnossen et al., 2014).

9.3.6 *ACUTE LIVER FAILURE (ALF)*

Acute liver failure (ALF) is a syndrome in which the patients sustain a rapid loss of hepatic functions, without any previous encounter of live diseases. ALF is accompanied with a cascade of complications such as hepatic encephalopathy. There are two different pathways that can lead to ALF. These include the direct damage and immune-mediated liver injury. In the first scenario, pathogen or toxic substance directly damages cellular organelles or triggers the cellular cascade pathway, thereby, disturbing the intracellular homeostasis (Wu et al., 2010). While on the other hand, the innate and adaptive immune responses eventually converge on liver cell death pathways including apoptosis, autophagy, necrosis, and necroptosis leading to immune-mediated liver injury (Wang et al., 2014). Among the advanced treatments in all levels, the transplantation of liver became one of the most promising treatments for ALF (Olivo et al., 2018).

9.3.7 *HEPATOCELLULAR CARCINOMA (HCC)*

The most common primary liver malignancy is the HCC which is a leading cause of cancer-related death worldwide (O'Connor et al., 2010). Chronic liver disease and cirrhosis remain the most important risk factors for the development of HCC, while on the other hand, viral hepatitis and excessive alcohol intake are the leading risk factors worldwide (Loprieno, 1975). Apart from this, the aflatoxin produced by Aspergillus species serves as a potent hepatocarcinogen (Kew et al., 2013). In addition, many metabolic and genetic diseases such as hemochromatosis, Wilson's disease, α–1 antitrypsin disease, tyrosinemia, glycogen-storage disease types I and II, and porphyria are associated with HCC (Hoffmann et al., 2017).

9.4 LIVER DISEASE PROGRESSION

The progressions of liver diseases are seen to be associated with liver complications like fibrosis and its final stage liver cirrhosis (Figure 9.2). This process may further be classified by dividing into major four stages like inflammation, fibrosis/scarring, cirrhosis, and end-stage liver disease (Sharma et al., 2021).

- **Stage 1: Inflammation:** It is the primary stage of the on-going liver damage caused due to a number of toxic agents, and diseases which eventually causes enlargement of the liver. As a result, the individual may experience abdominal pain in the upper right area.

- **Stage 2: Fibrosis/Scarring:** If it is occupied in 1st stage during the inflammation without diagnosis, the tissues of liver slowly deteriorate the condition and leads to scar in the liver, instead of regenerating the healthy liver tissues. In addition, if the toxins, fat, and scarred tissue start accumulating in the liver, the fibrosis can distort the blood flow as well as the shape of the liver. As a result of while, the tissues that don't get enough blood supply and starts becoming necrotic in nature.
- **Stage 3: Cirrhosis:** At this stage, the liver is completely scarred and since there is no possible cure for it, the only potential alternative is a liver transplantation. The individuals are shown to have signs and symptoms such as discomfort, and tenderness in the liver area of the abdomen, loss of appetite, bloating of abdomen, as well as jaundice, with the eyes and skin turning yellowish.
- **Stage 4: End-Stage of Liver Disease (ESLD):** It is also known as chronic hepatic failure also called the phenomenon ESLD, result of cirrhosis condition in turn exhibit as scar tissue without replacement of hepatic cells until the redevelopment of liver function adequately, progression in days, months, years, or decades. Patients with the diseased, abnormal state of liver who develop ascites in hepatic cells, variceal hemorrhage, hepatic encephalopathy, or renal impairment observed most often chronic liver failure. Liver failure classified as mainly two types. The ALF happens quickly within the time period of 48–72 hours due to alcohol addiction and drug abuse. Whereas other type is chronic liver failure, it takes a maximum time to develop and is often caused by alcohol abuse (Sharma et al., 2021; Singal et al., 2013).

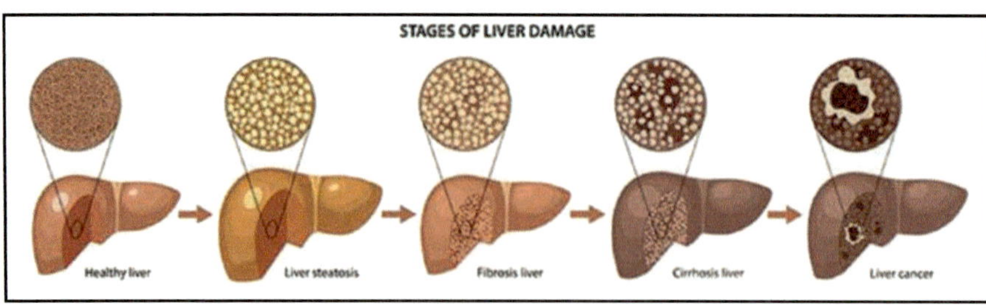

FIGURE 9.2 Diagrammatic representation of liver progression.

9.5 CURRENT TREATMENT FOR LIVER DISEASES

Despite the curative treatments for liver diseases such as surgical resection, liver transplantation and local ablation, it is effective only if the patients are diagnosed at early stages (Wong et al., 2002). In recent years, the FDA approved some novel drugs such as sofosbuvir and simeprevir for the treatment of liver hepatitis (Ende et al., 2015). The current treatment of choice for HBV/HCV infection is basically an interferon-based combination therapy with oral antiviral drugs which is associated with improving the therapeutic success and also preventing the development of resistances (Uhl et al., 2014). Apart from this, diet that limits the intake of fats and calories, avoid alcohol, and increase of fiber intakes plays a major role in the control of liver diseases.

Treatments are unique according to nature of liver issues; the types of treatments are limited and therapy with modern medicine may lack in efficacy. The effective treatments of liver issues such as corticosteroids administration and interferons are inconsistent due to the risk of adverse effects and too expensive (Stickel et al., 2007). Hence, there is a need for and alternative effective therapeutic agents with a low incidence of side effects. Thus, the plants metabolites which are traditionally used as hepatoprotective agents (Scott et al., 1998), serves as a potential drug as they are easily accessibility, cost effective with very less risk of adverse events.

9.6 PLANT METABOLITES

Since times immemorial, plants serve as a tremendous source for the discovery of new products that have medicinal importance in the drug discovery process. A plant produces a huge array of specialized metabolites such as primary metabolites which are associated with basic functions of cell such as cell growth, food synthesis and storage, respiration, and reproduction. While on the other hand, the secondary metabolites are produced to make them competitive in their own environment. These metabolites are natural constituents of fruits and vegetables, preventive against liver damage if adopted in our diet and free from exposure of xenobiotics, alcohol, and various drugs. Phytochemicals possess various roles apart from such impact as secondary metabolites exhibit potential role in various biological effect such as antibiotic, antifungal, antiviral, antioxidant, anti-inflammatory, antimutagenic, and anti-fibrosis activities. Thus, serves as economically important and unique sources for pharmaceuticals dugs, food additives, flavors, and other industrial values (David, 1998; Hussein et al., 2019).

9.6.1 CLASSIFICATION OF PLANT METABOLITES

Secondary metabolites are classified based on the chemical composition and structure, functional group, solubility, stability, various physic-chemical properties interacting their biosynthetic pathways (Harborne et al., 1999; Kabera et al., 2014). Secondary metabolites classified in three main common groups:

1. **Phenolic Compounds:** These which are developed it as simple sugars, possessing benzene rings with the stronger hydrogen bonds and oxygen. They are further classified on the bases of its structure as coumarins, chromones, flavonoids, lignans, stilbenes, tannins, xanthones, etc. (Kabera et al., 2014).
2. **Terpenoids:** These which are made from mevalonic acid (MVA) and composed almost entirely of carbon and hydrogen and may include the hemiterpenes, monoterpenes, sesquiterpenes, diterpenes, sesterterpenes, triterpenes (Kabera et al., 2014).
3. **Alkaloids:** These which are nitrogen-containing compounds and may include nicotine, caffeine, and vinblastine (Kabera et al., 2014).

9.7 APPLICATION OF PLANT METABOLITES AND VEGETABLES IN LIVER DISEASES

9.7.1 CAFFEIC ACID

Caffeic acid a bioactive phenolic component found in coffee beans and wines serves as potential antioxidant and anti-inflammatory responses against to the treatments of liver diseases. In an experimental setup, it has shown to modulate the expression of hepatic carcinoma factors by restraining its binding to the Keap1, thereby elevating the expressions of vital antioxidants (Kim et al., 2018). Along with caffeine, it also serves as a beneficial protective agent for various liver functions due to its ability to improve the liver triglyceride metabolism, oxidative stress, and fibrotic status (Kennedy et al., 2016) (Figure 9.3).

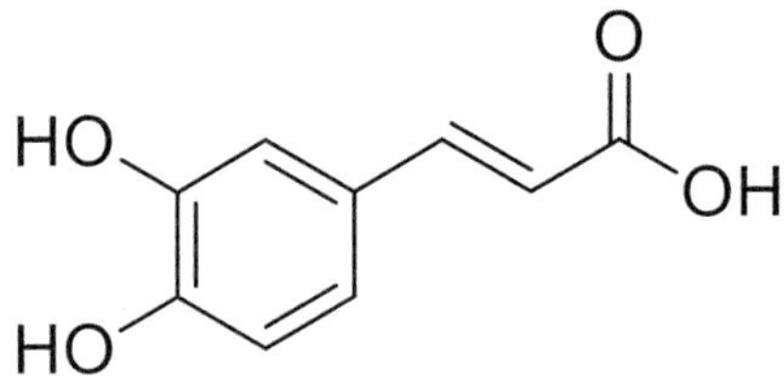

FIGURE 9.3 Structure of caffeic acid.

9.7.2 *CURCUMIN*

Curcumin is the principal curcuminoid obtained from *Curcuma longa* (Turmeric) which is characterized functional drug to be antitumor, antioxidant, antiarthritic, anti-myloid, anti-ischemic, and anti-inflammatory properties (Aggarwal et al., 2006; Hatcher et al., 2008). In experimental animal studies, curcumin have shown to mitigate the oxidative stress and prevent further liver cell damages (Bao et al., 2010). In another *in vitro* study, it has shown the significantly decrease lipid peroxidation, reduce the release of LDH and attenuate the antioxidant enzymes like superoxide dismutase (SOD), catalase, and GSH peroxidase in liver slice cultures (Naik et al., 2004). Furthermore, in a separate study with rat hepatocytes, curcumin have also decreased the ethanol-induced increase in malondialdehyde, LDH, and AST levels and induce the hemeoxygenase in the liver cells. It also has the ability to inhibit several factors like nuclear factor-kappa B and suppress the expression of cytokines, chemokines, COX-2, and iNOS in Kupffer cells of liver in rats having alcohol induced liver diseases (Nanji et al., 2003) (Figure 9.4).

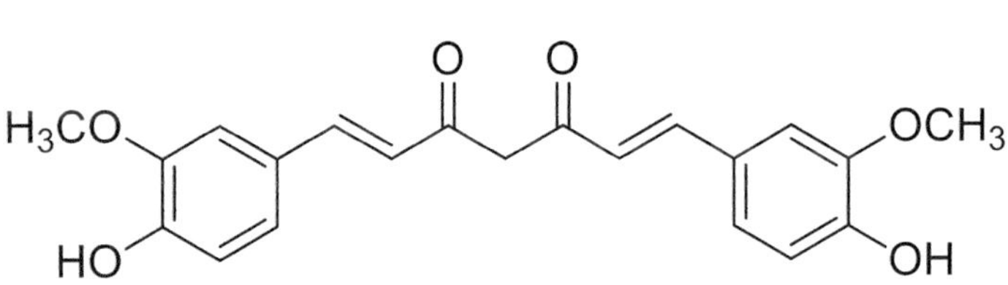

FIGURE 9.4 Structure of curcumin.

9.7.3 *ELLAGIC ACID*

Ellagic Acid potential polyphenol present in blackberries, cranberries, pecans, pomegranate, strawberries, and walnuts have been reported for their various biological properties such as anticancer, antimutagenic, and antioxidant properties. A rapid increase of serum level of aminotransferase and lipid peroxidase was observed when the rat with ethanol-induced hepatotoxicity was treated with ellagic acid (Devipriya et al., 2008). Various other studies have demonstrated the mechanism of how ellagic acid reduces the alcohol abused hepatotoxicity by modifying the mechanism of protein expression in matrix of the metalloproteinases as well as attenuate the lipid levels in the blood (Devipriya et al., 2007). Thereby, improving the body weight, immune response restores the antioxidant status and modulating the micronutrient levels (Figure 9.5).

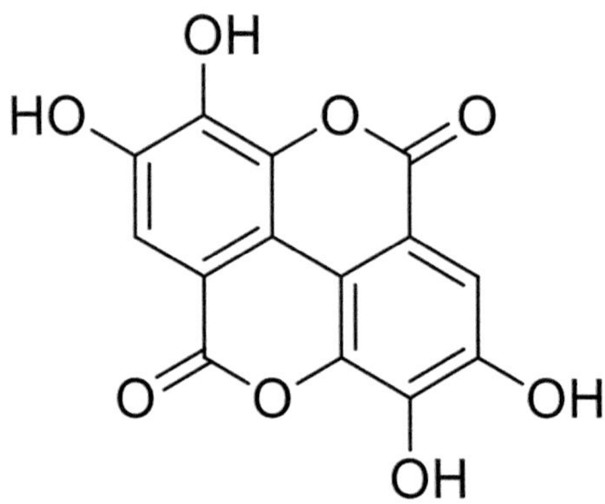

FIGURE 9.5 Structure of ellagic acid.

9.7.4 *EPIGALLOCATECHIN-3-GALLATE (EGCG)*

Epigallocatechin-3-gallate (EGCG) the main constituent of *Camellia sinensis* (green tea) (Yang et al., 2009) is polyphenoliccatechin plays a vital role in the treatment of various liver diseases such as immune-medicated liver injury, chronic alcohol-induced liver injury and NaF-induced oxidative hepatic stress (Wang et al., 2006; Thangapandiyan et al., 2013). It serves as a potential free radical scavenger and antioxidant. Dietary intake of EGCG have shown to prevent the oxidative stress, reduce expression of pro-inflammatory molecules as well as increase the serum levels of aminotransferases in ethanol-fed mice with overload iron (Yuan et al., 2006; Ren et al., 2011). In another study, the consumption of EGCG can promote the β-oxidation of fatty acids by enhancing the activities of carnitine palmitoyltransferase 1 and phospho-acetyl-CoA carboxylase, thereby reducing the elevated expression of TNF-a, COX-2 and iNOS in rats with ethanol-induced liver injuries (Yun et al., 2007; Yuan et al., 2006) (Figure 9.6).

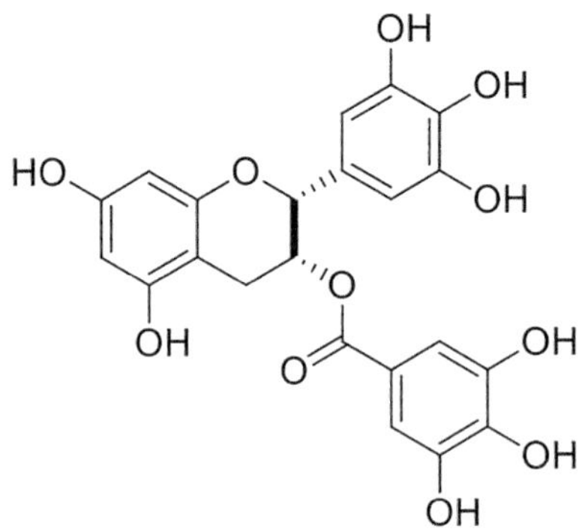

FIGURE 9.6 Structure of epigallocatechin-3-gallate.

9.7.5 *LUTEIN*

Lutein is a tetraterpenoid, a naturally occurring carotenoid is the most common in vegetables includes kale and spinach green leafy vegetables and certain fruits such as avocado (Sommerburg et al., 1998). It is shown the exhibit antioxidant as well as strong free radical scavenging effects. In a study carried out on rats treated with alcohol, lutein has tremendously reduced the serum levels of aminotransferase, bilirubin, alkaline phosphatase as well as decrease the lipid peroxidation in the liver (Shivashankara et al., 2013). In addition is also serves as a hepatoprotective agent against various xenobiotics that can damage to the liver (Firdous et al., 2011). Furthermore, lutein has shown to reduce the inflammatory proteins and cytokines and offered significant cryoprotection against alcohol-induced liver injury (Du et al., 2015) (Figure 9.7).

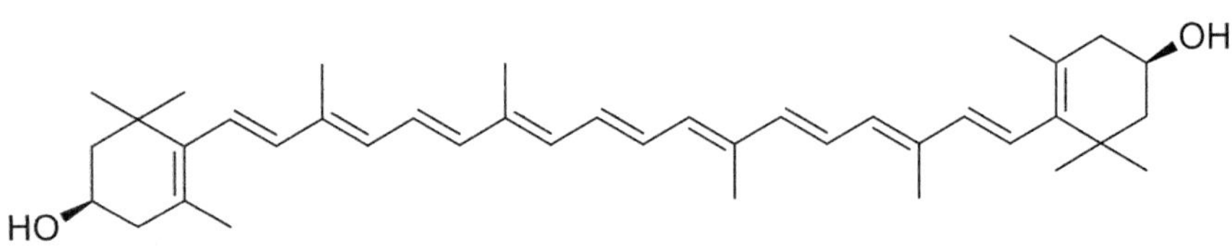

FIGURE 9.7 Structure of lutein.

9.7.6 *GENIPOSIDE*

Geniposide is a naturally occurring iridoid glycoside present in the fruits of *Gardenia jasminoides ELLIS* has a great potential against FLDs, hyperlipidemia, and can inhibit liver fibrosis (Kojima et al., 2011; Ma et al., 2011).

However, in a study carried out on rats fed with high fatty diet, geniposide have shown to exert protective effects against hepatic steatosis, mainly due to its antioxidant actions and regulation of the adipocytokine release as well as increases the presence of SOD in the liver (Zhang et al., 2013). Apart from this, the aglycone of geniposide can effectively express the anti-angiogenic and anti-inflammatory properties as well as induce apoptosis in human hepatocarcinoma Hep3B cells (Kim et al., 2005) (Figure 9.8).

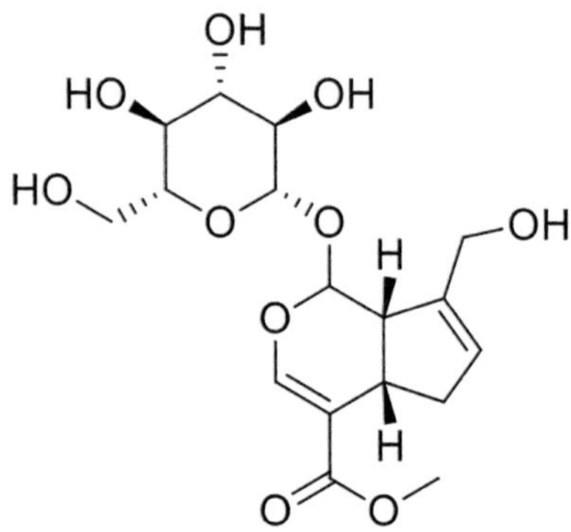

FIGURE 9.8 Structure of geniposide.

9.7.7 *GENISTEIN*

Genistein is an active isoflavone that is present in soybean that has characteristic properties such as antioxidant, anti-inflammation, antifibrosis, and aids in the prevention of NAFLD (Kim et al., 2004). In another study, the administration of genistein has reduced the levels of ALT as well as various inflammation biomarkers such as TNF-α and increase the GSH levels (Leelananthakul et al., 2019). Apart from this, genistein is shown to reduce the hepatic oxidative stress and demonstrated to prevent NAFLD-related liver cell tumorigenesis (Lee et al., 2006; Li et al., 2017) (Figure 9.9).

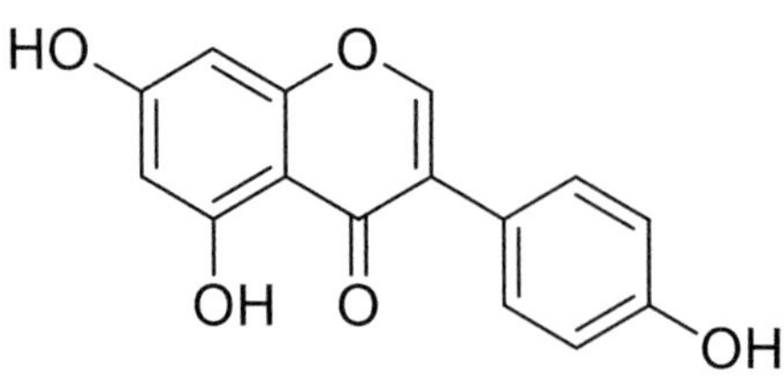

FIGURE 9.9 Structure of genistein.

9.7.8 GLYCYRRHIZIN

Glycyrrhizin is a major biological active constituent of *Glycyrrhiza glabra* roots which have predominant anti-inflammatory and anti-viral triterpene properties that is used for the treatment for chronic hepatitis. In a study carried out, glycyrrhizin shown to significantly inhibit the hepatocyte apoptosis by down-regulating the expression of caspase-3 (Tang et al., 2007) as well as shown anti-inflammatory activity by inhibiting the release of tumor necrosis factor-a and nuclear factor-kappa B (Ogiku et al., 2011). Another molecule of such plant namely, Glycyrrhetinic acid shows protective nature against to liver issues due to hepatotoxicity induced by the other harmful molecule like alkaloid, galactosamine, resorcin, etc. (Lin et al., 1999) (Figure 9.10).

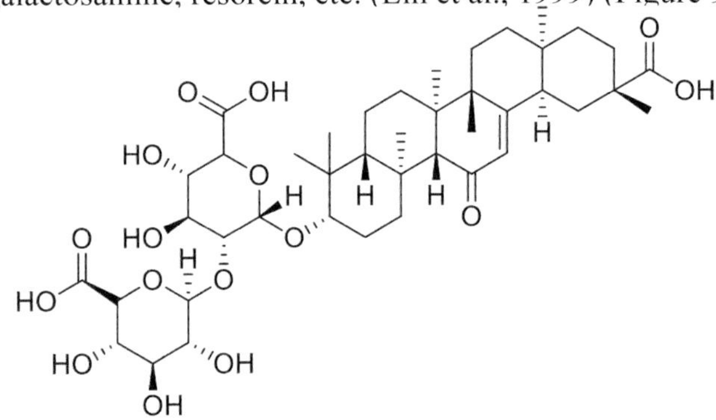

FIGURE 9.10 Structure of glycyrrhizin.

9.7.9 NARINGENIN

Naringenin is a citrus flavonoid generally present in oranges, grapefruits, tangerines, and tomatoes that possess antioxidant, anticancer, and anti-inflammatory activities and can exhibit *in vivo* hepatoprotective and anti-fibro genic effects against DMN-induced liver injuries. Along with this, it also exhibited various properties like antioxidant, redox potential, oxidative stress and decrease the lipid peroxidation in the liver, thereby, preventing the formation of hepatic fibrosis (Pari et al., 2006). A study on rats with alcohol induced liver injury suggested that on administration of naringenin can significantly decrease the levels of serum aspartate and alanine transaminase, gamma-glutamyl transpeptidase, conjugated dienes, lipid hydroperoxide (LOOH) while on the other hand increases the SOD, GSH peroxidase, GSH reductase and catalases activities (Jayaraman et al., 2009) (Figure 9.11).

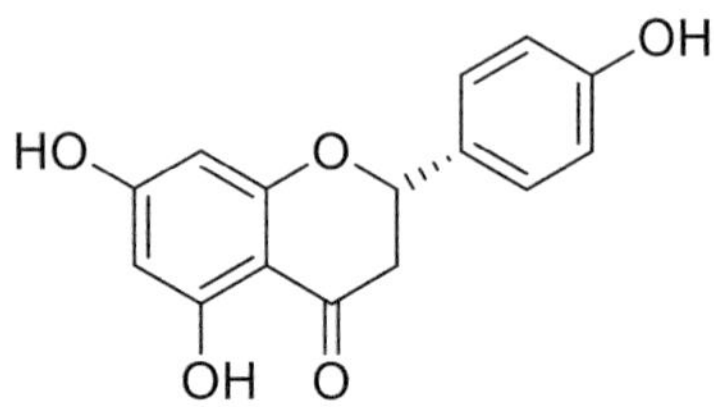

FIGURE 9.11 Structure of naringenin.

9.7.10 PICROLIV

Picroliv is a *Picrorhiza kurroa* plant molecule and a mixture of iridoid glycoside derivatives namely Picroside-1 and Kutkoside that has been demonstrated to have antioxidant and anti-inflammatory effects (Zhang et al., 2012). Administration of picroliv biomolecule have shown various biological properties, in that reducing ethanol-induced chronic hepatotoxicity in rats as well as reduces the serum levels of f AST, ALT. In addition, it has also shown to possess potential hepatoprotective effects against various xenobiotic hepatotoxins (Rastogi et al., 1996; Saraswat et al., 1999). Furthermore, it was observed that picroliv can elevate the leases of bile volume, bile alts and bile acids as well as restore the altered levels of lipids and glycogen in the liver (Saraswat et al., 1999), hence, it aids in potential preventive measures in hepatocarcinogenesis (Figure 9.12).

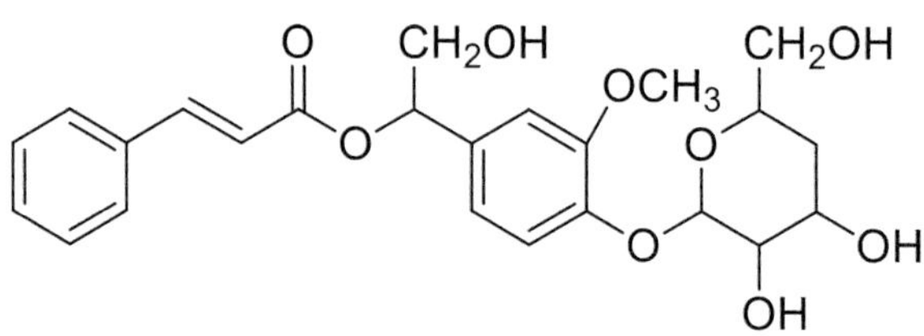

FIGURE 9.12 Structure of picroliv.

9.7.11 QUERCETIN

Quercetin, a flavanol that naturally occurs in onions, broccoli, and leafy vegetables is known to demonstrate free radical scavenging, anti-inflammatory, and antioxidant properties. *In vivo* studies on ethanol-induced liver cell injury, have shown that quercetin played an eminent role in decreasing the lipid peroxidation, the GSH content and upregulated the heme oxygenase-1 (HO-1) in human hepatocytes (Liu et al., 2010). In another experimental study, it can successfully reverse the ethanol-induced hepatotoxicity by decreasing the levels of prominent biomarkers of liver diseases such as AST, ALT, AD, GGT, and TNF-a (Vidhya et al., 2009). Furthermore, quercetin can significantly regulate the cholesterol metabolism as well as elevates the expression of liver X receptors to increase the pace of transforming cholesterol into bile acids (Zhang et al., 2016) (Figure 9.13).

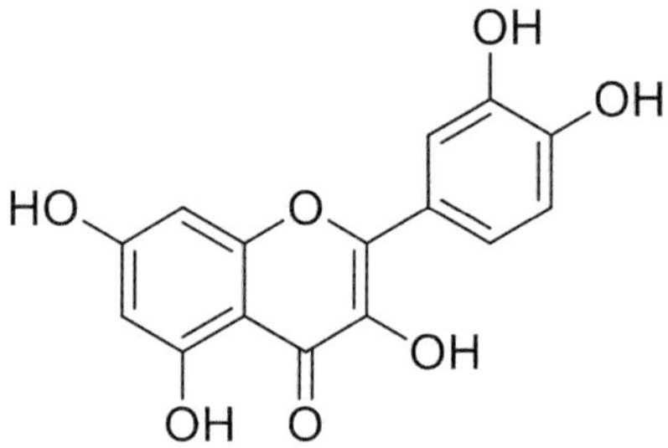

FIGURE 9.13 Structure of quercetin.

9.7.12 RESVERATROL (RES)

Resveratrol (RES) is a polyphenolic secondary metabolite most commonly produces in berries, grapes, and wine producing fruits. In an experiment carried out on BDL mice, RES has shown to tremendously reduce the tumor necrosis factor-α and IL-6 mRNA as well as decrease fibrosis and promoted hepatocyte regeneration. Thereby, playing an important role in the treatments of cholestatic liver injury (Chan et al., 2011). If administered at an early or advanced stage of liver diseases, RES has shown to effectively protect the liver cells by suppressing the oxidative stress and activated the apoptotic pathway in male Wistar rats (Rajasekaran et al., 2011). Apart from this, in one of the preclinical studies on RES on rats treated with ethanol, it played a vital role in the reduction of lipid peroxidation as well as restored the levels of the antioxidant enzymes SOD, CAT, and GSH-Px (Kasdallah-Grissa et al., 2007). Thus, RES treatment can significantly reduce the lipid synthesis, increase the rate of fatty acid oxidation, and prevent alcoholic liver steatosis in mice (Ajmo et al., 2008) (Figure 9.14).

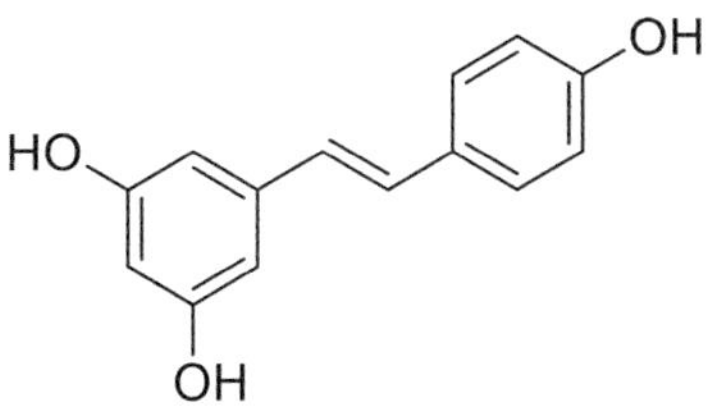

FIGURE 9.14 Structure of resveratrol.

9.7.13 RHEIN

Rhein is an anthraquinone naturally found in medicinal herbs, such as *Rheum palmatum* L. that exhibits various pharmacological properties such as anti-inflammatory, antitumor, and hepatoprotective effect. In a study carried out on hepatic N virus-transgenic mice with NASH, Rhein significantly attenuated the ameliorating glucose and lipid metabolism as well as decreased the serum levels of total cholesterol (Bian et al., 2013). Furthermore, it is seen to prevent the progress of hepatic fibrosis in rats due to its ability to inhibit the expression of TGF-beta 1 and suppressing the activation of hepatic stellate cells. In addition, Rhein has also shown to inhibit liver fibrosis that is inducing due to CCl_4 or alcohol (Wang et al., 2011) (Figure 9.15).

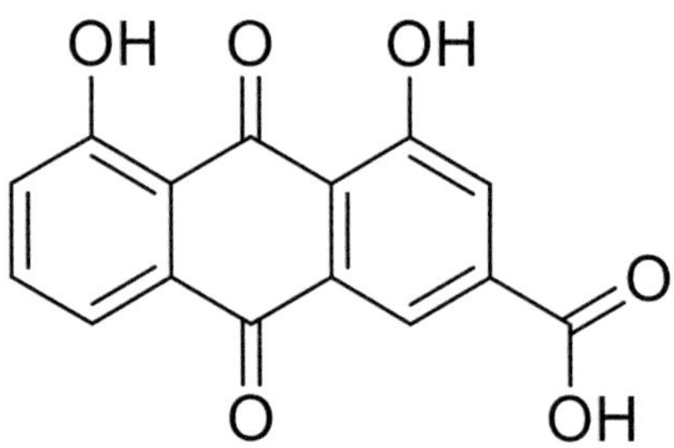

FIGURE 9.15 Structure of Rhein.

9.7.14 SILYMARIN

Silymarin, a flavonolignan present in *Silybum marianum* commonly known as milk thistle, is a cocktail of different constituents such as silibnins A and B, isosilibinins A and B, silicristin, and silidianin which have shown to protect the hepatic cells from various kinds of toxins such as ethanol, CCl_4, acetaminophen, and D-galactosamine (Habib-Ur-Rehman et al., 2009; Rasool et al., 2014). This is mainly due to its antioxidant effects, anti-lipid peroxidation which enhances the detoxification of the liver cells (Pradhan et al., 2006). Furthermore, it exerts membrane-stabilizing activity, promotes hepatocyte regeneration, reduces inflammatory reactions, and inhibits the fibrogenesis of the liver (Fehér et al., 2012). Apart from this, silymarin can efficiently reduce the proliferation of tumor cells, angiogenesis, and insulin resistance, thereby increasing the survival time of patients with alcohol induced liver cirrhosis (Zhang et al., 2013) (Figure 9.16).

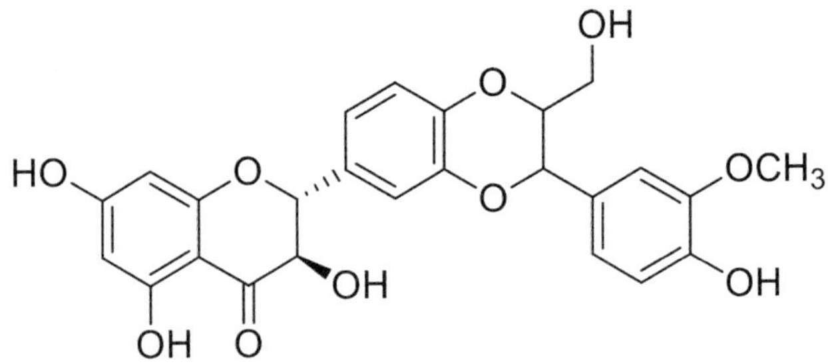

FIGURE 9.16 Structure of silymarin.

9.7.15 *URSOLIC ACID*

Ursolic acid, a pentacyclic triterpene acid mostly found in apples, bilberries, cranberries, peppermint, lavender, elder flower, oregano, thyme, and hawthorn have shown to have prominent properties such as antioxidant, anti-inflammatory, antifungal, and antibacterial. Apart from this, in study on ethanol-fed rats, ursolic acid has effectively reduced the lipid peroxidation, in addition to which it has also increase the circulatory antioxidants GSH, ascorbic acid and alpha-tocopherol (Saravanan et al., 2006). Apart from this, it can also increase the bile excretion as well as reduce the serum levels of ALT, AST, bilirubin, and restored serum albumin level in liver diseases (Xiong et al., 2003; Shankari et al., 2010) (Figure 9.17).

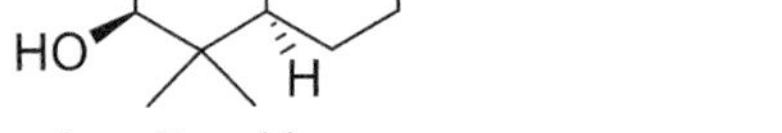

FIGURE 9.17 Structure of ursolic acid.

9.7.16 *WOGONIN*

Wogonin, a monoflavonoid isolated from *Scutellaria radix* have been used for 1,000 years for treatment of hepatitis and inflammatory related diseases (Guo et al., 2007). Due to its ability to suppress the hepatitis B surface antigen (HBsAg) secretion in cell cultures, wogonin has a potential activity against to anti-hepatitis B virus. In an experimental study carried on ducks, the plasma HBsAg level reduced as well as the considerable histopathological improvement of the liver on treating them with wogonin. It has also shown to be possess effective cytotoxic effects in the HCC cells SK-HEP-1 by inducing apoptosis, activation of caspase 3 cascades and induction of p53 proteins (Chen et al., 2002) (Figure 9.18).

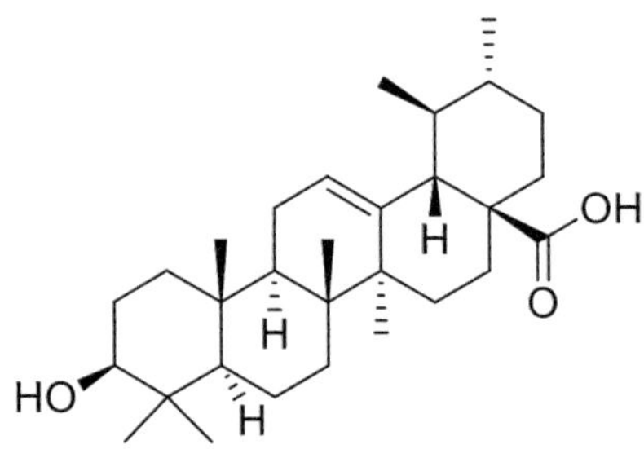

FIGURE 9.18 Structure of wogonin.

9.8 CONCLUSION AND FUTURE PROSPECTS

Since time immoral, for thousands of years, natural products from plant-derived foods or traditional medicine have been used to treat liver diseases; hence, it serves as promising lead compounds for prevention and treatment of various liver diseases. Numerous preclinical *in vivo* and *in vitro* studies on plant metabolites have demonstrated its hepatoprotective properties such as antioxidant, anti-inflammatory, antimutagenic, and lipotropic actions. With the rapid advancement in metabolic engineering and plant genomics, the 21st century has seen a paradigm shift towards therapeutic evaluation of plant-based products to aid in the isolation or production of potential bioactive metabolites. In addition, these natural agents can be used alone as single molecules or in combination with other agents to enhanced therapeutic effects against liver disease. However, there is still a need for well-designed clinical studies and large-scale trials to understand the biological function, composition, and therapeutic effects of various other metabolites that could, prevent, adverse effects from long-term administration and development of health promoting properties.

ACKNOWLEDGMENTS

The authors are thankful to Dr. S. M. Basavarajaiah, Assistant Professor, P. G. Department of Chemistry, Organic Chemistry Research Lab, Vijaya

College, Bangalore – 560004, Karnataka, India for chemical structural work and drawing in ChemSketch Chemical Software 12.0 version.

KEYWORDS

- **biomolecules**
- **hepatitis**
- **liver diseases**
- **plant secondary metabolites**
- **therapeutic agent**
- **treatment**
- **vegetables**

REFERENCES

Aggarwal, B. B., & Shishodia, S., (2006). Molecular targets of dietary agents for prevention and therapy of cancer. *Biochem. Pharmacol., 71*(10), 1397–1421.

Ajmo, J. M., Liang, X., Rogers, C. Q., Pennock, B., & You, M., (2008). Resveratrol alleviates alcoholic fatty liver in mice. *Am. J. Physiol. Gastroint. Liver Physiol., 295*(4), G833–842.

Asrani, S. K., Devarbhavi, H., Eaton, J., &Kamath, P. S., (2019). Burden of liver diseases in the world. *J. Hepatol., 70*(1), 151–171.

Bacon, B. R., Powell, L. W., Adams, P. C., Kresina, T. F., &Hoofnagle, J. H., (1999). Molecular medicine and hemochromatosis: At the crossroads. *Gastroenterol., 116*(1), 193–207.

Balachandran, P., & Govindarajan, R., (2003). Hepatic disorders. In: *Scientific Basis for Ayurvedic Therapies* (pp. 255–278). Routledge.

Bandmann, O., Weiss, K. H., & Kaler, S. G., (2015). Wilson's disease and other neurological copper disorders. *Lancet Neurol., 14*(1), 103–113.

Bao, W., Li, K., Rong, S., Yao, P., Hao, L., Ying, C., et al., (2010). Curcumin alleviates ethanol-induced hepatocytes oxidative damage involving heme oxygenase-1 induction. *J. Ethnopharmacol., 128*(2), 549–553.

Bhawna, S., & Kumar, S. U., (2009). Hepatoprotective activity of some indigenous plants. *Int. J. Pharm Tech Res., 4,* 1330–1334.

Bian, D. X., Liu, J., Lu, L., Liu, H., Guo, J. C., Yang, W. J., Liu, Y. L., et al., (2013). Protective effects of Rhein on hepatic progression in HBV-transgenic mice with nonalcoholic steatohepatitis induced by a high-fat diet. *Zhonghuashiyan he lin Chuang Bing du xuezazhi= ZhonghuaShiyan he LinchuangBingduxueZazhi= Chinese Journal of Experimental and Clinical Virology, 27*(5), 328–331.

Bowlus, C. L., & Gershwin, M. E., (2014). The diagnosis of primary biliary cirrhosis. *Autoimmunity Reviews, 13*(4, 5), 441–444.

Chan, C. C., Cheng, L. Y., Lin, C. L., Huang, Y. H., Lin, H. C., & Lee, F. Y., (2011). The protective role of natural phytoalexin resveratrol on inflammation, fibrosis, and regeneration in cholestatic liver injury. *Molecular Nutrition & Food Research, 55*(12), 1841–1849.

Chen, Y. C., Shen, S. C., Lee, W. R., Lin, H. Y., Ko, C. H., Shih, C. M., & Yang, L. L., (2002). Wogonin and fisetin induction of apoptosis through activation of caspase 3 cascade and alternative expression of p21 protein in hepatocellular carcinoma cells SK-HEP-1. *Archives of Toxicology, 76*(5), 351–359.

Cnossen, W. R., & Drenth, J. P., (2014). Polycystic liver disease: An overview of pathogenesis, clinical manifestations, and management. *Orphanet Journal of Rare Diseases, 9*(1), 1–3.

Cornelius, C. E., (1980). Liver function. In: *Clinical Biochemistry of Domestic Animals* (pp. 201–257). Academic Press.

Członkowska, A., Litwin, T., Dusek, P., Ferenci, P., Lutsenko, S., Medici, V., Rybakowski, J. K., et al., (2018). Wilson disease. *Nature Reviews Disease Primers, 4*(1), 1–20.

David, S. S., (1998). *Plant Secondary Metabolism*. Kluwier Acad. Publ., Massachusetts.

Devipriya, N., Sudheer, A. R., & Menon, V. P., (2007). Dose-response effect of ellagic acid on circulatory antioxidants and lipids during alcohol-induced toxicity in experimental rats. *Fundamental & Clinical Pharmacology, 21*(6), 621–630.

Devipriya, N., Sudheer, A. R., Vishwanathan, P., & Menon, V. P., (2008). Modulatory potential of ellagic acid, a natural plant polyphenol on altered lipid profile and lipid peroxidation status during alcohol-induced toxicity: A pathohistological study. *Journal of Biochemical and Molecular Toxicology, 22*(2), 101–112.

Dey, P. R., Saha, M. R., & Sen, A. R., (2013). An overview on drug-induced hepatotoxicity. *Asian J. Pharm. Clin. Res., 6*(4), 1–4.

Du, S. Y., Zhang, Y. L., Bai, R. X., Ai, Z. L., Xie, B. S., & Yang, H. Y., (2015). Lutein prevents alcohol-induced liver disease in rats by modulating oxidative stress and inflammation. *International Journal of Clinical and Experimental Medicine, 8*(6), 8785.

Ende, A. R., Kim, N. H., Yeh, M. M., Harper, J., & Landis, C. S., (2015). Fulminant hepatitis B reactivation leading to liver transplantation in a patient with chronic hepatitis C treated with simeprevir and sofosbuvir: A case report. *Journal of Medical Case Reports, 9*(1), 1–4.

Farrell, G. C., & Larter, C. Z., (2006). Nonalcoholic fatty liver disease: From steatosis to cirrhosis. *Hepatology, 43*(S1), S99–112.

Fehér, J., & Lengyel, G., (2012). Silymarin in the prevention and treatment of liver diseases and primary liver cancer. *Current Pharmaceutical Biotechnology, 13*(1), 210–217.

Ferrell, L., (2000). Liver pathology: Cirrhosis, hepatitis, and primary liver tumors. Update and diagnostic problems. *Modern Pathology, 13*(6), 679–704.

Firdous, A. P., Sindhu, E. R., & Kuttan, R., (2011). Hepato-protective potential of carotenoid meso-zeaxanthin against paracetamol, CCl_4 and ethanol induced toxicity. *Indian Journal of Experimental Biology, 49*(1), 44–49.

Gao, B., & Tsukamoto, H., (2016). Inflammation in alcoholic and nonalcoholic fatty liver disease: Friend or foe. *Gastroenterology, 150*(8), 1704–1709.

Gao, B., Jeong, W. I., & Tian, Z., (2008). Liver: An organ with predominant innate immunity. *Hepatology, 47*(2), 729–736.

Ghany, M. G., & Hoofnagle, J. H., (2014). Approach to the patient with liver disease. In: Kasper, D., Fauci, A., Hauser, S., Longo, D., Jameson, J., & Loscalzo, J., (eds.), *Harrison's Principles of Internal Medicine* (p. 19e). McGraw.

Guo, Q., Zhao, L., You, Q., Yang, Y., Gu, H., Song, G., Lu, N., & Xin, J., (2007). Anti-hepatitis B virus activity of wogonin *in vitro* and *in vivo*. *Antiviral Research, 74*(1), 16–24.

Habib-ur-Rehman, M., Mahmood, T., Salim, T., Afzal, N., Ali, N., Iqbal, J., Tahir, M., & Khan, A., (2009). Effect of silymarin on serum levels of ALT and GGT in ethanol induced hepatotoxicity in albino rats. *Journal of Ayub Medical College Abbottabad, 21*(4), 73–75.

Harborne, J. B., (1999). Classes and functions of secondary products from plants. *Chemicals from Plants, 26*, 1–25.

Hatcher, H., Planalp, R., Cho, J., Torti, F. M., & Torti, S. V., (2008). Curcumin: From ancient medicine to current clinical trials. *Cellular and Molecular Life Sciences, 65*(11), 1631–1652.

Hoffmann, G. F., & McKiernan, P., (2017). Liver disease. In: *Inherited Metabolic Diseases* (pp. 203–226). Springer, Berlin, Heidelberg.

Hussein, R. A., & El-Anssary, A. A., (2019). Plants secondary metabolites: The key drivers of the pharmacological actions of medicinal plants. *Herbal Medicine, 1*, 13.

Jayaraman, J., Veerappan, M., & Namasivayam, N., (2009). Potential beneficial effect of naringenin on lipid peroxidation and antioxidant status in rats with ethanol-induced hepatotoxicity. *Journal of Pharmacy and Pharmacology, 61*(10), 1383–1390.

Kabera, J. N., Semana, E., Mussa, A. R., & He, X., (2014). Plant secondary metabolites: Biosynthesis, classification, function, and pharmacological properties. *J. Pharm. Pharmacol., 2*(7), 377–392.

Kasdallah-Grissa, A., Mornagui, B., Aouani, E., Hammami, M., El May, M., Gharbi, N., Kamoun, A., & El-Fazaâ, S., (2007). Resveratrol, a red wine polyphenol, attenuates ethanol-induced oxidative stress in rat liver. *Life Sciences, 80*(11), 1033–1039.

Kennedy, O. J., Roderick, P., Poole, R., & Parkes, J., (2016). Coffee, caffeine, and non-alcoholic fatty liver disease? *Therapeutic Advances in Gastroenterology, 9*(3), 417–418.

Kew, M. C., (2013). Aflatoxins as a cause of hepatocellular carcinoma. *Journal of Gastrointestinal & Liver Diseases, 22*(3).

Kim, B. C., Kim, H. G., Lee, S. A., Lim, S., Park, E. H., Kim, S. J., & Lim, C. J., (2005). Genipin-induced apoptosis in hepatoma cells is mediated by reactive oxygen species/c-Jun NH2-terminal kinase-dependent activation of mitochondrial pathway. *Biochemical Pharmacology, 70*(9), 1398–1407.

Kim, H. M., Kim, Y., Lee, E. S., Huh, J. H., & Chung, C. H., (2018). Caffeic acid ameliorates hepatic steatosis and reduces ER stress in high fat diet–induced obese mice by regulating autophagy. *Nutrition, 55*, 63–70.

Kim, S., Shin, H. J., Kim, S. Y., Kim, J. H., Lee, Y. S., Kim, D. H., & Lee, M. O., (2004). Genistein enhances expression of genes involved in fatty acid catabolism through activation of PPARα. *Molecular and Cellular Endocrinology, 220*(1, 2), 51–58.

Köhnlein, T., & Welte, T., (2008). Alpha-1 antitrypsin deficiency: Pathogenesis, clinical presentation, diagnosis, and treatment. *The American Journal of Medicine, 121*(1), 3–9.

Kojima, K., Shimada, T., Nagareda, Y., Watanabe, M., Ishizaki, J., Sai, Y., Miyamoto, K. I., & Aburada, M., (2011). Preventive effect of geniposide on metabolic disease status in spontaneously obese type 2 diabetic mice and free fatty acid-treated HepG2 cells. *Biological and Pharmaceutical Bulletin, 34*(10), 1613–1618.

Lai, C. L., Ratziu, V., Yuen, M. F., & Poynard, T., (2003). Viral hepatitis B. *The Lancet, 362*(9401), 2089–2094.

Lee, W. M., (1997). Hepatitis B virus infection. *New England Journal of Medicine, 337*(24), 1733–1745.

Lee, Y. M., & Kaplan, M. M., (1995). Primary sclerosing cholangitis. *New England Journal of Medicine, 332*(14), 924–933.

Lee, Y. M., Choi, J. S., Kim, M. H., Jung, M. H., Lee, Y. S., & Song, J., (2006). Effects of dietary genistein on hepatic lipid metabolism and mitochondrial function in mice fed high-fat diets. *Nutrition, 22*(9), 956–964.

Leelananthakul, W., Werawatganon, D., Klaikeaw, N., Chayanupatkul, M., & Siriviriyakul, P., (2019). Protective effects of genistein alleviate alcohol-induced liver injury in rats. *Pharmacognosy Magazine, 15*(61), 342.

Levrero, M., (2006). Viral hepatitis and liver cancer: The case of hepatitis C. *Oncogene, 25*(27), 3834–3847.

Li, H. Y., Gan, R. Y., Shang, A., Mao, Q. Q., Sun, Q. C., Wu, D. T., Geng, F., et al., (2021). Plant-based foods and their bioactive compounds on fatty liver disease: Effects, mechanisms, and clinical application. *Oxidative Medicine and Cellular Longevity, 2021*.

Li, S., Li, J., Dai, W., Zhang, Q., Feng, J., Wu, L., Liu, T., et al., (2017). Genistein suppresses aerobic glycolysis and induces hepatocellular carcinoma cell death. *British Journal of Cancer, 117*(10), 1518–1528.

Li, S., Tan, H. Y., Wang, N., Zhang, Z. J., Lao, L., Wong, C. W., & Feng, Y., (2015). The role of oxidative stress and antioxidants in liver diseases. *International Journal of Molecular Sciences, 16*(11), 26087–26124.

Lin, G., Nnane, I. P., & Cheng, T. Y., (1999). The effects of pretreatment with glycyrrhizin and glycyrrhetinic acid on the retrorsine-induced hepatotoxicity in rats. *Toxicon., 37*(9), 1259–1270.

Liu, S., Hou, W., Yao, P., Zhang, B., Sun, S., Nüssler, A. K., & Liu, L., (2010). Quercetin protects against ethanol-induced oxidative damage in rat primary hepatocytes. *Toxicology in Vitro, 24*(2), 516–522.

Loprieno, N., (1975). International Agency for Research on Cancer (IARC) monographs on the evaluation of carcinogenic risk of chemicals to man: Relevance of data on mutagenicity. *Mutation Research, 31*(3), 210.

Ma, T., Huang, C., Zong, G., Zha, D., Meng, X., Li, J., & Tang, W., (2011). Hepatoprotective effects of geniposide in a rat model of nonalcoholic steatohepatitis. *Journal of Pharmacy and Pharmacology, 63*(4), 587–593.

Mieli-Vergani, G., Vergani, D., Czaja, A. J., Manns, M. P., Krawitt, E. L., Vierling, J. M., Lohse, A. W., & Montano-Loza, A. J., (2018). Autoimmune hepatitis. *Nature Reviews Disease Primers, 4*(1), 1–21.

Mitra, S., De, A., & Chowdhury, A., (2020). Epidemiology of non-alcoholic and alcoholic fatty liver diseases. *Translational Gastroenterology and Hepatology, 5*.

Mohamed, S. T. S., MadhusudhanaChetty, C., Ramkanth, S. V., Rajan, V. S., Mahesh, K. K., & Gauthaman, K., (2010). Hepatoprotective herbs: A review. *International Journal of Research in Pharmaceutical Sciences, 1*(1), 1–5.

Mokdad, A. A., Lopez, A. D., Shahraz, S., Lozano, R., Mokdad, A. H., Stanaway, J., Murray, C. J., & Naghavi, M., (2014). Liver cirrhosis mortality in 187 countries between 1980 and 2010: A systematic analysis. *BMC Medicine, 12*(1), 1–24.

Naik, R. S., Mujumdar, A. M., & Ghaskadbi, S., (2004). Protection of liver cells from ethanol cytotoxicity by curcumin in liver slice culture *in vitro*. *Journal of Ethnopharmacology, 95*(1), 31–37.

Nanji, A. A., Jokelainen, K., Tipoe, G. L., Rahemtulla, A., Thomas, P., & Dannenberg, A. J., (2003). Curcumin prevents alcohol-induced liver disease in rats by inhibiting the expression of NF-κB-dependent genes. *American Journal of Physiology-Gastrointestinal and Liver Physiology, 284*(2), G321–327.

O'Connor, S., Ward, J. W., Watson, M., Momin, B., & Richardson, L. C., (2010). Hepatocellular carcinoma-United States, 2001–2006. *Morbidity and Mortality Weekly Report, 59*(17), 517–520.

Ogiku, M., Kono, H., Hara, M., Tsuchiya, M., & Fujii, H., (2011). Glycyrrhizin prevents liver injury by inhibition of high-mobility group box 1 production by Kupffer cells after ischemia-reperfusion in rats. *Journal of Pharmacology and Experimental Therapeutics, 339*(1), 93–98.

Olivo, R., Guarrera, J. V., & Pyrsopoulos, N. T., (2018). Liver transplantation for acute liver failure. *Clinics in Liver Disease, 22*(2), 409–417.

Pari, L., & Gnanasoundari, M., (2006). Influence of naringenin on oxytetracycline mediated oxidative damage in rat liver. *Basic & Clinical Pharmacology & Toxicology, 98*(5), 456–461.

Pietrangelo, A., (2010). Hereditary hemochromatosis: Pathogenesis, diagnosis, and treatment. *Gastroenterology, 139*(2), 393–408.

Polimeni, L., Del Ben, M., Baratta, F., Perri, L., Albanese, F., Pastori, D., Violi, F., & Angelico, F., (2015). Oxidative stress: New insights on the association of non-alcoholic fatty liver disease and atherosclerosis. *World Journal of Hepatology, 7*(10), 1325.

Pradhan, S. C., & Girish, C., (2006). Hepatoprotective herbal drug, silymarin from experimental pharmacology to clinical medicine. *Indian Journal of Medical Research, 124*(5), 491.

Rajasekaran, D., Elavarasan, J., Sivalingam, M., Ganapathy, E., Kumar, A., Kalpana, K., & Sakthisekaran, D., (2011). Resveratrol interferes with N-nitrosodiethylamine-induced hepatocellular carcinoma at early and advanced stages in male Wistar rats. *Molecular Medicine Reports, 4*(6), 1211–1217.

Rasool, M., Iqbal, J., Malik, A., Ramzan, H. S., Qureshi, M. S., Asif, M., Qazi, M. H., et al., (2014). Hepatoprotective effects of *Silybum marianum* (silymarin) and *Glycyrrhiza glabra* (glycyrrhizin) in combination: A possible synergy. *Evidence-Based Complementary and Alternative Medicine, 2014.*

Rastogi, R., Saksena, S., Garg, N. K., Kapoor, N. K., Agarwal, D. P., & Dhawan, B. N., (1996). Picroliv protects against alcohol-induced chronic hepatotoxicity in Rats1. *Plantamedica, 62*(03), 283–285.

Raynaud, P., Tate, J., Callens, C., Cordi, S., Vandersmissen, P., Carpentier, R., Sempoux, C., et al., (2011). A classification of ductal plate malformations based on distinct pathogenic mechanisms of biliary dysmorphogenesis. *Hepatology, 53*(6), 1959–1966.

Ren, Y., Deng, F., Zhu, H., Wan, W., Ye, J., & Luo, B., (2011). Effect of epigallocatechin-3-gallate on iron overload in mice with alcoholic liver disease. *Molecular Biology Reports, 38*(2), 879–886.

Rietman, A., Sluik, D., Feskens, E. J., Kok, F. J., & Mensink, M., (2018). Associations between dietary factors and markers of NAFLD in a general Dutch adult population. *European Journal of Clinical Nutrition, 72*(1), 117–123.

Sánchez-Valle, V., Chavez-Tapia, N. C., Uribe, M., & Méndez-Sánchez, N., (2012). Role of oxidative stress and molecular changes in liver fibrosis: A review. *Current Medicinal Chemistry, 19*(28), 4850–4860.

Saraswat, B., Visen, P. K., Patnaik, G. K., & Dhawan, B. N., (1999). Ex vivo and *in vivo* investigations of picroliv from *Picrorhiza kurroa* in an alcohol intoxication model in rats. *Journal of Ethnopharmacology, 66*(3), 263–269.

Saravanan, R., Viswanathan, P., & Pugalendi, K. V., (2006). Protective effect of ursolic acid on ethanol-mediated experimental liver damage in rats. *Life Sciences, 78*(7), 713–718.

Schiff, L., & Schiff, E. R., (1993). *Diseases of the Liver*. Philadelphia.

Scott, L. N. D., (1998). A review of plants used in the treatment of liver disease: Part 1. *Alternative Medicine Review, 3*(6), 410–421.

Shankari, S. G., Karthikesan, K., Jalaludeen, A. M., & Ashokkumar, N., (2010). Hepatoprotective effect of morin on ethanol-induced hepatotoxicity in rats. *Journal of Basic and Clinical Physiology and Pharmacology, 21*(4), 277–294.

Sharma, A., & Nagalli, S., (2021). *Chronic Liver Disease*. In StatPearls [Internet]. StatPearls Publishing.

Shivashankara, A. R., Sunitha, V., Bhat, H. P., Palatty, P. L., & Baliga, M. S., (2013). Phytochemicals are effective in the prevention of ethanol-induced hepatotoxicity: Preclinical observations. In: *Bioactive Food as Dietary Interventions for Liver and Gastrointestinal Disease* (pp. 743–758). Academic Press.

Singal, A. K., & Kamath, P. S., (2013). Model for end-stage liver disease. *Journal of Clinical and Experimental Hepatology, 3*(1), 50–60.

Singh, B., Saxena, A. K., Chandan, B. K., Anand, K. K., Suri, O. P., Suri, K. A., & Satti, N. K., (1998). Hepatoprotective activity of verbenalin on experimental liver damage in rodents. *Fitoterapia-Milano, 69*, 135–140.

Sivakrishnan, S., & Pharm, M., (2019). Liver disease overview. *World Journal of Pharmacy and Pharmaceutical Sciences, 8*(1), 1385–1395.

Sommerburg, O., Keunen, J. E., Bird, A. C., & Van, K. F. J., (1998). Fruits and vegetables that are sources for lutein and zeaxanthin: The macular pigment in human eyes. *British Journal of Ophthalmology, 82*(8), 907–910.

Stickel, F., & Schuppan, D., (2007). Herbal medicine in the treatment of liver diseases. *Digestive and Liver Disease, 39*(4), 293–304.

Sumida, Y., Niki, E., Naito, Y., & Yoshikawa, T., (2013). Involvement of free radicals and oxidative stress in NAFLD/NASH. *Free Radical Research, 47*(11), 869–880.

Szabo, G., & Bala, S., (2010). Alcoholic liver disease and the gut-liver axis. *World Journal of Gastroenterology: WJG., 16*(11), 1321.

Tang, B., Qiao, H., Meng, F., & Sun, X., (2007). Glycyrrhizin attenuates endotoxin-induced acute liver injury after partial hepatectomy in rats. *Brazilian Journal of Medical and Biological Research, 40*(12), 1637–1646.

Thangapandiyan, S., & Miltonprabu, S., (2013). Epigallocatechin gallate effectively ameliorates fluoride-induced oxidative stress and DNA damage in the liver of rats. *Canadian Journal of Physiology and Pharmacology, 91*(7), 528–537.

Tilg, H., Moschen, A. R., & Kaneider, N. C., (2011). Pathways of liver injury in alcoholic liver disease. *Journal of Hepatology, 55*(5), 1159–1161.

Tsimogiannis, D., & Oreopoulou, V., (2019). Classification of phenolic compounds in plants. In: *Polyphenols in Plants* (pp. 263–284). Academic Press.

Uhl, P., Fricker, G., Haberkorn, U., & Mier, W., (2014). Current status in the therapy of liver diseases. *International Journal of Molecular Sciences, 15*(5), 7500–7512.

Vidhya, A., & Indira, M., (2009). Protective effect of quercetin in the regression of ethanol-induced hepatotoxicity. *Indian Journal of Pharmaceutical Sciences, 71*(5), 527.

Wang, X., & Ning, Q., (2014). Immune mediated liver failure. *EXCLI Journal, 13*, 1131.

Wang, X., Sun, H., Zhang, A., Jiao, G., Sun, W., & Yuan, Y., (2011). Pharmacokinetics screening for multi-components absorbed in the rat plasma after oral administration traditional Chinese medicine formula Yin-Chen-Hao-Tang by ultra-performance liquid chromatography-electrospray ionization/quadrupole-time-of-flight mass spectrometry combined with pattern recognition methods. *Analyst, 136*(23), 5068–5076.

Wang, Y., Mei, Y., Feng, D., & Xu, L., (2006). (–)-epigallocatechin-3-gallate protects mice from concanavalin A-induced hepatitis through suppressing immune-mediated liver injury. *Clinical & Experimental Immunology, 145*(3), 485–492.

Williams, R., (2006). Global challenges in liver disease. *Hepatology, 44*(3), 521–526.

Wong, L. L., (2002). Current status of liver transplantation for hepatocellular cancer. *The American Journal of Surgery, 183*(3), 309–316.

Worm, H. C., Van, D. P. W. H., & Brandstätter, G., (2002). Hepatitis E: An overview. *Microbes and Infection, 4*(6), 657–666.

Wu, Z., Han, M., Chen, T., Yan, W., & Ning, Q., (2010). Acute liver failure: Mechanisms of immune-mediated liver injury. *Liver International., 30*(6), 782–794.

Xiao, J., Ho, C. T., Liong, E. C., Nanji, A. A., Leung, T. M., Lau, T. Y., Fung, M. L., & Tipoe, G. L., (2014). Epigallocatechingallate attenuates fibrosis, oxidative stress, and inflammation in non-alcoholic fatty liver disease rat model through TGF/SMAD, PI3 K/Akt/FoxO1, and NF-kappa B pathways. *European Journal of Nutrition, 53*(1), 187–199.

Xiong, X., Chen, W., Cui, J., Yi, S., Zhang, Z., & Li, K., (2003). Effects of ursolic acid on liver-protection and bile secretion. *Zhongyaocai= Zhongyaocai= Journal of Chinese Medicinal Materials., 26*(8), 578–581.

Yan, F. J., Wang, X., Wang, S. E., Hong, H. T., Lu, J., Ye, Q., Zheng, Y. L., & Wang, Y. J., (2020). C-Jun/C7ORF41/NF-κB axis mediates hepatic inflammation and lipid accumulation in NAFLD. *Biochemical Journal, 477*(3), 691–708.

Yang, C. S., Wang, X., Lu, G., & Picinich, S. C., (2009). Cancer prevention by tea: Animal studies, molecular mechanisms, and human relevance. *Nature Reviews Cancer, 9*(6), 429–439.

Yuan, G., Gong, Z., Zhou, X., Zhangq, P., Sun, X., & Li, X., (2006). Epigallocatechin-3-gallate ameliorates alcohol-induced liver injury in rats. *International Journal of Molecular Sciences, 7*(7), 204–219.

Yun, J. W., Kim, Y. K., Lee, B. S., Kim, C. W., Hyun, J. S., Baik, J. H., Kim, J. J., & Kim, B. H., (2007). Effect of dietary epigallocatechin-3-gallate on cytochrome P450 2E1-dependent alcoholic liver damage: Enhancement of fatty acid oxidation. *Bioscience, Biotechnology, and Biochemistry, 71*(12), 2999–3006.

Zhang, A., Sun, H., & Wang, X., (2013). Recent advances in natural products from plants for treatment of liver diseases. *European Journal of Medicinal Chemistry, 63*, 570–577.

Zhang, D. K., Yu, J. J., Li, Y. M., Wei, L. N., Yu, Y., Feng, Y. H., & Wang, X., (2012). A *Picrorhiza kurroa* derivative, picroliv, attenuates the development of dextran-sulfate-sodium-induced colitis in mice. *Mediators of Inflammation, 2012*.

Zhang, M., Xie, Z., Gao, W., Pu, L., Wei, J., & Guo, C., (2016). Quercetin regulates hepatic cholesterol metabolism by promoting cholesterol-to-bile acid conversion and cholesterol efflux in rats. *Nutrition Research, 36*(3), 271–279.

Zhou, J., Zhou, F., Wang, W., Zhang, X. J., Ji, Y. X., et al., (2020). Epidemiological features of NAFLD from 1999 to 2018 in China. *Hepatology, 71*(5), 1851–1864.

CHAPTER 10

PLANT METABOLITES AND VEGETABLES FOR BRAIN DISEASE PREVENTION AND TREATMENT

MADHURI VERMA

Assistant Professor, Naraina Group of Institutions of Pharmacy, Kanpur, Dr. A.P.J. Abdul Kalam Technical University, Uttar Pradesh, India

ABSTRACT

Brain diseases are developed in population due stress, unhealthy lifestyle, metal toxicity, aging, environmental factors, etc., causes various neurodegenerative diseases (NDs). Currently, available research results show that secondary metabolites of vegetables and herbs are promising in preventing and managing neurodegenerative disorders. These metabolites elucidate their activity on mitochondrial changes in ND, also has antioxidant, anti-aggregation, memory enhancing, revert neuronal function, neuroprotective activity in animal model and clinical trials. Metabolites obtained from plant sources are more safe than chemically synthesized molecules, also, they downregulate various pathways of ND mechanism, regulate neurotransmitter level. Researchers develop new methods for delivery of these metabolites across BBB such as nanoparticles or targeted drug delivery so metabolites can bind more effectively and show good results. There are various clinical studies are going choosing plant metabolite for research to evaluate their effect on ND. In the present study, we are focusing on various plant metabolites acting on various NDs.

Plant Metabolites and Vegetables as Nutraceuticals.
Raj K. Keservani, Bui Thanh Tung, Rajesh K. Kesharwani, and Eknath D. Ahire (Eds.)

10.1 INTRODUCTION

Millions of people around the world suffer from brain disorders, the most common of which are Alzheimer's and Parkinson's amyotrophic lateral sclerosis (ALS), Huntington's disease (HD), spinocerebellar ataxia (SCA), spinal muscular atrophy (SMA), vascular cognitive impairment (VCI) prion disease, Frontotemporal disease, Pick's disease, etc. (Chen et al., 2015). In ND, both central and peripheral nervous system nerve cells eventually cease to function and perish. Alzheimer's disease (AD) is a severe, complex illness that causes irreversible cognitive impairment due to environmental and hereditary factors (Farkhondeh et al., 2019). Amyloid plaques, neurofibrillary tangles, and neuronal loss are the primary neuropathological characteristics of AD in the limbic and neocortical regions (Macdonald et al., 2014). Amyloid deposition, tau hyperphosphorylation, and aggregation, loss of the cholinergic system, oxidative stress, inflammation, apoptosis, glutamatergic excitotoxicity, and a decrease in neurogenesis and neurotrophic factors are just a few of the mechanisms linked to the disorder's progression (Farkhondeh et al., 2019; Webber et al., 2005; Ganguli et al., 2000). After AD, PD is the condition that affects older people most frequently. It is characterized by dopaminergic (DA) neuronal loss in the striatum, locus coeruleus, raphe nuclei, nucleus basalis of meynert, and hippocampus. Resting tremors, rigidity, bradykinesia, muscular impairment, postural instability, difficulty walking, soft speech, sleep issues, and sluggish handwriting are the main signs and symptoms of PD patients (Leverenz et al., 2009). Lewy Bodies (LBs, an aggregation of -synuclein), DNA mutations, and neuronal cytoplasmic inclusions in the cortex and brain stem are the pathological findings in PD patients' brains (Zecca et al., 1996; Hoang et al., 2014; Gibb et al., 1988).

Another neurodegenerative disease (ND) that is heterogeneous and causes motor symptoms is ALS, also known as motor neuron disease (MND) or Lou Gehrig's disease. Individuals with spinal-onset disease may experience limb muscle weakness, while those with bulbar-onset disease may initially experience dysarthria and dysphagia (Hardiman et al., 2017; According to Boillée et al. (2006), the majority of patients will die from respiratory failure within three to five years of the first manifestation of the disease. The exact cause of this disease is unknown, but a small number of cases are linked to a family history with gene mutations as a major contributor (Al-Chalabi et al., 2016) and associated mechanisms of oxidative stress (Simpson et al., 2003), inflammation (McGeer et al., 2002), chronic viral infections (Swash et al., 1992), and excitotoxicity (Niebroj-Dobosz et al., 1999). Spasticity, sialorrhea, pain, muscle cramps, liver issues, deep vein thrombosis, mood swings, respiratory insufficiency, fever, vomiting, exhaustion, and nausea are additional symptoms. Because of mutations in SOD1, ALS patients' brains' capacity to transform harmful radicals into harmless compounds is compromised (Semmler et al., 2020). Although this type of mutation does not cause the disease to advance, it does make SOD1 more susceptible to protein misfolding, which results in a buildup of protein and a consequent decrease in motor neuron function.

HD is a fatal, familial, genetic disorder that causes progressive loss of brain and muscular function. In this single-gene condition, neurons in various parts of the brain are programmed to degenerate. Exon 1 of the Huntingtin (Htt) gene on chromosomal number 4 has 6–35 CAG repeats in the wild type gene, but it has more than 36 repeats in the affected individuals (Szlachcic et al., 2015). As a result, the neuron begins to accumulate Htt protein, which eventually kills it. The cortex, thalamus, and hypothalamus nuclei atrophy as nearly 95% of GABAergic medium spiny neurons (MSNs) projecting to the substantia nigra (SN) and globus pallidus are lost. Short-term memory loss, falling, lack of attention, clumsiness, sadness, weight loss, food issues, difficulties speaking, uncontrollable facial movements, itching, and other symptoms are among the symptoms (Manoharan et al., 2016).

According to O'Brien et al. (2003), VCI is characterized by a particular cognitive profile that includes retained memory along with deficits in attentional and executive functioning. Broadening the definition of vascular dementia is necessary to acknowledge the significant role that cerebrovascular disease plays in a number of cognitive disorders, such as hereditary vascular dementia, multi-infarct dementia, post-stroke dementia, subcortical ischemic vascular disease and dementia, mild cognitive impairment, and degenerative dementias (such as AD, frontotemporal dementia, and dementia with Lewy bodies) (Jeong et al., 2020). Although a stroke is not a neurodegenerative illness in and of itself, it can result in one. It is well acknowledged to be a significant factor in VCI. It is the third leading global cause of death from brain illnesses (Hicks et al., 2009). An area of the brain has an abrupt drop in blood flow that may be temporary or permanent, depriving it of oxygen, glucose, and energy. *Cornu ammonis* 1 (CA1) of the hippocampus is the primary area of the brain injured in the case of Ischemia (Kirino et al., 1982). The term "delayed neuronal death" refers to the loss of pyramidal neurons in the CA1 area that occurs 4 to 5 days after ischemia (DND). After ischemia/reperfusion, it is followed by a number of pathological processes that cooperate, such as excitotoxicity, oxidative stress, apoptosis, inflammation, gliosis of astrocytes, and microglia in the CA1 region. There is a significant load from the loss of mitochondrial function and depletion of endogenous antioxidants, which leads to neuronal death via intrinsic programmed cell death and causes behavioral and histological changes (Abas et al., 2010). Following all of this, the BBB breaks down, resulting in edema. Proinflammatory cytokines like TNF-, IL-1, and IL-6 are released by activated immune cells in response to edema (Jin et al., 2017). Along with the inflammatory and apoptotic pathways, oxidative stress leads to additional loss of neurons and brain tissues, enlarging the infarct area (Jin et al., 2017).

Prion disease is the only naturally occurring infectious protein misfolding disorder. It is also known as transmissible spongiform encephalopathies (TSE). Normal prion protein (PrPc), which is a protein that stands for pertinacious infectious particles, is unnaturally converted into an abnormal form of a protein called prions (PrPsc), which causes human prion disease (Barron et al., 2017). Although PrPc is a surface glycoprotein that is expressed in many tissues, it is most prominent in the central nervous system (CNS) (Marn-Moreno et al., 2017). The "protein only-hypothesis," which explains how a protein may reproduce itself without a nucleic acid, provides an explanation for the initiation and course of the prions disease. It asserts that the infectious agent is PrPc itself (Ma et al., 2014). The infectious protein PrPsc clumps and develops resistance to proteinase digestion as the prion illness progresses. Following this, PrPc is converted into PrPsc by PrPsc-seeded processes, resulting in a high concentration of PrPsc in an individual and causing neurodegeneration and the disease's clinical manifestation (Ma et al., 2020). In terms of both its course and appearance, this illness is regarded as unusual (Baiardi et al., 2020).

The term "FTD" refers to a large spectrum of diverse progressive neurological illnesses. These non-Alzheimer dementias are characterized by frontal and temporal lobe atrophy (Bott et al., 2014). As the frontal and temporal lobes are severely impacted by this illness, FTD results in changes in behavior and language that are controlled by these areas. While some FTD sufferers lose their language skills, others experience dramatic behavioral and personality changes that make them impulsive and socially inappropriate (Kelley et al., 2016). Other typical symptoms include a lack of restraint, judgment, and apathy, compulsive conduct that repeats itself, a reduction in personal hygiene, eating inedible objects, and losing empathy for the suffering of other people (Mohandas et al., 2009). Behavioral variation FTD (bvFTD), non-fluent/agrammatic variant primary progressive aphasia (nfvPPA), and semantic variant PPA (svPPA) are the three common variant syndromes that make up FTD (Miller et al., 2014). Pick's disease (PiD), a neurodegenerative disorder that causes permanent dementia and is characterized by an excessive number of tau inclusions known as Pick bodies, is a subset of FTD (Kertesz et al., 1998). Round argyrophilic neuronal inclusions are called pick bodies. PiD results in cortical atrophy, which affects the limbic systems, neocortex, and dentate granular cells of the hippocampus, as well as the frontal and temporal poles. Such neural degenerations lead to behavioral modifications, personality disorders, and other problems (Frederick et al., 2006). A pragmatic attitude and mutism are the early symptoms of PiD.

The term "spinocerebellar ataxia" (SCA) is used to describe a group of inherited (autosomal dominant) ataxias that affect the cerebellum, which controls movement, and the spinal cord (Klockgether et al., 2019). It is a sizable category of neurodegenerative disorders characterized by dysarthria, cerebellar ataxia, and brain stem involvement. The cerebellum's atrophy is what causes the resultant ataxia (Schöls et al., 2004). Loss of coordination in eye and hand movements, including speaking, is a symptom of all kinds of SCA. SCA often develops in adults, and because its advancement is progressive, it gets worse with time (Groth et al., 2018).

SMA is a hereditary condition characterized by skeletal muscular atrophy. The loss of motor neurons, which control muscle movement, causes the development of SMA. It causes weakness and makes movement or any physical activity difficult; over time, this condition gets worse (Kolb et al., 2011). This condition is brought on by insufficient survival motor neuron (SMN) synthesis, which is mostly produced by the SMN1 gene. SMN is also produced by SMN2, albeit in a much less significant way (Farrar et al., 2014).

10.2 MEDICINAL PLANT EFFECTIVE IN BRAIN DISEASE

10.2.1 ALZHEIMER'S DISEASE (AD)

1. ***Withania somnifera***: Withanolide A, withanoside IV, and withanoside VI are present and produce pre- and postsynaptic restoration in cortical neurons' axons and dendrites (Tohda, 2008; Kuboyama et al., 2005). In a clinical trial, people who are under a lot of stress are given an extract containing withanolide glycosides, withaferin A, oligosaccharides, alkaloids, and polysaccharides without scopolamine to see if they can reduce the levels of biochemical

stress markers by affecting the hypothalamic-pituitary-adrenal axis (Auddy et al., 2008).

2. ***Bacopa monnieri***: In contrast to glutamate-induced excitotoxicity, the plant extract was found to protect neurons from -amyloid-induced cell death. This neuroprotection was made feasible by the extract's capacity to inhibit cellular acetylcholinesterase activity. Additionally, this extracts enhanced cell survival and shown lowering and lipid peroxidation inhibitory effects (Limpeanchob et al., 2008). Additionally, it exhibits antioxidant properties in young rats and lessens brain oxidative stress (Priyanka et al., 2013; Shinomol & Muralidhara, 2011; Saini et al., 2012). Additionally, it increases the amount of acetylcholine in the hippocampus and cerebral blood flow while also raising the expression or activity of Na^+/K^+-ATPase, which suggests that Brahmi may have cognitive enhancing and neuroprotective properties in AD (Uabundit et al., 2010; Zhang et al., 2013). A poly-herbal formulation containing Brahmi was administered for a 12-month period in clinical research comprising healthy human volunteers and SDAT patients aged 60 to 75 (Sharma et al., 2020; Rane et al., 2020a–c). In these healthy volunteers and SDAT patients, the authors assessed cognitive capabilities, inflammatory markers, and oxidative stress. They discovered an improvement in cognitive assessments and a decrease in inflammation and oxidative stress levels (Sadhu et al., 2014). Bacosides reduce anterograde experimental amnesia in mice and help them remember things from the past (Kishore & Singh, 2005). In healthy elderly adults 65 years of age and over, a standardized whole plant dry extract of *B. monnieri* was tested for its effect and function on cognition as well as for safety and tolerability. The outcomes demonstrated that the standardized extract has the ability to improve cognitive function in aging (Calabrese et al., 2008).

3. ***Centella asiatica***: It is used in Ayurveda and Unani to heal ailments like asthma, skin conditions, ulcers, and bodily aches as well as to enhance memory and as an energizer (Singh et al., 2010). Leaf extract promotes CA3 neuron proliferation in the hippocampal region, and in B103 cell cultures and hippocampal slices, its active ingredients asiatic acid, asiaticoside, and SM2 have neuroprotective properties that prevent -amyloid and free radical-induced cell death (Rao et al., 2006; Mook-Jung et al., 1999; Lee et al., 2000). Pretreatment with

asiatic acid shown to reduce apoptotic cell death, reduce reactive oxygen species (ROS), stabilize mitochondrial membrane potential (MMP), and attenuate neuronal damage of the pyramidal layer in the CA1 and CA3 regions was studied *in vitro* (human neuroblastoma SH-SY5Y cells) and *in vivo* (neonatal mice) (Xu et al., 2012).

4. ***Ginkgo biloba*:** The ginkgolides and bilobalides found in ginkgo biloba have been categorized as nootropic substances (Kumar et al., 2006). According to studies, it boosts the body's oxygen supply and aids in the removal of free radicals, which enhances memory (Chandrasekaran et al., 2001). By preventing PAF and causing blood vessels to produce nitric oxide, it improves cerebral blood flow (Hirsch et al., 2017). Gb extract reduces A1-42 caused hippocampus neuron dysfunction and death, A induced abnormal behavior, and A aggregation in animal models of AD. It also increases neurogenesis (Suliman et al., 2016).

5. ***Vitis vinifera*:** RES, a phenolic compound found in Vitis vinifera, reduces the amount of intracellular ROS produced by A (Koukoulitsa et al., 2016) by increasing the expression of phase 2 enzymes and cellular antioxidants (such as glutathione), preventing oxidative and electrophilic injury (Cao et al., 2004), and supporting the HO-1 pathway (Kwon et al., 2011). Inhibiting the expression of TNF-, IL-1, and IL-6 as well as the phosphorylation of STAT1 and STAT3 has been demonstrated to have anti-inflammatory effects (Huang et al., 2011; Yao et al., 2015). It has also been proven to reverse the A-induced iNOS upregulation (Capiralla et al., 2012).

6. ***Curcuma longa*:** The active ingredient, curcumin ($C_{21}H_{20}O_6$), has been identified from Curcuma longa. We may highlight this substance's antioxidant, anti-inflammatory, and amyloid disaggregating properties among its known biological effects (Ullah et al., 2017; Ganjali et al., 2017). Additionally, it reduces the inflammation brought on by A while just slightly impairing g-secretase and acetylcholinesterase activity (Lim et al., 2001; Yang et al., 2005). Curcumin nanoparticles were created in another study to enhance cognition by increasing BDNF levels via the Akt/GSK-3 signaling pathways (Hoppe et al., 2013; Nam et al., 2014; Kim et al., 2004). According to Rogers et al. (2001), curcumin reduces the expression of inflammatory mediators (IL-6 and TNF-), which lead to inflammation in the brain. It also restores phagocytosis and is able to

remove amyloid plaques from the microglial cell line N_9 (He et al., 2016).

7. ***Crocus sativus*:** The primary component is crocin ($C_{44}H_{64}O_{24}$), which is also known to have antidepressant and anti-inflammatory properties as well as to enhance learning and memory (Kongkeaw et al., 2014; Schmidt et al., 2007). It primarily works on its antioxidant characteristics and suppresses acetyl-cholinesterase activity (Naghizadeh et al., 2013).

10.2.2 PARKINSON'S DISEASE (PD)

1. ***Mucuna pruriens*:** Long-term use as a potent aphrodisiac, for the treatment of Parkinson's disease (PD), neurological disorders, and arthritis (Amin et al., 1996; Jeyaweera et al., 1981; Sathiyanarayanan & Arulmozhi, 2007). In this mouse model, seed extract decreased oxidative damage, physiological abnormalities, and immunohistochemistry alterations to minimize paraquat-induced neurotoxicity (Yadav et al., 2013). The SN and striatum extract restored the number of tyrosine hydroxylase-positive cells, decreased the expression of iNOS and glial fibrillary acidic protein (GFAP), and boosted dopamine levels. Additionally, the extract reduced the formation of nitric oxide, neuroinflammation, and microglial activation, all of which may have contributed to its neuroprotective effects (Yadav et al., 2014). *Mucuna pruriens* cotyledon powder significantly improves SN endogenous levodopa, dopamine, norepinephrine, and serotonin content (Manyam et al., 2004a). HP-200 was given to rats over the course of a lengthy research (52 weeks), and it had a considerable impact on the amount of dopamine in the brain (Manyam et al., 2004b). When *Mucuna* and *Withaina* extracts are administered together, behavioral and antioxidant recovery is seen, indicating that the combination is neuroprotective. It is well known that *M. pruriens* has antiparkinsonian effects without causing drug-induced dyskinesias (Lieu et al., 2010). Additionally, dopa decarboxylase inhibitors, when administered in combination, produced long-term antiparkinsonian effects without resulting in drug-induced dyskinesias (Lieu et al., 2012). It was discovered that *M. pruriens*, when administered orally to 60 PD patients for a period of 12 weeks in the form of a powder (HP-200), was a successful treatment for these individuals (Parkinson's Disease Study Group, 1995).

2. ***Bacopa monnieri***: In Caenorhabditis elegans models of PD, the extract was tested alone or in combination with other herbs. It was discovered that *B. monnieri* reduced α-synuclein aggregation, prevented DA neurodegeneration, and restored the lipid content in *C. elegans*, demonstrating *B. monnieri* as a potential antiparkinsonian agent (Jadiya et al., 2011). PD model fruit flies (Drosophila melanogaster) treated with *B. monnieri* alone or in combination have improved walking abilities (Jansen et al., 2014). Brahmi supplementation improved the striatal dopamine levels and cholinergic enzyme activity in mice treated with paraquat (Hosamani et al., 2014). Different brain areas of young mice were investigated for acute paraquat-induced oxidative stress using a standardized Brahmi extract. In this study, mice treated with paraquat had their striatal dopamine levels and cholinergic enzyme activity restored by Brahmi supplementation (Hosamani et al., 2014).
3. ***Curcuma longa***: Additionally, it has been demonstrated that curcumin promotes neuronal regeneration in a PD model by enhancing BDNF levels through activating Trk/PI3K signaling pathways (Yang et al., 2014). Curcumin is demonstrated to block the JNK pathway in animal models of MPTP- and MPP-induced neurotoxicity (Yu et al., 2010).
4. ***Glycine max***: In PD flies exposed to genistein derived from Glycine max, a substantial dose-dependent reduction in oxidative stress indicators, an increase in dopamine content, and an increase in life expectancy were seen (Siddique et al., 2018).

10.2.3 AMYOTROPHIC LATERAL SCLEROSIS (ALS)

1. ***Panax***: Genistein (derived from Panax) reduced ALS-related symptoms and marginally increased the lifetime of SOD1-G93A mice. It improves onset on sign and survival in transgenic mice in comparison to control (Jiang et al., 2000; Zhao et al., 2019).
2. ***Withania somnifera***; Changes in the expression of several cytokines and chemokines provided additional evidence of WS's overall immunomodulatory effect. Additionally, WS acted as an autophagy inducer, which may be helpful when the disease is still in its early stages. These findings imply that WS extracts may represent

therapies with promise for the treatment of ALS involving misfolded SOD1 (Dutta et al., 2018).

3. ***Tripterygium wilfordii*** **and** ***Celastrus regelii*****:** Increasing cell survival, lowering MDA, upregulating mRNA levels of GST and GCLC, and activating the ERK1/2 and Akt signaling pathways (Li et al., 2017).

10.2.4 HUNTINGTON'S DISEASE (HD)

1. ***Withania somnifera*****:** Increases in body weight, behavioral activity, SOD, catalase, mitochondrial complex (I, II, III) levels, and decreases in lipid peroxidation, nitrite, and LDH are all demonstrated by root extract (Kumar et al., 2009).
2. ***Vitis vinifera*****:** boost mitochondrial genesis, SIRT3-dependent AMPK activation, decrease cell toxicity, oxidative stress, and mitochondrial malfunction (Fu et al., 2012).
3. ***Panax ginseng*****:** Weight gain, increased physical activity, antioxidant enzymes, HO-1, NQO1, and a reduction in ROS (Gao et al., 2015).
4. **Resveratrol:** Rise in autophagy, autophagic degradation of mutant Htt, and reduction in ROS (Vidoni et al., 2018).
5. **Melatonin:** Reduce protein carbonyl, LPO, and succinate dehydrogenase levels while raising SOD and SOD (Tunez et al., 2004).

10.2.5 VASCULAR COGNITIVE IMPAIRMENT (VCI)

1. ***Aster ageratoides*****:** Extract exhibits decreased neuronal excitotoxicity *in vitro*, decreased hippocampal structures, and decreased memory impairment *in vivo* (Jeong et al., 2020).
2. ***Ginkgo biloba*****:** Bilobalide, a compound derived from *G. biloba*, exhibits memory improvement reduction, decreased nuclear condensation, increased SOD and GSH, decreased NOS and MDA, and decreased TNF (Li et al., 2013).

10.2.6 STROKE (CAUSES SECONDARY NEURODEGENERATION)

1. ***Piper nigrum***: The primary component, piperine, decreases infarct volume, increases behavioral impairments and histological abnormalities, and increases expression of iNOS, COX-2, TNF-, and IL-1 and IL-6 (Vaibhav et al., 2012).
2. ***Zingiber officinale***: Zingerone enhances behavioral impairments, decreases infarct size, LPO, Caspase-3 and -9, Apaf-1, Bax, increases GSH and related enzymes, Catalase, and Bcl-2 (Vaibhav et al., 2013).
3. ***Magnolia officinalis***: Honokiol reduces levels of cerebral inflammation, p65, NO, TNF-β, RANTES, and CCL5 (Zhang et al., 2013).
4. ***Sinomenium acutum***: Sinomenine causes apoptosis, activation of astrocytes and microglia, the production of IL-1β, IL-6, IL-18, and TNF-α, an increase in behavioral impairments, a decrease in infarct volume, and changes in the NLRP3 inflammasome (Qiu et al., 2016).
5. ***Angelica sinensis***: Ligustilide, increase behavior deficits, decrease infarct volume, increase HSP-70 and MAPK (mitogen activated protein kinase) activation (Yu et al., 2015).
6. ***Scutellaria baicalensis***: Microglia/macrophage M2 markers CD206 and CD 163 are upregulated, infarct volume is downregulated, microglia/macrophage M1 markers CD86 and CD16 are upregulated, MAPK and NF-B are upregulated (Yang et al., 2019).
7. **Green Tea:** Extract increases cell survival, SOD, migration, and tube formation, VEGF (vascular endothelial growth factor), and Bcl2 mRNA expression, and decreases ROS, LDH, MDA, death, and autophagy, as well as Bax and caspase-3 mRNA expression (Fu et al., 2019).
8. ***Curcuma longa***: Increased cell viability, COX-2, TNF-expression, DOPAC, HVANE, and 5-HT, as well as curcumin, DA, and its metabolites (De et al., 2016).
9. ***Ginkgo biloba***: Ginkgolides increase NSE, GFAP, and SOCS2-positive cell sizes as well as cell survival, process duration, and cell body area (Zheng et al., 2018).
10. **Grapes and Other Plants:** Resveratrol (RES) increases cell survival and stimulates synaptogenesis, neurite outgrowth, and SIRT1 activation (Tang et al., 2017).

10.2.7 PRION DISEASE

1. ***Scutellari biacalenesis*:** Baicalein, decrease ROS production, mitochondrial dysfunction, apoptosis, and JNK signaling (Moon et al., 2014).
2. **Resveratrol (RES):** Increase autophagy, decrease mitochondrial dysfunction and apoptosis (Jeong et al., 2012).
3. **Hinokitiol:** Increase HIF-1α, and autophagy, increase p62/SQSTM1, Decrease apoptosis (Moon et al., 2016).

10.2.8 PICK'S DISEASE

1. **δ-Tocopherol:** Decrease cholesterol accumulation, and lysosomal size, increase intracellular Ca^{2+}, increase Ca^{2+} deficiency (Xu et al., 2012).
2. **Quercetin:** Decrease cholesterol uptake in Coca-2 cells and male Wistar rats (Xu et al., 2012).
3. **Luteolin:** Decrease cholesterol uptake in Coca-2 cells and male Wistar rats (Nekohashi et al., 2014).

10.2.9 SPINAL MUSCULAR ATROPHY (SMA)

1. **Melibiose:** Reduced expression of caspase 3 and ROS, increased autophagy, and decreased polyQ aggregation (Chen et al., 2020).

10.2.10 SPINOCEREBELLAR ATAXIA (SCA)

1. ***Brucea javanica:*** Bruceine D, Improving SMN phenotypic deficits by repairing the SMN2 splice defect (Baek et al., 2019).
2. ***Tripterygium wilfordii*:** Triptolide increases SMN transcription, SMN expression, Gemin2 and Gemin3 levels, survival rate, and SMA-related abnormalities (Hsu et al., 2012).

10.3 CLINICAL STUDIES (TABLE 10.1)

TABLE 10.1 Phytochemicals Used for Clinical Studies on Neurodegenerative Diseases

Phytochemical/ Drug Candidate	AD Model	Outcomes	References
Dimebon	Mild-to-moderate AD patients.	Improvement of cognition and memory function.	Doody et al. (2008)
Melatonin	157 AD patients	Increase nocturnal total sleep time, decrease wake after sleep.	Zhang et al. (2017)
Grapes and other plants.	20PD patients double blind, crossover, randomized, placebo-controlled phase 1 study.	To investigate BIA 6-512's impact on the pharmacokinetics of levodopa when combined with benserazide.	ClinicalTrials.gov. Identifier: NCT03091543
	BM-MSCs derived from ALS patients.	Nestin, NF-M, Tuj-1, and Map-2, as well as AMPK/SIRT1 signaling and neuroprogenitor markers.	Yun et al. (2019)
	Mild to moderate AD-related dementia in 199 cases.	Increase CSF (cerebro spinal fluid) and plasma Aβ (amyloid β) 40 in comparison to placebo, maintaining the integrity of the brain barrier.	Sawda et al. (2017)
	Double blind, randomized controlled 102 early affected HD patients.	To assess the impact of resveratrol on HD patients' caudate volume.	ClinicalTrials.gov Identifier: NCT02336633
α-lipoic acid	AD patients	Increase in cognition function, protection against Aβ toxicity.	Sancheti et al. (2013)
N-acetylcysteine (NAC)	A double-blind AD patient.	Improvement of cognitive and behavioral functions.	Remington et al. (2015)
G. biloba	Older adults and AD patients.	Prevention of cognition and memory decline.	Vellas et al. (2012)
	90 VCI patients, randomized, double-blind, placebo-controlled trial.	Increase CGI scores, no effect on trans-cranial doppler ultrasound, more adverse reactions in placebo group.	Demarin et al. (2017)
Green tea	Randomized, double-blind, 480 PD patients' blind.	Progression of motor function is delayed, while cognition, mood, and quality of life are all improved.	ClinicalTrials.gov Identifier: NCT00461942
Huperzia serrata	Randomized, double-blinded, research with 78 individuals that was placebo-controlled with minor to average VCI.	Increase MMSE, CDR, and ADL scores.	Xu et al. (2012); Xing et al. (2014)
Curcuma longa	42 randomized ALS patients.	Decrease disease progression, AOPPs, oxidative damage, increase aerobic metabolism.	Chico et al. (2018)

Source: Modified from: Khan et al. (2020).

10.4 CHALLENGES AND PROSPECTS

In the upcoming year, it should need to more investigated to find out unique phytochemicals, exactly mimic the biological molecules acting on human brain. Also, we know the cases of brain diseases were increase during past decades, i.e., the burden of brain disease increases in the world population, we should conduct more clinical trial to evaluate the efficacy, potency of more phytochemicals.

10.5 CONCLUSION

The secondary plant metabolites are effective in prevention and management of various brain diseases, which was reflected in *in-vivo*, *in-vitro*, and in clinical studies by various investigators to establish safety and efficacy of plant metabolites.

KEYWORDS

- **amyotrophic lateral sclerosis**
- **neurodegenerative disease**
- **neuroprotective**
- **plant metabolite**
- **prion disease**
- **spinal muscular atrophy**
- **vascular cognitive impairment**

REFERENCES

Abas, F., Alkan, T., Goren, B., Taskapilioglu, O., Sarandol, E., & Tolunay, S., (2010). Neuroprotective effects of postconditioning on lipid peroxidation and apoptosis after focal cerebral ischemia/reperfusion injury in rats. *Turk. Neurosurg., 20*, 1–8.

Al-Chalabi, A., Berg, L. H. V. D., & Veldink, J., (2016). Gene discovery in amyotrophic lateral sclerosis: Implications for clinical management. *Nat. Rev. Neurol., 13*, 96–104.

Amin, K. M. Y., Khan, M. N., Zillur-Rehman, S., et al., (1996). Sexual function improving effect of *Mucuna pruriens* in sexually normal male rats. *Fitoterapia Milano, 67*, 53–56.

Auddy, B., Hazra, J., Mitra, A., et al., (2008). *Withania somnifera* extract significantly reduces stress-related parameters in chronically stressed humans. A double blind, randomized, placebo-controlled study. *J. Am. Nutraceut. Assoc., 11*, 5056.

Baek, J., Jeong, H., Ham, Y., Jo, Y. H., Choi, M., Kang, M., Son, B., Choi, S., Ryu, H. W., Kim, J., et al., (2019). Improvement of spinal muscular atrophy via correction of the SMN2 splicing defect by *Brucea javanica* (L.) Merr. extract and *Bruceine*, D. *Phytomedicine, 65*, 153089.

Baiardi, S., Rossi, M., Capellari, S., & Parchi, P., (2019). Recent advances in the histo-molecular pathology of human prion disease. *Brain Pathol., 29*, 278–300.

Barron, R., (2017). Infectious prions and proteinopathies. *Prion, 11*, 40–47.

Boillée, S., Yamanaka, K., Lobsiger, C. S., Copeland, N. G., Jenkins, N. A., Kassiotis, G., Kollias, G., & Cleveland, D. W., (2006). Onset and progression in INHERITED ALS determined by motor neurons and microglia. *Science, 312*, 1389–1392.

Bott, N., Radke, A., Stephens, M. L., & Kramer, J. H., (2014). Frontotemporal dementia: Diagnosis, deficits, and management. *Neurodegener. Dis. Manag., 4*, 439–454.

Calabrese, C., Gregory, W. L., Leo, M., et al., (2008). Effects of a standardized *Bacopa monnieri* extract on cognitive performance, anxiety, and depression in the elderly: A randomized, double-blind, placebo-controlled trial. *J. Altern. Complement Med., 14*(6), 707–713.

Cao, Z., & Li, Y., (2004). Potent induction of cellular antioxidants and phase 2 enzymes by resveratrol in cardiomyocytes: Protection against oxidative and electrophilic injury. *Eur. J. Pharmacol., 489*, 39–48.

Capiralla, H., Vingtdeux, V., Zhao, H., Sankowski, R., Al-Abed, Y., Davies, P., & Marambaud, P., (2012). Resveratrol mitigates lipopolysaccharide- and Aβ-mediated microglial inflammation by inhibiting the TLR4/NF-κB/STAT signaling cascade. *J. Neurochem., 120*, 461–472.

Chandrasekaran, K., Mehrabian, Z., Spinnewyan, B., Drieu, K., & Kiskum, G., (2001). Neuroprotective effects of bilobalide, a component of Ginkgo biloba extract (EGb 761), in gerbil global brain ischemia. *Brain Res., 922*, 282–292.

Chen, C. M., Lin, C. H., Wu, Y. R., Yen, C. Y., Huang, Y. T., Lin, J. L., Lin, C. Y., Chen, W. L., Chao, C. Y., Lee-Chen, G. J., et al., (2020). Lactulose and melibiose inhibit α-synuclein aggregation and up-regulate autophagyto reduce neuronal vulnerability. *Cells, 9*, 1230.

Chen, X., & Pan, W., (2015). The treatment strategies for neurodegenerative diseases by integrative medicine. *Integr. Med. Int., 1*, 223–225.

Chico, L., Caldarazzo-Ienco, E., Bisordi, C., LoGerfo, A., Petrozzi, L., Petrucci, A., Mancuso, M., & Siciliano, G., (2018). Amyotrophic lateral sclerosis and oxidative stress: A double-blind therapeutic trial after curcumin supplementation. *CNS Neurol. Disord. Drug Targets, 17*, 767–779.

De Alcântara, G. F. T., Simões-Neto, E., Da Cruz, G. M. P., Nobre, M. E. P., Neves, K. R. T., De Andrade, G. M., Brito, G., & Viana, G. S. D. B., (2016). Curcumin reverses neurochemical, histological and immuno-histochemical alterations in the model of global brain ischemia. *J. Tradit. Complement. Med., 7*, 14–23.

Demarin, V., Kes, V. B., Trkanjec, Z., Budišic´, M., Pasic´, M. B., Cˇ rnac, P., & Budincˇevic, H., (2017). Efficacy and safety of Ginkgo biloba standardized extract in the treatment of vascular cognitive impairment: A randomized, double-blind, placebo-controlled clinical trial. *Neuropsychiatr. Dis. Treat., 13*, 483–490.

Doody, R. S., Gavrilova, S. I., Sano, M., Thomas, R. G., Aisen, P. S., Bachurin, S. O., Seely, L., & Hung, D., (2008). Effect of dimension on cognition, activities of daily living, behavior, and global function in patients with mild-to-moderate Alzheimer's disease: A randomized, double-blind, placebo-controlled study. *Lancet, 372*, 207–215.

Dutta, K., Patel, P., & Julien, J. P., (2018). Protective effects of *Withania somnifera* extract in SOD1G93A mouse model of amyotrophic lateral sclerosis. *Exp. Neurol., 309*, 193–204.

Farkhondeh, T., Samarghandian, S., Pourbagher-Shahri, A. M., & Sedaghat, M., (2019). The impact of curcumin and its modified formulations on Alzheimer's disease. *J. Cell. Physiol., 234*, 16953–16965.

Farrar, M. A., & Kiernan, M. C., (2014). The genetics of spinal muscular atrophy: Progress and challenges. *Neurotherapeutics, 12*, 290–302.

Frederick, J., (2006). Pick disease: A brief overview. *Arch. Pathol. Lab. Med., 130*, 1063–1066.

Fu, B., Zeng, Q., Zhang, Z., Qian, M., Chen, J., Dong, W. L., & Li, M., (2019). Epicatechin gallate protects HBMVECs from ischemia/reperfusion injury through ameliorating apoptosis and autophagy and promoting neovascularization. *Oxidative Med. Cell. Longev., 2019*, 7824684.

Fu, J., Jin, J., Cichewicz, R. H., Hageman, S. A., Ellis, T. K., Xiang, L., Peng, Q., Jiang, M., Arbez, N., Hotaling, K., et al., (2012). Trans-(-)-€_-Viniferin increases mitochondrial sirtuin 3 (SIRT3), activates AMP-activated protein kinase (AMPK), and protects cells in models of Huntington disease. *J. Biol. Chem., 287*, 24460–24472.

Ganguli, M., Chandra, V., Kamboh, M. I., Johnston, J. M., Dodge, H. H., Thelma, B. K., Juyal, R. C., et al., (2000). Apolipoprotein E polymorphism and Alzheimer disease. *Arch. Neurol., 57*, 824–830.

Ganjali, S., Blesso, C. N., Banach, M., Pirro, M., Majeed, M., & Sahebkar, A., (2017). Effects of curcumin on HDL functionality. *Pharmacol Res., 119*, 208–218. doi: 10.1016/j.phrs.2017.02.008.

Gibb, W. R., & Lees, A. J., (1988). The relevance of the Lewy body to the pathogenesis of idiopathic Parkinson's disease. *J. Neurol. Neurosurg. Psychiatry, 51*, 745–752.

Grosskreutz, J., Bosch, L. V. D., & Keller, B. U., (2010). Calcium dysregulation in amyotrophic lateral sclerosis. *Cell Calcium, 47*, 165–174.

Groth, C. L., & Berman, B. D., (2018). Spinocerebellar ataxia 27: A review and characterization of an evolving phenotype. *Tremor Other Hyperkinetic Mov., 8*, 534.

Hardiman, O., Al-Chalabi, A., Chio, A., Corr, E. M., Logroscino, G., Robberecht, W., Shaw, P. J., Simmons, Z., & Van, D. B. L. H., (2017). Erratum: Amyotrophic lateral sclerosis. *Nat. Rev. Dis. Primers, 3*, 1–19.

He, G. L., Luo, Z., Yang, J., Shen, T. T., Chen, Y., & Yang, X. S., (2016). Curcumin ameliorates the reduction effect of PGE(2) on fibrillar beta-amyloid peptide (1-42)-induced microglial phagocytosis through the inhibition ofEP2-PKA signaling in N9 microglial cells. *PLoS One, 11*, e0147721.

Hicks, A., & Jolkkonen, J., (2009). Challenges and possibilities of intravascular cell therapy in stroke. *Acta Neurobiol. Exp., 69*, 1–11.

Hirsch, G. E., Viecili, P. R., De Almeida, A. S., Nascimento, S., Porto, F. G., Otero, J., Schmidt, A., Da Silva, B., Parisi, M. M., & Klafke, J. Z., (2017). Natural products with antiplatelet action. *Curr. Pharm. Des., 23*, 1228–1246.

Hoang, Q. Q., (2014). Pathway for Parkinson's disease. *Proc. Natl. Acad. Sci. USA, 111*, 2402–2403.
Hoppe, J. B., Coradini, K., Frozza, R. L., Oliveira, C. M., Meneghetti, A. B., Bernardi, A., Pires, E. S., et al., (2013). Free and nanoencapsulated curcumin suppress beta-amyloid-induced cognitive impairments in rats: Involvement of BDNF and Akt/GSK-3 beta signaling pathway. *Neurobiol. Learn. Mem., 106*, 134–144.
Hosamani, R., Krishna, G., & Muralidhara, (2014). Standardized *Bacopa monnieri* extract ameliorates acute paraquat-induced oxidative stress, and neurotoxicity in prepubertal mice brain. *Nutr. Neurosci.* doi: http://dx.doi.org/ 10.1179/1476830514Y.0000000149.
HP-200 in Parkinson's Disease Study Group, (1995). An alternative medicine treatment for Parkinson's disease: Results of a multicenter clinical trial. HP-200 in Parkinson's disease study group. *J. Altern. Complement Med., 1*(3), 249–255.
Hsu, Y. Y., Jong, Y. J., Tsai, H. H., Tseng, Y. T., An, L. M., & Lo, Y. C., (2012). Triptolide increases transcript and protein levels of survival motor neurons in human SMA fibroblasts and improves survival in SMA-like mice. *Br. J. Pharmacol., 166*, 1114–1126.
Huang, T. C., Lu, K. T., Wo, Y. Y., Wu, Y. J., & Yang, Y. L., (2011). Resveratrol protects rats from Aβ-induced neurotoxicity by the reduction of iNOS expression and lipid peroxidation. *PLoS One, 6*, e29102.
im, G. P., Chu, T., Yang, F., Beech, W., Frautschy, S. A., & Cole, G. M., (2001). The curry spice curcumin reduces oxidative damage and amyloid pathology in an Alzheimer transgenic mouse. *J. Neurosci., 21*, 8370–8377.
Jadiya, P., Khan, A., Sammi, S. R., et al., (2011). Anti-Parkinsonian effects of *Bacopa monnieri:* Insights from transgenic and pharmacological *Caenorhabditis elegans* models of Parkinson's disease. *Biochem. Biophys. Res. Commun., 413*(4), 605–610.
Jansen, R. L. M., Brogan, B., Whitworth, A. J., & Okello, E. J., (2014). Effects of five ayurvedic herbs on locomotor behavior in a *Drosophila melanogaster* Parkinson's disease model. *Phytotherapy Research, 28*(12), 1789–1795.
Jeong, J. H., Lee, S. E., Lee, J. H., Kim, H. D., Seo, K. H., Kim, D. H., & Han, S. Y., (2020). *Aster ageratoides* Turcz. Extract attenuates Alzheimer's disease-associated cognitive deficits and vascular dementia-associated neuronal death. *Anat. Cell Boil., 53*, 216–227.
Jeong, J. K., Moon, M. H., Bae, B. C., Lee, Y. J., Seol, J. W., Kang, H. S., Kim, J. S., et al., (2012). Autophagy induced by resveratrol prevents human prion protein-mediated neurotoxicity. *Neurosci. Res., 73*, 99–105.
Jeyaweera, D. M. A., (1981). *Medicinal Plants Used in Ceylon*. Colombo, Sri Lanka: National Science Council of Sri Lanka.
Jiang, F., DeSilva, S., & Turnbull, J., (2000). Beneficial effect of ginseng root in SOD-1 (G93A) transgenic mice. *J. Neurol. Sci., 180*(1, 2), 52–54. doi: 10.1016/s0022-510x(00)00421-4. PMID: 11090864.
Jin, W. N., Shi, S. X. Y., Li, Z., Li, M., Wood, K., Gonzales, R. J., & Liu, Q., (2017). Depletion of microglia exacerbates post ischemic inflammation and brain injury. *Br. J. Pharmacol., 37*, 2224–2236.
Kelley, R. E., & El-Khoury, R., (2016). Frontotemporal dementia. *Neurol. Clin., 34*, 171–181.
Kertesz, A., & Muñoz, D., (1998). Pick's disease, frontotemporal dementia, and pick complex. *Arch. Neurol., 55*, 302–304.
Khan, A., Jahan, S., Imtiyaz, Z., Alshahrani, S., Antar, M. H., Mohammed, A. B., Kumar, A., et al., (2020). Neuroprotection: Targeting multiple pathways by naturally occurring

phytochemicals. *Biomedicines, 8*(8), 284. doi: 10.3390/biomedicines8080284. PMID: 32806490; PMCID: PMC7459826.

Kim, H., Park, B. S., Lee, K. G., Choi, C. Y., Jang, S. S., Kim, Y. H., & Lee, S. E., (2005). Effects of naturally occurring compounds on fibril formation and oxidative stress of beta-amyloid. *J. Agric. Food Chem., 53*, 8537–8541.

Kirino, T., (1982). Delayed neuronal death in the gerbil hippocampus following ischemia. *Brain Res., 239*, 57–69.

Kishore, K., & Singh, M., (2005). Effect of bacosides, alcoholic extract of *Bacopa monniera* Linn. (Brahmi), on experimental amnesia in mice. *Indian J. Exp. Biol., 43*(7), 640–645.

Klockgether, T., Mariotti, C., & Paulson, H. L., (2019). Spinocerebellar ataxia. *Nat. Rev. Dis. Primers, 5*, 24.

Kolb, S. J., & Kissel, J. T., (2011). Spinal muscular atrophy. *Arch. Neurol., 68*, 979.

Kongkeaw, C., Dilokthornsakul, P., Thanarangsarit, P., Limpeanchob, N., & Norman, S. C., (2014). Meta-analysis of randomized controlled trials on cognitive effects of *Bacopa monnieri* extract. *J. Ethnopharmacol., 151*(1), 528–535. doi: 10.1016/j.jep.2013.11.008.

Koukoulitsa, C., Villalonga-Barber, C., Csonka, R., Alexi, X., Leonis, G., Dellis, D., Hamelink, E., Belda, O., Steele, B. R., Micha-Screttas, M., et al., (2016). Biological and computational evaluation of resveratrol inhibitors against Alzheimer's disease. *J. Enzyme Inhib. Med. Chem., 31*, 67–77.

Kuboyama, T., Tohda, C., & Komatsu, K., (2005). Neuritic regeneration and synaptic reconstruction induced by with anolide A. *Br J Pharmacol., 144*, 961–971.

Kumar, P., & Kumar, A., (2009). Possible neuroprotective effect of *Withania somnifera* root extract against 3-nitropropionic acid-induced behavioral, biochemical, and mitochondrial dysfunction in an animal model of Huntington's disease. *J. Med. Food, 12*, 591–600.

Kumar, V., (2006). Potential medicinal plants for CNS disorders: Anoverview. *Phytother. Res., 20*, 1023–1035.

Kwon, K. J., Kim, J. N., Kim, M. K., Lee, J., Ignarro, L. J., Kim, H. J., Shin, C. Y., & Han, S. H., (2011). Melatonin synergistically increases resveratrol induced heme oxygenase-1 expression through the inhibition of ubiquitin-dependent proteasome pathway: A possible role in neuroprotection. *J. Pineal. Res., 50*, 110–123.

Lee, M. K., Kim, S. R., Sung, S. H., et al., (2000). Asiatic acid derivatives protect cultured cortical neurons from glutamate-induced excitotoxicity. *Res. Commun. Mol. Pathol. Pharmacol., 108*, 75–86.

Leverenz, J. B., Quinn, J. F., Zabetian, C., Zhang, J., Montine, K. S., & Montine, T. J., (2009). Cognitive impairment and dementia in patients with Parkinson's disease. *Curr. Top. Med. Chem., 9*, 903–912.

Li, W. Z., Wu, W. Y., Huang, H., Yin, Y. Y., & Wu, Y. Y., (2013). Protective effect of bilobalide on learning and memory impairment in rats with vascular dementia. *Mol. Med. Rep., 8*, 935–941.

Li, Y. H., Liu, S. B., Zhang, H. Y., Zhou, F. H., Liu, Y. X., Lu, Q., & Yang, L., (2017). Antioxidant effects of celastrol against hydrogen peroxide-induced oxidative stress in the cell model of amyotrophic lateral sclerosis. *Sheng Li XueBao Acta Physiol. Sin., 69*, 751–758.

Lieu, C. A., Kunselman, A. R., Manyam, B. V., et al., (2010). A water extract of *Mucuna pruriens* provides long-term amelioration of parkinsonism with reduced risk for dyskinesias. *Parkinsonism Relat. Disord., 16*(7), 458–465.

Lieu, C. A., Venkiteswaran, K., Gilmour, T. P., et al., (2012). The antiparkinsonian and anti-dyskinetic mechanisms of Mucuna pruriens in the MPTP-treated nonhuman primate. *Evid. Based Complement Alternat. Med.*, 10. Article ID 840247.

Limpeanchob, N., Jaipan, S., Rattanakaruna, S., et al., (2008). Neuroprotective effect of *Bacopa monnieri* on beta-amyloid-induced cell death in primary cortical culture. *J. Ethnopharmacol., 120*(1), 112–117.

Ma, J., & Ma, J., (2014). Prion disease and the 'protein-only hypothesis. *Essays Biochem., 56*, 181–191.

Ma, Y., & Ma, J., (2020). Immunotherapy against prion disease. *Pathogens, 9*, 216.

Macdonald, I. R., Rockwood, K., Martin, E., & Darvesh, S., (2014). Cholinesterase inhibition in Alzheimer's disease: Is specificity the answer? *J. Alzheimer's Dis., 42*, 379–384.

Manoharan, S., Guillemin, G. J., Abiramasundari, R. S., Essa, M. M., & Akbar, M., (2016). The role of reactive oxygen species in the pathogenesis of Alzheimer's disease, Parkinson's disease, and Huntington's disease: A mini review. *Oxidative Med. Cell. Longev., 2016*, 1–15.

Manyam, B. V., Dhanasekaran, M., & Hare, T. A., (2004a). Neuroprotective effects of the anti-Parkinson drug *Mucuna pruriens*. *Phytother. Res., 18*(9), 706–712.

Manyam, B. V., Dhanasekaran, M., & Hare, T. A., (2004b). Effect of anti-Parkinson drug HP-200 (*Mucuna pruriens*) on the central monoaminergic neurotransmitters. *Phytother. Res., 18*(2), 97–101.

Marín-Moreno, A., Fernández-Borges, N., Espinosa, J. C., Andréoletti, O., & Torres, J. M., (2017). Transmission and replication of prions. *Prog. Mol. Biol. Transl. Sci., 150*, 181–201.

McGeer, P. L., & McGeer, E. G., (2002). Inflammatory processes in amyotrophic lateral sclerosis. *Muscle Nerve, 26*, 459–470.

Miller, B. L., (2014). Frontotemporal dementia. *Front. Dement., 35*, 339–374.

Mohandas, E., & Rajmohan, V., (2009). Frontotemporal dementia: An updated overview. *Indian J. Psychiatry, 51*, S65–S69.

Mook-Jung, I., Shin, J. E., Yun, S. H., et al., (1999). Protective effects of asiaticoside derivatives against beta amyloid neurotoxicity. *J. Neurosci. Res., 58*, 417–425.

Moon, J. H., Lee, J. H., Lee, Y. J., & Park, S. Y., (2016). Hinokitiol protects primary neuron cells against prion peptide-induced toxicity via autophagy flux regulated by hypoxia inducing factor-1. *Oncotarget., 7*, 29944–29957.

Naghizadeh, B., Mansouri, M. T., Ghorbanzadeh, B., Farbood, Y., & Sarkaki, A., (2013). Protective effects of oral crocin against intracerebroventricular streptozotocin-induced spatial memory deficit and oxidative stress in rats. *Phytomedicine, 20*(6), 537–542. doi: 10.1016/j.phymed.2012.12.019.

Nam, S. M., Choi, J. H., Yoo, D. Y., Kim, W., Jung, H. Y., Kim, J. W., Yoo, M., Lee, S., Kim, C. J., Yoon, Y. S., et al., (2014). Effects of curcumin (*Curcuma longa*) on learning and spatial memory as well as cell proliferation and neuroblast differentiation in adult and aged mice by upregulating brain-derived neurotrophic factor and creb signaling. *J. Med. Food, 17*, 641–649.

Nekohashi, M., Ogawa, M., Ogihara, T., Nakazawa, K., Kato, H., Misaka, T., Abe, K., & Kobayashi, S., (2014). Luteolin and quercetin affect the cholesterol absorption mediated by epithelial cholesterol transporter Niemann–pick C1-like 1 in Caco-2 cells and rats. *PLoS One, 9*, e97901.

Niebroj-Dobosz, I., & Janik, P., (1999). Amino acids acting as transmitters in amyotrophic lateral sclerosis (ALS). *Acta Neurol. Scand., 100*, 6–11.

O'Brien, J., Erkinjuntti, T., Reisberg, B., Román, G., Sawada, T., Pantoni, L., Bowler, J. V., Ballard, C., DeCarli, C., Gorelick, P. B., et al., (2003). Vascular cognitive impairment. *Lancet Neurol., 2*, 89–98.

Priyanka, H. P., Bala, P., Ankisettipalle, S., et al., (2013). *Bacopa monnieri* and L-deprenyl differentially enhance the activities of antioxidant enzymes and the expression of tyrosine hydroxylase and nerve growth factor via ERK 1/2 and NF-κB pathways in the spleen of female Wistar rats. *Neurochem. Res., 38*, 141–152.

Rane, B. R., Bharath, M. S., Patil, R. R., Keservani, R. K., & Jain, A. S., (2020b). Novel approaches in nutraceuticals; Chapter 11. In: Rajesh, K. K., Raj, K. K., & Anil, K. S., (eds.), *Enhancing the Therapeutic Efficacy of Herbal Formulations Through Novel Drug Delivery Systems* (pp. 241–266). IGI Global International Publisher, Pennsylvania, USA. ISBN13: 9781799844532.

Rane, B. R., Patil, A. S., Keservani, R. K., & Jain, A. S., (2020a). Novel approaches in herbal formulation; Chapter 2. In: Rajesh, K. K., Raj, K. K., & Anil, K. S., (eds.), *Enhancing the Therapeutic Efficacy of Herbal Formulations Through Novel Drug Delivery Systems* (pp. 43–68). IGI Global International Publisher, Pennsylvania, USA. ISBN13: 9781799844532.

Rane, B. R., Tadavi, S. A., & Keservani, R. K., (2020c). Naturopathy; Chapter 12. In: Anil, K. S., Raj, K. K., & Surya, P. G., (eds.), *Herbal Product Development* (pp. 321–347). Apple Academic Press, CRC Press, Taylor & Francis Group. ISBN: 9781771888776.

Rao, M. K. G., Rao, M. S., & Rao, G. S., (2006). *Centella asiatica* (L.) leaf extract treatment during the growth spurt period enhances hippocampal CA3 neuronal dendritic arborization in rats. *Evid. Based Complement Alternat Med., 3*, 349–357.

Remington, R., Bechtel, C., Larsen, D., Samar, A., Doshanjh, L., Fishman, P., Luo, Y., Smyers, K., Page, R., Morrell, C., et al., (2015). APhase II randomized clinical trial of a nutritional formulation for cognition and mood in Alzheimer's disease. *J. Alzheimer's Dis., 45*, 395–405.

Rogers, J., & Lue, L. F., (2001). Microglial chemotaxis, activation, and phagocytosis of amyloid beta-peptide as linked phenomena in Alzheimer's disease. *Neurochem. Int., 39*, 333–340.

Sadhu, A., Upadhyay, P., Agrawal, A., et al., (2014). Management of cognitive determinants in senile dementia of Alzheimer's type: Therapeutic potential of a novel polyherbal drug product. *Clin. Drug Investig., 34*(12), 857–869.

Sancheti, H., Akopian, G., Yin, F., Brinton, R. D., Walsh, J. P., & Cadenas, E., (2013). Age-dependent modulation of synaptic plasticity and insulin mimetic effect of lipoic acid on a mouse model of Alzheimer's disease. *PLoS One, 8*, e69830.

Sathiyanarayanan, L., & Arulmozhi, S., (2007). *Mucuna pruriens*. A comprehensive review. *Pharmacognosy Rev., 1*, 157–162.

Sawda, C., Moussa, C., & Turner, R. S., (2017). Resveratrol for Alzheimer's disease. *Ann. NY Acad. Sci., 1403*, 142–149.

Schmidt, M., Betti, G., & Hensel, A., (2007). Saffron in phytotherapy: Pharmacology and clinical uses. *Wien Med. Wochenschr., 157*(13, 14), 315–319.

Schöls, L., Bauer, P., Schmidt, T., Schulte, T., & Riess, O., (2004). Autosomal dominant cerebellar ataxias: Clinical features, genetics, and pathogenesis. *Lancet Neurol., 3*, 291–304.

Semmler, S., Gagné, M., Garg, P., Pickles, S. R., Baudouin, C., Hamon-Keromen, E., Destroismaisons, L., Khalfallah, Y., Chaineau, M., Caron, E., et al., (2020). TNF receptor-associated factor 6 interacts with ALS-linked misfolded superoxide dismutase 1 and promotes aggregation. *J. Boil. Chem., 295*, 3808–3825.

Sharma, A. K., Keservani, R. K., & Gautam, S. P., (2020). *Herbal Product Development*. Apple Academic Press, CRC Press, Taylor & Francis Group. ISBN: 9781771888776.

Shinomol, G. K., & Muralidhara, (2011). *Bacopa monnieri* modulated endogenous cytoplasmic and mitochondrial oxidative markers in prepubertal mice brain. *Phytomedicine, 18*, 317–326.

Siddique, Y. H., Naz, F., Jyoti, S., Ali, F., & Rahul, (2019). Effect of genistein on the transgenic *Drosophila* model of Parkinson's disease. *J. Diet. Suppl., 16*(5), 550–563. doi: 10.1080/19390211.2018.1472706. Epub 2018 Jul 3. PMID: 29969325.

Simpson, E. P., Yen, A. A., & Appel, S. H., (2003). Oxidative stress: A common denominator in the pathogenesis of amyotrophic lateral sclerosis. *Curr. Opin. Rheumatol., 15*, 730–736.

Singh, S., Gautam, A., Sharma, A., et al., (2010). *Centella asiatica* (L.): A plant with immense medicinal potential but threatened. *Int. J. Pharm. Sci., 4*(2), Article 003. ISSN 0976-044X.

Suliman, N. A., Mat, T. C. N., Mohd, M. M. A., Adenan, M. I., Hidayat, B. M. T., & Basir, R., (2016). Establishing natural nootropics: Recent molecular enhancement influenced by natural nootropic. *Evid. Based Complement Alternat Med., 2016*, 4391375. doi: 10.1155/2016/4391375.

Swash, M., & Schwartz, M. S., (1992). What do we really know about amyotrophic lateral sclerosis? *J. Neurol. Sci., 113*, 4–16.

Szlachcic, W. J., Switonski, P. M., Krzyzosiak, W. J., Figlerowicz, M., & Figiel, M., (2015). Huntington disease iPSCs show early molecular changes in intracellular signaling, the expression of oxidative stress proteins and the p53pathway. *Dis. Model. Mech., 8*, 1047–1057.

Tang, F., Guo, S., Liao, H., Yu, P., Wang, L., Song, X., Chen, J., & Yang, Q., (2017). Resveratrol enhances neurite outgrowth and synaptogenesis via sonic hedgehog signaling following oxygen-glucose deprivation/reoxygenation injury. *Cell. Physiol. Biochem., 43*, 852–869.

Tohda, C., (2008). Overcoming several neurodegenerative diseases by traditional medicines—The development of therapeutic medicine and unraveling pathophysiological mechanisms. *Yakugaku Zasshi., 128*(8), 1159–1167.

Túnez, I., Montilla, P., Muñoz, M. D. C., Feijóo, M., & Salcedo, M., (2004). Protective effect of melatonin on 3 nitropropionic acid-induced oxidative stress in synaptosomes in an animal model of Huntington's disease. *J. Pineal Res., 37*, 252–256.

Uabundit, N., Wattanathorn, J., Mucimapura, S., et al., (2010). Cognitive enhancement and neuroprotective effects of *Bacopa monnieri* in Alzheimer's disease model. *J. Ethnopharmacol., 127*, 26–31.

Ullah, F., Liang, A., Rangel, A., Gyengesi, E., Niedermayer, G., & Münch, G., (2017). High bioavailability curcumin: An anti-inflammatory and neuro supportive bioactive nutrient for neurodegenerative diseases characterized by chronic neuroinflammation. *Arch. Toxicol., 91*, 1623–1634.doi: 10.1007/s00204-017-1939-4.

Vaibhav, K., Shrivastava, P., Javed, H., Khan, A., Ahmed, E., Tabassum, R., Khan, M. M., Khuwaja, G., Islam, F., Siddiqui, M. S., et al., (2012). Piperine suppresses cerebral ischemia–reperfusion-induced inflammation through the repression of COX-2, NOS-2, and NF-_B in middle cerebral artery occlusion rat model. *Mol. Cell. Biochem., 367*, 73–84.

Vaibhav, K., Shrivastava, P., Tabassum, R., Khan, A., Javed, H., Ahmed, E., Islam, F., Safhi, M. M., & Islam, F., (2013). Delayed administration of zingerone mitigates the behavioral and histological alteration via repression of oxidative stress and intrinsic programmed cell death in focal transient ischemic rats. *Pharmacol. Biochem. Behav., 113*, 53–62.

Vellas, B., Coley, N., Ousset, P. J., Berrut, G., Dartigues, J. F., Dubois, B., & Touchon, J., (2012). Grp, long-term use of standardized *Ginkgo biloba* extract for the prevention of Alzheimer's disease (GuidAge): A randomized placebo-controlled trial. *Lancet Neurol., 11*, 851–859.

Vidoni, C., Secomandi, E., Castiglioni, A., Melone, M. A. B., & Isidoro, C., (2018). Resveratrol protects neuronal-like cells expressing mutant huntingtin from dopamine toxicity by rescuing ATG4-mediated autophagosome formation. *Neurochem. Int., 117*, 174–187.

Webber, K. M., Raina, A. K., Marlatt, M. W., Zhu, X., Prat, M. I., Morelli, L., Casadesus, G., et al., (2005). The cell cycle in Alzheimer's disease: A unique target for neuropharmacology. *Mech. Ageing Dev., 126*, 1019–1025.

Xing, S. H., Zhu, C. X., Zhang, R., & An, L., (2014). Huperzine a in the treatment of Alzheimer's disease and vascular dementia: A meta-analysis. *Evid. Based Complement. Altern. Med., 2014*.

Xu, M. F., Xiong, Y. Y., Liu, J. K., et al., (2012). Asiatic acid, a pentacyclic triterpene in *Centella asiatica*, attenuates glutamate-induced cognitive deficits in mice and apoptosis in SH-SY5Y cells. *Acta. Pharmacol. Sin., 33*(5), 578–587.

Xu, M., Liu, K., Swaroop, M., Porter, F. D., Sidhu, R., Finkes, S., Ory, D. S., Marugan, J. J., Xiao, J., Southall, N. T., et al., (2012). α-tocopherol reduces lipid accumulation in Niemann-pick type C1 and Wolman cholesterol storage disorders. *J. Boil. Chem., 287*, 39349–39360.

Xu, Z. Q., Liang, X. M., Wu, J., Zhang, Y. F., Zhu, C. X., & Jiang, X. J., (2011). Treatment with huperzine a improves cognition in vascular dementia patients. *Cell Biophys., 62*, 55–58.

Yadav, S. K., Prakash, J., & Chouhan, S., (2014). Comparison of the neuroprotective potential of *Mucuna pruriens* seed extract with estrogen in 1-methyl-4-phenyl-1,2,3,6-tetrahydropyridine (MPTP)-induced PD mice model. *Neurochem. Int., 65*, 1–13.

Yadav, S. K., Prakash, J., Chouhan, S., et al., (2013). *Mucuna pruriens* seed extract reduces oxidative stress in nigrostriatal tissue and improves neurobehavioral activity in paraquat-induced Parkinsonian mouse model. *Neurochem. Int., 62*(8), 1039–1047.

Yang, F., Lim, G. P., Begum, A. N., Ubeda, O. J., Simmons, M. R., Ambegaokar, S. S., Chen, P. P., et al., (2005). Curcumin inhibits formation of amyloid beta oligomers and fibrils, binds plaques, and reduces amyloid *in vivo*. *J. Biol. Chem., 280*, 5892–5901.

Yang, J. Q., Song, S. L., Li, J., & Liang, T., (2014). Neuroprotective effect of curcumin on hippocampal injury in 6-ohda-induced Parkinson's disease rat. *Pathol. Res. Pract., 210*, 357–362.

Yang, S., Wang, H., Yang, Y., Wang, R., Wang, Y., Wu, C., & Du, G. H., (2019). Baicalein administered in the subacute phase ameliorates ischemia-reperfusion-induced brain injury by reducing neuroinflammation and neuronal damage. *Biomed. Pharm., 117*, 109102.

Yao, Y., Li, J., Niu, Y., Yu, J. Q., Yan, L., Miao, Z. H., Zhao, X. X., Li, Y. J., Yao, W. X., Zheng, P., et al., (2015). Resveratrol inhibits oligomeric Aβ-induced microglial activation via NADPH oxidase. *Mol. Med. Rep., 12*, 6133–6139.

Yu, J., Jiang, Z., Ning, L., Zhao, Z., Yang, N., Chen, L., Ma, H., Li, L., Fu, Y., Zhu, H., et al., (2015). Protective HSP70 induction by Z-ligustilide against oxygen–glucose deprivation injury via activation of the MAPK Pathway but not of HSF1. *Boil. Pharm. Bull., 38*, 1564–1572.

Yu, S., Zheng, W., Xin, N., Chi, Z. H., Wang, N. Q., Nie, Y. X., Feng, W. Y., & Wang, Z. Y., (2010). Curcumin prevents dopaminergic neuronal death through inhibition of the c-Jun N-terminal kinase pathway. *Rejuv. Res., 13,* 55–64.

Yun, Y. C., Jeong, S. G., Kim, S. H., & Cho, G. W., (2019). Reduced sirtuin 1/adenosine monophosphate-activated protein kinase in amyotrophic lateral sclerosis patient-derived mesenchymal stem cells can be restored by resveratrol. *J. Tissue Eng. Regen. Med., 13*, 110–115.

Zecca, L., Shima, T., Stroppolo, A., Goj, C., Battiston, G., Gerbasi, R., Sarna, T., & Swartz, H., (1996). Interaction of neuromelanin and iron in substantia nigra and other areas of human brain. *Neuroscience, 73*, 407–415.

Zhang, L. N., Sun, Y. J., Pan, S., et al., (2013). Na^+-K^+-ATPase, a potent neuroprotective modulator against Alzheimer's disease. *Fundam. Clin. Pharmacol., 27*, 96–103.

Zhang, N., Hu, Z., Zhang, Z., Liu, G., Wang, Y., Ren, Y., Wu, X., & Geng, F., (2017). Protective role of naringenin against Aβ25-35-caused damage via ER and PI3K/Akt-mediated pathways. *Cell. Mol. Neurobiol., 38*, 549–557.

Zhang, P., Liu, X., Zhu, Y., Chen, S., Zhou, D., & Wang, Y. Y., (2013). Honokiol inhibits the inflammatory reaction during cerebral ischemia reperfusion by suppressing NF-_B activation and cytokine production of glial cells. *Neurosci. Lett., 534*, 123–127.

Zhao, Z., Fu, J., Li, S., & Li, Z., (2019). Neuroprotective effects of genistein in a SOD1-G93A transgenic mouse model of amyotrophic lateral sclerosis. *J. Neuroimmune Pharm., 14*, 688–696.

Zheng, J. S., Zheng, P. D., Mungur, R., Zhou, H. J., Hassan, M., & Jiang, S. N., (2018). Ginkgolide B promotes the proliferation and differentiation of neural stem cells following cerebral ischemia/reperfusion injury, both *in vivo* and *in vitro*. *Neural Regen. Res., 13*, 1204–1211.

CHAPTER 11

PLANT METABOLITES AND VEGETABLES FOR GASTROINTESTINAL DISEASE PREVENTION AND TREATMENT

VIDISHA MISHRA

Department of Home Science, R.B.B.M. College, B.R.A. Bihar University, Muzaffarpur, Bihar, India

ABSTRACT

A variety of therapeutic chemicals can be found in plant metabolites, and their multidimensional chemical diversity makes them the best agents for treating severe ailments. The therapeutic benefits of herbs are based on the chemistry of plants. Understanding the chemical makeup of plants helps one better appreciate their potential medical usefulness. Plant metabolite levels are greatly influenced by genetic, environmental, transportation, and storage circumstances. The physiological components of plants include things like light, temperature, humidity, soil type, fertilizer application, damage from insects and pathogens, stress brought on by heavy metals, UV radiation, and pesticides, among other growth-promoting elements. The present chapter discusses the nutritional importance of vegetables, their key source in plants, and their therapeutic significance in gastrointestinal illnesses. Primary metabolites include carbohydrates, fatty acids, amino acids, and organic acids have a significant role in protein and hormone production, growth, respiration, and development. Flavonoids, sterols, carotenoids, phenolic acids, glucosinolates (GSs), and alkaloids are secondary metabolites that affect vegetable color,

Plant Metabolites and Vegetables as Nutraceuticals.
Raj K. Keservani, Bui Thanh Tung, Rajesh K. Kesharwani, and Eknath D. Ahire (Eds.)

defend plants against herbivores and microbes, seed dispersion animals and draw pollinators, and function as signals in stressful situations.

11.1 INTRODUCTION

There are many different digestive system illnesses that have a significant global impact on morbidity and mortality rates. According to the World Health Organization (WHO), digestive system illnesses, notably diarrhea, were the fifth largest cause of death worldwide in 2012, accounting for around 100 million deaths. Additionally, in Southeast Asia, diarrhea has been responsible for 10% of mortality in children under the age of five (WHO, 2014). In 2019, 3,70,000 children died from diarrheal illness, the second greatest reason of death in children below the age of five years (WHO, 2019). There has been a widespread tendency in recent years toward a resurgence of interest in a conventional system of medical care. The rise of health care and conservation initiatives around the world has increased the value of plant studies, which have grown to be of particular interest (Balick, 1996). The WHO has acknowledged the value of conventional medicine in the system of basic healthcare (Akerele, 1984). Medicinal plants continue to be the primary source of medication in poor nations. According to estimates, traditional medicine is the major source of medical care for about 88% of people living in poor nations (Balick & Cox, 1997). Numerous therapeutic chemicals can be found in plant metabolites, and because to their variety of multidimensional chemical structures, they are more effective in treating serious disorders. Some of them have already been identified as promising complementary therapies and starting points for the development of new drugs (Bhuiyan et al., 2020).

The therapeutic benefits of herbs are based on the chemistry of plants The pharmacology activities of medicinal plants are based on the bioactive compounds of plants (Sharma et al., 2020; Rane et al., 2020a). Understanding the bioactive compounds of plants will provide the better appreciate their using. The importance of primary plant metabolites in basic biological processes such as development and cell division, respiration, reproduction, and storage has been defined by modern chemistry. The parts of procedures like the Krebs or citric acid cycle, glycolysis, photosynthesis, and related pathways are included in them. Small molecules including amino acids, sugars, polysaccharides, protein, nucleic acid tricarboxylic acids, or Krebs cycle intermediates are examples of primary metabolites. Eventually, every living cells produce comparable basic metabolites (Seigler, 1995).

Numerous chemical substances created through metabolic pathways by the plant cell descended from the basic metabolic pathways are known as secondary plant metabolites. Albrecht Kossel, who won the Nobel Prize (1910) in physiology or medicine, was the first to develop the term "secondary metabolite" (Jones & Kossel, 1953). Czapek referred to them as end-products 30 years later (Bourgaud et al., 2001). Czapek claimed that these compounds come from metabolism of nitrogen through what he termed "secondary alterations" like deamination. The development in the field of phytochemistry was based on the increasing recovery of these chemical compounds by the mid of the 20th century thanks to advancements in analytical techniques like chromatography. Many ancient civilizations have used herbs in traditional medicine (Rane et al., 2020b, c), and secondary metabolites have been proven to have a variety of biological effects. They can defend plants from diseases because they have been described as antibiotics, antifungals, and antivirals. The chemical structures of secondary plant metabolites are divided into various kinds.

The nutritional makeup of plants and vegetables is extremely intricate and challenging to analyze. Plant metabolite levels are greatly influenced by genetic, environmental, transportation, and storage circumstances. Light, temperature, humidity, soil type, fertilizer application, insect and microbial damage, stress brought by heavy metals, UV radiation, pesticides, and other growth conditions can affect the polite composition of the plants (Orcutt & Nilsen, 2000). Greens have been handled by farmers, packagers, transporters, storage operators, processors, and distributors before they end up on a supermarket shelf. Vegetables undergo chemical and physical changes during these stages that may cause the loss of certainly healthy constituents (MacEvilly & Peltola, 2003).

The recognition, biochemical characterization, localization, and health effects of plant metabolites have attracted a lot of research during the past 10 years. The particulars regarding metabolites, which affect the nutritional content of greens and their prevalence in plants, and their therapeutic character in gastrointestinal illnesses, is attempted to be condensed in this chapter.

11.2 PRIMARY PLANT METABOLITES

Vegetables include a wide range of organic substances that are the end result of primary and secondary plant metabolism. Primary metabolic processes like respiration, photosynthesis, hormone, and protein production, as well as growth and development, depend on substances like carbohydrates, fatty acids, amino acids, and organic acids. All species across wide evolutionary groups manufacture primary metabolites through the same (or very similar) metabolic pathways. Flavonoids (Keservani & Sharma, 2014; Keservani et al., 2016a, b), sterols, carotenoids, phenolic acids, GSs, and alkaloids are secondary metabolites that affect vegetable color, defend plants against herbivores and microbes, seed dispersion animals and draw pollinators and function as signal in stressful situation (Seiger, 1998) (Figure 11.1).

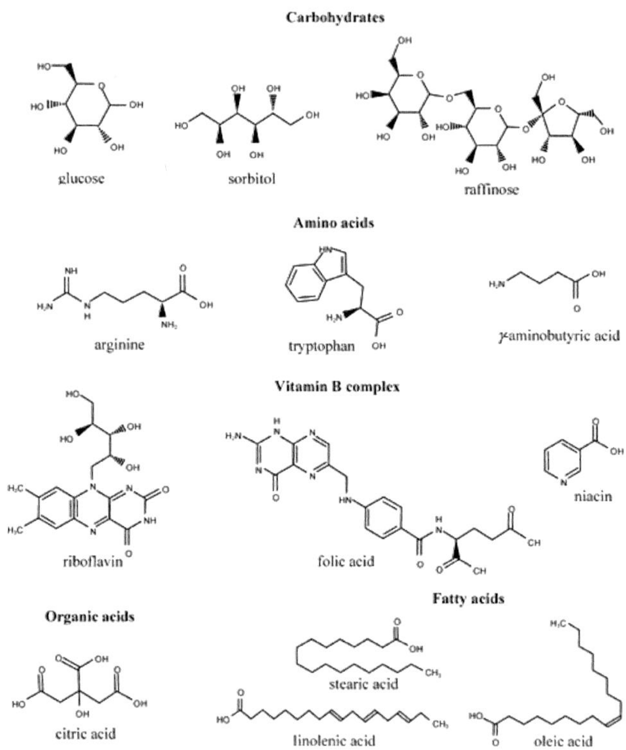

FIGURE 11.1 The chemical composition of indicative primary metabolites.

11.2.1 CARBOHYDRATES

Vegetables contain sugar monosaccharides such as fructose, glucose, galactose, rhamnose, and arabinose as well as disaccharides such as maltose, sucrose, trehalose, and sugar alcohols like mannitol, sorbitol, and xylitol, oligosaccharides such as raffinose and stachyose, and polysaccharides like (cellulose, starch, pectins, hemicellulose). All plants contain monosaccharides, disaccharides, and polysaccharides. According to Kuo et al. (1988); Obendorf et al. (1998); Frias et al. (1999); and Peterbauer et al. (2001), stachyose and raffinose are present in peas, beans, and lentils. Asparagus, beans, peas, leeks, chicory, and onions accumulate fructans (fructo-oligosaccharides) (Shiomi, 1992; Shiomi et al., 2005). Celery, carrots, and parsley contain mannitol, while corn, maize, and strawberries contain xylitol (Mäkinen & Söderling, 1980; Lewis, 1984; Stoop et al., 1996). Carbohydrates are frequently categorized as available and unavailable based on their biological or nutritive purpose. Available carbohydrates are those that the human gastrointestinal system's enzymes hydrolyze into monosaccharides like sucrose and digestible starch. Monosaccharides can enter the bloodstream without needing to be digested first. Endogenous human enzymes do not hydrolyze unavailable carbohydrates, such as sugar alcohols, many oligosaccharides, and non-starch polysaccharides. They may undergo variable degrees of fermentation by bacteria in the large intestine before being absorbed (Asp, 1996). Dietary fiber (Keservani et al., 2010, 2017, 2010b, 2020) contains essential substances including fructo-oligosaccharides and non-starch polysaccharides. The body's most readily available energy source is carbohydrate, and a diet with an adequate amount of them prevents protein from being used for energy. Since the body can produce all monosaccharides, humans don't have any particular sugar needs. Intestinal microbiota activity, food fermentation, and blood glucose and insulin metabolism are all regulated by sugars. Cell signaling involves monosaccharides attached to lipid and protein molecules (glycolipids and glycoproteins). Advanced glycation end products, which are created by the non-enzymatic binding of proteins to sugars (also known as glycation), have been linked to a number of age-associated chronic diseases, including cardiovascular disease (CVD), type 2 diabetes, Alzheimer's disease (AD), peripheral neuropathy, and cancer (Kudláková et al., 2002).

Non-starch polysaccharides, lignin, resistant oligosaccharides, and related compounds such as resistant starch, cutin, waxes, and suberin are few examples of plant materials referred to as dietary fiber (De Vries, 2003). All of the substances move by the digestive system as mass fiber, where they are altered and digested by colon microbes (Blaut, 2002). The body may absorb substances that gut bacteria create. Some items, including fatty acids, vitamin K and biotin can be advantageous. In addition to hydrogen in gaseous form produced by colon fermentation, other compounds such alcohols, lactic acid, and formate may also be undesired (Flamm et al., 2001; Tungl & Meyer, 2002; McGarr et al., 2005). Consuming foods rich in dietary fiber has been shown to lessen the symptoms of diverticular disease, persistent constipation, and few kinds of colitis (Stollman & Raskin, 2004). Low-fiber diets have been linked to a greater risk of obesity, CVDs, and colon cancer (Marlett, 2001; McGarr et al., 2005; Slavin, 2005). Since high-fiber veggies don't always have palatable tastes, increasing fiber consumption has proven to be a challenging task (Tungland & Meyer, 2002). Hence, in spite of its beneficial effects on well-being, dietary fiber consumption is still inadequate in many populations around the world.

11.2.2 AMINO ACID

In both plant and animal metabolism, amino acids function as intermediaries and combine to make proteins. The human body is made up of proteins, which also serve as hormones, antibodies, and enzymes. The primary source of amino acids is food, specifically proteins. Enzymes convert the majority of proteins into amino acids, which are then take up by the small intestine. Numerous amino acids, such as aspartic acid, alanine, asparagine, cystine, cysteine, glutamine, glutamic acid, proline, glycine, tyrosine, and serine can be produced by humans. The nine essential amino acids— histidine, arginine, leucine, isoleucine, methionine, lysine, valine, tryptophan, and phenylalanine must be obtained from food. Because the rate of synthesis is insufficient to meet the body's needs, the amino acids arginine, phenylalanine, and methionine are regarded as necessary for reasons other than the absence of a metabolic pathway (Spallholz et al., 1999). Histidine is regarded as an essential amino acid in youngsters. All essential amino acids are present in vegetables, though some may be present in lesser amounts than those needed by humans (Young & Pellett, 1994). Brussels sprouts, asparagus, potatoes, and watercress consist of high levels of arginine; broccoli, brussels sprouts, cauliflower, and tomatoes consist of high levels of histidine; parsley, beets, tomatoes, carrots, and spinach consist of high levels of phenylalanine; and cauliflower, cabbage, kale, radish, and watercress consist of high levels of methionine (FAO, 1970). There are about 250 non-protein amino acids that have been found in plants, including carnitine, homoarginine, citrulline, amino-butyric acid and taurine. Beans, spinach, potatoes, and kale contain aminobutyric acid (GABA), an inhibitory neurotransmitter found in the retina of the human central nervous system (CNS) (Oh et al., 2003). Carnitine plays an important role metabolism of lipid in the skeletal muscle and heart and can be found in potatoes, peas, and zucchini (Demarquoy et al., 2004). In human cells, taurine, which is found in foods including peanuts, lentils, and maize, is involved in membrane stabilization and detoxification (Stapleton et al., 1997). A few non-protein amino acids, such as – canavanine and cyano-alanine found in beans, are thought to be hazardous to people because they impair immunological function and protein synthesis (Bell, 2003). Free amino acids play a part in human metabolism and the flavor of plants. Glutamate and aspartic acid have a sour taste, while leucine and valine are bitter and glycine and alanine are sweet.

11.2.3 VITAMIN B COMPLEX

The water-soluble vitamins thiamine (B1), riboflavin (B2), nicotinic acid (B3), niacin, pantothenic acid (B5), pyridoxine (B6), biotin (B7), and folic acid (B9) are all part of the vitamin B complex found in plants (Bender, 2003). These substances are essential cofactors for enzymes in plants that are involved in respiration (thiamine, biotin, riboflavin), the synthesis of organic and amino acids (folic acid, thiamine), the control of cell division and blooming (niacin), and photosynthesis (riboflavin) (Heldt, 2005). Vitamin B complex play a role in the carbohydrate metabolism, amino acids, and fatty acids in humans. Thiamine, nicotinic acid, riboflavin, glossitis, seborrhea, peripheral neuropathies, dermatitis, nausea, anemia, and angular stomatitis are all symptoms of vitamin B insufficiency. Other symptoms include cheilosis, polyneuritis (thiamine), dermatitis, and angular stomatitis (Combs, 1998). B vitamins are abundant in leafy vegetables like Brussels sprouts, asparagus, spinach, turnips, cauliflower, and lettuce (USDA, 2005).

11.2.4 ORGANIC ACID

A class of organic molecules known as organic acids include carboxylic groups. Protons are released by organic acids in solution, and these protons give them their "acidic taste." Plants contain a variety of organic acids, including acetic, citric, oxalic, malic, fumaric, succinic, tartaric, quinic, aconitic, shikimic, malonic, and ascorbic (Heldt, 2005). Malic and citric acids are the most common. There is widespread use of succinic, fumaric, and quinic acids. Lettuce, carrots, chicory, endives, and celery all contained tartaric acid (Ruhl & Herrmann, 1985). Organic acids are typically found in small amounts in green vegetables. However, they are crucial as flavor enhancers and organic antibacterial agents. Organic acids give veggies their acidity and change flavor by changing how sweet things taste (Fisher & Scott, 1997). The sugar/acid ratio is frequently used to technologically characterize the maturity of vegetables (Bartz & Brecht, 002). For instance, the sugar/acid ratio in tomatoes is typically regarded to be around 7.5, while ranges as wide as 3.3 to 21.7 have been documented (Kmiecik & Lisiewska, 2000). Vegetables' colors are affected by organic acids as several plant pigments serve as pH indicators naturally (Davies, 2004). For instance, some anthocyanins, which are present in red lettuce and cabbage, turn from red to blue as the pH rises. The organic acid ascorbic acid, also referred to as vitamin C, has potent antioxidant effects. Green onions, cabbage, spinach, cress, bell pepper, broccoli, beans, cauliflower, and peas are some vegetables high in ascorbic acid (USDA, 2005). The formation of steroid hormones, neurotransmitters, and collagen, as well as the transformation of cholesterol into bile acids and the absorption of calcium and iron, all depend on vitamin C. It promotes the recovery of burns and wounds prevents bruising and blood clotting and fortifies the capillary walls (Combs, 1998). Due to its potent biological antioxidant properties, vitamin C has also been associated with the protection of degenerative illnesses such cataracts, cardiovascular conditions, and certain malignancies (Carr & Frei, 1999). Development conditions and the use of nitrogen fertilizers have an impact on the ascorbic acid concentration of vegetables (Mozafar, 1993).

11.2.5 FATTY ACID

A hydrophilic carboxylate group is joined to a substantial hydrocarbon chain to form fatty acids. Prime parts of fats are fatty acids. They give the body its energy as well as the building blocks for organ cushioning and cell membranes. According to Shahidi & Miraliakbari (2005), fatty acids play a part in the blood coagulation, immune system, and the absorption of specific vitamins (Nettleton, 1995). Some of them are the chemical building blocks for various hormones (Yehuda et al., 1999). The amount of double bonds is used to categorize fatty acids. Double bonds are absent from saturated fatty acids like palmitic, capric, stearic, and myristic. Oleic and palmitoleic are examples of monounsaturated acids, while docosahexaenoic acid, linolenic acid, eicosapentaenoic acid, and arachidonic acid are examples of polyunsaturated acids (Gurr et al., 2002). Alpha-linolenic acid and linoleic acid and are two necessary fatty acids that necessarily obtained through diet because they cannot be produced by the body (Innis, 1991). Soybeans and the majority of vegetable oils, such as corn, sunflower, and safflower oil, all contain alpha-linolenic acid (Connor, 1999). Pumpkin seeds, soybeans, and wheat germ all contain linolenic acid (Jakab et al., 2002). It is well known that green vegetables including parsley, watercress, Chinese cabbage, and Brussels sprouts have a disproportionately high concentration of polyunsaturated fatty acids, especially in the form of alpha-linolenic acid (Pereira et al., 2001). According to studies (Simopoulos, 1999; Connor, 2000), consuming monounsaturated fatty acids lowers cholesterol levels and has positive effects on several of the established threat factors for type 2 diabetes and CVDs.

The following are details on the key metabolites and their primary vegetable sources:

Bioactive Compounds	**Plants**
Glucose	Onion
	Cabbage
	Tomato
	Pumpkin
Fructose	Broccoli
	Asparagus
	Cucumber
	Cabbage
	Brussels sprouts
Sucrose	Cabbage
	Beetroot
	Carrot
	Broccoli
	Green corn
Raffinose	Cabbage
	Cauliflower
	Broccoli
Stachyose	Broccoli
	Beetroot
	Leeks
	Onion
	Cabbage
Mannitol	Artichoke
	Endive
	Asparagus
	Pumpkin
	Onion
Xylitol	Spinach
	Cauliflower

Bioactive Compounds	Plants
	Carrot
	Lettuce
Sorbitol	Beetroot
	Cabbage
Fiber	Lettuce
	Carrot
	Tomato
	Potato
	Non-starch polysaccharides
	Carrot
	Cabbage
	Tomato
	Lettuce
	Potato
Lignin	Carrot
	Cabbage
	Lettuce
	Tomato
	Potato
	Amino acids
Arginine	Cabbage
	Broccoli
	Carrot
	Peas
	Cucumber
	Spinach
Histidine	Brussels sprouts
	Endive
	Cauliflower
	Onion
	Lettuce
	Peas
Isoleucine	Carrot
	Brussels sprouts
	Peas
	Tomato
	Lettuce

Bioactive Compounds	Plants
	Spinach
Leucine	Cabbage
	Broccoli
	Spinach
	Peas
	Onion
	Tomato
Lysine	Cauliflower
	Brussels sprouts
	Endive
	Cucumber
	Peas
	Carrot
Methionine	Cucumber
	Broccoli
	Peas
	Onion
	Endive
	Tomato
Phenylalanine	Cabbage
	Brussels sprouts
	Carrot
	Endive
	Cucumber
	Peas
Tryptophan	Cauliflower
	Brussels sprouts
	Onion
	Cauliflower
	Cucumber
Valine	Cabbage
	Cauliflower
	Peas
	Onion
	Tomato
	Spinach
	Vitamin B complex

Bioactive Compounds	Plants
Thiamine	Cucumber
	Broccoli
	Spinach
	Leeks
	Garlic
Riboflavin	Lettuce
	Brussels sprouts
	Tomato
	Onion
	Spinach
Niacin	Cabbage
	Broccoli
	Capsicum
	Peas
	Leeks
	Endive
Pantothenic acid	Cabbage
	Broccoli
	Capsicum
	Endive
	Tomato
	Spinach
Pyridoxine	Cauliflower
	Broccoli
	Carrot
	Spinach
	Cucumber
Folic acid	Celery
	Broccoli
	Spinach
	Endive
	Tomato
	Organic acids
Ascorbic acid	Cauliflower
	Cabbage
	Capsicum
	Spinach

Bioactive Compounds	Plants
	Tomato
	Onion
Oxalic acid	Broccoli
	Cabbage
	Brussels sprouts
	Onion
	Lettuce
Aconitic acid	Soybean
	Maize
	Fatty acids
Saturated	Cauliflower
	Broccoli
	Endive
	Carrot
	Capsicum
	Tomato
Monounsaturated	Cucumber
	Onion
	Broccoli
	Celery
	Spinach
	Onion
Polyunsaturated	Carrot
	Broccoli
	Capsicum
	Lettuce
	Tomato
	Spinach

11.3 SECONDARY PLANT METABOLITES

Plants create a wide range of organic materials that are indirectly occupied in growth and development. Within the kingdom of flora, these materials—traditionally known as secondary metabolites—are frequently distributed differently within taxonomic groups. Their roles, many of which are still unclear, are increasingly being explained. Secondary plant metabolites can be classified into four primary groups using the terminology used by the British Nutrition Foundation: phenolic and polyphenolic compounds (approximately 8,000 compounds), terpenoids (about 25,000 compounds), alkaloids (about 12,000 compounds), and sulfur-containing compounds (Goldberg, 2003) (Figure 11.2).

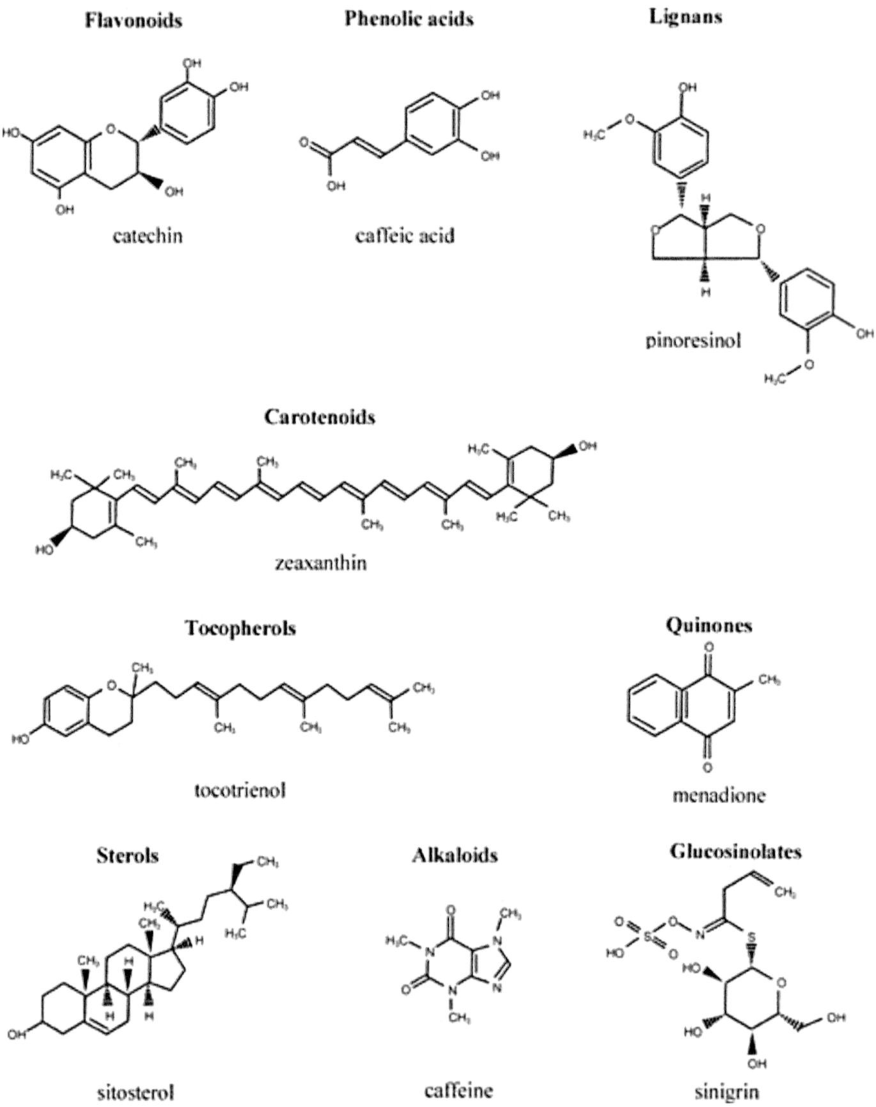

FIGURE 11.2 The chemical composition of indicative secondary metabolites.

11.3.1 COMPOUNDS OF PHENOL AND POLYPHENOL

Flavonoids, lignans, phenolic acids are examples of phenolic and polyphenolic chemicals, which are identified by an aromatic or phenolic ring structure. Elevated levels of phenolic compounds are frequently associated to greater protection to fungal plant diseases. Phenolic compounds are feeding deterrents for many parasites and other creatures (Nicholson & Hammerschmidt, 1992). A few phenolics affect how plants look and smell, drawing pollinators in. Phenolics aid in cold acclimatization and UV radiation defense. Most phenolic chemicals are combined with sugars inside plant cells to lessen their inherent toxicity. External factors such bacterial illness, UV rays, temperature, and the most abundant phenolic compound class is flavonoids, which include flavonols (quercetin, isorhamnetin, and kaempferol) found in leeks, endives, onion, and broccoli, flavones (apigenin, chrysoberyl, and luteolin) found in celery, parsley, and thyme, anthocyanidins (cyanidin, malvidin, and delphinidin) found Flavonoids are plant pigments that give vegetables their color, including anthocyanidins, chalcones, and flavones (Hou et al., 2004). Flavonoids found in food have antioxidant, anti-inflammatory, antiviral, and antihistamine effects. According to studies (Heim et al., 2002; Rice-Evans & Packer, 2003), they have been exhibited to reduce lipid peroxidation, chelate iron, and copper ions (which can catalyze the generation of free radicals), scavenge free radicals and modify cell signaling pathways. Atherosclerosis, cancer, aging, Parkinson's disease (PD), and AD have all been related to the production of peroxides and free radicals, which harm DNA, lipids, and proteins. Low-density lipoprotein cholesterol is shielded from oxidation by flavonoids, inhibiting the development of atherosclerotic plaques in the artery wall. They increase the activity of enzymes that help remove cancer-causing chemicals from the body and reduce inflammation linked to the local development of free radicals (exposed by Hollman & Katan, 1999; McCord, 2000). The majority of flavonoids taste bitter, astringent, or bitter with a sweet aftertaste (Drewnowski & Gomez-Carneros, 2000).

Chemicals known as phenolic acids have at least one aromatic ring and one or more hydroxyl groups. According to Rice-Evans & Packer (2003); and Crozier et al. (2006), phenolic acids, such as caffeic and gallic acid, can be found in lettuce and Pac Choi, cinnamic, and vanillic acid in onions, spinach, and parsley and coumaric acid in tomatoes, garlic, and carrots. According to Nicholson & Hammerschmidt (1992; Chou, 1999), these substances have antipathogen, antiherbivore, and allelopathic properties in plants. Under stressful circumstances, salicylic acid plays a critical function in cell signaling

(Klessig & Malamy, 1994). Nourishing phenolic acids like caffeic, benzoic, vanillic, and hydroxybenzoic have been established possess antibacterial and antifungal properties. This is likely because the oxidized molecules inhibit certain enzymes (Cowan, 1999). Due to their ability to suppress oxidation of lipid and scavenge reactive oxygen species (ROS), hydroxycinnamic acid derivatives such chlorogenic, caffeic, p-coumaric, ferulic, and sinapic acid have significant antioxidant activity (Sroka & Cisowski, 2003; Cheng et al., 2007). Syringic acid is one phenolic that may help explain why some plants have a bitter and astringent flavor (Drewnowski & Gomez-Carneros, 2000).

Lignans are diphenolic substances that are used in the production of lignin, a hydrophobic element for the cell wall of plants. High levels of lignans are found in foods like kale, broccoli, garlic, Brussels sprouts, and beans (Milder et al., 2005). Lignans operate as key of insect feeding and development (molting), which helps plants defend themselves against insects (Harmatha & Dinan, 2003). Few plant lignans, including secoisolariciresinol, pinoresinol, matairesinol, and lariciresinol, may be transformed into "enterolignans" by the intestinal microbiota and ingested by people (Heinonen et al., 2001). According to Adlercreutz & Mazur (1997); Arts & Hollman (2005), lignans have numerous biological actions in humans, including antioxidant and (anti) estrogenic qualities, which can lower the threat of some malignancies and cardiovascular illnesses. According to a theory, the Western culture diet can affect metabolism, hormone production, or action at the cellular level differently than the semi-vegetarian diet common in many Asian nations, leading to an increase in the incidence of colorectal, breast, and prostate cancer. Higher urine lignan excretion was linked to lower cancer risk (Webb & McCullough, 2005).

11.3.2 TERPENOIDS

A vast group of chemical compounds known as terpenoids is created when branched 5-carbon isoprene units repeatedly fuse. Terpenoids play a variety of physiological roles in plants, including those of hormones (gibberellins, abscisic acid), electron carriers (ubiquinone, plastoquinone), and structural elements of membranes (sterols, carotenoids, phytol) (Seiger, 1998). Sterols, carotenoids, quinones, tocopherols, and tocotrienols are some of the most important dietary terpenoids.

Plant carotenoids are lipid-soluble pigments that are orange, red, and yellow. They include – carotenes, carotenes, xanthophylls, and lycopene.

Any green leafy vegetables, including sweet potatoes, pumpkin, carrots, and tomatoes as well as all other vegetables whose color is obscured by chlorophyll, contain them (Granado et al., 1992; Crozier et al., 2006). According to Demmig-Adams & Adams (1996); and Taylor et al. (2000), carotenoids in plants shield photosynthetic tissues from photooxidative damage and are precursors of the phytohormone abscisic acid, which regulates stress and developmental processes. In the human diet, carotenoids with provitamin A activity are crucial nutrients. Vitamin A plays a role in the production of hormones, the control of cell division and growth, and immunological responses (Combs, 1998; Bender, 2003). Certain carotenoids, particularly carotene (found in carrots, sweet potatoes, and spinach) and carotene (located in carrots, red, and yellow peppers, and pumpkin), can be used by the body to make it (Bureau & Bushway, 1986). A deficiency in the human dietary supply of carotenoids can cause xerophthalmia (night blindness), which can cause early mortality. Diets high in carotenoids are linked to a significantly lower risk of developing some coronary heart diseases (CADs), cancers, and several deteriorating diseases like CADs and cataracts. By serving as biological antioxidants and defending cells and tissues against oxidative damage, carotenoids play a significant role (Combs, 1998; Bender, 2003).

Broccoli, kale, turnips, brussels sprouts, cauliflower, lettuce, cabbage, sweet potatoes, asparagus, spinach, and tomatoes all contain tocotrienols and tocopherols, sometimes familiar as vitamin E (Piironen et al., 1986; Eitenmiller & Lee, 2004). Tocopherols offer an ideal environment for the photosynthetic machinery in plants by preventing photooxidation of chloroplast membranes (Munné-Bosch & Alegre, 2002). All human cell membranes and plasma lipoproteins contain vitamin E (specially in red blood cells). Vitamin E prevents free radical-induced oxidation of DNA, low-density lipoproteins, and polyunsaturated fatty acids since this is the main fat-soluble chain-breaking antioxidant in human beings. It helps in maintaining the integrity of membranes, produce hemoglobin, and control immunological response (Brigelius-Flohe & Traber, 1999).

Quinones have two ketone substitutions on their aromatic rings. In spinach, lettuce, asparagus, cabbage, broccoli, cauliflower, and kale, you'll find phylloquinone, often known as vitamin K1 (Bolton-Smith et al., 2000). Phylloquinone is a compound found in plant kingdom that is involved in both the electron transport process during photosynthesis and the production of the active oxygen species that are seen as a response to pathogen attack or stress (Lochner et al., 2003). In people, vitamin K1 regulates blood clotting, bone growth, and bone healing. In addition to haematuria, surgical bleeding, muscular hematomas, and intercranial hemorrhages in adults, a shortage of this vitamin causes hemorrhagic illness in new-born neonates (Combs, 1998; Bender, 2003). Menadione, often called vitamin K3, has been found to have cytotoxic properties and to stop the development of human malignancies (Taper et al., 2004). Quinones, which are very reactive, are what cause sliced or damaged vegetables to turn brown (Cowan, 1999).

Soy, broccoli, Brussels sprouts, cauliflower, and spinach are high in plant sterols, including – sitosterol, campesterol, brassicasterol, and stigmasterol (Piironen et al., 2003). Sterols in plant membranes control the permeability and fluidity of phospholipid bilayers (Hartmann, 1998). Some sterols serve as building blocks for the plant hormones brassinosteroids, which play a role in fertilization, cell division, and embryonic development (Clouse & Sasse, 1998). These sterols produce calciferol (vitamin D2) when exposed to UV light, which is essential for calcium absorption and bone development. Prostaglandins and leukotrienes, crucial elements of the immune system, are synthesized from sterols. Plant sterols prevent the absorption of cholesterol because they resemble cholesterol structurally. Plant sterols may also have antioxidant, anti-inflammatory, anti-cancer, and anti-atherosclerosis properties in addition to their ability to decrease cholesterol (Awad & Fink, 2000; Ostlund, 2002; Dutta, 2003).

11.3.3 ALKALOIDS

Alkaloids are a class of primarily produced from amino acids basic nitrogen-containing molecules. Because of the strong physiological and therapeutic effects of alkaloids, they have historically attracted a lot of attention (for example nicotine, caffeine, morphine, quinine, and atropine). Common chemical antecedents of alkaloids, like pyridine (nicotine, coniine), tropane (cocaine, atropine) isoquinone (codeine, morphine), purine (caffeine), and steroids (solanine), are typically used to classify them (Facchini, 2001). Role of Alkaloids in plant metabolism is controversial. Since most alkaloids are extremely poisonous, they may serve as a chemical barrier against herbivores and microbes (Wittstock & Gershenzon, 2002). Additionally, it has been proposed that these substances act as UV light protectors (Jansen et al., 1998). Plant alkaloids with cytotoxic and antibacterial properties include berberine (BBR), palmatine, and mahanine (Facchini, 2001). Alkaloids found in plants can be hazardous to people. Solanine and chaconine, two significant alkaloid fractions, are present in potatoes and raw tomatoes that have been exposed to light (Jadhav et al., 1981). As a cholinesterase inhibitor, solanine may result in gastrointestinal disorders and neurological disorders, including a decrease in the function of CNS (Dalvi & Bowie, 1983). It hasn't been demonstrated, nevertheless, specifically eating these veggies could be harmful unless they made up an unusually large amount of the diet. The majority of alkaloids have a bitter or acrid flavor, such as lactucin and lactucopicrin, which are found in lettuce and chicory (Van Beek et al., 1990).

Because these substances exhibit surfactant qualities and produce foam in aqueous solutions, the plant glycoalkaloids tomatine, solanine, and chaconine are known as saponins (Hostettmann & Marston, 2005). Beans, peas, spinach, tomatoes, asparagus, potatoes, onions, and garlic all contain saponins (Sparg et al., 2004). Because of their insecticidal and molluscicidal activity, saponins protect plants from attack by microbes and herbivores. They also have allelopathic effects on several weeds (Haralampidis et al., 2001). Dietary saponins lower blood cholesterol, stop cancer cells from growing, and boost the immunity (Francis et al., 2002). Vegetables have an acidic, bitter flavor because to saponins (Drewnowski & Gomez-Carneros, 2000). On, the other hand, sweet saponins like periandrin V are 100 times sweeter in comparison to sucrose (Kinghorn et al., 1998). Sapotoxin, one kind of saponin, is hazardous to humans. Red blood cell membrane permeability is increased, and urticaria (a skin rash) and respiratory and digestive tract membranes are irritated (Francis et al., 2002).

11.3.4 COMPOUND CONTAINING SULFUR

A sizable functional group of amino acid derivatives with sulfur that includes a group produced from glucose is known as GSs. Brussels sprouts, red, and white cabbage, broccoli, cauliflower, daikon, and red radish all contain the GSs progoitrin and sinigrin, whereas mustard and horseradish include sinigrin and gluconasturtiin (Fahey et al., 2001; Johnson, 2002). GS-derived compounds serve as both natural insecticides and a defense against herbivores. The enzyme myrosinase quickly hydrolyzes GSs after tissue damage, cleaving off the glucose group. The remaining molecule swiftly transforms into the allelochemicals thiocyanate, isothyocyanate, or nitrile, which defend plants against pathogens, herbivores, and pests (Bennett & Wallsgrove, 1994; Bones & Rossiter, 1996). Some GSs play a crucial role as taste component precursors. Isothiocyanates, sometimes known as "mustard oils," have an unpleasant, lachrymatory taste and a strong fragrance. Examples include allyl isothiocyanate and benzyl isothiocyanate (Drewnowski & Gomez-Carneros, 2000). Other GSs are undesirable because the disintegration of products they produce have unfavorable physiological or sensory properties. Sinigrin has a harsh taste, as does the breakdown product. Though tasteless, progoitrin, and gluconapoleiferin's hydrolytic by products are bitter (Van Doorn et al., 1998). Excessive consumption of foods high in GSs can interfere with the production of thyroid hormone and aggravate stomach mucous membrane inflammation (Fenwick & Heaney, 1983). But these incidents are uncommon. There has been evidence that several GSs (glucoraphanin (Grn), glucobrassin, and glucotropaeolin) and the disintegration of products of these compounds may reduce the incidence of some cancers (Das et al., 2000). The inhibition of enzymes affecting the metabolism of steroid hormones, the activation of enzymes involved in the detoxification of carcinogens, and protection against oxidative damage all contribute to the anticarcinogenic activity of GSs (Johnson, 2002).

As was previously said, the nutritional value of vegetables is determined by a vast range of plant metabolites that are advantageous to human health. Consumers, however, are unable to accurately estimate the nutritional value of food because it requires specialized knowledge. As a result, the consumer's access to nutritional information will have an impact on how well the product is received. The provision of high-quality vegetables and fruits that have attained an acceptable level of ripeness and display their actual organoleptic flaws is required by law and public concern. According to Shewfelt & Brückner (2000), internal factors include flavor, texture, nutritional components, and internal flaws such cavities, water cores, frost damage, and decay. Vegetable texture, shape, size, color, pathogens, internal defects, maturity, and freshness have all been estimated using a variety of techniques, including mechanical and electrical impedance measurements, soft X-ray imaging, visible, and near-infrared spectroscopy, nuclear magnetic resonance imaging, acoustic (sonic and ultrasonic), and chlorophyll fluorescence (Abbott, 1999; Butz et al., 2005). Both destructive and non-destructive methods can be used to evaluate the nutritional value of vegetables. According to Buchgraber & Karaali (2005), the majority of standardized procedures for food chemical analysis used today are destructive and based on techniques like gas chromatography, liquid chromatography, thin-layer chromatography and spectrophotometry.

Major secondary metabolites and their primary vegetable sources are described as follows:

Flavonoids	Plants
Flavonoids	
Quercetin	Cabbage
	Broccoli
	Lettuce
	Endive
	Tomato
	Onion
Apigenin	Lettuce
	Celery
	Cabbage
Luteolin	Spinach
	Cauliflower
	Capsicum
	Celery
Myricetin	Spinach
	Lettuce
Cyanidin	Red lettuce
	Green lettuce
	Phenolic acids
	Chlorogenic acid Carrot
	Beans
	Lettuce
	Cauliflower
	Tomato
	Soya beans
p-Coumaric acid	Cabbage
	Cauliflower
Caffeic acid	Carrot
Ferulic acid	Turnip
Protocatechuic acid	Beans
	Carrot
Sinapic acid	Cauliflower
	Turnip
Vanillic acid	Soya beans
	Lignans
Lariciresinol	Cauliflower
	Broccoli

Flavonoids	Plants
	Kale
	Onion
	Lettuce
	Capsicum
Pinoresinol	Cabbage
	Endive
	Kale
	Broccoli
	Capsicum
	Leeks
Secoisolariciresinol	Brussels sprout
	Broccoli
	Carrot
	Leeks
	Tomato
	Lettuce
Matairesinol	Kale
	Carotenoids
α-Carotene	Capsicum
	Tomato
	Peas
	Broccoli
	Carrot
β-Carotene	Brussels sprout
	Broccoli
	Tomato
	Carrot
	Peas
β-Cryptoxanthin	Capsicum
Zeaxanthin	Kale
	Celery
	Lettuce
	Carrot
	Spinach
Lycopene	Tomato
	Tocopherols and tocotrienols
α-Tocopherol	Spinach

Flavonoids	Plants
	Broccoli
	Tomato
	Carrot
	Cabbage
	Onions
	Celery
β-Tocopherol	Lettuce
	Carrots
	Cucumber
γ-Tocopherol	Broccoli
	Cauliflower
	Lettuce
	Spinach
	Tomato
α-T3	Cabbage
	Cauliflower
	Onions
γ-T3	Corn
	Peas
	Quinones
Phylloquinone	Capsicum
	Broccoli
	Celery
	Onion
	Cucumber
	Carrot
	Lettuce
	Sterols
Campesterol	Brussels sprouts
	Broccoli
	Onion
	Cauliflower
	Carrot
	Toma
Campestanol	Tomato
	Broccoli
	Kale

Flavonoids	Plants
	Tomato
Stigmasterol	Carrot
	Celery
	Brussels sprouts
	Carrot
	Cauliflower
	Broccoli
	Tomato
β-Sitosterol	Celery
	Tomato
	Cauliflower
	Broccoli
	Kale
	Brussels sprouts
β-Sitostanol	Tomato
	Carrot
	Broccoli
	Celery
	Cauliflower
	Alkaloids
α-Tomatine	Tomato
α-Solanine	Potato
Dehydrotomatine	Tomato
α-Chaconine	Potato
	Lactucin Chicory
	α-Solanine Potato
Lactucopicrin	Chicory
	Glucosinolates
Glucoiberin	Green cabbage
	Cauliflower
	Broccoli
	Brussels sprouts
Glucoraphanin	Brussels sprouts
	Cauliflower
	Broccoli
	Green cabbage
Glucoalyssin	Broccoli

Flavonoids	Plants
	Brussels sprouts
Sinigrin	Green cabbage
	Cauliflower
	Broccoli
	Brussels sprouts
Gluconapin	Broccoli
	Brussels sprouts
	Cauliflower
	Green cabbage
Progoitrin	Cauliflower
	Broccoli
	Brussels sprouts
	Green cabbage
Gluconasturtiin	Cauliflower
	Broccoli
	Brussels sprouts

11.4 CONCLUSION

Nearly every nutrient required for human nutrition is present in plant meals, along with a significant number of chemical compounds that are beneficial to health. Over 2,00,000 metabolites are thought to be present in the plant kingdom (Pichersky & Gang, 2000; Fiehn, 2002). The particular substances are all involved in the primary metabolism and secondary metabolism of plants. These substances are crucial for growth and development of plants, defense in case of pests and more creatures that feed on plants, disease resistance, and stress reactions brought on by heavy metals, UV radiation and pesticides. Understanding the biochemical and molecular mechanisms of metabolite formation in plants as well as the development of analytical tools for metabolite identification are necessary for the development of methods to increase the nutritional content of plant foods. Through conventional breeding and the use of genomics technology, major efforts have been made over the past 10 years to increase the production of plant metabolites (Della Penna, 2001; Capell & Christou, 2004; Morandini et al., 2005). Enhancement in oleic acid content in soybeans (Kinney, 1996), carotenoids, and flavonoids in tomatoes (Fraser et al., 2002; Mehta et al., 2002), and amino acid and starch content in potatoes (Kinney, 1996) are instances of successful changes (Regierer et al., 2002). However, due to background levels of more metabolites are mostly unfamiliar, the introduction of high phytochemical-containing plants necessitates rigorous metabolite control. A variety of therapeutic chemicals can be found in plant metabolites, and their multidimensional chemical diversity makes them the best agents for treating severe ailments (Hounsome et al., 2008). Dietary modifications, hygiene, and inclusion of these metabolites in appropriate amount can be useful to target to the specific gastro-intestinal diseases and can be guaranteed to reduce disease advancement and enhancement in quality of life. Though, the consciousness and understanding about the relationship between plant metabolites and the digestive system disorders is still finite.

KEYWORDS

- **alkaloids**
- **carbohydrates**
- **constipation**
- **diarrhea**
- **irritable bowel disease**
- **primary metabolites**
- **secondary metabolites**

REFERENCES

Abbott, J. A., (1999). Quality measurement of fruits and vegetables. *Postharv. Biol. Technol., 15*(3), 207–225.

Adlercreutz, H., & Mazur, W., (1997). Phyto-oestrogens and Western diseases. *Ann. Med., 29*(2), 95–120.

Akerele, O., (1984). WHO's traditional medicine program, progress, and perspectives. *WHO Chron., 38*, 76–81.

Arts, I. C. W., & Hollman, P. C. H., (2005). Polyphenols and disease risk in epidemiologic studies. *Am. J. Clin. Nutr., 81*(Suppl), 317S–25S.

Asp, N. G., (1996). Dietary carbohydrates: Classification by chemistry and physiology. *Food Chem., 57*(1), 9–14.

Awad, A. B., & Fink, C. S., (2000). Phytosterols as anticancer dietary components: Evidence and mechanism of action. *J. Nutr., 130*, 2127–2130.

Balick, M. J., & Cox, P. A., (1997). Ethnobotanical research and traditional healthcare in developing countries. In: Bodeker, G., Bhat, K. K. S., Burley, J., & Vantomme, P., (eds.), Medicinal *Plant for Forest Conservation and HealthCare*. Rome: Food and Agriculture Organization of the United Nation.

Balick, M. J., (1996). Transforming ethnobotany for the new millennium. *Ann. Missouri Bot. Gard., 83,* 58–66.

Bartz, J. A., & Brecht, J. K., (2002). *Postharvest Physiology and Pathology of Vegetables* (2nd edn., p. 744). CRC Press.

Bell, A., (2003). Nonprotein amino acids of plants: Significance in medicine, nutrition, and agriculture. *J. Agric. Food Chem., 51*(10), 2854–2865.

Bender, D. A., (2003). *Nutritional Biochemistry of the Vitamins* (2nd edn., p. 512). Cambridge, U.K.: Cambridge Univ. Press.

Bennett, R. N., & Wallsgrove, R. M., (1994). Tansley review no 72. Secondary metabolites in plant defense mechanisms. *New Phytol., 127*, 617–633.

Blaut, M., (2002). Relationship of prebiotics and food to intestinal microflora. *Eur. J. Nutr., 41*(Suppl 1), I/11–I/16.

Bolton-Smith, C., Price, R. J. G., Fenton, S. T., Harrington, D. J., & Shearer, M. J., (2000). Compilation of a provisional UK database for the phylloquinone (vitamin K1) content of foods. *Br. J. Nutr., 83*(4), 389–399.

Bones, A. M., & Rossiter, J. T., (1996). The myrosinase-glucosinolate system: Its organization and biochemistry. *Physiol. Plant, 97*(1), 194–208.

Bourgaud, F., Gravot, A., Milesi, S., & Gontier, E., (2001). Production of plant secondary metabolites: A historical perspective. *Plant Science, 161*, 839–851.

Brigelius-Flohe, R., & Traber, M. G., (1999). Vitamin E: Function and metabolism. *FASEB J., 13*, 1145–1155.

Buchgraber, M., & Karaali, A., (2005). *Compilation of Standardized Analytical Methods for the Analysis of Active Ingredients in Functional Foods*. Brussels: European Commission Directorate-General Joint Research Centre Inst. for Reference Materials and Measurements. Available from: http://www.irmm.jrc.be/html/publications/technical_reports/publications/EUR21831ENcomplete.pdf (accessed on 15 November 2023).

Bureau, J. L., & Bushway, R. J., (1986). HPLC determination of carotenoids in fruits and vegetables in the United States. *J. Food Sci., 51*(1), 128–130.

Butz, P., Hofmann, C., & Tauscher, B., (2005). Recent developments in noninvasive techniques for fresh fruit and vegetable internal quality analysis. *J. Food Sci., 70*(9), 131–141.

Capell, T., & Christou, P., (2004). Progress in plant metabolic engineering. *Curr. Opin. Biotechnol., 15*(2), 148–154.

Carr, A. C., & Frei, B., (1999). Toward a new recommended dietary allowance for vitamin C based on antioxidant and health effects in humans. *Am. J. Clin. Nutr., 69*, 1086–1107.

Cheng, J. C., Dai, F., Zhou, B., Yang, L., & Liu, Z. L., (2007). Antioxidant activity of hydroxycinnamic acid derivatives in human low-density lipoprotein: Mechanism and structure–activity relationship. *Food Chem., 104*(1), 132–139.

Clouse, S. D., & Sasse, J. M., (1998). Brassinosteroids: Essential regulators of plant growth and development. *Annu. Rev. Plant Physiol. Plant Mol. Biol., 49*, 427–451.

Combs, G. F. Jr., (1998). *The Vitamins: Fundamental Aspects in Nutrition and Health* (p. 618). San Diego, Calif.: Academic Press.

Connor, W. E., (1999). α-linolenic acid in health and disease. *Am. J. Clin. Nutr., 69*, 827–828.

Connor, W. E., (2000). Importance of n-3 fatty acids in health and disease. *Am. J. Clin. Nutr., 71*(1 Suppl), 171S–175S.

Cowan, M. M., (1999). Plant products as antimicrobial agents. *Clin. Microbiol. Rev., 12*(4), 564–582.

Crozier, A., Clifford, M., & Ashihara, H., (2006). *Plant Secondary Metabolites: Occurrence Structure and Role in the Human Diet* (p. 384). Oxford, U.K.: Blackwell Publishing Ltd.

Dalvi, R. R., & Bowie, W. C., (1983). Toxicology of solanine: An overview. *Vet. Human Toxicol., 25*(1), 13–15.

Das, S., Tyagi, A. K., & Kaur, H., (2000). Cancer modulation by glucosinolates: A review. *Curr. Sci., 79*(12), 1665–1671.

Davies, K., (2004). Plant pigments and their manipulation. *Annual Plant Reviews* (Vol. 14, p. 368). Oxford, U.K.: Blackwell Publishing Ltd.

De Vries, J. W., (2003). On defining dietary fiber. *Proc. Nutr. Soc., 62*(1), 37–33.

DellaPenna, D., (2001). Plant metabolic engineering. *Plant Physiol., 125*, 160–163.

Demarquoy, J., Georges, B., Rigault, C., Royer, M. C., Clairet, A., Soty, M., Lekounoungou, S., & Le Borgne, F., (2004). Radioisotopic determination of L-carnitine content in foods commonly eaten in Western countries. *Food Chem., 86*(1), 137–142.

Demmig-Adams, B., & Adams, W. W., (1996). The role of xanthophyll cycle carotenoids in the protection of photosynthesis. *Trends Plant Sci., 1*, 21–26.

Drewnowski, A., & Gomez-Carneros, C., (2000). Bitter taste phytonutrients and the consumer: A review. *Am. J. Clin. Nutr., 72*, 1424–1435.

Dutta, P. C., (2003). *Phytosterols as Functional Food Components and Nutraceuticals* (p. 450). New York: Marcel Dekker Inc.

Eitenmiller, R., & Lee, J., (2004). *Vitamin E: Food Chemistry Composition and Analysis* (p. 540). New York: Marcel Dekker Inc.

Facchini, P. J., (2001). Alkaloid biosynthesis in plants: Biochemistry cell biology molecular regulation and metabolic engineering applications. *Ann. Rev. Plant Physiol. Plant Mol. Biol., 52*, 29–66.

Fahey, J. W., Zalcmann, A. T., & Talalay, P., (2001). The chemical diversity and distribution of glucosinolates and isothiocyanates among plants. *Phytochemistry, 56*(1), 5–51.

Farhana, R. B., Sabbir, H., Topu, R., & Mahmudul, H., (2020). Plants metabolites: Possibility of natural therapeutics against the COVID-19 pandemic/ *Front. Med.* https://doi.org/10.3389/fmed.2020.00444.

Fenwick, G. R., & Heaney, R. K., (1983). Glucosinolates and their breakdown products in cruciferous crops foods and feeding stuffs. *Food Chem., 11*(4), 249–271.

Fiehn, O., (2002). Metabolomics—The link between genotypes and phenotypes. *Plant Mol. Biol., 48*, 155–171.

Fisher, C., & Scott, T. R., (1997). *Food Flavors: Biology and Chemistry* (p. 176). Cambridge, U.K.: The Royal Society of Chemistry.

Flamm, G., Glinsmann, W., Kritchevsky, D., Prosky, L., & Roberfroid, M., (2001). Inulin and oligofructose as dietary fiber: A review of the evidence. *Crit. Rev. Food Sci. Nutr., 41*(5), 353–362.

Food and Agriculture Organization (FAO) of the United Nations, (1970). *The Amino Acid Content of Foods and Biological Data on Proteins*. Rome: Food Policy and Food Science Service Nutrition Division, FAO. Available from: http://www.fao.org/docrep/005/AC854T/AC854T00.HTM (accessed on 15 November 2023).

Francis, G., Kerem, Z., Makkar, H. P. S., & Becker, K., (2002). The biological action of saponins in animal systems: A review. *Br. J. Nutr., 88*(6), 587–605.

Fraser, P. D., Romer, S., Shipton, C. A., Mills, P. B., Kiano, J. W., Misawa, N., Drake, R. G., et al., (2002). Evaluation of transgenic tomato plants expressing an additional phytoene synthase in a fruit-specific manner. *Proc. Natl. Acad. Sci. U S A., 99*, 1092–1097.

Frias, J., Bakhsh, A., Jones, D. A., Arthur, A. E., Vidal-Valverde, C., Rhodes, M. J. C., & Hedley, C., (1999). Genetic analysis of the raffinose oligosaccharide pathway in lentil seeds. *J. Exp. Bot., 50*, 469–476.

Goldberg, G., (2003). *Plants: Diet and Health* (p. 347). The report of a British nutrition foundation task force. Oxford, U.K.: Blackwell Publishing Ltd.

Granado, F., Olmedilla, B., Blanco, I., & Rojas-Hidalgo, E., (1992). Carotenoid composition in raw and cooked Spanish vegetables. *J Agric. Food Chem., 40*(11), 2135–2140.

Gurr, M., Harwood, J., & Frayn, K., (2002). *Lipid Biochemistry: An Introduction* (5th edn., p. 368). Oxford, U.K.: Blackwell Science Ltd.

Haralampidis, K., Trojanowska, M., & Osbourn, A. E., (2001). Biosynthesis of triterpenoid saponins in plants. *Adv. Biochem. Eng. Biotechnol., 75*, 31–49.

Harmatha, J., & Dinan, L., (2003). Biological activities of lignans and stilbenoids associated with plant-insect chemical interactions. *Phytochem. Rev., 2*(3), 321–330.

Hartmann, M. A., (1998). Plant sterols and the membrane environment. *Trends Plant Sci., 3*(5), 170–175.

Heim, K. E., Tagliaferro, A. R., & Bobilya, D. J., (2002). Flavonoid antioxidants: Chemistry metabolism and structure-activity relationships. *J. Nutr. Biochem., 13*(10), 572–584.

Heinonen, S., Nurmi, T., Liukkonen, K., Poutanen, K., Wähälä, K., Deyama, T., Nishibe, S., & Adlercreutz, H., (2001). *In vitro* metabolism of plant lignans: New precursors of mammalian lignans enterolactone and enterodiol. *J. Agric. Food Chem., 49*(7), 3178–3186.

Heldt, H. W., (2005). *Plant Biochemistry* (3rd edn., p. 656). London, U.K.: Elsevier Academic Press.

Hollman, P. C. H., & Katan, M. B., (1999). Dietary flavonoids: Intake health effects and bioavailability. *Food Chem. Toxicol., 37*(9, 10), 937–942.

Hostettmann, K., & Marston, A., (2005). *Saponins* (p. 560). Cambridge, U.K.: Cambridge Univ. Press.

Hou, D. X., Fujii, M., Terahara, N., & Yoshimoto, M., (2004). Molecular mechanisms behind the chemopreventive effects of anthocyanidins. *J. Biomed. Biotechnol., 5*, 321–325.

Hounsome, N., Hounsome, B., Tomos, D., & Edwards-Jones, G., (2008). Plant metabolites and nutritional quality of vegetables. *Journal of Food Science, 73*, R48–R65. https://doi.org/10.1111/j.1750-3841.2008.00716.x.

Innis, S. M., (1991). Essential fatty acids in growth and development. *Progr. Lipid Res., 30*(1), 39–103.

Jadhav, S. J., Sharma, R. P., & Salunkhe, D. K., (1981). Naturally occurring toxic alkaloids in foods. *Crit. Rev. Toxicol., 9*(1), 21–104.

Jakab, A., Nagy, K., Héberger, K., Vékey, K., & Forgács, E., (2002). Differentiation of vegetable oils by mass spectrometry combined with statistical analysis. *Rapid Commun. Mass Spectrom., 16*(24), 2291–2297.

Jansen, M. A. K., Gaba, V., & Greenberg, B. M., (1998). Higher plants and UV-B radiation: Balancing damage repair and acclimation. *Trends Plant Sci., 3*(4), 131–135.

Jha, S. N., & Matsuoka, T., (2000). Non-destructive techniques for quality evaluation of intact fruits and vegetables. *Food Sci. Technol. Res., 6*(4), 248–251.

Johnson, I. T., (2002). Glucosinolates in the human diet. Bioavailability and implications for health. *Phytochem. Rev., 1*(2), 183–188.

Jones, M. E., & Kossel, A., (1953). A biographical sketch. *Yale Journal of Biology and Medicine, 26*(1), 80–97.

Keservani, R. K., & Sharma, A. K., (2014). Flavonoids: Emerging trends and potential health benefits. *Journal of Chinese Pharmaceutical Sciences*, *23*(12), 815.

Keservani, R. K., Kesharwani, R. K., Sharma, A. K., Vyas, N., & Chadoker, A., (2010b). Nutritional supplements: An overview. *International Journal of Current Pharmaceutical Review and Research, 1*(1), 59–75.

Keservani, R. K., Kesharwani, R. K., Vyas, N., Jain, S., Raghuvanshi, R., & Sharma, A. K., (2010a). Nutraceutical and functional food as future food: A review. *Der Pharmacia Lettre*, *2*(1), 106–116.

Keservani, R. K., Sharma, A. K., & Kesharwani, R. K., (2016a). Nutraceutical and functional foods for cardiovascular health. In: *Food Process Engineering* (pp. 257–278, 291–312). Apple Academic Press, CRC Press. ISBN: 9781771884020.

Keservani, R. K., Sharma, A. K., & Kesharwani, R. K., (2016b). Medicinal effect of nutraceutical fruits for the cognition and brain health. *Scientifica*, 10. Article ID 3109254. doi: 10.1155/2016/3109254.

Keservani, R. K., Sharma, A. K., & Kesharwani, R. K., (2017). An overview and therapeutic applications of nutraceutical and functional foods. *Recent Advances in Drug Delivery Technology*, 160–201. ISBN:9781522507543.

Keservani, R. K., Sharma, A. K., & Kesharwani, R. K., (2020). *Nutraceuticals and Dietary Supplements: Applications in Health Improvement and Disease Management*. CRC Press. ISBN: 9781771888738.

Kinghorn, A. D., Kaneda, N., Baek, N. I., Kennelly, E. J., & Soejarto, D. D., (1998). Noncariogenic intense natural sweeteners. *Med. Res. Rev., 18*(5), 347–360.

Kinney, A. J., (1996). Development of genetically engineered soybean oils for food applications. *J. Food Lipids, 3*, 273–292.

Klessig, D. F., & Malamy, J., (1994). The salicylic acid signal in plants. *Plant Mol. Biol., 26*(5), 1439–1458.

Kmiecik, W., & Lisiewska, Z., (2000). Studies on the morphological traits and chemical composition of the fruit of six tomato cultivars recommended as raw material for freezing. *Nahrung/Food, 44*(5), 394–353.

Krajčovičová-Kudláčková, M., Šebeková, K., Schinzel, R., & Klvanová, J., (2002). Advanced glycation end products and nutrition. *Physiol. Res., 51*(3), 313–316.

Kuo, T. M., Van, M. J. F., & Wolf, W. J., (1988). Content of raffinose oligosaccharides and sucrose in various plant seeds. *J. Agric. Food Chem., 36*, 32–36.

Lewis, D. H., (1984). Physiology and metabolism of alditols. In: Lewis, D. H., (ed.), *Storage Carbohydrates in Vascular Plants* (pp. 157–179). Cambridge, U.K.: Cambridge Univ. Press.

Lochner, K., Döring, O., & Böttger, M., (2003). Phylloquinone, what can we learn from plants? *Biofactors, 18*(1–4), 73–78.

MacEvilly, C., & Peltola, K., (2003). The effect of agronomy storage processing and cooking on bioactive substances in food. In: Goldberg, G., (ed.), *Plants: Diet and Health* (pp. 226–39). Oxford, U.K.: Blackwell Publishing Ltd.

Mäkinen, K. K., & Söderling, E., (1980). A quantitative study of mannitol sorbitol xylitol and xylose in wild berries and commercial fruits. *J. Food Sci., 45*(2), 367–371.

Marlett, J. A., (2001). Dietary fiber and cardiovascular disease. In: Cho, S. S., & Dreher, M. L., (eds.), *Handbook of Dietary Fiber* (pp. 17–30). New York: Marcel Dekker Inc.
McCord, J. M., (2000). The evolution of free radicals and oxidative stress. *Am. J. Med., 108*(8), 652–659.
McGarr, S. E., Ridlon, J. M., & Hylemon, P. B., (2005). Diet anaerobic bacterial metabolism and colon cancer: A review of the literature. *J. Clin. Gastroenterol., 39*(2), 98–109.
Mehta, R. A., Cassol, T., Li, N., Ali, N., Handa, A. K., & Mattoo, A. K., (2002). Engineered polyamine accumulation in tomato enhances phytonutrient content, juice quality, and vine life. *Nat. Biotechnol., 20*, 613–618.
Milder, I. E. J., Arts, I. C. W., Van De, P. B., Venema, D. P., & Hollman, P. C. H., (2005). Lignan contents of Dutch plant foods: A database including lariciresinol, pinoresinol, secoisolariciresinol and matairesinol. *Brit. J. Nutr., 93*, 393–402.
Morandini, P., Salamini, F., & Gantet, P., (2005). Engineering of plant metabolism for drug and food. *Curr. Med. Chem. Immunol Endocr. Metab. Agents, 5*(2), 103–112.
Mozafar, A., (1993). Nitrogen fertilizers and the amount of vitamins in plants: A review. *J. Plant Nutr., 16*(12), 2479–2506.
Munné-Bosch, S., & Alegre, L., (2002). The function of tocopherols and tocotrienols in plants. *Crit. Rev. Plant Sci., 21*(1), 31–57.
Nettleton, J. A., (1995). *Omega-3 Fatty Acids and Health* (p. 384). New York: Chapman Hall.
Nicholson, R. L., & Hammerschmidt, R., (1992). Phenolic compounds and their role in disease resistance. *Annu. Rev. Phytopathol., 30*, 369–389.
Obendorf, R. L., Horbowicz, M., Dickerman, A. M., Brenac, P., & Smith, M. E., (1998). Soluble oligosaccharides and galactosyl cyclitols in maturing soybean seeds in planta and *in vitro*. *Crop Sci., 38*, 78–84.
Oh, S. H., Moon, Y. J., & Oh, C. H., (2003). γ-Aminobutyric acid (GABA) content of selected uncooked foods. *Nutraceuticals Food, 8*(1), 75–78.
Orcutt, D. M., & Nilsen, E. T., (2000). *The Physiology of Plants Under Stress Soil and Biotic Factors* (p. 680). New York: John Wiley and Sons Inc.
Ostlund, R. E., (2002). Phytosterols in human nutrition. *Annu. Rev. Nutr., 22*, 533–549.
Parr, A. J., & Bolwell, G. P., (2000). Phenols in the plant and in man. The potential for possible nutritional enhancement of the diet by modifying the phenols content or profile. *J. Sci. Food Agric., 80*(7), 985–1012.
Pereira, C., Li, D., & Sinclair, A. J., (2001). The alpha-linolenic acid content of green vegetables commonly available in Australia. *Int. J. Vit. Nutr. Res., 71*(4), 223–228.
Peterbauer, T., Lahuta, L. B., Blöchl, A., Mucha, J., Jones, D. A., Hedley, C. L., Gòrecki, R. J., & Richter, A., (2001). Analysis of the raffinose family oligosaccharide pathway in pea seeds with contrasting carbohydrate composition. *Plant Physiol., 127*, 1764–1772.
Pichersky, E., & Gang, D., (2000). Genetics and biochemistry of secondary metabolites: An evolutionary perspective. *Trends Plant Sci., 5*(10), 439–445.
Piironen, V., Syväoja, E. L., Varo, P., Salminen, K., & Koivistoinen, P., (1986). Tocopherols and tocotrienols in Finnish foods: Vegetables fruits and berries. *J. Agric. Food Chem., 34*(4), 742–746.
Piironen, V., Toivo, J., Puupponen-Pimiä, R., & Lampi, A. M., (2003). Plant sterols in vegetables fruits and berries. *J. Sci. Food Agric., 83*(4), 330–337.
Rane, B. R., Bharath, M. S., Patil, R. R., Keservani, R. K., & Jain, A. S., (2020b). Novel approaches in nutraceuticals; Chapter 11. In: Rajesh, K. K., Raj, K. K., & Anil, K. S., (eds.), *Enhancing the Therapeutic Efficacy of Herbal Formulations Through Novel Drug Delivery*

Systems (pp. 241–266). IGI Global International Publisher, Pennsylvania, USA. ISBN13: 9781799844532.

Rane, B. R., Patil, A. S., Keservani, R. K., & Jain, A. S., (2020a). Novel approaches in herbal formulation; Chapter 2. In: Rajesh, K. K., Raj, K. K., & Anil, K. S., (eds.), *Enhancing the Therapeutic Efficacy of Herbal Formulations Through Novel Drug Delivery Systems* (pp. 43–68). IGI Global International Publisher, Pennsylvania, USA. ISBN13: 9781799844532.

Rane, B. R., Tadavi, S. A., & Keservani, R. K., (2020c). Naturopathy; Chapter 12. In: Anil, K. S., Raj, K. K., & Surya, P. G., (eds.), *Herbal Product Development* (pp. 321–347). Apple Academic Press, CRC Press, Taylor & Francis Group. ISBN: 9781771888776.

Regierer, B., Fernie, A. R., Springer, F., Perez-Melis, A., Leisse, A., Koehl, K., Willmitzer, L., et al., (2002). Starch content and yield increase as a result of altering adenylate pools in transgenic plants. *Nat. Biotechnol., 20*, 1256–1212.

Rice-Evans, C. A., & Packer, L., (2003). *Flavonoids in Health and Disease* (p. 504). New York: Marcel Dekker Inc.

Ruhl, I., & Herrmann, K., (1985). Organic acids in vegetables: *Brassica* leaf and bulb vegetables as well as carrots and celery. *Zeitschrift Lebensm Forsch A, 180*(3), 215–220.

Seiger, D. S., (1998). *Plant Secondary Metabolism* (p. 592). Dordrecht, The Netherlands: Kluwer Academic Publishers.

Seigler, D. S., (1995). *Plant Secondary Metabolism*. New York: Springer Science: Business Media; ISBN: 978-1-4613-7228-8; ISBN: 978-1-4615-4913-0 (eBook). doi: 10.1007/978-1-4615-4913-0.

Shahidi, F., & Miraliakbari, H., (2005). Omega-3 fatty acids in health and disease: Part 2—Health effects of omega-3 fatty acids in autoimmune diseases, mental health, and gene expression. *J. Med. Food, 8*(2), 133–148.

Sharma, A. K., Keservani, R. K., & Gautam, S. P., (2020). *Herbal Product Development*. Apple Academic Press, CRC Press, Taylor & Francis Group. ISBN: 9781771888776.

Shewfelt, R. L., & Brückner, B., (2000). *Fruit and Vegetable Quality: An Integrated View* (p. 664). Boca Raton, Fla.: CRC Press.

Shiomi, N., (1992). Content of carbohydrate and activities of fructosyltransferase and invertase in asparagus roots during the fructo-oligosaccharide- and fructo-polysaccharide-accumulating season. *New Phytol., 122*(3), 421–432.

Shiomi, N., Benkeblia, N., & Onodera, S., (2005). The metabolism of the fructooligosaccharides in onion bulbs: A comprehensive review. *J. Appl. Glycosci., 52*, 121–127.

Simopoulos, A. P., (1999). Essential fatty acids in health and chronic disease. *Am. J. Clin. Nutr., 70*(3 Suppl), 560S–569S.

Slavin, J. L., (2005). Dietary fiber and body weight. *Nutrition, 3*, 411–418.

Spallholz, J. E., Boylan, L. M., & Driskel, J. A., (1999). *Nutrition: Chemistry and Biology* (2nd edn., p. 345). Boca Raton, Fla.: CRC Press.

Sparg, S. G., Light, M. E., & Van, S. J., (2004). Biological activities and distribution of plant saponins. *J. Ethnopharmacol., 94*(2, 3), 219–243.

Sroka, Z., & Cisowski, W., (2003). Hydrogen peroxide scavenging antioxidant and anti-radical activity of some phenolic acids. *Food Chem. Toxicol., 41*(6), 753–758.

Stapleton, P. P., Charles, R. P., Redmond, H. P., & Bouchier-Hayes, D. J., (1997). Taurine and human nutrition. *Clin. Nutr., 16*(3), 103–108.

Stollman, N., & Raskin, J. B., (2004). Diverticular disease of the colon. *Lancet, 363*, 631–639.

Stoop, J. M. H., Williamson, J. D., & Mason, P. D., (1996). Mannitol metabolism in plants: A method for coping with stress. *Trends Plant Sci., 1*, 139–144.

Taper, H. S., Jamison, J. M., Gilloteaux, J., Summers, J. L., & Calderon, P. B., (2004). Inhibition of the development of metastases by dietary vitamin C: K3 combination. *Life Sci., 75*(8), 955–967.

Taylor, I. B., Burbidge, A., & Thompson, A. J., (2000). Control of abscisic acid synthesis. *J. Exp. Bot., 51*(350), 1563–1574.

Tungland, B. C., & Meyer, D., (2002). Nondigestible oligo- and polysaccharides (dietary fiber): Their physiological role in human health and food. *Comp. Rev. Food Sci. Food Saf., 1*(3), 90–109.

U.S. Dept. of Agriculture Agricultural Research Service (USDA), (2005). *USDA National Nutrient Database for Standard Reference Release 18.* Available from: http://www.nal.usda.gov/fnic/foodcomp/Data/ (accessed on 15 November 2023).

Van, B. T. A., Maas, P., King, B. M., Leclercq, E., Voragen, A. G. J., & De Groott, A., (1990). Bitter sesquiterpene lactones from chicory roots. *J. Agric. Food Chem., 38*(4), 1035–1038.

Van, D. H. E., Van, D. K. G. C., Van, H. G. J., Raaijmakers-Ruijs, N. C. M. E., Postma, E., Groeneweg, B., & Jongen, W. H. F., (1998). The glucosinolates sinigrin and progoitrin are important determinants for taste preference and bitterness of Brussels sprouts. *J. Sci. Food Agric., 78*(1), 30–38.

Webb, A. L., & McCullough, M. L., (2005). Dietary lignans: Potential role in cancer prevention. *Nutr. Cancer, 51*(2), 117–131.

WHO, (2014). *World Health Statistics 2014*. Switzerland: WHO Press.

WHO, (2019). https://www.who.int/news-room/fact-sheets/detail/diarrhoeal-disease (accessed on 15 November 2023).

Wittstock, U., & Gershenzon, J., (2002). Constitutive plant toxins and their role in defense against herbivores and pathogens. *Cur. Opin. Plant Biol., 5*(4), 300–307.

Yao, L. H., Jiang, Y. M., Shi, J., Tomás-Barberán, F. A., Datta, N., Singanusong, R., & Chen, S. S., (2004). Flavonoids in food and their health benefits. *Plant Foods Human Nutr., 59*(3), 113–122.

Yehuda, S., Rabinovitz, S., & Mostofsky, D. I., (1999). Essential fatty acids are mediators of brain biochemistry and cognitive functions. *J. Neurosci. Res., 56*, 565–570.

Young, V. R., & Pellett, P. L., (1994). Plant proteins in relation to human protein and amino acid nutrition. *Am. J. Clin. Nutr., 59*, 1203S–1212S.

CHAPTER 12

PLANT METABOLITES AND VEGETABLES FOR DIABETES PREVENTION AND TREATMENT

AMRUTA BALEKUNDRI,[1] EKNATH D. AHIRE,[2] and RAJ K. KESERVANI[3]

[1]*Department of Pharmacognosy, KLE College of Pharmacy Belagavi, KLE University of Higher Education and Research, Belagavi, Karnataka, India*

[2]*Department of Pharmaceutics, METs, Institute of Pharmacy, BKC, Adgaon, Nashik, Maharashtra, India*

[3]*Faculty of B. Pharmacy, CSM Group of Institutions, Prayagraj, Uttar Pradesh, India*

ABSTRACT

Diabetes Mellitus is one of the most prevailing disorders which is caused when the body is not able to regulate the blood glucose level leading to hyperglycemic conditions. The regulation of the blood glucose level is very important for the normal functioning of the body. There are various classes of drugs which are available in treatment and prevention of diabetes. Apart from medicines diet also plays a very important role in management of diabetes. Plant metabolites like flavonoids, alkaloids, terpenes, and phenolics are very much potential in management of diabetes. These various plant metabolites can be used as an adjuvant in treatment of diabetes. Everyday diet contains a lot of vegetables which are helpful in diabetes management

Plant Metabolites and Vegetables as Nutraceuticals.
Raj K. Keservani, Bui Thanh Tung, Rajesh K. Kesharwani, and Eknath D. Ahire (Eds.)

which are bitter gourd, okra, lettuce, cluster bean, drumstick, etc. These vegetables can play an important role in management of diabetes by protecting the beta cells of pancreas and inhibiting the enzymes involved in hyperglycemic process. So, these vegetables can be made an important part of diet of a diabetic patient.

12.1 INTRODUCTION

About 422 million people are diabetic worldwide in which the majority of population belongs to countries with low and middle income. In accordance with WHO diabetes is a chronic disease which is marked by an increase in the level of blood glucose level than that of the normal, which further leads to secondary complications with time. The secondary complication results in damage to heart, blood vessels, kidneys, and nerves (Platel & Srinivasan, 1997; Asif, 2011; Franz et al., 2002). Diabetes cannot be completely cured it's a long-lasting chronic health condition were in the body is unable to convert the food we consume to the energy. The food that we consume is broken down into small sugar molecules (glucose blocks) which are further released into blood flow. The increase in the blood sugar level signals the pancreas, so that the pancreas releases the insulin hormone. Insulin is responsible to keep the blood glucose level in check and required for the conversion of blood glucose into energy. But when the person is diabetic his body is unable to produce enough of insulin as per the requirement or either the released insulin is not completely utilized, as the insulin isn't in required amount the glucose in the blood flow is increased which further leads to secondary complications over time like kidney, heart and vision diseases, etc. (Ghorbani et al., 2019; Vinayagam et al., 2015; Rasouli et al., 2020). Diabetes is one of the most leading causes of death in the world. Diabetes is also one of the major reasons for the kidney failure, adult blindness, limb-amputation, heart diseases. The number of diabetes patients have been doubled in last two decades (Varshney et al., 2019; Forouhi & Wareham, 2010; Ahire et al., 2021; Surana et al., 2021).

12.1.1 TYPES OF DIABETES

The diabetes is broadly divided into three domains, they are:

1. **Type I Diabetes:** The diabetes caused due to the auto immune response of the body is said to be the type I diabetes. In this type of the body itself attacks the insulin producing cells which results

in reduction of the insulin production which results in the diabetic condition. There is no treatment currently available to prevent this type of diabetes so external insulin is given on daily doses for survival to the people suffering from this type. The type I diabetes is found in about 5–10% of entire diabetes population (Chen et al., 2015; Panche et al., 2016).

2. **Type II Diabetes:** It is found in 90–95% of the diabetes population. In this type II diabetes, the body is not able to utilize the insulin released by the cells ultimately resulting in the increase of glucose in the blood flow. This type II diabetes can be prevented/delayed by modifying the diet and healthy lifestyle (Coskun et al., 2005; Babu et al., 2013).
3. **Gestational Diabetes:** The diabetes developed by the women who is pregnant and who never had this condition before is called gestational diabetes. After the women delivers the baby, the gestational diabetes goes away. The baby born to the mother with gestational diabetes is at higher risk of health issues, obesity as well as more likely to develop diabetes type II in life later (Rasouli et al., 2020; Shi et al., 2014; Kumar et al., 2019).
4. **Pre-Diabetes:** It is the condition where in the person has blood sugar level which is slightly higher than that of the normal blood sugar level but that is not sufficient to be diagnosed for type II types. The person who is pre-diabetic has a higher risk of developing type II diabetes with time. The pre-diabetic condition can be reversed by lifestyle and diet changes up to certain level (Adhikari et al., 2021; Yang et al., 2020).

12.2 ROLE OF DIET IN DIABETES

In the management of diabetes, the major role is played by the lifestyle and diet habits. The diet can help patients to control the blood glucose level to great extent before using the therapeutic insulin. The diet includes the intake of vegetables and fruits which help in controlling the blood glucose level. Apart from the normal vegetable diet additional sources of plant metabolites are given which are also used as traditional medicines for controlling type II diabetes (Teoh & Das, 2018; Cox-Georgian et al., 2019). About one third population of the world depends on the traditional medicines for the prevention, treatment, and cure of diseases. So as a primary measure WHO recommends the utilization of the traditional medicines in prevention and treatment of type II diabetes (Lin et al., 2016; Demir et al., 2019).

Diet plays a critical role in overall diabetes management which includes the diet along with insulin, diet along with oral hypoglycemic and sometimes diet alone (pre-diabetes). In accordance with age, weight, and other secondary complications the diet is individualized for every patient, based on individual requirements. Along with healthy eating habits, it is advisable to do exercise, walking, and other activities that increase physical activity as physical inactivity is the major reason for the obesity and burdening of diabetes. Diet which is rich in vegetables, fruits, and fibers which reduces the type II diabetes. The World Health Organization (WHO), National Institute of Health (NIH), Canadian Diabetes Association (CDA), and United States Diabetes Association (USDA) all these associations recommend the diet therapy for the prevention of type II diabetes which includes vegetables, fruits, whole grains, and plant metabolites. They also suggest five servings of fruits and vegetables on daily basics (Albuquerque et al., 2021; Attanayake et al., 2016; Mukhtar et al., 2006).

12.3 DIABETES PREVENTION AND TREATMENT

The diet plays an important role in management, prevention, and treatment of diabetes. The Individualization of the diet based on the cultural preferences and with respect to personal willingness and wish. The health improvement must be kept in check with healthy food habits/choices and physical exercises. In case the patient has secondary complications (obesity, cardiovascular diseases (CVD), kidney disorders, and hypertension) the lifestyle modification is required for prevention and treatment of diabetes. There must be proper check on the metabolic parameters of the patients (blood glucose level, blood pressure, body weight, renal functions, and lipid profiles) so that essential modification of the nutrition and diet therapy can be done for further prevention and treatment of diabetes (Bazzano et al., 2003; Gupta et al., 2012; Wang et al., 2016).

12.3.1 PLANT METABOLITES FOR DIABETES PREVENTION AND TREATMENT

The substances which directly influence the growth and development of the plant are called primary plant metabolites, whereas the substances that are important but not essential in plant functioning directly are called secondary metabolites. Secondary metabolites are responsible for defense mechanism of plants and also responsible for metabolic regulation.

The small organic molecules produced by the plants by modification of primary metabolite synthase are called secondary metabolites which aren't concerned with the growth, reproduction, and development (Watanabe et al., 2021; Tian & Deng, 2020; Platel & Srinivasan, 1997). At times prior to the era of anti-diabetic drugs/medicines the diabetic patients were treated with traditional medicinal plants. Mechanisms like insulin kinases, inhibition of insulinase activity and increase the pancreatic beta cell reconstruction are followed by the plant metabolites. There are many plant metabolites that are responsible for prevention and treatment of diabetes. It is not possible to specify which phytoconstituent is responsible for the activity as there are number of other constituents present in the medicinal plants. But the plant metabolites can play a significant role when it comes to discovery of the new class of drugs as safe and potent anti-diabetic agents (Gandhi et al., 2014; Jaiswal et al., 2013; Psaltopoulou et al., 2010; Joseph & Jini, 2013).

There are many different classes of compounds and metabolites that are responsible for the anti-diabetic activity, but the most common classes of plant metabolite are flavonoids, alkaloids, terpenoids, and phenolic (Ali et al., 2015; Gaddam et al., 2015).

12.3.2 FLAVONOIDS

The naturally occurring compound in plants which have poly-phenolic structure is called flavonoids. The structure of flavonoids comprises of three ring structure where in there are two aromatic rings which are joined to one another by heterocyclic pyran ring (Haber & Keonavong, 2013; Abdul-rahman et al., 2020).

The flavonoids include sub-classes like:

1. **Flavones:** The flavones are chemically 2-phenylchromen-4-one, the common examples of this class are Luteolin, Apigenin whereas the natural sources of flavones are green pepper, Parsley, Oregano, etc. (Jardine et al., 2021).
2. **Flavanonols:** 3-hydroxy-2,3-dihydro-2-phenylchromen-4-one is the chemical name of flavanonols. Aromadendrin, Taxifolin are the examples of flavanonols sub-class, and the natural resources are Deodar cedar, Chir pine, etc. (Roberts & Moreau, 2016).

3. **Flavonols:** Also called 3-hydroxy-2-phenylchromen-4-one chemically. Quercetin and Kaempferol are the most common examples of this sub-class. The natural sources of this sub-class are Parsely, Radish, Onion skin, etc. (Pandey et al., 2020).

4. **Isoflavones:** Chemically isoflavones are 3-phenylchromen-4-one, where in Genistein and Daidzein are the common examples. The natural sources of isoflavones are Soybean, Green bean, cowpea, etc. (Karkute et al., 2018).

5. **Flavanones:** 2,3-dihydro-2-phenylchromen-4-one is the chemical name of flavanones. Naringenin and Hesperetin are examples of flavanones subclass. Lemons and Oranges are the natural sources of flavanones (Uuh-Narvaez and Segura-Campos, 2021).

6. **Anthocyanidins:** These are chemically known as 2-phenyl chromenylium. This sub-class has natural resources like bilberries, elderberry, etc., and the most common compounds of this sub-class are cyaniding and pelargonidin (Gupta et al., 2022).

7. **Flavanols:** Also known as 2-phenyl-3, 4-dihydro-2H-chromen-3-ol. The common compounds of this sub-class are epicatechin and the natural source are green tea, blackberries, etc. The diabetes can be treated in early stage by protecting the beta cells mass and function. Flavonoids are the secondary metabolites that can protect the function and survival of beta cells of pancreas. The flavonoids have antioxidant and anti-inflammatory properties, due to these properties the beta cell survival and functioning can be benefited. The flavonoids help in protecting the beta cell by increasing their survival index and their functioning capacity which is beneficial in the prevention and treatment of diabetes (Esmaeilzadeh et al., 2020). Enhancing insulin secretion, promoting beta cell proliferation, and decreasing insulin resistance are the probable mechanisms by which flavonoids are responsible in prevention and treatment of diabetes (Ravikumar, 2015; Bahadoran et al., 2011).

7. **Alkaloids:** These are the secondary metabolites which have large nitrogen-compounds of phytochemicals. The Alkaloids are broadly divided into three categories like true alkaloids, pseudo-alkaloids, and proto-alkaloids. The Alkaloids can also be classified into various sub-class based on the amino acid backbone the subs classes are indole alkaloids, piperidine alkaloid, quinolizidine alkaloid, indolizidine

alkaloid, benzylisoquinoline alkaloid, isoquinoline alkaloid, pyrrolizidine alkaloid, pyrrolidine alkaloid, histamine alkaloid (Khan et al., 2021). Anti-diabetic activity is shown by extracts having alkaloid compounds. Although alkaloids have some toxic effect on both animals and humans, but the advantage of anti-diabetes property of the alkaloid will outweigh the toxicity of the alkaloids (Bahadoran et al., 2012; Basharat et al., 2021). The anti-diabetic activity of the alkaloid is achieved by different mechanisms. They are inhibiting the digestive enzymes, increasing the insulin secretion, antioxidant activity, antioxidant activity, and inhibiting aldose reductase.

12.3.3 INHIBITING THE DIGESTIVE ENZYMES

The digestive enzymes are required for breaking down the dietary polysaccharides into glucose units which will increase the level of blood glucose in blood flow. The Alkaloids interfere with the enzyme binding site both competitive and non-competitive site and inhibits the digestive enzymes. The binding of the alkaloids to the enzyme will automatically lower the blood glucose level which is responsible for the anti-diabetic activity (Uthpala et al., 2020; Thanh et al., 2021; Adnan et al., 2017). The examples of plant which exhibit this mechanism are *Murraya koenigii* L., *Adhatoda vasica, Oriciopsis glaberrima, Catharanthus roseus* (L.), *Sarcococa saligna, Combretum dolichopetalum*, etc. (Banihani, 2017).

12.3.4 INCREASING THE INSULIN SECRETION

The glucagon (alpha-cells) and insulin (beta-cells) are secreted by islets of Langerhans. The glucagon and insulin exert antagonistic effect on the organs, the glucose level is lowered by insulin whereas the glucose level is increased by glucagon. The increase in the insulin secretion is required for anti-diabetic activity. The secretion of the insulin is increased by two ways first is by increasing the secretion of insulin producing stimulators by the presence of alkaloid compound. When the insulin secretion increases the glucose uptake increases in turn decreasing the blood glucose level. Examples of alkaloids which follow this type of mechanism are *Xanthocercis zambesiaca, Berberis aristata, Stephania glabra, Cissampelos pareira* (Shukla et al., 2011). Another mechanism is Enhancing the regeneration of beta cells of pancreas. The alkaloids will enhance the regeneration of beta cells and that will lead to increase in insulin production which will decrease the blood glucose level, which is required for the anti-diabetic activity (Cao et al., 2021; Zeb, 2015; Sadeghi et al., 2014). Examples are *Trigonella foenumgraecum, Tabernaemontana bufalina* Lour.

12.3.5 ANTIOXIDANT ACTIVITY

When there is increase in the glucose level in blood flow that results in hyperglycemic environment where the oxidative stress increases and causes micro-vascular and macro-vascular complications. The antioxidants react to the radicals in such a way that they counter act the oxidative stress that is resulted due to hyperglycemia which will in turn reduce the glucose level in the blood and act as anti-diabetic agents (Salem et al., 2015; Lattanzio et al., 2009; Mirmiran et al., 2020). The examples are *Rhizoma coptidis, Nasturtium officinale.*

12.3.6 INHIBITING THE ADVANCED GLYCATION END-PRODUCT

The Advanced glycation end-products are formed when reaction takes place between the carbonyl group of reducing sugar and amino groups (proteins/nucleic acid, lipids), The advanced glycation products that are formed in this reaction are responsible for the secondary complications (like retinopathy, nephropathy, cardiomyopathy, and neuropathy) which take place in diabetic patients. Natural alkaloids use different mechanisms like degradation, protection of amino groups, balancing of oxygen species, removal or reducing the carbonyl group to inhibit the production of advanced glycation end-products, hence when the products are not formed the secondary complications will not occur (Ninfali & Angelino, 2013; Kwon et al., 2010; Ghahari et al., 2017). The plant alkaloids showing this mechanism are *Octaea paranapiacabensis, Rhizoma coptidis, Herba leonuri.*

12.3.7 INHIBITING ALDOSE REDUCTASE

In the environment where there is increase in blood glucose level various complications occur due to excess of polyol metabolism, this further leads to increase in the osmotic stress and sorbitol levels. This polyol metabolism can be reduced by Aldose reductase which is the key enzyme in polyol pathway which will lead to excess production of reactive oxygen species (ROS). So, inhibiting the Aldose reductase by the plant alkaloid can be used for reducing the blood glucose levels (Kwon et al., 2008). The plant alkaloid examples are *Coptis japonica, Tinospora cardifolia, Catharanthus roseus.*

12.3.8 TERPENES

Terpenes are also called terpenoids or isoprenoids. Terpenes are the compounds which are naturally occurring in plants, they are also responsible for the pigmentation, taste, and fragrance of plants. Terpenes are derived naturally from mevalonic acid (MVA) pathway. Terpenes also have medicinal uses apart for taste and fragrance properties. The basic unit of terpenoids is called isoprene unit which has formula C_5H_8. The terpenes are classified based on the number of isoprene units present, they are Monoterpenes: The terpenes are said to be monoterpenes when they have presence of C_{10}, with chemical formula $C_{10}H_{16}$. Sesquiterpenes: In Sesquiterpenes there are C_{15}, with chemical formula $C_{15}H_{24}$. Diterpenes: There are C_{20} in the diterpenes, with chemical formula $C_{20}H_{36}$. Triterpenes: Triterpenes have presence of C_{30}, with chemical formula $C_{30}H_{48}$. Tetraterpenes: They are also known as carotenoids, with chemical formula $C_{40}H_{56}$ (Gürbüz et al., 2018).

There are various terpenes which show effect against diabetes treatment. *Stevia rebaudiana* has presence of active principle called stevia which is a diterpenoid. It is found that the stevia shows significant effect in treatment of hyperglycemia. Arteminisin which is widely known for anti-malarial effect also shows potential in treatment of type 1 diabetes. Terpenes, terpenediols derived from plant showed anti-diabetic properties along with alkaloids and flavonoids. Diterpenoid lactone Andrograholide is one of the promising terpenes which shows potential against diabetes treatment which is the major component of *Andrographis paniculata.* This compound acts by utilizing the glucose available in turn decreasing the level of plasma glucose (Mohamed et al., 2014).

Another terpene curcumin commonly known as turmeric which is obtained from Curcuma longa also shows potential anti-diabetic activity. It acts by the mechanism of counteracting with the oxidative stress and inflammation in reducing the blood glucose level. *Buddleja saligna* Willd. with terpenoids (betulonic acid, beutlone, and spinasterol) shows anti-diabetic by significantly inhibiting the alpha-glucosidase. *Cichorium intybus* L. also acts by inhibiting the alpha-glucosidase where in the terpenoids present are lactucin, dihydrolactucin, jacquinelin, and lactucopicrin. *Citrus sinensis* L. and Citrus limon with terpenoid limonene shows effect against diabetes by inhibiting alpha-amylase and alpha-glucosidase activities.

Panax ginseng (Ginsenoside Rb2, Re, Rg1, Rd) acts by promoting human pancreatic beta cell function. *Momordica charantia* L. (curcurbitane triterpenoids) acts against diabetes by showing alpha-glucosidase inhibition. *Prosopis juliflora* (24-methylencycloartan-3-one) acts by reducing the glucose level in the blood flow. *Salvia urmiensis* Bunge mechanism of action is by inhibiting the activities of alpha-amylase and alpha-glucosidase (Zou et al., 2021).

12.3.9 PHENOLICS

The Shikimic acid pathways in the plants are responsible for the production of phenolic compounds. Phenolic are secondary metabolites they vary from range of simple phenolic compounds to highly polymerized phenolic compounds. Chemically phenolic compounds have basic benzene rings which are substituted with hydroxyl groups. Phenolic compounds are responsible for defense response in plants, accelerating the process of pollination, pigmentation, astringency, defense against herbivores, antifungal, and antibacterial activities.

The Phenolic compounds have different variety they are phenolic acids (Hydroxybenzoic acids and Hydroxycinnamic acids), simple and complex flavonoids (flavonols, flavones, flavanols, flavanones, and isoflavones) and colored anthocyanins. The diabetic activity is shown by phenolic compounds by inhibiting the alpha-amylase and alpha-glucosidase which was concentration dependent. The secondary diabetic complications can also be prevented by phenolic compounds like nephropathy, neuropathy, retinopathy, and CVDs.

Few examples of phenolic acting as anti-diabetic agents along with the mechanism of action – *B. ciliate* (galloylepicatechin, galloylcatechin) acts on pancreatic alpha-amylase and maltase to give anti-diabetic effect (Shidfar et al., 2015). *E. coracana* (apigenin, caffeic acid, ferulic acid, syringic acid) acts on pancreatic alpha-amylase and maltase to give anti-diabetic effect. *Diospyros kaki* (proanthocyanidin, epigallocatechin) shows anti-diabetic effect by acting on pancreatic alpha-amylase. *C. anthelminticum* (Caffeic acid, ellagic acid ferulic acid) shows anti-diabetic effect by acting on maltase, sucrose, salivary alpha-amylase. *Rubus idaeus* (ellagic acid, catechin) acts on alpha-glucosidase to give anti-diabetic effect. *T. chebula* (chebulagic acid, chebulanin, chebulinic acid) acts on maltase to give effect against diabetes (Liu et al., 2007).

12.3.10 VEGETABLES FOR DIABETES PREVENTION AND TREATMENT

As we know that the diet is very much important for the diabetes patients because the diet is also potential in fluctuating the blood glucose levels. There are vegetables and fruits which may increase the blood glucose level whereas other may decrease the blood glucose level. Vegetables (cruciferous vegetables, green leafy vegetable, yellow vegetables) and Fruits (berries) can be used as the tool for prevention of diabetes and secondary complications caused by diabetes. Vegetables are defined as the edible/eatable part of the plant. For the treatment of diabetes vegetables can be considered as an adjuvant therapy. Nutrition can be considered as the crucial part of the prevention approach. The diet with low glycemic index and high fiber can play critical role in prevention of diabetes. The various examples of vegetables that have hypoglycemic action are discussed in subsections (Wang et al., 2017).

12.3.10.1 MOMORDICA CHARANTIA

M. charantia commonly called bitter gourd or bitter melon or karela belonging to the family Cucurbitaceae. This plant is widely found in Asia, India, few parts of Africa and South America. Bitter gourd has chemical constituents like triterpenes, steroids, alkaloids, and phenolic compounds which are accountable for anti-diabetic activity. The mechanism of action like protection of pancreatic beta cells and insulin secretion, stimulation of glucose metabolism is effective against the prevention of diabetes. This vegetable has been used as remedy for treatment of diabetes in traditional medicine. The pre-clinical studies also show promising results and beneficiary evidence towards the treatment of diabetes. The bitter gourd has been used as ethnomedicine and dietary supplement in treatment of diabetes (El-Saber et al., 2020).

12.3.10.2 COCCINIA GRANDIS

C. indica commonly called Ivy gourd or scarlet gourd or tindora or kowai, belonging to the family Cucurbitaceae. The plant is found in Asian countries and widely in India and Sri Lanka. The main chemical constituents present in *C. indica* are triterpenoids, flavonoids, alkaloids, beta-sitosterol which are having potential anti-diabetic. The mechanism of action in treatment of diabetes is suppressing the key enzymes involved in gluconeogenesis and stimulating the key enzymes involved in HMP (hexose monophosphate shunt) pathway. By many different traditional medicine literatures, it is seen that this vegetable is being used as dietary supplement in treatment of type II diabetes. There are also studies carried out that show effectiveness of ivy gourd towards the diabetes treatment so it can be recommended as one of the potential nutraceuticals in treating the patients with diabetes (Keservani et al., 2010a, b; Eidi et al., 2006).

12.3.10.3 BRASSICA JUNCEA

B. juncea commonly called mustard or brown mustard or oriental mustard, belonging to the family Brassicaceae (Cruciferae). The plant is found in Asia, North America, Europe, and Africa. The main active principles responsible for anti-diabetic are sinapic acid, sinapine, gallic acid, ferulic acid, rutin, caffeic acid. The mode of action is by inhibiting the glycogen phosphorylase and gluconeogenic effect in prevention and treatment of diabetes. The Mustard leaves are consumed as vegetables, which was found to be hypoglycemic. The evidence also show that the mustard leaves vegetable can be recommended as potential dietary supplement in prevention of diabetes (Kim et al., 2016).

12.3.10.4 CYAMOPSIS TETRAGONOLOBA

C. tetragonoloba commonly called cluster bean or Indian bean cluster, belonging to the family Fabaceae. The plant is found in Asia, South Africa, and United states. The chemical composition of Cluster bean is gallic acid, caffeic acid, ellagic acid, quercetin, rutin, kaempferol. Cluster bean possesses hypoglycemic effect and anti-hyperglycemic property. The mode of action in this case is by showing protective effect on the beta cell of pancreatic tissue. *C. tetragonoloba* vegetables is consumed as food, the gum which is accumulated from the endosperm of the seeds is beneficial for the activity. Cluster bean can be used as adjuvant in the diabetes therapy (Gopal et al., 2017).

12.3.10.5 MORINGA OLEIFERA LAM.

M. oleifera commonly called horseradish or drumstick or miracle tree, belonging to the family Moringaceae. This plant is found in sub-tropic and tropic regions. This plant is rich sources of antioxidants and also possess activity against diabetes. The major constituent responsible to show anti-diabetic activity are Quercetin, moringinine, chlorogenic acid. The presence of the above-mentioned ingredients is responsible to show the anti-diabetic activity by protecting the beta cells against the ROS mediated damage by enhancing the antioxidant effect. Due to the excellent antioxidant activity of this vegetable, it can be considered as one of the potential supplements in diabetes treatment (Jaafar et al., 2019).

12.3.10.6 TRIGONELLA FOENUMGRAECUM

T. foenumgraecum commonly called Fenugreek or methi, belonging to the family Leguminosae. This plant is cultivated in Europe, Western Asia, India, Northern Africa regions. This plant contains alkaloids, flavonoids, steroids, saponins. There is enhancement of in serum level of insulin because of insulinotropic effect which is the reason this plant shows hypoglycemic effect. The increase in the level of serum insulin is due to the presence of alkaloids which gives anti-diabetic effect. In pre-diabetes treatment and prevention fenugreek powder is used to lower the blood glucose level. Fenugreek can be used comprehensively in the treatment and prevention of diabetes as a part of diet therapy. Fenugreek is available as nutraceutical product in diabetes management (Hichri et al., 2019).

12.3.10.7 PHASEOLUS VULGARIS L.

P. vulgaris is commonly called Haricot beans or navy beans or Boston beans, belonging to the family Fabaceae. This vegetable is consumed in various parts of the world as staple food in Central America, African countries, and Asian countries. *P. vulgaris* is rich source of polyphenols, flavonoids, anthocyanins, and tannins. The constituents like ferulic acid, sinapic acid, vanillic acid, chlorogenic acid, kaempferol, quercetin, etc., are the constituents which are present in *P. vulgaris*. The *P. vulgaris* shows hypoglycemic effect by antagonist action on the key enzymes' alpha-amylase and alpha-glucosidase. The *P. vulgaris* also shows promising antioxidant activity which also adds up to the treatment of diabetes. Due to the anti-diabetic and antioxidant action of *P. vulgaris* it is being used as functional food and/or nutraceuticals in various parts of the world (Hichri et al., 2019).

12.3.10.8 SPINACIA OLERACEA L.

S. oleracea is commonly called spinach or palak, belonging to the family Amaranthaceae. This vegetable is consumed in different part of Southeast Asia in countries like Japan, Indonesia, China Korea, Taiwan, India, and Malaysia. *S. oleracea* contains quercetin, kaempferol, ferulic, p-coumaric, vanillic acid. *S. oleracea* shows hypoglycemic action by protecting the beta cells of pancreas by which the efficiency of insulin also increases. It also shows very good antioxidant activity. Due to anti-diabetic and antioxidant potential this vegetable can be used as adjuvant therapy in treatment of diabetes patient. This is also used as prevention in pre-diabetes (Sharma et al., 2020).

12.3.10.9 BRASSICA OLERACEA VAR. ITALICA

B. oleracea has common name Broccoli, belongs to the cabbage family namely Brassicaceae. The vegetable Broccoli is native to Eastern Mediterranean and Asia. It is now even cultivated in America and England. Broccoli contains phytochemical constituents like alkaloids, flavonoids, polyphenols, phenolics, and polypeptides. The vegetable Broccoli possesses anti-diabetic potential by improving the insulin resistance and antioxidant potential. The phytochemical compounds that are present in Broccoli have been showing anti-diabetic potentials so can be considered as supplementary diet therapy in treatment and prevention of diabetes (Dixit & Kar, 2010).

12.3.10.10 DAUCUS CAROTA L.

D. carota was commonly called black carrot or Kaali Gajar, which belongs to the family of Umbellifers. They are commonly eaten in Asian countries like India, Turkey Afghanistan, and Pakistan. The black carrot has presence of chemical constituents like flavonoids, phenolic acids (ferulic acid, gallic acid, sinapic acid), anthocyanins. Black carrot has significant antioxidant property which plays a significant role in oxidative stress. The phenolic compounds which are present in black carrot are very much useful in management of diabetes. Due to various nutraceuticals potentials, it can be used as dietary supplement in treatment and prevention of diabetes (Fikselová et al., 2008).

12.3.10.11 BRASSICA OLERACEA VAR CAPITATA

B. oleracea is commonly known as Cabbage or celery cabbage, belonging to the family Brassicaceae or Cruciferae. The geographical origin of cabbage is Asia, Middle east, costal area of southeastern Europe. The Phyto-chemicals present in Cabbage are phenolic, flavonoids, anthocyanins, and glucosinolates (GSs). The mechanism of action which show anti-diabetic action is through alpha-glucosidase inhibition and alpha amylase inhibition. Cabbage is being widely cultivated worldwide and also being consumed by many people all around the world as an important part of diet. Cabbage is having valuable bioactive compounds which make cabbage a potential function food in treatment and prevention of diabetes (Pinilla et al., 1995).

12.3.10.12 ABELMOSCHUS ESCULENTUS

A. esculentus is commonly called okra or gumbo or ladies finger, belonging to the family Malvaceae. Bioactive compounds include polyphenol, flavonoids (Keservani & Sharma, 2014; Keservani et al., 2016a, b) and catechins. Due to the presence of quercetin, fiber, and polysaccharides okra is considered as one of the valuable functional foods. The mode of action in diabetes is by inhibition of alpha-glucosidase and alpha-amylase. Okra also possesses very good antioxidant activity. Both the actions mentioned above are responsible to keep in check the blood glucose level which is the basic requirement of anti-diabetic prevention and treatment. Okra has been widely used as functional food in various regions of world in prevention of diabetes (Sengkhamparn et al., 2010).

12.3.10.13 CUCUMIS SATIVUS L.

C. sativus commonly called cucumber or kheera, belonging to the family Cucurbitaceae. Cucumber originated in South Asia, but now cultivated in different parts of world in Asia, Europe. Major bioactive principles present in cucumber polyphenols, Anthocyanins, Flavonoids, and Alkaloids. Oxidative and carbonyl stress plays very critical role in diabetes progression which further leads to development of free radical generation and weakening of antioxidant defense mechanism so to overcome this condition compounds which protect ROS are required. In case of cucumber there are many compounds like beta-carotene, alpha-carotene, and Vitamin A and C which are responsible for giving promising antioxidant activity, which can help in prevention of diabetes. So, Cucumber can be used as one of the dietary supplements in diabetes management (Tatlioglu, 1993).

12.3.10.14 CUCURBITA PEPO L.

C. pepo is widely known as Pumpkin or field pumpkin or kadoo, belonging to the family Cucurbitaceae. In Asia and North America pumpkin is used in diabetes management. The chemical class presents in Pumpkin are proteins, polysaccharides, peptides, vitamins, and sterols. The major bioactive principles available are p-coumaric acid, caffeic acid and derivatives, vanillic acid, quercetin, and linoleic acid. Flavonoids and polyphenols are responsible for antioxidant and hypoglycemic effect from pumpkin. By improving the pancreas tissue damage, inhibiting alpha-amylase, and alpha-glucosidase enzymes the blood glucose level is reduced in patients with diabetes, hence giving hypoglycemic effect. Pumpkin is widely used as traditional medicine and can be used as dietary supplement in prevention and treatment of diabetes (Andolfo et al., 2017).

12.3.10.15 RAPHANUS SATIVUS

R. sativus commonly called Radish, belongs to family Brassicaceae. Radish was found in anciently in Mediterranean region. Later it was spread to China and Japan and now widely used all over the world. In Indian system as well as Unani system of medicine Radish is used in treatment and prevention of Diabetes. The bioactive constituents present in Radish are catechins, polyphenolics, anthocyanins, and flavonoids. Radish has the ability to improve the antioxidant defense mechanism, promote glucose uptake and improve hormonal induced glucose hemostasis, which is useful in treatment and prevention of diabetes. This vegetable is already being used in traditional medicine for diabetes treatment, so it can be considered as adjuvant dietary supplement in prevention of diabetes (Han et al., 1996).

12.3.10.16 BRASSICA RAPA SUBSP. RAPA

Brassica rapa subsp. Rapa commonly known as Turnip belonging to the family Cruciferae Brassica. It has originally been cultivated in Asia. Turnip has multiple bioactive constituents like isothiocyanate, GSs, flavonoids, and phenolics. Turnip has abundant antioxidant components which decrease the oxidative stress resulted by the excess ROS. It shows effect on alpha-glucosidase enzyme by inhibiting it. Turnip can be used as function food in treatment and prevention of diabetes. Due to the benefits of the Turnip, it is considered as one of the potential diet supplements in adjuvant therapy (Li et al., 2003).

12.3.10.17 NASTURTIUM OFFICINALE

N. officinale commonly known as watercress or yellowcress, belonging to the family Brassicaceae. Watercress is found in some part of Asia and Europe. The leaves are being extensively used as diuretic, anti-diabetic, and expectorant traditionally. The bioactive compounds which are present in watercress are coumaric acid derivatives, quercetin derivatives and gallic acid derivatives. The phenolic components have very potential antioxidant effect which is helpful in decreasing the oxidative stress during the diabetes treatment. The antioxidant benefit can be helpful in prevention and treatment of diabetes; hence watercress can be used as the part of diet of diabetic patient (Yazdanparast et al., 2008).

12.3.10.18 CYNARA CARDUNCULUS VAR. SCOLYMUS

C. cardunculus is commonly called artichoke or green artichoke or French artichoke or Globe artichoke, belonging to the family Asteraceae. The Greek and the Romans from centuries utilize artichoke as food as well as medicine. The chemical components present in artichoke are Phenolic acids (caffeic acid and quinic acid), flavonoids (caffeoylquinic acid and luteolin) and Polyphenols. Artichoke is one among the richest source of dietary antioxidant which makes it potential phytopharmaceutical product. The presence of flavonoids helps in reducing the oxidative stress. Artichoke can be used as one of the nutraceutical products and can be used as a part of diet in prevention and treatment of diabetes patients (Pandino et al., 2011).

12.3.10.19 BETA VULGARIS SUBSP. VULGARIS

B. vulgaris commonly called beet root or common beet or garden beet, Belongs to family Amaranthaceae. Beet root originated in Middle East, which was further spread to America, Europe, and Asia. The phytoconstituents present in beet root are essential oils (EOs), polyphenols, Phenolic acids, Flavonoids, Betalains, and phenolic amides. The hypoglycemic action is associated with the inhibition of gluconeogenesis and glycogenolysis whereas flavonoids inhibit the glucose transporters. The above mechanism is responsible for showing anti-diabetic effect. Beetroot is widely used as functional food as it has various potential therapeutic properties, hence can be used as diet therapy in prevention and treatment of diabetes (Pink, 1993).

12.3.10.20 GLYCINE MAX L. MERRIL

G. max commonly called soya bean or soja bean, belonging to the family Fabaceae. Soya bean is being used for years in Asian countries as functional staple food. Soybean contains isoflavones, phenol, proteins, volatile compounds, tocopherols, and sterols. Many studies suggest that the presence of isoflavones provides protection against diabetes by improving the glucose tolerance. As the structure of isoflavones is similar to that of estrogen they act as agonist or competitive antagonist by which the insulin secretion is improved along with increase in beta mass cells, which help in anti-diabetic action. Due to this mechanism the functional food soya bean can be considered as an important diet supplement in prevention and treatment of diabetes (Chiari et al., 2005).

12.3.10.21 SOLANUM MELONGENA

S. melongena commonly called eggplant or aubergine or brinjal or berenjena, belongs to family of Solanaceae. This plant is cultivated in Asia, Africa, and Europe for centuries. The bioactives present in eggplant are phenolics, flavonoids, hydrocinnamic acid derivatives, anthocyanins, carotenoids, and glycoalkaloids. The eggplant has good antioxidant and inhibitory activity against the carbohydrate modulating enzymes (alpha glucosidase). Eggplant is potential in managing the glucose induced hyperglycemia. The diabetes related complications can be sorted by glucose absorption and cellular antioxidant protection. The Eggplant has strong rationale to be used as functional food in improving the condition of diabetes patients as well as in diabetes prevention (Kalloo, 1993).

12.3.10.22 BRASSICA RAPA SUBSP. CHINENSIS

B. rapa commonly called Bokchoy or pak choi or bak choi Chinese cabbage or spoon cabbage, belonging to the family Brassicaceae. The Bokchoy is widely found in China, East Asia, and some parts of Europe. Bok choy contains carotenoids, GSs, phenolic acid (and their derivatives), glycosides, and flavonoids. The major chemical constituents are ascorbic acid, sinigrin, sinapic, ferulic p-coumaric, caffeic acids, quercetin, and kaempferol. The phytoconstituents reduce the oxidative stress by antioxidant activity and inhibiting the alpha-glucosidase enzyme can help in reducing the blood glucose level. This mechanism can help in treatment of diabetes by acting as hypoglycemic agents. Bokchoy can be used as functional food in the diet of diabetes patient in prevention and treatment of diabetes (Liao et al., 2005).

12.3.10.23 ZINGIBER OFFICINALE

Z. officinale commonly called Ginger or rhizomes, belongs to family Zingiberaceae. The plant is widely seen in tropical regions of Asia and Africa. Ginger has major class of compounds are phenolic and terpene, whereas chemical constituents namely gingerol, shojail, paradol, and zingerone. Inhibition of alpha-amylase and alpha-glucosidase enzymes which are associated diabetes like carbohydrate metabolism and hyperglycemia is done by virtue of phenolic compounds. Ginger stimulates the insulin secretion by pancreatic beta cells protection. The side effects of ginger can be neglected which make ginger potential food supplemental. Ginger can be used as one of the food supplements in diabetes treatment and prevention (Mbaveng & Kuete, 2017).

12.3.10.24 ALLIUM SATIVUM

A. sativum commonly called Garlic belongs to family Alliaceae. Garlic is originated in Mediterranean region but now it has been cultivated worldwide. Garlic has been used for ages by many traditional cultures in treatment as well as part of diet. Garlic chemically contains allicin, allyl propyl disulfide, cysteine sulfoxide and S-allyl cysteine sulfoxide. The above-mentioned chemical constituents are responsible for reducing the glucose level in the blood by enhancing the insulin secretion by preventing the beta cells of pancreas. Garlic also possesses very effective antioxidant property. The mechanism noted here is responsible to give anti-diabetic action and garlic can be used in prevention of diabetes. So, garlic can be considered as one of the potential candidates for the diet of diabetic patients (Takagi, 2020).

12.3.10.25 LACTUCA SATIVA L

L. sativa commonly called lettuce or garden lettuce, belonging to the family Asteraceae. Lettuce is cultivated mostly in China, US, and Western Europe. The lettuce contains high amount of water due to which it is not considered as nutritional food, whereas lettuce is actually nutritious. The bioactive compounds present in Lettuce are carotenoids, phenolic compounds, and flavonoids. Lactucaxanthin is the chemical constituent belonging to carotenoid class which is present in Lettuce is responsible for giving anti-diabetic activity. The approach by which the level of blood glucose is decreased is by inhibiting the intestinal and pancreatic alpha-amylase and alpha-glucosidase enzymes and also by controlling the breakdown of dietary starch. The nutritional value of Lettuce is not appreciated but it can be one of the most potential vegetables in prevention of diabetes, so it can be considered as a part of diet of diabetic patients (Chon et al., 2005).

12.3.10.26 ERUCA SATIVA L.

E. sativa commonly known as Arugula or jarjeer, belonging to the family Brassicaceae. Arugula originated in the coastal region of Mediterranean Sea, grown in South Asia and Middle East. The phytochemical which are present in Arugula are flavonoids, phenolic acid, carotenoids, terpenes, glycosides, sterols, and alkaloids. Arugula shows effect against hyperglycemia and hyperlipidemia by modifying the pancreatic islet cells. The vegetable also shows effect against metabolism of carbohydrates by inhibiting carbohydrate hydrolyzing enzymes such as alpha-amylase, alpha-glucosidase, and beta-glucosidase. This vegetable can be considered as one of the important nutraceuticals (Keservani et al., 2017, 2020) which can be added in the diet of patients with diabetes which can help them in treatment and prevention (Zhi et al., 2015).

12.3.10.27 PRAECITRULLUS FISTULOSUS

P. fistulosus commonly called Indian squash or Indian baby pumpkin or Tinda or Indian round gourd, belonging to the family of Cucurbitaceae. This vegetable is widely grown as well as popular in India and South Asian countries. Indian squash possesses higher quantity of phytoconstituents like phenolic compounds, flavonoids, saponins, terpenoids, cardiac glycoside and tannins. Indian squash is storehouse to natural antioxidant, which has very high therapeutic value in treatment of diabetes. The antioxidant potential of Indian squash can reduce the oxidative stress build in the patient with diabetes. So Indian squash can be taken in consideration as one of the important vegetables in the diet plan during the treatment and prevention of diabetes (Gautam et al., 2011).

12.4 CONCLUSION

Lifestyle and food habits can play a critical role when it comes to management of blood glucose level in diabetic patients. Hence Diet plan can be used a key in prevention and treatment of diabetes. The vegetables which show effect on beta cells of pancreas and inhibit the glucose production can be considered as a part of diet plan. This adjuvant therapy can be beneficial in the management of diabetes Mellitus.

ACKNOWLEDGMENT

The author acknowledges the kind support and guidance of the Department of Pharmacognosy, KLE College of Pharmacy, KLE University of Higher Education and Research, Belagavi, Karnataka, India.

KEYWORDS

- **aldose reductase**
- **diabetes**
- **glycation**
- **metabolites**
- **plant**
- **prevention**
- **terpenes**
- **vegetables**

REFERENCES

Abdulrahman, B. O., Bala, M., & Bello, O. M., (2020). Bioactive compounds of black bean (*Phaseolus vulgaris* L.). *Bioactive Compounds in Underutilized Vegetables and Legumes*, 1–20.

Adhikari, B., (2021). Roles of alkaloids from medicinal plants in the management of diabetes mellitus. *Journal of Chemistry, 2021*.

Adnan, M., Gul, S., Batool, S., Fatima, B., Rehman, A., Yaqoob, S., Shabir, H., et al., (2017). A review on the ethnobotany, phytochemistry, pharmacology and nutritional composition of *Cucurbita pepo* L. *The Journal of Phytopharmacology, 6*(2), 133–139.

Ahire, E. D., Sonawane, V. N., Surana, K. R., & Talele, G. S., (2021). Drug discovery, drug-likeness screening, and bioavailability: Development of drug-likeness rule for natural products. In: *Applied Pharmaceutical Practice and Nutraceuticals* (pp. 191–208). Apple Academic Press.

Albuquerque, B. R., Heleno, S. A., Oliveira, M. B., Barros, L., & Ferreira, I. C., (2021). Phenolic compounds: Current industrial applications, limitations, and future challenges. *Food & Function, 12*(1), 14–29.

Ali, F. T., Hassan, N. S., & Abdrabou, R. R., (2015). Potential activity of *Moringa oleifera* leaf extract and some active ingredients against diabetes in rats. *Int. J. Sci. Eng. Res., 6*(5), 1490.

Andolfo, G., Di Donato, A., Darrudi, R., Errico, A., Aiese, C. R., & Ercolano, M. R., (2017). Draft of zucchini (*Cucurbita pepo* L.) proteome: A resource for genetic and genomic studies. *Frontiers in Genetics*, *8*, 181.

Asif, M., (2011). The role of fruits, vegetables, and spices in diabetes. *International Journal of Nutrition, Pharmacology, Neurological Diseases, 1*(1), 27.

Attanayake, A. P., Jayatilaka, K. A., & Mudduwa, L. K., (2016). Anti-diabetic potential of ivy gourd (*Coccinia grandis*, family: *Cucurbitaceae*) grown in Sri Lanka: A review. *Journal of Pharmacognosy and Phytochemistry, 5*(6), 286–289.

Babu, P. V., Liu, D., & Gilbert, E. R., (2013). Recent advances in understanding the anti-diabetic actions of dietary flavonoids. *The Journal of Nutritional Biochemistry, 24*(11), 1777–1789.

Bahadoran, Z., Mirmiran, P., Hosseinpanah, F., Hedayati, M., Hosseinpour-Niazi, S., & Azizi, F., (2011). Broccoli sprouts reduce oxidative stress in type 2 diabetes: A randomized double-blind clinical trial. *European Journal of Clinical Nutrition., 65*(8), 972–977.

Bahadoran, Z., Tohidi, M., Nazeri, P., Mehran, M., Azizi, F., & Mirmiran, P., (2012). Effect of broccoli sprouts on insulin resistance in type 2 diabetic patients: A randomized double-blind clinical trial. *International Journal of Food Sciences and Nutrition, 63*(7), 767–771.

Banihani, S. A., (2017). Radish (*Raphanus sativus*) and diabetes. *Nutrients, 9*(9), 1014.

Basharat, S., Junaid, A., Masood, I., Azhar, N., Imran, S., Basit, A. A., & Saleem, M., (2021). Beneficial effects of okra in diabetes mellitus. *Asian Journal of Allied Health Sciences (AJAHS),* 67–77.

Bazzano, L. A., Serdula, M. K., & Liu, S., (2003). Dietary intake of fruits and vegetables and risk of cardiovascular disease. *Current Atherosclerosis Reports, 5*(6), 492–499.

Cao, Q., Wang, G., & Peng, Y., (2021). A critical review on phytochemical profile and biological effects of turnip (*Brassica rapa* L.). *Frontiers in Nutrition*, 459.

Chen, J., Mangelinckx, S., Adams, A., Wang, Z. T., Li, W. L., & De Kimpe, N., (2015). Natural flavonoids as potential herbal medication for the treatment of diabetes mellitus and its complications. *Natural Product Communications, 10*(1), 1934578X1501000140.

Chiari, W. C., Toledo, V. D. A. A. D., Ruvolo-Takasusuki, M. C. C., Oliveira, A. J. B. D., Sakaguti, E. S., Attencia, V. M., Costa, F. M., & Mitsui, M. H., (2005). Pollination of soybean (*Glycine max* L. Merril) by honeybees (*Apis mellifera* L.). *Brazilian Archives of Biology and Technology, 48*, 31–36.

Chon, S. U., Jang, H. G., Kim, D. K., Kim, Y. M., Boo, H. O., & Kim, Y. J., (2005). Allelopathic potential in lettuce (*Lactuca sativa* L.) plants. *Scientia Horticulturae, 106*(3), 309–317.

Coskun, O., Kanter, M., Korkmaz, A., & Oter, S., (2005). Quercetin, a flavonoid antioxidant, prevents and protects streptozotocin-induced oxidative stress and β-cell damage in rat pancreas. *Pharmacological Research, 51*(2), 117–123.

Cox-Georgian, D., Ramadoss, N., Dona, C., & Basu, C., (2019). Therapeutic and medicinal uses of terpenes. In: *Medicinal Plants* (pp. 333–359). Springer, Cham.

Demir, Y., Durmaz, L., Taslimi, P., & Gulçin, İ., (2019). Antidiabetic properties of dietary phenolic compounds: Inhibition effects on α-amylase, aldose reductase, and α-glycosidase. *Biotechnology and Applied Biochemistry, 66*(5), 781–786.

Dixit, Y., & Kar, A., (2010). Protective role of three vegetable peels in alloxan induced diabetes mellitus in male mice. *Plant Foods for Human Nutrition, 65*(3), 284–289.

Eidi, A., Eidi, M., & Esmaeili, E., (2006). Antidiabetic effect of garlic (*Allium sativum* L.) in normal and streptozotocin-induced diabetic rats. *Phytomedicine, 13*(9, 10), 624–629.

El-Saber, B. G., Magdy, B. A., Wasef, L. G., Elewa, Y. H., Al-Sagan, A. A., Abd El-Hack, M. E., Taha, A. E., et al., (2020). Chemical constituents and pharmacological activities of garlic (*Allium sativum* L.): A review. *Nutrients, 12*(3), 872.

Esmaeilzadeh, D., Razavi, B. M., & Hosseinzadeh, H., (2020). Effect of *Abelmoschus esculentus* (okra) on metabolic syndrome: A review. *Phytotherapy Research, 34*(9), 2192–2202.

Fikselová, M., Šilhár, S., Mareček, J., & Frančáková, H., (2008). Extraction of carrot (*Daucus carota* L.) carotenes under different conditions. *Czech Journal of Food Sciences, 26*(4), 268–274.

Forouhi, N. G., & Wareham, N. J., (2010). Epidemiology of diabetes. *Medicine, 38*(11), 602–606.

Franz, M. J., Bantle, J. P., Beebe, C. A., Brunzell, J. D., Chiasson, J. L., Garg, A., Holzmeister, L. A., et al., (2002). Evidence-based nutrition principles and recommendations for the treatment and prevention of diabetes and related complications. *Diabetes Care, 25*(1), 148–198.

Gaddam, A., Galla, C., Thummisetti, S., Marikanty, R. K., Palanisamy, U. D., & Rao, P. V., (2015). Role of fenugreek in the prevention of type 2 diabetes mellitus in prediabetes. *Journal of Diabetes & Metabolic Disorders, 14*(1), 1–10.

Gandhi, G. R., Vanlalhruaia, P., Stalin, A., Irudayaraj, S. S., Ignacimuthu, S., & Paulraj, M. G., (2014). Polyphenols-rich *Cyamopsis tetragonoloba* (L.) Taub. beans show hypoglycemic and β-cells protective effects in type 2 diabetic rats. *Food and Chemical Toxicology, 66*, 358–365.

Gautam, S., Singh, P., & Shivhare, Y., (2011). *Praecitrullus fistulosus*: A miraculous plant. *Asian Journal of Pharmacy and Technology, 1*(1), 9–12.

Ghahari, S., Alinezhad, H., Nematzadeh, G. A., Tajbakhsh, M., & Baharfar, R., (2017). Chemical composition, antioxidant and biological activities of the essential oil and extract of the seeds of glycine max (soybean) from North Iran. *Current Microbiology, 74*(4), 522–531.

Ghorbani, A., Rashidi, R., & Shafiee-Nick, R., (2019). Flavonoids for preserving pancreatic beta cell survival and function: A mechanistic review. *Biomedicine & Pharmacotherapy, 111*, 947–957.

Gopal, S. S., Lakshmi, M. J., Sharavana, G., Sathaiah, G., Sreerama, Y. N., & Baskaran, V., (2017). Lactucaxanthin–a potential anti-diabetic carotenoid from lettuce (*Lactuca sativa*) inhibits α-amylase and α-glucosidase activity *in vitro* and in diabetic rats. *Food & Function, 8*(3), 1124–1131.

Gupta, R., Mathur, M., Bajaj, V. K., Katariya, P., Yadav, S., Kamal, R., & Gupta, R. S., (2012). Evaluation of antidiabetic and antioxidant activity of *Moringa oleifera* in experimental diabetes. *Journal of Diabetes, 4*(2), 164–171.

Gupta, S., Burman, S., Nair, A. B., Chauhan, S., Sircar, D., Roy, P., Dhanwat, M., et al., (2022). *Brassica oleracea* extracts prevent hyperglycemia in type 2 diabetes mellitus. *Preventive Nutrition and Food Science, 27*(1), 50.

Gürbüz, N., Uluişik, S., Frary, A., Frary, A., & Doğanlar, S., (2018). Health benefits and bioactive compounds of eggplant. *Food Chemistry, 268*, 602–610.

Haber, S. L., & Keonavong, J., (2013). Fenugreek use in patients with diabetes mellitus. *American Journal of Health-System Pharmacy, 70*(14), 1196–1203.

Han, D. H., & Lee, J. H., (1996). Effects of liming on uptake of lead and cadmium by *Raphanus sativa*. *Archives of Environmental Contamination and Toxicology*, *31*(4), 488–493.

Hichri, F., Omri, A., Hossan, A. S., & Ben, J. H., (2019). Alpha-glucosidase and amylase inhibitory effects of *Eruca vesicaria* subsp. *longirostris* essential oils: Synthesis of new 1, 2, 4-triazole-thiol derivatives and 1, 3, 4-thiadiazole with potential inhibitory activity. *Pharmaceutical Biology, 57*(1), 564–570.

Hichri, F., Omri, H. A., Maha, M., Saad, M. H. A., Flamini, G., & Ben, J. H., (2019). Chemical composition, antibacterial, antioxidant and *in vitro* antidiabetic activities of essential oils from Eruca vesicaria. *Chemistry & Biodiversity, 16*(8), e1900183.

Jaafar, N. S., & Jaafar, I. S., (2019). *Eruca sativa* Linn.: Pharmacognostical and pharmacological properties and pharmaceutical preparations. *Asian J. Pharm. Clin. Res., 12*(3), 39–45.

Jaiswal, D., Rai, P. K., Mehta, S., Chatterji, S., Shukla, S., Rai, D. K., Sharma, G., et al., (2013). Role of *Moringa oleifera* in regulation of diabetes-induced oxidative stress. *Asian Pacific Journal of Tropical Medicine, 6*(6), 426–432.

Jardine, M. A., Kahleova, H., Levin, S. M., Ali, Z., Trapp, C. B., & Barnard, N. D., (2021). Perspective: Plant-based eating pattern for type 2 diabetes prevention and treatment: Efficacy, mechanisms, and practical considerations. *Advances in Nutrition, 12*(6), 2045–2055.

Joseph, B., & Jini, D., (2013). Antidiabetic effects of *Momordica charantia* (bitter melon) and its medicinal potency. *Asian Pacific Journal of Tropical Disease, 3*(2), 93–102.

Kalloo, G., (1993). Eggplant: *Solanum melongena* L. *Genetic Improvement of Vegetable Crops*, 587–604.

Karkute, S. G., Koley, T. K., Yengkhom, B. K., Tripathi, A., Srivastava, S., Maurya, A., & Singh, B., (2018). Anti-diabetic phenolic compounds of black carrot (*Daucus carota Subspecies sativus* var. atrorubens Alef.) inhibit enzymes of glucose metabolism: An in silico and *in vitro* validation. *Medicinal Chemistry, 14*(6), 641–649.

Keservani, R. K., & Sharma, A. K., (2014). Flavonoids: Emerging trends and potential health benefits. *Journal of Chinese Pharmaceutical Sciences, 23*(12), 815.

Keservani, R. K., Kesharwani, R. K., Sharma, A. K., Vyas, N., & Chadoker, A., (2010b). Nutritional supplements: An overview. *International Journal of Current Pharmaceutical Review and Research, 1*(1), 59–75.

Keservani, R. K., Kesharwani, R. K., Vyas, N., Jain, S., Raghuvanshi, R., & Sharma, A. K., (2010a). Nutraceutical and functional food as future food: A review. *Der Pharmacia Lettre, 2*(1), 106–116.

Keservani, R. K., Sharma, A. K., & Kesharwani, R. K., (2016a). Nutraceutical and functional foods for cardiovascular health. In: *Food Process Engineering* (pp. 257–278, 291–312). Apple Academic Press, CRC Press. ISBN: 9781771884020.

Keservani, R. K., Sharma, A. K., & Kesharwani, R. K., (2016b). Medicinal effect of nutraceutical fruits for the cognition and brain health. *Scientifica*, 10. Article ID 3109254. doi: 10.1155/2016/3109254.

Keservani, R. K., Sharma, A. K., & Kesharwani, R. K., (2017). An overview and therapeutic applications of nutraceutical and functional foods. *Recent Advances in Drug Delivery Technology*, 160–201. ISBN:9781522507543.

Keservani, R. K., Sharma, A. K., & Kesharwani, R. K., (2020). *Nutraceuticals and Dietary Supplements: Applications in Health Improvement and Disease Management*. CRC Press. ISBN: 9781771888738.

Khan, A., Mishra, A., Hasan, S. M., Usmani, A., Ubaid, M., Khan, N., & Saidurrahman, M., (2021). Biological and medicinal application of *Cucumis sativus* Linn. – review of current status with future possibilities. *Journal of Complementary and Integrative Medicine*.

Kim, M. J., Moon, Y., Tou, J. C., Mou, B., & Waterland, N. L., (2016). Nutritional value, bioactive compounds, and health benefits of lettuce (*Lactuca sativa* L.). *Journal of Food Composition and Analysis., 49*, 19–34.

Kumar, A., Aswal, S., Semwal, R. B., Chauhan, A., Joshi, S. K., & Semwal, D. K., (2019). Role of plant-derived alkaloids against diabetes and diabetes-related complications: A mechanism-based approach. *Phytochemistry Reviews, 18*(5), 1277–1298.

Kwon, D. Y., Daily, III. J. W., Kim, H. J., & Park, S., (2010). Antidiabetic effects of fermented soybean products on type 2 diabetes. *Nutrition Research, 30*(1), 1–3.

Kwon, Y. I., Apostolidis, E., & Shetty, K. J., (2008). *In vitro* studies of eggplant (*Solanum melongena*) phenolics as inhibitors of key enzymes relevant for type 2 diabetes and hypertension. *Bioresource Technology, 99*(8), 2981–2988.

Lattanzio, V., Kroon, P. A., Linsalata, V., & Cardinali, A., (2009). Globe artichoke: A functional food and source of nutraceutical ingredients. *Journal of Functional Foods, 1*(2), 131–144.

Li, H., Sivasithamparam, K., & Barbetti, M. J., (2003). Breakdown of a *Brassica Rapa* subsp. sylvestris single dominant blackleg resistance gene in B. napus rapeseed by *Leptosphaeria maculans* field isolates in Australia. *Plant Disease*, *87*(6), p.752.

Liao, C. S., Yen, J. H., & Wang, Y. S., (2006). Effects of endocrine disruptor di-n-butyl phthalate on the growth of Bok choy (*Brassica Rapa* subsp. chinensis). *Chemosphere, 65*(10), 1715–1722.

Lin, D., Xiao, M., Zhao, J., Li, Z., Xing, B., Li, X., Kong, M., et al., (2016). An overview of plant phenolic compounds and their importance in human nutrition and management of type 2 diabetes. *Molecules*, *21*(10), 1374.

Liu, C. T., Sheen, L. Y., & Lii, C. K., (2007). Does garlic have a role as an antidiabetic agent? *Molecular Nutrition & Food Research, 51*(11), 1353–1364.

Mbaveng, A. T., & Kuete, V., (2017). Zingiber officinale. In: *Medicinal Spices and Vegetables from Africa* (pp. 627–639). Academic Press.

Mirmiran, P., Houshialsadat, Z., Gaeini, Z., Bahadoran, Z., & Azizi, F., (2020). Functional properties of beetroot (*Beta vulgaris*) in management of cardio-metabolic diseases. *Nutrition & Metabolism, 17*(1), 1–5.

Mohamed, S., (2014). Functional foods against metabolic syndrome (obesity, diabetes, hypertension, and dyslipidemia) and cardiovasular disease. *Trends in Food Science & Technology., 35*(2), 114–128.

Mukhtar, H. M., Ansari, S. H., Bhat, Z. A., & Naved, T., (2006). Antihyperglycemic activity of *Cyamopsis tetragonoloba*. Beans on blood glucose levels in alloxan-induced diabetic rats. *Pharmaceutical Biology, 44*(1), 10–13.

Ninfali, P., & Angelino, D., (2013). Nutritional and functional potential of *Beta vulgaris* cicla and rubra. *Fitoterapia, 89*, 188–199.

Panche, A. N., Diwan, A. D., & Chandra, S. R., (2016). Flavonoids: An overview. *Journal of Nutritional Science, 5*.

Pandey, P., & Grover, K., (2020). Characterization of black carrot (*Daucus carota* L.), role in health promotion and disease prevention: An overview. *J. Pharmacogn. Phytochem., 9*(5), 2784–2792.

Pandino, G., Lombardo, S., Mauromicale, G., & Williamson, G., (2011). Profile of polyphenols and phenolic acids in bracts and receptacles of globe artichoke (*Cynara cardunculus* var. scolymus) germplasm. *Journal of Food Composition and Analysis*, *24*(2), 148–153.

Pinilla, M. G., Pérez-Alonso, M. J., & Velasco-Negueruela, A., (1995). Volatile constituents from fruits of *Daucus carota* L., subsp. *Gummifer* Hooker Fil. *Journal of Essential Oil Research*, *7*(4), 433–435.

Pink, D. A. C., (1993). Beetroot: *Beta vulgaris* subsp. vulgaris. In: *Genetic Improvement of Vegetable Crops* (pp. 473–477). Pergamon.

Platel, K., & Srinivasan, K., (1997). Plant foods in the management of diabetes mellitus: Vegetables as potential hypoglycaemic agents. *Food/Nahrung, 41*(2), 68–74.

Psaltopoulou, T., Ilias, I., & Alevizaki, M., (2010). The role of diet and lifestyle in primary, secondary, and tertiary diabetes prevention: A review of meta-analyses. *The Review of Diabetic Studies: RDS, 7*(1), 26.

Rasouli, H., Yarani, R., Pociot, F., & Popović-Djordjević, J., (2020). Anti-diabetic potential of plant alkaloids: Revisiting current findings and future perspectives. *Pharmacological Research, 155*, 104723.

Ravikumar, C., (2015). Therapeutic potential of *Brassica oleracea* (broccoli): A review. *Int. J. Drug Dev. Res., 7*(7), 9–10.

Roberts, J. L., & Moreau, R., (2016). Functional properties of spinach (*Spinacia oleracea* L.) phytochemicals and bioactives. *Food & Function, 7*(8), 3337–3353.

Sadeghi, H., Mostafazadeh, M., Sadeghi, H., Naderian, M., Barmak, M. J., Talebianpoor, M. S., & Mehraban, F., (2014). *In vivo* anti-inflammatory properties of aerial parts of *Nasturtium officinale*. *Pharmaceutical Biology, 52*(2), 169–174.

Salem, M. B., Affes, H., Ksouda, K., Dhouibi, R., Sahnoun, Z., Hammami, S., & Zeghal, K. M., (2015). Pharmacological studies of artichoke leaf extract and their health benefits. *Plant Foods for Human Nutrition, 70*(4), 441–453.

Sengkhamparn, N., Sagis, L. M., De Vries, R., Schols, H. A., Sajjaanantakul, T., & Voragen, A. G., (2010). Physicochemical properties of pectins from okra (*Abelmoschus esculentus* (L.) Moench). *Food Hydrocolloids*, *24*(1), 35–41.

Shaheer, M. K., Finose, A., Shafi, K. M., & Hashim, K. M. (2019). *Arugula: The Green Solution for Diabetics and Cholesterol. 6*, 179–183.

Sharma, R., Kour, J., Nayik, G. A., Alam, M. S., & Anand, N., (2020). Tinda (*Praecitrullus fistulosus*). In: *Antioxidants in Vegetables and Nuts-Properties and Health Benefits* (pp. 127–141). Springer, Singapore.

Shi, Q. I., Hui, S. U., Zhang, A. H., Hong-Ying, X. U., Guang-Li, Y. A., Ying, H. A., & Xi-Jun, W. A., (2014). Natural alkaloids: Basic aspects, biological roles, and future perspectives. *Chinese Journal of Natural Medicines., 12*(6), 401–406.

Shidfar, F., Rajab, A., Rahideh, T., Khandouzi, N., Hosseini, S., & Shidfar, S., (2015). The effect of ginger (*Zingiber officinale*) on glycemic markers in patients with type 2 diabetes. *Journal of Complementary and Integrative Medicine, 12*(2), 165–170.

Shukla, S., Chatterji, S., Mehta, S., Rai, P. K., Singh, R. K., Yadav, D. K., & Watal, G., (2011). Antidiabetic effect of *Raphanus sativus* root juice. *Pharmaceutical Biology, 49*(1), 32–37.

Surana, K. R., Ahire, E. D., Sonawane, V. N., & Talele, S. G., (2021). Biomolecular and molecular docking: A modern tool in drug discovery and virtual screening of natural products. In: *Applied Pharmaceutical Practice and Nutraceuticals* (pp. 209–223). Apple Academic Press.

Takagi, H., (2020). Garlic *Allium sativum* L. In: *Onions and Allied Crops* (pp. 109–146). CRC press.

Tatlioglu, T., (1993). Cucumber: *Cucumis sativus* L. In: *Genetic Improvement of Vegetable Crops* (pp. 197–234). Pergamon.

Teoh, S. L., & Das, S., (2018). Phytochemicals and their effective role in the treatment of diabetes mellitus: A short review. *Phytochemistry Reviews, 17*(5), 1111–1128.

Thanh, T. T., Quach, T. T., Yuguchi, Y., Nguyen, N. T., Van, N. Q., Van, B. N., Kawashima, S., & Ho, C. D., (2021). Molecular structure and anti-diabetic activity of a polysaccharide extracted from pumpkin *Cucurbita pepo*. *Journal of Molecular Structure, 1239*, 130507.

Tian, Y., & Deng, F., (2020). Phytochemistry and biological activity of mustard (*Brassica juncea*): A review. *CyTA-Journal of Food, 18*(1), 704–718.

Uthpala, T. G., Marapana, R. A., Lakmini, K., & Wettimuny, D. C., (2020). Nutritional bioactive compounds and health benefits of fresh and processed cucumber (*Cucumis sativus* L.). *Sumerianz Journal of Biotechnology, 3*(9), 75–82.

Uuh-Narvaez, J. J., & Segura-Campos, M. R., (2021). Cabbage (*Brassica oleracea* var. capitata): A food with functional propertics aimed to type 2 diabetes prevention and management. *Journal of Food Science., 86*(11), 4775–4798.

Varshney, R., Mishra, R., Das, N., Sircar, D., & Roy, P., (2019). A comparative analysis of various flavonoids in the regulation of obesity and diabetes: An *in vitro* and *in vivo* study. *Journal of Functional Foods, 59*, 194–205.

Vinayagam, R., & Xu, B., (2015). Antidiabetic properties of dietary flavonoids: A cellular mechanism review. *Nutrition & Metabolism, 12*(1), 1–20.

Wang, J., Zhang, X., Lan, H., & Wang, W., (2017). Effect of garlic supplement in the management of type 2 diabetes mellitus (T2DM): A meta-analysis of randomized controlled trials. *Food & Nutrition Research, 61*(1), 1377571.

Wang, P. Y., Fang, J. C., Gao, Z. H., Zhang, C., & Xie, S. Y., (2016). Higher intake of fruits, vegetables or their fiber reduces the risk of type 2 diabetes: A meta-analysis. *Journal of Diabetes Investigation, 7*(1), 56–69.

Watanabe, S., Okoshi, H., Yamabe, S., & Shimada, M., (2021). *Moringa oleifera* Lam. in diabetes mellitus: A systematic review and meta-analysis. *Molecules, 26*(12), 3513.

Yang, W., Chen, X., Li, Y., Guo, S., Wang, Z., & Yu, X., (2020). Advances in pharmacological activities of terpenoids. *Natural Product Communications, 15*(3), 1934578X20903555.

Yazdanparast, R., Bahramikia, S., & Ardestani, A., (2008). Nasturtium officinale reduces oxidative stress and enhances antioxidant capacity in hypercholesterolaemic rats. *Chemico-Biological Interactions*, *172*(3), 176–184.

Zeb, A., (2015). *Phenolic Profile and Antioxidant Potential of Wild Watercress (Nasturtium Officinale L.)* (Vol. 4, No. 1, pp. 1–7). Springer Plus.

Zhi, Y., Deng, Z., Luo, M., Ding, W., Hu, Y., Deng, J., Li, Y., et al., (2015). Influence of heavy metals on seed germination and early seedling growth in *Eruca sativa* mill. *American Journal of Plant Sciences*, *6*(5), 582.

Zou, L., Tan, W. K., Du, Y., Lee, H. W., Liang, X., Lei, J., Striegel, L., et al., (2021). Nutritional metabolites in *Brassica Rapa* subsp. chinensis var. parachinensis (choy sum) at three different growth stages: Microgreen, seedling, and adult plant. *Food Chemistry*, *357*, 129535.

CHAPTER 13

PLANT METABOLITES AND VEGETABLES FOR KIDNEY DISEASE PREVENTION AND TREATMENT

AMRUTA BALEKUNDRI,[1] SHALNI SATYA,[2] EKNATH D. AHIRE,[3] and RAJ K. KESERVANI[4]

[1]*Department of Pharmacognosy, KLE College of Pharmacy Belagavi, KLE University of Higher Education and Research, Belagavi, Karnataka, India*

[2]*Department of Environmental Science, Central University of Rajasthan, Rajasthan, India*

[3]*Department of Pharmaceutics, METs, Institute of Pharmacy, BKC, Adgaon, Nashik, Maharashtra, India*

[4]*Faculty of B. Pharmacy, CSM Group of Institutions, Prayagraj, Uttar Pradesh, India*

ABSTRACT

The kidney and the entire excretion system is very much essential in keeping the human body in balance. Many of the major functions of the body are directly or indirectly related to the kidney functions. The disturbance in the kidney functions leads to an imbalance in many other functions. It is very much essential that the kidney function properly. Kidney disease can be prevented and treated by proper management of the diet to a large extent. The intake of carbohydrates, proteins, vitamins,

Plant Metabolites and Vegetables as Nutraceuticals.
Raj K. Keservani, Bui Thanh Tung, Rajesh K. Kesharwani, and Eknath D. Ahire (Eds.)

fats, and micro-nutrients (sodium, potassium, and phosphorus) must be monitored properly as these play roles in the functioning of the kidney. The disturbance in the amount of intake might lead to several complications like proteinuria, hyperkalemia, and also fasten the progression of the kidney diseases. Plant-based diets are preferred over animal-based diets in patients suffering from chronic kidney diseases. The individualization of the diet therapy is also recommended to the patients suffering from kidney diseases.

13.1 INTRODUCTION

The bean-shaped organs responsible for filtration and removal of waste from the body are kidney. Kidneys are required to maintain the acid-base balance or electrolyte balance. The waste, toxins, and excess of water is removed out of the body via kidney. When the Kidney are unable to functional properly due to the damage in kidney lead to kidney diseases. Diabetes and High blood pressure are also responsible for kidney diseases as secondary complication. There are different types of kidney diseases. When the Kidney is dysfunctional for short period of time, where in there is temporary loss of Kidney function for not more than three months. The cause for acute kidney injury is due to reduced blood flow to kidney, low blood pressure, heavy blood and fluid loss or shock. When the Kidney is damaged to that extend where in reversing back to normal is not possible then it is been termed as long term or chronic kidney injury. The most possible causes for chronic kidney injuries are high blood pressure, diabetes, genetics, infections, and immune kidney disorder (Surana et al., 2021; Ahire et al., 2021). The clumps of crystals which are formed by the waste chemicals in the urine which vary in size and shape whereas hard in nature are called kidney stones. The cause for the kidney stones is dehydration or less water uptake. There are different types of Kidney stones: calcium stones, struvite stones, uric acid (UA) stones, cystine stones (Rapa et al., 2019; Carrero et al., 2020; Stenvinkel et al., 2018; Chen et al., 2022).

13.1.1 KIDNEY CANCER

When the cells of the Kidney mutate/change and the growth is out of control is termed as renal cell carcinoma or Kidney cancer. Kidney cancer can be broadly divided into three types which are discussed in subsections (Chow et al., 2010; Motzer et al., 2011).

13.1.2 LOCALIZED KIDNEY CANCER

The cancer which is formed at one site and doesn't spread is termed as localized cancer which is also sometimes termed as tumor. There can be presence of one or multiple tumors on the same or both the kidney (Lenis et al., 2018).

13.1.3 ADVANCED KIDNEY CANCER

The cancer that is spread from the Kidney to other parts of the body is termed as advanced kidney cancer. Genetic or hereditary, smoking, Alcohol consumption, blood pressure, Hepatitis C infection, Obesity, etc., are factors which increase the risk of Kidney cancer, whereas the exact cause of cancer is unknown (Graham et al., 2018).

13.1.4 URINARY TRACT INFECTIONS (UTI)

The infection that occurs in nay part of the urinary tract is called UTI. The cause of UTI is the microorganism which enter the urinary tract and multiply. There are different infections depending on the site where they multiply (Behzadi et al., 2019; Kolawole et al., 2009; Aiyegoro et al., 2007):

- If the infection occurs in the urinary bladder is called cystitis;
- If the infection occurs in the Kidney is called pyelonephritis;
- If the infection occurs in vagina it is termed as vaginitis;
- If the infections occur in urethra is termed as urethritis.

13.2 ROLE OF DIET IN KIDNEY DISEASES

The role of diet is very crucial in kidney diseases. The waste which is found in the body is the result of the food and fluids that are consumed, when the kidney is not in proper condition or having chronic kidney disease the kidney functions will be compromised. If the waste is not excreted due to the co promised functioning, then it will negatively affect the levels of electrolytes in the blood stream. To promote the proper functioning of the Kidney diet is very much responsible. Diet can also help in prolonging the life of chronic kidney disease patient. Diet can play role in delaying the progression of the chronic kidney diseases. Kidney is responsible for regulation of sodium, phosphates, potassium content in body as well as their serum levels, so it is very necessary to take in consideration of this while designing the diet plan for kidney disease patient. The diet which has optimum amount of sodium, potassium, and phosphate is considered. The vegetables, plants, and meat which have high amount of these salts are avoided. Apart from diet the fluid intake of the patient suffering from kidney disease is also considered as crucial aspect (Naber & Purohit, 2021; Vervloet et al., 2017; Iseki, 2008; Kramer, 2019).

13.3 KIDNEY DISEASES PREVENTION AND TREATMENT

13.3.1 PLANT METABOLITES

Primary plant metabolites which are consumed as the part of the diet are responsible in altering the conditions of the kidney in the patients having kidney diseases. The kidney in the disease conditions cannot perform their fullest, which might lead to further complications by accumulations of these nutrients or metabolites in the body further worsening the situation. So, modifications are supposed to be done in the diet of the kidney disease patients to decrease the further complications. In accordance with the requirement and conditions of the patients the intake of the dietary plant metabolites is to be consumed. The different metabolites are carbohydrates, proteins, fats, and vitamins. The mineral and nutrients like Sodium, Phosphorus, and potassium also play crucial role in the kidney diseases so the intake of these is also required to be in optimum amount. The details about the major plant metabolites which are commonly consumed on regular bases are noted in further subsections (Eknoyan et al., 2003; Takashi & Kawanami, 2022).

13.3.2 CARBOHYDRATES

Carbohydrates are the major source of energy and also called building blocks of the body. Chemically carbohydrates are poly-hydroxy aldehydes and ketones. Carbohydrates can be classified into categories based on the chains Single molecule carbohydrates are monosaccharaides. Two molecules' carbohydrates are disaccharides. More than two molecules of carbohydrates are polysaccharides. Carbohydrates are the fuel source for the body, they are major nutrients found in food that are consumed. They are also considered as essential nutrients. The energy is gained from carbohydrates by breaking them down in to simple sugars. The risk of chronic kidney diseases increases with intake of higher dietary carbohydrates, as it leads to unfavorable metabolic processes. The carbohydrates combined with high fiber content are favored in kidney disease patients, as they reduce the dietary protein and phosphorus. The plant-based diets are preferred which are low in refined carbohydrates and have higher fiber content. Starch is the major source of carbohydrates that is been consumed which are absorbable in nature (Rumessen, 1992; Jequier, 1994).

13.3.3 PROTEINS

Proteins are the chains formed by large number of amino acids; they are responsible for structural components of body. Proteins are essential for maintenance and repair of the body organs, muscles, etc. Proteins are classified in different types: Primary proteins where there is single sequence of amino acids. Secondary proteins are when there is linking of hydrogen bonds between the amino acid sequences. Tertiary proteins there is presence of attraction between the pleated sheets and alpha helices. Quaternary proteins in this type there is presence of more than one amino acid chains (Whitford, 2013). Proteins when consumed by body are further converted to waste, so higher quantity of protein intake will lead to stress on the kidneys, where in patients suffering from chronic kidney disease might lead to further complications. The low protein diet helps in dilating the efferent arterioles of angiotensin pathway as post glomerular effect and as result lowers the intraglomerular pressure. By reducing the dietary protein intake, the proportion of generation of urea also reduces. The conditions like azotemia can also be avoided by consumption of low-protein diet. The nitrogenous compounds which may further result into uremic toxins can also be reduced by intake of low-protein diet (Damodaran, 2017).

The normal body requires about 0.8 g/kg of protein as daily intake, whereas in chronic kidney disease this quantity can be reduced up to 0.66 g/kg of protein intake daily. The bioavailability of plant proteins is low when compared with animal proteins. The low protein diets can be achieved by plant-based diets, which are rich in alanine and glycine. Sources of animal proteins are chicken, meat, eggs, meat, and dairy products, whereas plant-based protein sources are beans, grains, nuts, and vegetables (Barik, 2006; Efimov, 1993).

13.3.4 FATS

Fats (Singh et al., 2017b) are type of lipids consisting of triglycerides and fatty acids. Fats play a crucial role in the cholesterol control of the body. Dietary fats are also one of the major sources of energy. Fats contain almost double the energy than that of the carbohydrates. The fats are classified into different categories: Monounsaturated fats are the having one unsaturated carbon bond in the molecule. Polyunsaturated fats are the fats having more than two carbon bonds in the molecule Saturated fats are the fats having saturated carbon bonds in the molecule. Tran's fats are obtained after hydrogenation process where in the liquid fats are converted to solid fats. The excess of fats gets accumulated in the renal parenchyma which leads to decrease in the efficacy of the kidney functions. Excess of saturated and trans fats are responsible to raise the cholesterol in blood stream and further leads to clogging of the blood vessels. When Kidney disease patients consume saturated and trans fats in higher quantity then it increases the load on the kidney which might worsen the condition of the kidneys (O'brien, 2008). The fats namely monounsaturated and polyunsaturated fats are good for chronic kidney disease as it helps in lagging the progression of chronic kidney disease. The modifications in lifestyle and diet help in improvement of the condition of the kidney disease patients. The dietary fats from animal products such as chicken, red meat, butter, and whole milk are of saturated fats. The intake of animal fats increases the risk and complications in the kidney disease. The monounsaturated, polyunsaturated, and low saturated fatty acids are found in plants, so plant-based diets are essential for kidney disease. Vegetable oil, nuts, avocado, walnuts, flaxseeds, and chia seeds, etc. (Stauffer, 1996).

13.3.5 VITAMINS

Vitamins are the micro-nutrients that are essential for proper functioning of the body. Vitamins are derived from the dietary sources (Keservani et al., 2010a, b, 2017, 2020). Vitamins are classified broadly into two categories: Fat-soluble vitamins are the vitamins which are soluble in fats. Vitamin A, D, E, and K. Water-soluble vitamins are the vitamins which are soluble in water. Vitamin B group and C (Halver, 2003; Zempleni et al., 2013).

13.3.6 VITAMIN C

Vitamin C belongs to water soluble vitamin category. The common source of Vitamin C citrus fruits and vegetables. Cherries, Oranges, lemon, chives, and guavas are the potential sources of Vitamin C. The anti-inflammatory activity and lowering the cardiovascular mortality and morbidity are the potential benefits of the micronutrient consumption of vitamin C (Combs & McClung, 2016).

13.3.7 VITAMIN K

Vitamin K belongs to fat-soluble vitamin category. The common source of Vitamin K is ssprouts, broccoli, and raw spinach. The potential benefits of Vitamin K are reduction vessel calcification and lowering the risk related to heart (Friedrich, 2013).

13.3.8 VITAMIN B_{12}

This belongs to the group of water-soluble vitamins. The common sources of vitamin B_{12} are cucumber, spinach, mushroom, kiwi, seaweed. The Vitamin B_{12} helps in reducing the progression of chronic kidney diseases and also is potential in lowering the cardiovascular risks (Friedrich, 2013).

13.3.9 VITAMIN D

Vitamin D is the fat-soluble vitamin. There are few plant sources like mushrooms and yeast that have Vitamin D, whereas in deficiency the vitamin D is given by supplementary medicine. Vitamin D is very helpful in treating proteinuria and also renal osteodystrophy healing (Bender, 2003).

13.3.10 MINERALS AND NUTRIENTS

The natural occurring element in soil and air are termed as minerals which are absorbed by the plants and consumed by animals and human beings. The minerals which are essential for human body are calcium, phosphorus, potassium, sodium, copper, chloride, etc. (Miller, 2017). These minerals are required in very low amount, so they are also called micronutrients. They are very much essential for developing the structure and function of the human body (Schwalfenberg & Genuis, 2015).

13.3.11 PHOSPHORUS

Naturally, phosphorus is found in protein rich food. The mineral phosphorus is required in proper working of muscles and blood vessels. In processed food phosphorus has been added. In Chronic kidney diseases the high intake of phosphorus can worsen the condition by getting into the blood stream as well as making the bones thin and weaker. The intake of phosphorus must be reduced from the normal intake approximately the intake should not be more than 800 mg/day. In case in the patients intake of phosphorus is higher that might lead to condition named hyperphosphatemia. Low-protein diet helps in achieving the reduced levels of phosphorus intake, the processed food intake must be reduced which has higher phosphorus to protein ratio. The plant diet can be a good alternative for the diet as the plant sources phosphorus absorption is less efficient than that from the animal sources. So, plant resources can be better alternatives. This will further help in controlling the disorder related to mineral bone in chronic kidney disease patients (Miller, 2017).

13.3.12 POTASSIUM

Potassium is also one of the essential micro-nutrients in human body it is helpful in functioning of nerves and muscles. In chronic kidney disease patient's higher intake of potassium is not recommended as the kidney may not be able to throw out the potassium from the blood which is excess in amount which might lead to further complications. The diet and the medicines we take contribute to the intake of potassium (Miller, 2017). The sources of potassium are both animal and plant based, as the absorption of the plant-based potassium is lower compared to that of the animal based. This also counts to be advantageous as it helps in reducing the risk of hyperkalemia in patients with chronic kidney diseases. The higher potassium output via urine will worsen the condition of the kidney disease patients. Another potential benefit of the plant-based potassium intake is the improved control on metabolic acidosis. The plant-based diets lower in potassium are apricots, berries, watermelons, okra, cabbage, etc.

13.3.13 SODIUM

Sodium commonly found in sodium chloride that is salt. The fast foods which are canned, packaged are processed with sodium. The blood pressure is associated with the sodium intake. Lower the intake of sodium lower is the blood pressure which also results in slowing down the progression in chronic kidney disease patients. The low sodium intake can be achieved by plant-diets as they are usually low in sodium content which also benefits in improving the control over the blood pressure. The risk of hypertension is also reduced by the lower ratio of sodium to potassium. The sodium intake in the patients suffering from the kidney disease must be less than that of 4 g/day (Miller, 2017).

13.3.14 VEGETABLES

The studies recently have shown the evidence towards the prevention and treatment of the patients in chronic kidney diseases by changing the diet. The plant based/vegetable-based diet has shown improvement in the health condition and nutritional quality of the patients. The vegetables which have nephroprotective activity, anti-inflammatory, and antioxidant potential have shown positive outcome towards the prevention from kidney disease. When it comes to risk benefit ratio of the plant-based diet it is more tilted towards the benefits (Ammerman et al., 1995). The high intake of meat and refined cereals leads to increase in the dietary acidity load which ends up in metabolic acidosis in chronic kidney disease patients, whereas the intake of vegetables is having potentials to neutralize the acidosis which avoids further harmful consequences. The toxins formed by uremic acid can be decreased by the consumption of vegetable-based diet as they help in improvement of the gut microbial flora and reducing the pathobionts and protein-fermenting species. The vegetable diet also helps in reducing hypophosphatemia as the bioavailability and absorption of phosphorus is low in the vegetables. The deficiency of the Vitamin K and magnesium can be compensated by vegetable diet. The slowing down of the chronic kidney disease progression, prevention, and treatment to some extent can be done by the vegetables consumed as they have chemicals that play an important role in reducing the oxidative stress by antioxidant potentials and inflammation by anti-inflammatory activity. The antioxidant and anti-inflammatory activity benefit the renal patients by reducing the further complications which results in slow progression of the disease. Even though they are many benefits of the vegetable diet in the chronic kidney disease patients there are also harmful effects that can be caused by vegetables to these patients. One such event is hyperkalemia which is been favored by the consumption of the vegetable diet. Different cooking techniques are being utilized to reduce the potassium and phosphorus content in the vegetables, which can be helpful in decreasing the risk of hyperkalemia in the kidney disease patients. The commonly consumed vegetables which can be consumed by the patients suffering from kidney disease are given in subsections along with the chemical constituents and mechanism of action of these vegetables (Grubben, 2004; Slavin & Lloyd, 2012).

13.3.14.1 TELFAIRIA OCCIDENTALIS

T. occidentalis is commonly called fluted gourd or fluted pumpkin, belonging to the family Cucurbitaceae. This leafy vegetable is found in Western Africa, especially in Nigeria where it is been consumed as food. The chemical composition of fluted pumpkin contains polyphenolic compounds (Keservani & Sharma, 2014; Keservani et al., 2016a, b), flavonoids, saponins, steroids, and terpenoids. The key role is played by renin-angiotensinogen system in kidney dysfunction, the important regulation is the formation of angiotensin-II in management of renal injury. The inhibition of angiotensin converting enzyme which is responsible in production of angiotensin-II can play important role in management of chronic kidney disease. The above-mentioned mechanism is shown by fluted pumpkin so it can be considered as one of the potential candidates in the diet regimen of kidney disease patients (Okoli & Mgbeogu, 1983).

13.3.14.2 BRASSICA OLERACEA VAR ITALICA PLENCK

B. oleracea var italica Plenck is commonly called Broccoli, belonging to the family Brassicaceae. Broccoli is widely consumed across the world but were cultivated in Asia and East Mediterranean. The chemical constituents of broccoli are phenolic compounds, flavonoids, alkaloids, vitamins (Singh et al., 2017a), mineral, fibers, and glucosinolates (GSs). The Phenolic compounds present in Broccoli play an important role in prevention and/or progression of chronic kidney disease by antioxidant and anti-inflammatory action as oxidative stress and inflammation are hallmark in chronic kidney diseases. As this vegetable plays an essential role in hindering the progression of the disease it can be considered as a part of diet of patients who are suffering from chronic kidney diseases (Li et al., 2021).

13.3.14.3 BRASSICA OLERACEA L. VAR. CAPITATA F. RUBRA

Commonly known as red cabbage, purple cabbage, blaukraut, which belongs to family Brassicaceae. The red cabbage is native to southern part of Europe continent, whereas now widely grown in entire Europe. The bioactive constituents of red cabbage include flavonoids, anthocyanin, and GSs. The potent antioxidants present in red cabbage are due to presence of the chemical bioactive namely anthocyanin. These antioxidants can be utilized in combatting the oxidative stress in the body. The antioxidant potential helps in hindering renal damage in the kidney diseases patient. The vegetable red cabbage can be added as one of the potential candidates for the diet plan of kidney disease patients (Haddar et al., 2018).

13.3.14.4 RAPHANUS SATIVUS

R. sativus commonly known as Radish, belonging to the family Cruciferae. This vegetable was originally found in Asian and European countries. Radish is a very commonly used vegetable in variety of dishes. The phytochemical compounds present in this vegetable are alkaloids, phenols, and flavonoids. The presence of ferulic acid, vanillic acid and quercetin chemicals are responsible for the nephroprotective activity of this vegetable. The activity might be due to the antioxidant effect of the phytochemicals present in the radish. Due to the nephroprotective activity of this vegetable it can be considered as one of the vegetables in the diet of patients suffering from chronic kidney diseases (Banihani, 2017).

13.3.14.5 BRASSICA RAPA

B. rapa commonly called turnip is widely found across the world and is one of the oldest used vegetables. Turnip has also been used as medicinal plant in various traditional medicines. Turnip belongs to family of Brassicaceae. The phytochemicals present in turnip are GSs, isothiocyanates, flavonoids, phenolics indoles and volatiles. The flavonoids and phenolic present show very good antioxidant effect whereas indole derivatives are responsible for the anti-inflammatory activity. The nephroprotective property is by the virtue of reducing oxidative stress, lipid peroxidation, reactive oxidation species production and glutathione (GSH) depletion. Due to this nephroprotective activity turnip can be added as one of the good choices in the diet regimen of the patients suffering from kidney diseases (Suwabe et al., 2002).

13.3.14.6 DAUCUS CAROTA

D. carota is commonly called carrot, which belongs to family Apiaceae. This vegetable has been widely consumed by people all around the globe was originated in Iran and Afghanistan. Chemically carrots contain secondary metabolites namely tannins, steroids, flavonoids, carotenoids, and volatile oils. The antioxidant property is by the virtue of carotenoids and anthocyanins. Depending on the various chemical compounds available in the carrot are responsible for anti-nephrotoxic potential and nephron-curative properties. So, carrots can be used as one of the vegetables in the diet plan of patients suffering through kidney diseases or disorders (Smallwood et al., 1999).

13.3.14.7 APIUM GRAVEOLANS L.

A. greveolans is commonly called celery or ajamoda, belonging to the family Apiaceae. Celery is found in Sweden, Algeria, Egypt, and Asian countries. This vegetable has carbohydrates, terpenes, alkaloids, saponins, steroids, and flavonoids as the chemical composition. The presence of antioxidant potential in Celery is due to presence active constituent apigenin and the anti-inflammatory activity is due to presence of umbelliferone chemical. The actions associated with the chemicals mentioned above are responsible for the nephroprotective activity of this vegetable. The celery has also been used since ages as nephroprotective agent in the traditional system of Unani. So, we can consider celery as one of the potential vegetables in the diet of the patient suffering from kidney diseases (Kooti & Daraei, 2017).

13.3.14.8 CUCUMUS SATIVUS

C. sativus is commonly called cucumber or Kheera, which belongs to the family Cucurbitaceae. The cucumber vegetable was originally grown in South Asia, whereas now widely found across the globe. The phytochemicals present in cucumber are alkaloids, flavonoids, polyphenols, anthocyanins, phenolics, and tannins. In Kidney diseases there is increase in oxidative stress which further leads to nephrotoxicity due to various other factors. The antioxidant property of this vegetable is responsible for showing the nephroprotective activity. The antioxidant property of cucumber is due to the presence of chemicals like beta carotene, alpha carotene and vitamin C. Cucumber can be considered as one of the important vegetables in diet of kidney disease patients (Tatlioglu, 1993).

13.3.14.9 NASTURTIUM OFFICINALE

N. officinale is commonly called yellowcress or watercress. This vegetable has origin from Europe whereas now even cultivated in Asia. Watercress is the family member of Brassicaceae. Phytochemicals like Alkaloids, Flavonoids, saponins, terpenoids, Vitamin E, tannins, and glycosides are present in this vegetable. Notable anti-inflammatory and antioxidant activity is shown by watercress. The renoprotective potential of watercress is by the virtue of anti-inflammatory and antioxidant effects. The vegetable has also been used in treatment of various urinary disorders in traditional medicines. By understanding the effect of watercress, we can add up this vegetable in the diet of patients who are suffering from kidney or urinary disorders (Al-Snafi, 2020).

13.3.14.10 BETA VULGARIS RUBRA

B. vulgaris is commonly known as beetroot or common beat or garden beat, belonging to the family Amaranthaceae. Beetroots are extensively grown in Asia, North America, and Europe. The class of compounds found in beetroot are polyphenols, flavonoids, betalains, and phenolic amides. The chemical constituents like betaine, betalain, and nitrate are responsible for anti-inflammatory effects, reducing the blood pressure and also antioxidant effect. The combine effect of all the above-mentioned mechanism helps in treatment of patients suffering from chronic kidney diseases. So, beetroot can be included as an important vegetable for treating the chronic kidney disease patients (Ninfali & Angelino, 2013).

13.3.14.11 TRICHOSANTHES DIOICA ROXB.

T. dioica commonly known as parwal or pointed gourd, belonging to the family Cucurbitaceae. The vegetable is widely grown in Asia and Australia. The various chemical constituents present in pointed gourd are vitamin A, vitamin C, tannins, saponins, tri terpenoids and alkaloids. Antiproliferative activity which is shown by the triterpenoid constituents and the anti-inflammatory effect of pointed gourd are responsible for nephroprotective activity. The nephroprotective activity of pointed gourd makes it helpful in treatment and prevention of chronic kidney disease, so it can be used in the diet of patients with kidney disease (Kumar et al., 2012).

13.3.14.12 TRICHOSANTHES CUCUMERINA

T. cucumerina commonly called snake gourd or serpent gourd, belonging to family Cucurbitaceae. The vegetable is widely grown in southern Asia. The chemical constituents which are present in snake gourd are flavonoids, terpenoids, carbohydrates, saponins, and tannins. The nephroprotective activity is due to the presence of phytochemicals such as flavonoids, saponins, and terpenoids which are responsible for activating the enzymes for antioxidant effect. The nephroprotective activity makes snake gourd vegetable one of the choices in diet of chronic kidney disease patient (Arawwawala et al., 2010).

13.3.14.13 SECHIUM EDULE

S. edule is commonly known as vegetable pear or choko or chow-chow, belonging to the family Cucurbitaceae. This vegetable is native to Mexico. The phytochemicals present are sterol, triterpenes, saponins, alkaloids, and flavonoids. The vegetable has nephroprotective effect by the virtue of anti-inflammatory and antioxidant potential. The damage to kidney can be prevented by utilizing the nephroprotective vegetable in the diet. *S. edule* can be used as one of the vegetables in the diet as it has the renal protective effect in the patients with kidney diseases (Flores, 1989).

13.3.14.14 MOMORIDICA DIOICA ROXB

M. dioica commonly known as spiny gourd or spine gourd, belongs to family Cucurbitaceae. It has been consumed in India and some other parts of South Asia as vegetable. The major chemical class presents in spiny gourd are phenolics and flavonoids, whereas there is other chemicals present are beta-sitosterol, saponins glycosides, triterpenes, lectins, etc. The free radical activity shown by spine gourd is due to the presence of high content of flavonoids and phenolic compounds. Due to the promising antioxidant activity of this vegetable, it shows nephroprotective and curative activity which makes it one of the good options in prevention and treatment of kidney disease patients. Spiny gourd can be utilized as a vegetable in the diet regimen of patients suffering from kidney diseases (Jain & Singhai, 2010).

13.3.14.15 ABELMOSCHUS ESCULENTUS

A. esculentus commonly known as okra or ladies' finger, belonging to family Malvaceae. Okra originated in South-East Asia, whereas now been consumed all over the world. Many countries consume okra as nutritional enhancer. The phytochemicals present in okra are flavonoids, polyphenol, catechin, quercetin, procyanidin. The nephroprotective and antioxidant potentials are shown by okra vegetable. The above-mentioned nephroprotective and antioxidant activity makes it an important vegetable in prevention and treatment of kidney diseases. Okra can be used in the diet of patients who suffering from the kidney disease as a preventive measure (Jain et al., 2012).

13.3.14.16 CUCURBITA PEPO

C. pepo commonly known as summer squash or pumpkin, belonging to the family Cucurbitaceae. This vegetable originates in Northeastern Mexico. The chemical compounds categories present are alkaloids, flavonoids, palmitic, oleic, and linoleic acids. This vegetable has widely been consumed all over the world. The consumption of roasted seeds of pumpkin also shows very good protective effect against nephrotoxicity. The protective activity is by the virtue of improvement in antioxidant property, lipid peroxidation is inhibited and protection of glomerular injuries. So, the Pumpkin as well as pumpkin seeds can be consumed in the diet by the patients having kidney diseases (Ratnam et al., 2017).

13.3.14.17 BETA VULGARIS L. VAR. CICLA

B. vulgaris leaves commonly known as beta root leaves or edible Swiss chard or chard leaves, belonging to family Amaranthaceae. Chard is consumed in Asia, America, and European continent. The chemical compounds present in chard are fatty acids, glycosides, polysaccharides, flavonoids, saponins, phenolic acids and betalains. The anti-inflammatory and antioxidant effects are shown by the chard by the virtue of the phytochemicals. The nephroprotective property of this chard makes it one of the potential one in the diet regime of kidney disease patients (Sacan & Yanardag, 2010).

13.3.14.18 ERUCA SATIVA

E. sativa commonly called jarjeer or argugula, belonging to the family Brassicaceae. Due to the nutraceutical properties of jarjeer it has been used as medicinal product traditionally as well as consumed as food. The phytochemicals that are present are flavonoids, phenolic acids, terpenes, glycosides, alkaloids, saponins, and sterols. During the kidney disease the kidney have increased oxidative stress which is being reduced or maintained by jarjeer as it has antioxidant potentials. The free radical scavenging potential is due to the presence of polyphenols, which helps in the nephroprotective activity and as the vegetable has nephroprotective activity it can be added into the diet of patients suffering from the kidney disease. The diet of the patients which kidney disease must have reduced levels of sodium, potassium, phosphorus, and proteins. So, the vegetables which are having high content of any of this must be avoided in the diet of kidney disease patients. The vegetables can be boiled in case to reduce the higher content of potassium which may lead to hyperkalemia (Miyazawa et al., 2002; Alqasoumi et al., 2009). The function of the kidney is to throw the waste out of the body by conversion, whereas the patients suffering from kidney disease have compromised kidney functions. It is very much required that these patients don't consume any such food which can further strain their kidney and lead to complications. Depending on the stage in which the patient is the diet is to be decided, even other factors like age, other diseases, lifestyle, and geographical conditions also affect the efficacy of the diet. The diet is being designed after considering all the factors which can affect the kidneys. The diet plays a very crucial role in the prevention and management of the kidney disease so the diet must be properly planned by the dietician. If the patients are suffering from Cardiovascular or Diabetes along with the kidney disease then care must be taken while designing the diet for such patients, if neglected might lead to complications or faster progression of the disease. In some cases, the DASH (Dietary Approaches to stop Hypertension) diet is followed by the dietician for the kidney disease patients. Individualization of the diet can be a better option in kidney disease patients, based on the requirements and conditions of the individual patients the diet plan can be designed. This will give good results in prevention and also help in treating the patients with kidney diseases. It will also help the patients in complying with the diet plan. The diet plans only work when the patients follow them on regular bases otherwise there is a very little effect seen, whereas the patients who follow the diet strictly have been observed with benefiting results (Yaniv et al., 1998).

13.4 CONCLUSION

Diet plays an important role in management of many diseases. The intake of plant-based diets, micronutrients are all responsible for condition of the patients with chronic kidney disease. The proper diet plan can help in management as well as slowing down the progression of the kidney diseases. It can also help in improving the conditions of the kidney functions. So proper individualized diet can help in prevention and treatment of kidney disease patients and is recommended.

ACKNOWLEDGMENTS

The author acknowledges the kind support and guidance of the Department of Pharmacognosy, KLE College of Pharmacy, KLE University of Higher Education and Research, Belagavi, Karnataka, India.

KEYWORDS

- **carbohydrates**
- **diet**
- **kidney diseases**
- **plant metabolites**
- **prevention**
- **treatment**
- **urinary tract infections**
- **vegetables**

REFERENCES

Ahire, E. D., Sonawane, V. N., Surana, K. R., & Talele, G. S., (2021). Drug discovery, drug-likeness screening, and bioavailability: Development of drug-likeness rule for natural products. In: *Applied Pharmaceutical Practice and Nutraceuticals* (pp. 191–208). Apple Academic Press.

Aiyegoro, O., igbinosa, O., Ogunmwonyi, I., Odjadjare, E., Igbinosa, O., & Okoh, A., (2007). Incidence of urinary tract infections (UTI) among children and adolescents in Ile-Ife, Nigeria. *African Journal of Microbiology Research, 1*, 13–19.

Ammerman, C. B., Baker, D. P., & Lewis, A. J., (1995). *Bioavailability of Nutrients for Animals: Amino Acids, Minerals, Vitamins*. Elsevier.

Banihani, S. A., (2017). Radish (*Raphanus sativus*) and diabetes. *Nutrients, 9*, 1014.

Barik, S., (2006). Immunophilins: For the love of proteins. *Cellular and Molecular Life Sciences CMLS, 63*, 2889–2900.

Behzadi, P., Behzadi, E., & Pawlak-Adamska, E. A., (2019). Urinary tract infections (UTIs) or genital tract infections (GTIs)? It's the diagnostics that count. *GMS Hygiene and Infection Control,* 14.

Bender, D. A., (2003). *Nutritional Biochemistry of the Vitamins*. Cambridge University Press.

Carrero, J. J., González-Ortiz, A., Avesani, C. M., Bakker, S. J., Bellizzi, V., Chauveau, P., Clase, C. M., et al., (2020). Plant-based diets to manage the risks and complications of chronic kidney disease. *Nature Reviews Nephrology, 16*, 525–542.

Chen, Y., Wu, J., Yu, D., & Liu, M., (2022). Plant or animal-based or PLADO diets: Which should chronic kidney disease patients choose? *Journal of Renal Nutrition.*

Chow, W. H., Dong, L. M., & Devesa, S. S., (2010). Epidemiology and risk factors for kidney cancer. *Nature Reviews Urology, 7*, 245–257.

Combs, Jr. G. F., & Mcclung, J. P., (2016). *The Vitamins: Fundamental Aspects in Nutrition and Health*. Academic press.

Damodaran, S., (2017). Food proteins: An overview. *Food Proteins and Their Applications*, 1–24.

Efimov, A., (1993). Standard structures in proteins. *Progress in Biophysics and Molecular Biology, 60*, 201–239.

Eknoyan, G., Levin, A., & Levin, N. W., (2003). Bone metabolism and disease in chronic kidney disease. *American Journal of Kidney Diseases, 42*, 1–201.

Friedrich, W., (2013). *Vitamins*. Walter de Gruyter.

Graham, J., Heng, D. Y., Brugarolas, J., & Vaishampayan, U., (2018). Personalized management of advanced kidney cancer. *American Society of Clinical Oncology Educational Book, 38*, 330–341.

Grubben, G. J. H. (2004). *Vegetables. Vol. 2*. Prota.

Haddar, W., Ben, T. M., Meksi, N., & Guesmi, A., (2018). Application of anthocyanins as natural dye extracted from *Brassica oleracea* L. var. capitata f. rubra: Dyeing studies of wool and silk fibers. *Natural Product Research, 32*, 141–148.

Halver, John E. (2003). The vitamins. In: *Fish Nutrition*. Academic Press, 61–141.

Iseki, K., (2008). Gender differences in chronic kidney disease. *Kidney International, 74*, 415–417.

Jequier, E., (1994). Carbohydrates as a source of energy. *The American Journal of Clinical Nutrition, 59*, 682S–685S.

Keservani, R. K., & Sharma, A. K., (2014). Flavonoids: Emerging trends and potential health benefits. *Journal of Chinese Pharmaceutical Sciences, 23*(12), 815.

Keservani, R. K., Kesharwani, R. K., Sharma, A. K., Vyas, N., & Chadoker, A., (2010b). Nutritional supplements: An overview. *International Journal of Current Pharmaceutical Review and Research, 1*(1), 59–75.

Keservani, R. K., Kesharwani, R. K., Vyas, N., Jain, S., Raghuvanshi, R., & Sharma, A. K., (2010a). Nutraceutical and functional food as future food: A review. *Der Pharmacia Lettre, 2*(1), 106–116.

Keservani, R. K., Sharma, A. K., & Kesharwani, R. K., (2016a). Nutraceutical and functional foods for cardiovascular health. In: *Food Process Engineering* (pp. 257–278, 291–312). Apple Academic Press, CRC Press. ISBN: 9781771884020.

Keservani, R. K., Sharma, A. K., & Kesharwani, R. K., (2016b). Medicinal effect of nutraceutical fruits for the cognition and brain health. *Scientifica*, 10. Article ID 3109254. doi: 10.1155/2016/3109254.

Keservani, R. K., Sharma, A. K., & Kesharwani, R. K., (2017). An overview and therapeutic applications of nutraceutical and functional foods. *Recent Advances in Drug Delivery Technology*, 160–201. ISBN:9781522507543.

Keservani, R. K., Sharma, A. K., & Kesharwani, R. K., (2020). *Nutraceuticals and Dietary Supplements: Applications in Health Improvement and Disease Management*. CRC Press. ISBN: 9781771888738.

Kolawole, A., Kolawole, O., Kandaki-Olukemi, Y., Babatunde, S., Durowade, K., & Kolawole, C., (2009). Prevalence of urinary tract infections (UTI) among patients attending Dalhatu Araf Specialist Hospital, Lafia, Nasarawa state, Nigeria. *International Journal of Medicine and Medical Sciences, 1*, 163–167.

Kramer, H., (2019). Diet and chronic kidney disease. *Advances in Nutrition, 10*, S367–S379.

Lenis, A. T., Donin, N. M., Johnson, D. C., Faiena, I., Salmasi, A., Drakaki, A., Belldegrun, A., et al., (2018). Adjuvant therapy for high risk localized kidney cancer: Emerging evidence and future clinical trials. *The Journal of Urology, 199*, 43–52.

Li, H., Xia, Y., Liu, h. Y., Guo, H., HE, X. Q., Liu, Y., Wu, D. T., et al., (2021). Nutritional values, beneficial effects, and food applications of broccoli (*Brassica oleracea* var. italica Plenck). *Trends in Food Science & Technology*.

Miller, D. D., (2017). Minerals. *Fennema's Food Chemistry*. CRC Press.

Motzer, R. J., Agarwal, N., Beard, C., Bhayani, S., Bolger, G. B., Carducci, M. A., Chang, S. S., et al., (2011). Kidney cancer. *Journal of the National Comprehensive Cancer Network, 9*, 960–977.

Naber, T., & Purohit, S., (2021). Chronic kidney disease: Role of diet for a reduction in the severity of the disease. *Nutrients, 13*, 3277.

O'Brien, R. D., (2008). *Fats and Oils: Formulating and Processing for Applications*. CRC press.

Okoli, B. E., & Mgbeogu, C., (1983). Fluted pumpkin, *Telfairia occidentalis*: West African vegetable crop. *Economic Botany, 37*, 145–149.

Rapa, S. F., Di Iorio, B. R., Campiglia, P., Heidland, A., & Marzocco, S., (2019). Inflammation and oxidative stress in chronic kidney disease—Potential therapeutic role of minerals, vitamins, and plant-derived metabolites. *International Journal of Molecular Sciences, 21*, 263.

Rumessen, J. J., (1992). Fructose and related food carbohydrates: Sources, intake, absorption, and clinical implications. *Scandinavian Journal of Gastroenterology, 27*, 819–828.

Schwalfenberg, G. K., & Genuis, S. J., (2015). Vitamin D, essential minerals, and toxic elements: Exploring interactions between nutrients and toxicants in clinical medicine. *The Scientific World Journal, 2015*.

Singh, P., Kesharwani, R. K., & Keservani, R. K., (2017a). Antioxidants and vitamins: Roles in cellular function and metabolism. In: *Sustained Energy for Enhanced Human Functions and Activity* (pp. 385–407). Academic Press. ISBN:9780128054130.

Singh, P., Kesharwani, R. K., & Keservani, R. K., (2017b). Protein, carbohydrates, and fats: Energy metabolism. In: *Sustained Energy for Enhanced Human Functions and Activity* (pp. 103–115). Academic Press. ISBN:9780128054130.

Slavin, J. L., & Lloyd, B., (2012). Health benefits of fruits and vegetables. *Advances in Nutrition, 3*, 506–516.

Stauffer, C. E., (1996). *Fats and Oils*. Eagan Press.

Stenvinkel, P., Painer, J., Kuro-O, M., Lanaspa, M., Arnold, W., Ruf, T., Shiels, P. G., & Johnson, R. J., (2018). Novel treatment strategies for chronic kidney disease: Insights from the animal kingdom. *Nature Reviews Nephrology, 14*, 265–284.

Surana, K. R., Ahire, E. D., Sonawane, V. N., & Talele, S. G., (2021). Biomolecular and molecular docking: A modern tool in drug discovery and virtual screening of natural products. In: *Applied Pharmaceutical Practice and Nutraceuticals* (pp. 209–223). Apple Academic Press.

Takashi, Y., & Kawanami, D., (2022). The role of bone-derived hormones in glucose metabolism, diabetic kidney disease, and cardiovascular disorders. *International Journal of Molecular Sciences, 23*, 2376.

Vervloet, M. G., Sezer, S., Massy, Z. A., Johansson, L., Cozzolino, M., & Fouque, D., (2017). The role of phosphate in kidney disease. *Nature Reviews Nephrology, 13*, 27–38.

Whitford, D., (2013). *Proteins: Structure and Function*. John Wiley & Sons.

Zempleni, J., Suttie, J. W., Gregory, III. J. F., & Stover, P. J., (2013). *Handbook of Vitamins*. CRC Press.

CHAPTER 14

DELAYING THE ONSET OF PARKINSON'S DISEASE THROUGH PHYTOTHERAPY

DEVANGNA PAUL and MYRENE ROSELYN D'SOUZA

Department of Biochemistry, Mount Carmel College Autonomous, Karnataka, India

ABSTRACT

This chapter inspects evidence of plant-derived natural products and their components that have been demonstrated to reduce or reverse the underlying neuronal degeneration seen in Parkinson's disease (PD), with an emphasis on dopaminergic (DA) neurotransmission levels and motor function regulation. Several research papers investigating the anti-PD therapeutic potential of fruits, vegetables, herbs, spices, and parasitic plants, and fungi have been published in the recent decade. Phytochemicals may have a role in the prevention and moderation of some of the symptoms of PD. The ability of these natural products to modulate dopamine production, reactive oxygen species (ROS) generation, neuroinflammation, mitochondrial function, excitotoxicity, cellular signaling pathways, and metal homeostasis which are all disturbed in the PD brain, is thought to be responsible for their anti-PD effects. However, the specific neuroprotective mechanism of action of natural compounds in the treatment of PD is unknown.

14.1 INTRODUCTION

Neuroscience has been a topic of interest explored over centuries throughout civilizations. Although the works of Charles Sherrington and Edgar Adrian

Plant Metabolites and Vegetables as Nutraceuticals.
Raj K. Keservani, Bui Thanh Tung, Rajesh K. Kesharwani, and Eknath D. Ahire (Eds.)

among many other renowned neuroscientists significantly aided scientists to understand the functioning of the brain, an immense portion of its working is yet to be pioneered (Network Management, 2021). The brain yet remains one of the intricate organs whose unabridged mechanism is yet to be apprehended. The ongoing area of concern in neuroscience has shifted from neuroanatomy and neurophysiology to neurodegenerative diseases (NDs) in the past 4 decades, as a progressive increase in the percentage of the population affected has been reported. In the strict sense, neurodegeneration is the progressive atrophy and dysfunction of neuronal structures but in medical terms, it represents a wide spectrum of progressive neurological disorders of unknown origin affecting specific anatomic functional units in neurons (Przedborski et al., 2003). Amidst hundreds of neurodegenerative disorders, Parkinson's disease (PD), Alzheimer's disease (AD), Huntington's disease (HD), and amyotrophic lateral sclerosis (ALS), affects the population incredibly. Most of them being late onset indicates that aging plays an important role in the progression of neurodegeneration. Hence, shifting our focus on the development of therapy to delaying their onset and reversing the prodrome of neurodegeneration is imperative.

PD being one of the major progressive neurological disorders, affects over 1.7% male population and 1.2% female population over the age of 60 (GDB, 2018). In 1817, it was first elucidated by James Parkinson in his article named, "An essay on the shaking Palsy" (Parkinson, 2002). PD, as described by Tanner et al. (1996) is "A disorder of unknown cause, is a distinct clinical and neuropathologic entity, characterized clinically by bradykinesia, resting tremor, cogwheel rigidity, and postural reflex impairment. Loss of pigmented neurons, most prominently in the substantia nigra (SN), and presence of associated characteristic ubiquitin-positive cytoplasmic inclusion bodies (Lewy bodies, clear bodies) are the chief pathological features."

PD was primarily known to affect the movement control of the patient which occurs due to failure in impulse transmission by liberation of the neurotransmitter dopamine from neurons of SN pars compacta to the basal ganglia (Triarhou, 2013). This is a result of progressive degeneration of the neurons of the SN which in turn lowers the concentration of available dopamine for neurotransmission (Triarhou, 2013). Along with movement dysfunction the patients are also found to develop complications of non-motor origin like; depression, dementia, cognitive impairment, REM sleep disorders, dysautonomia, depression, anosmia, and constipation (Jankovic & Tan, 2020). Although neuronal death has been established as the primary cause, its progression remains to be fully understood. The onset of PD is generally observed during the 5th–6th decades of life, but early onsets have also been reported where genetic factors come into play. This indicates, how mitochondria and oxidative stress might have a significant role in its progression.

Considering how the current research is focused on the development of therapies for delaying the onset of neurodegeneration this chapter emphasizes a Phyto-therapeutic approach to delay the onset of PD. In phytotherapy, biologically active compounds of plant origin are used as either a treatment or a preventive measure for the disease. These active compounds are often toxic in higher quantities to protect plants from their predators, but in lower quantities, they show therapeutic potential. Although we have a plethora of plants available with potential therapeutic efficacies, only a few of them have been studied extensively for their neuroprotectant potential. This is what makes it a prospective area of future research for PD and many other NDs.

14.2 ETIOLOGY

Initially, the causes of PD were unclear as most of the early studies focused on the disease symptoms. Research emphasizing anatomical, pathological, and neurochemical characteristics of PD was found to be initiated after 1899 (Li & Le, 2017) which encouraged a better understanding of the approach on PD and its therapy. In most cases, the causes are still unknown but several risk factors of either genetic or sporadic origin associated with PD have been identified. Along with these factors, cross-cultural variations have also been observed with a higher risk of PD in western countries compared to eastern and middle eastern countries. This advancement might help elucidate the cause in near future. In this section, we will be discussing some of the common genetic, epigenetic, and environmental factors which have been discovered over the past several decades.

14.2.1 ENVIRONMENTAL FACTORS

The sporadic form of PD was said to be associated mainly with age as well as external factors some of which are due to exposure to specific elements while others are a result of certain lifestyle choices made by the individual which increase the risk of PD. Some of these factors which pose a great risk of developing parkinsonian symptoms are discussed in subsections, but by no means is this an exhaustive list.

14.2.1.1 SYNTHETIC NEUROTOXINS

1-methyl-4-phenyl-1,2,3,6-tetrahydropyridine (MPTP), a toxic by-product of an opioid analgesic has a neurotoxic effect on the dopaminergic (DA) neurons. The risk of developing parkinsonian symptoms posed by MPTP was accidentally discovered by Dr. J. W. Langston, a movement disorder specialist, when certain individuals from the San Francisco Bay brought in the medical facility were conscious but could not move or speak at all after "synthetic heroin" substance abuse (Langston et al., 1983). These frozen addicts had developed the features of advanced-stage PD. This discovery paved the way for etiopathogenesis and therapeutic developments of PD. Oxidopamine or 6-hydroxydopamine or 2,4,5-trihydroxydopamine (6-OHDA) is another synthetic neurotoxin mainly used to create PD animal models. It is a catecholaminergic neurotoxin exhibiting high affinity towards dopamine and epinephrine transporters due to similarities in their structure (Jonsson, 1983). It enters DA and noradrenergic neurons destroying the catecholaminergic structures and forming ROS which leads to a cascade of reactions ultimately leading to nigrostriatal lesions. PCBs which once gained fame as a coolant for electrical equipment, soon faced a ban in 1979 (USA), due to several of its health effects including slow metabolic rate along with lipophilic nature which allows it to readily pass through the blood-brain barrier and deposit in brain tissues (Hatcher-Martin et al., 2012). PCBs not only affect our neurological system, but it also has severe effects on the immune system, endocrine system, reproductive system, etc. Among the range of neurological effects PCBs have, significant reduction of dopamine unlike other neurotransmitters (serotonergic, glutamatergic, etc.) which remained unaffected (Seegal et al., 1990) has been observed in several *in vitro* and *in vivo* (Chishti et al., 1996; Richardson & Miller, 2004). Further studies confirmed the link between PD and loss of striatal DA due to PCB deposition in the brain (Caudle et al., 2006).

14.2.1.2 PESTICIDES, FUNGICIDES, AND INSECTICIDES

The individuals living in rural areas are frequently exposed to pesticides and fungicides which are washed into their local waterbodies. Some of these are potential neurotoxins associated with PD along with other health conditions. Rotenone, a lipophilic potent rotenoid is one insecticide and piscicide which has a neurotoxic effect. It impairs oxidative phosphorylation by binding to complex I of ETC in mitochondria (Schuler & Casida, 2001). Although due to its short half-life (1–3 days), PD cases caused solely by rotenone exposure is very low (Schuler & Casida, 2001). Another herbicide that has a neurotoxic effect is Paraquat (N,N′-dimethyl-4-4′-bipyridinium). It acts on the nitric oxide synthase releasing an excess of ROS, which causes oxidative stress resulting in mitochondrial dysfunction (Przedborski & Ischiropoulos, 2005; Shimizu et al., 2001). Although its exact mechanism of action remains unknown as it cannot cross the blood brain barrier (BBB) easily. There are also variances in the experimental results obtained, but there seems to be a link between oxidative stress and neuronal cell death. Maneb (manganese ethylene bis-dithiocarbamate), a fungicide is said to show a synergistic effect with Paraquat on DA neurons (Thiruchelvam et al., 2000). There is no adequate evidence to conclude it as a cause of PD but its concurrent exposure with Paraquat has a potentiation effect which makes it a risk factor (McCormack et al., 2002).

14.2.1.3 HEAVY METALS

Metals including iron, copper, zinc, and molybdenum play significant roles in our body although most of them are present in trace amounts (Singh et al., 2011). These function in maintaining the homeostasis in a cell. In contrast, exposure to certain metals like mercury, aluminum, have deleterious health. Heavy metal exposure usually comes in form of food like contaminated seafood and plant-based foods, certain medication, dental procedures involving amalgam, and even occupational exposure (Ullah et al., 2021). These can lead to the early onset of PD if high concentrations of the heavy metals accumulate in neurons. Iron has several significant roles in a cell, one of them being a component of Fe-S clusters for mitochondrial complex 1. In certain diseased conditions, Fe-S assembly is disrupted, and the iron starts getting accumulated in the mitochondria instead (Bjørklund et al., 2019). Iron metabolism aggravates oxidative stress when metabolized and accelerate the onset of PD (Galaris et al., 2019). Dietary iron on the other hand is important and is not involved with PD pathogenesis (Cheng et al., 2015). Copper resembles iron as an oxidative stress factor, and its accumulation in neurons leads to loss of DA neurons. The levels of copper in an individual rising to toxic levels is uncommon in everyday life unless it is an occupational exposure (Cheng et al., 2015). Mercury and manganese are heavy metals that get into our system upon ingestion of food contaminated by them either in form of the fungicides used in agriculture or biomagnification and bioaccumulation as in the case of meat products. Mercury toxicity can also be a result of dental procedures where amalgam is used (Mutter, 2011). Upon long term exposure, mercury accelerates the onset of PD (Ngim & Devathasan, 1989). Manganese exposure unless occupational is from Maneb, a fungicide discussed in the previous section. The essential heavy metals which are necessary for the proper functioning of the cell sometimes result in oxidative stress, neuroinflammation, and apoptosis upon disruption of equilibrium consequently leading to the development of PD. Organophosphates along with lead, calcium, zinc, aluminum, are some more metals whose association with PD have been extensively studied (Ullah et al., 2021). Although the exact mechanism of action remains a mystery, further knowledge can help infer potential therapies.

14.2.1.4 EXPOSURE TO PSYCHOMOTOR STIMULANTS

Unlike the common psychomotor stimulants such as caffeine, tobacco, and nicotine, methamphetamine is a synthetic drug with pronounced effects on the CNS and development of PD. Methamphetamine is a more addictive derivative of amphetamine, affecting monoamine neurotransmitters (Panenka et al., 2013). Although it was developed for use as a nasal decongestant and bronchial inhaler, its long-lasting euphoric effect and easy transport across the blood-brain barrier got it into illegal drug abuse (Panenka et al., 2013). Initially, scientists emphasized Parkinsonism as not being a predominant feature of drug abuse in methamphetamine users (Granado et al., 2013), until later studies inferred otherwise. The clinical evidence cumulated confirmed a 165% increased risk of developing PD in methamphetamine users (Callaghan et al., 2010). According to recent studies, the risk is identified as moderately increased as there is a long lag period along with other environmental factors (like nicotine intake) which may attenuate the progress (Lappin et al., 2018).

14.2.1.5 HEALTH CONDITIONS

Health conditions that sacrifice neuro-immunity or results in neuroinflammation is said to have a neurodegenerative effect. As PD is also an age-related disease, several health conditions have the potential to accelerate the onset of PD. Several epidemiological studies have delineated the link between melanoma and PD along with its co-occurrence. The common link between the two is melanin, specifical neuromelanin found in DA neurons. Any abnormalities in neuromelanin can cause motor impairment and oxidative stress. Whereas melanosis on the other hand, which is the excess production of the pigment melanin is said to have a protective effect against both PD and melanomas (Bose et al., 2018). Studies show that infections might also have a direct correlation and contribution to the onset of PD. The small immune response molecules like cytokines and chemokines are abundantly synthesized and released during viral infections. These immune response molecules easily cross over the blood-brain barrier and activate the microglial cells, resulting in neuroinflammation (Smeyne et al., 2021). As already stated, neuroinflammation is one of the causes of DA neuron loss. Influenza A virus, Herpes simplex virus-1, Epstein-Barr virus, Varicella zoster virus and Hepatitis C are some of the infections whose association has been established (Limphaibool et al., 2019).

Impaired glucose regulation in diabetes mellitus causes the brain to take up 20% of the total glucose resulting in disorders of the nervous system. Several studies have shown an increased risk of PD due to health complications other than type 2 diabetes mellitus (Cheong et al., 2020), like traumatic brain injury (Delic et al., 2020), and progressive supranuclear palsy (Alster et al., 2020). Further insight into the association of these diseases with PD can help discover a novel therapeutic target.

14.2.1.6 BIOLOGICAL NEUROTOXINS

Epidemiologic studies suggest that certain infections can act as a trigger and further advance PD manifestations. Initially, most of the researchers were trying to prove otherwise, but after Braak's hypothesis was stated more studies to deduce the association between infections and PD gained acceptance. The origin of these hypotheses lies in the influenza pandemic of 1918 and its association with postencephalitic parkinsonism (PEP) (Rail et al., 1981). Biological neurotoxins gain entry via such infections and trigger PD usually in immunocompromised individuals.

14.2.2 EPIGENETIC FACTORS

Epigenetic factors are gene regulation factors that turn off/on certain genes. These factors indirectly dictate protein synthesis and other cellular functions.

DNA methylation, histone modification, RNA post-transcriptional silencing are three systems responsible for epigenetic silencing.

14.2.2.1 DNA METHYLATION

It is a biochemical process in which the cytosine residues on DNA undergo methylation by the methyl group from the donor, S-adenosyl methionine. CpG islands are associated with promoters of 60% coding genes. Meta-analysis data suggests that inhibition of DNA methylation of the CpG islands on SNCA gene results in overexpression of the SNCA protein which in turn results in SNCA aggregation and Lewy body formation in PD (Egger et al., 2004). Several studies have found that hyper- or hypomethylation of the PD risk genes is associated with its pathophysiology. These dysregulations are a result of environmental factors and aging. Along with nuclear genes, mitochondrial genes are also found to increase the risk of PD due to their contribution to the pathophysiology of PD (Miranda-Morales et al., 2017).

1. **Histone Modification:** Studies show that upon exposure to certain environmental factors, the enzymes responsible for catalyzing histone modifications are depleted (Delgado-Morales, 2017). Since histone modification is an important element of regulation, any alteration can result in the manifestation of diseases. Ones that have shown histone modification with respect to PD are, paraquat (histone H3 acetylation), MPTP (reduced histone deacetylases), dieldrin (histone H3 and H4 acetylation), α-Syn (reduced histone H3 acetylation) (Qureshi & Mehler, 2011).
2. **miRNA:** In studies using toxin-induced models of PD, there is detectable downregulation of certain miRNA that are associated with pathways that result in DA cell death or neurotoxicity. For instance, miR-126 elevation results in inhibition of IGF-1/PI3K signaling pathway (Kim et al., 2014). These miRNAs are involved in the pathophysiology of PD and thereby can be used as PD biomarkers.

14.2.3 GENETIC FACTORS

Despite PD being characterized as idiopathic, there have been several cases where the Mendelian inheritance pattern has been observed. Once the pattern was recognized several studies were initiated to elucidate the genes responsible. By the year 2012, researchers had identified around 28 distinct chromosomal regions showing varying degrees of association with PD, out of which only 6 were known to have a monogenic effect (Klein & Westenberger, 2012) this count rose to over 20 monogenic forms with over 100 loci identified as risk factors within eight years (Jankovic & Tan, 2020), representing 16–36% of the inheritable component of PD. The name assigned to the potential PD causing genes is "PARK genes." However, only *SNCA*, *LRRK2*, *VPS35*, *PRKN*, *PINK1*, *GBA*, and *DJ-1* (Bandres-Ciga, 2020) out of all the genes identified have been confirmed to be the monogenic cause of PD, the rest of them have been designated as risk factors. Recent discoveries have also presented the involvement of genes; *TMEM230*, *LRP10*, *NUS1*, and *ARSA* in the onset of PD (Veeriah et al., 2010).

14.2.3.1 AUTOSOMAL RECESSIVE

A germline mutation in *PRKN* gene is one of the established origins of the early onset of PD called juvenile PD. This gene is associated with ubiquitination activity and regulation of proteasome-mediated degradation due to which the mutation affects neuronal cytoprotective potential (Kouli et al., 2018). Mutation in *PRKN* gene is one of the causes of familial autosomal recessive forms of PD along with; *PINK1*, *DJ-1*, *ATP13A2*, *HTRA2*, *PLA2G6*, *FBX07*, *DNAJC6*, *SYNJ1*, *VPS13C*. All the listed autosomal recessive genes are associated with either juvenile or early onset of PD.

14.2.3.2 AUTOSOMAL DOMINANT

One of the hallmarks of PD are Lewy bodies which are aggregates of misfolded α-synuclein, a small neuronal protein (Prusiner et al., 2015) caused by SNCA mutations (point mutations of A30P, A53T, E46K, H50Q, and G51D) (Siddiqui et al., 2016) although rare, plays a pivotal role in misfolding of α-synuclein. It is one of the autosomal dominant genes involved in the familial form of PD, other genes with autosomal dominant inheritance confirmed as monogenic causes are; *UCHL1*, *LRRK2*, *GIGYF2*, *VPS35*, *EIF4G1*, *DNAJC13* (Kouli et al., 2018).

14.3 OVERVIEW OF THE PATHOPHYSIOLOGY AND TREATMENT

PD is a multi-factorial neurodegenerative disorder having various possible pathophysiological mechanisms. Despite several elements, there have been recurrent reports of several molecular mechanisms recorded in human post-mortem tissues. According to Braak's hypothesis, which is based on Lewy bodies, the progression of PD is characterized by a progressive spreading pattern of αSyn starting from the olfactory bulb and neurons in the gut (Rietdijk et al., 2017). This hypothesis elucidates how pathological factors infiltrates the nervous system via enteral routes infections such as the intestinal mucosa and nasal mucosa.

Some of the pathogenic mechanisms which play a significant role development of PD have been elucidated. This occurs by misfolding and aggregation of α-synuclein, a major component of the Lewy bodies, due to hindrances in protein clearance mechanism caused by impaired ubiquitin-proteasome and autophagy along with the mitochondrial dysfunction, neuroinflammation, and oxidative stress ensued by genetic, epigenetic, and external factors (Cacabelos, 2017). The molecular pathogenic mechanisms involving PARK-SNCA (PARK1) mutation result in misfolding of α-synuclein protein due to which it forms inclusion aggregates. These aggregates hinder different cellular functions like fibril formation, and subcellular distribution. Another result of this point mutation is overexpression which impairs a wide spectrum of cellular functions including cytotoxicity, mitochondrial dysfunction (Rocha et al., 2018), and impaired autophagy (Fields et al., 2019).

Another pathological characteristic of PD is the progressive loss of DA neurons in the SN, which in turn lowers the concentration of available dopamine for neurotransmission (Tanner & Goldman, 1996). The dopamine neurotransmitter is released by the SN neurons to communicate with basal ganglia forming the biochemical basis of movement in the organism (Triarhou, 2013). The biochemical imbalance brought upon by neurodegeneration directly manifests as motor symptoms.

Currently, the usage of Levodopa for the medicinal treatment of PD along with surgical treatment is commonly prescribed. Although prior to medicinal treatment of PD with Levodopa doctors performed invasive surgical procedures like pallidotomies and thalamotomies (Groiss et al., 2009). The risk of severe side effects made these approaches less popular. Since then, researchers slowly shifted their focus from – *ectomy* surgical approaches to medicinal therapies and non-invasive deep brain stimulation procedures. The development of medicinal treatments was instigated based on Ehringer and Hornykiewicz, who reported, that the individuals suffering from PD had dopamine deficiency in their striatum (Ehringer & Hornykiewicz, 1960), along with Carlssone who substantiated the reversal of akinesia upon administration of levodopa (Carlssone et al., 1957). The incredible efficacy of levodopa to drastically improve the motor symptoms of PD was first demonstrated by Cotzias et al. (1967). These are some of the early milestones achieved in the field of medicine for PD.

Albeit L-dopa has been the superior monotherapy with incredible efficacy, its efficacy is progressively lost after long-term usage (5+ years) resulting in the development of dopa resistant motor symptoms (speech impairment, postural abnormalities, freezing episodes) and dopa resistant non-motor symptoms (autonomic dysfunctions, mood, cognitive impairment) along with side effects like psychosis, dyskinesia, toxicity, and motor fluctuations (Marsden, 1994; Thanvi & Lo, 2004). Levodopa also has a wearing-off effect known as the "on-and-off phenomenon" as described by a PD patient of Duvoisin (1974) where the "on-time" denotes the duration during which levodopa work and significantly reduces symptoms and the "off-time" denotes the downtime when levodopa does not show any therapeutic effect (Lees, 1989). The scientific data retrieved, depicts that extracerebral metabolism of monoamines accelerates and thereby reducing the sensitivity of DA neurons to levodopa after its continuous administration for 1–3 years (Petelin et al., 1977). Severity of this phenomenon increases significantly with long term continuous usage of high dose levodopa. This can be managed by coupling the treatment with an additional dose of levodopa and peripheral dopa decarboxylase inhibitor which reduces the inter-dose interval (Lees, 1989), but it might worsen the conditions in the long term with increased tolerance. This is an inevitable fate of levodopa therapy, as it primarily focuses to satiate the dopamine deficiency in the striatum and the continual loss of neurons of SN remains unaffected.

Recent elucidation on the complicity of noradrenergic, glutamatergic, serotonergic, adenosine pathways, along with DA pathways in the manifestation of non-motor PD symptoms has broadened our horizon for alternate therapeutic approaches way beyond conventional levodopa therapies. Since there are several repercussions of the medicinal treatment as well and the outcomes remain unsatisfactory with the involvement of new emerging complications as the treatment and disease progress, researchers are focusing on developing neurosurgery procedures to treat the disease. Stereotactic surgeries focusing on the motor complex of the brain to mitigate the motor symptoms have been the basis of developing those surgical procedures for advanced forms of PD.

The bilateral subthalamic nucleus deep brain stimulation is one of the standards and highly effective therapy for advanced forms of PD (Benabid et al., 2019). This procedure focuses on suppressing the tremors along with most of the primary motor symptoms of PD by an electrical impulse generated by tiny electrodes implants inserted in the motor complex. Since the exact mechanism of action is still unspecified, therefore hypothetically, it is said to induce blank noise which would hinder the off tuning of the looped feedback responsible for tremors (Krack et al., 2003). This therapeutic approach has shown tremendous improvement in individuals suffering from advanced PD. A five-year follow-up study shows improvement of about 50% in motor function (off medication) and dyskinesia (on medication) (Krack e al., 2003). However, this procedure has its own drawbacks with significant impact on the psychosocial aspect of the patient's life post-surgery (Kalteis et al., 2006).

Other antiparkinsonian agents are also commercially available, but they are likely to cause visual hallucinations and/or cognitive impairment in the patients. Anticholinergic treatment was the first treatment devised for PD patients, as it particularly focuses on the motor symptoms like tremors in the early stage of PD. As an antagonist of muscarinic receptors, they minimize the frequency of neurotransmission via neostriatal acetylcholine (Brocks, 1999). Some of the anticholinergic drugs prescribed to PD patients are trihexyphenidyl, benztropine, procyclidine, and biperiden. Dopamine agonists were introduced to moderate the levodopa dosage given to the patient, due to its antiparkinsonian effect. These agonists act directly upon the dopamine receptors (D2 type receptors) imitating the action of neurotransmitter (dopamine): ergoline (bromocriptine, pergolide, lisuride) and non-ergoline (ropinirole, pramipexole) are the two types of dopamine agonist used for treatment (Brooks, 2000). The COMT inhibitors (nitecapone, entacapone, tolcapone) are another class of drug prescribed along with levodopa therapy. It inhibits the action of the catechol-o-methyl-transferase enzyme thereby preventing peripheral metabolism of levodopa, this ensures high bioavailability of levodopa across the blood-brain barrier (Rivest et al., 1999). Amantadine is prescribed to patients that help ease the side effects of levodopa therapy. It palliates the levodopa-induced dyskinesia, as it is a weak non-competitive antagonist for NMDA receptors it inhibits the reuptake of dopamine, thereby maintaining the levels of dopamine in the brain (Chang & Ramphul, 2022). The adverse effects of using these drugs include chronic cognitive impairment (anticholinergic), compulsive behavior (dopamine agonist), insomnia (dopamine agonist, COMT inhibitors, Mao B inhibitors), Serotonin syndrome (MAO B inhibitors, dopamine agonists), confusion (Amatadine, anticholinergic) (Sposato, 2014).

As science and technologies are advancing, our grasp on neurochemical features of PD has expanded immensely over the past 200 years. Even with all these advances, we still do not have a cure. Consequently, further exploration framing pathogenic factors of PD is needed to enhance our knowledge on understanding the disease and develop a novel cure.

14.4 ROLE OF ABNORMAL MITOCHONDRIAL FUNCTION AND OXIDATIVE STRESS

Despite the unclear ideologies, the fact that neurons are highly dependent on oxidative energy metabolism has elucidated the underlying role of mitochondrial dysfunction in neurodegeneration. The energy requirements of a cell are predominantly furnished by oxidative phosphorylation, effectuated by the mitochondria. Any deterioration in the organelle's ability to produce energy would impair numerous cellular functions, including maintenance of homeostasis and primarily loss of ATP which would further lead to cellular deterioration. Both genetic and environmental factors influence a wide spectrum of mitochondrial function including apoptosis regulation, cell cycle, ATP biogenesis, Ca^{2+} homeostasis, pro-apoptotic gene activation, axonal transport, mitochondrial fission, and fusion.

Oxidative stress is one of the prime causes of NDs. Pathophysiological studies suggest a direct link between α-synuclein accumulation and oxidative stress, as mitochondrial antioxidants are known to reduce the soluble α-synuclein (Stoker, 2018). Oxidative stress can arise from environmental factors directly by ROS formation or indirectly by hampering mitochondrial function. Mitochondria being the organelle responsible for energy production is mostly exposed to oxidative stress. The proteins involved in ETC chain when defective becomes the root cause of several degenerative diseases. Mitochondrial complex I deficiency is a common characteristic of PD patients irrespective of the origin of causes. Neurotoxic molecules like maneb, MPTP, and rotenone also have a direct impact on the ETC, i.e., either impairing or inhibiting its function. ETC dysfunction aggravates ROS generation and diminishes ATP production along with altered mitochondrial functions. As for the familial cases, PD causing genes like PINK1, Parkin, α-synuclein, DJ-1, LRRK2 and $CHCH_2$ to have deleterious effects on mitochondrial functioning (Moon & Paek, 2015). PINK1 and Parkin are the product of PARK6 and PARK2 genes, respectively which upon mutation affects quality control of the mitochondria, and mitophagy (Stoker & Greenland, 2018). The hallmark of PD, α-synuclein misfolding and aggregation has deleterious effects on the mitochondrial complex I. Certain mutations in

PARK8 gene results in overexpression of LRRK2 resulting in mitochondrial fragmentation and reduced mitochondrial fusion (Martin et al., 2014).

Mitochondrial fission is the process by which larger mitochondria get fragmented, thereby enabling the clearance of damaged mitochondria via mitophagy. An amplified occurrence of this process gives rise to the accumulation of mitochondrial fragments that hampers ATP production, affecting ETC cycle and thereby resulting in cell death (Panchal & Tiwari, 2019). Mitochondrial fusion on the other hand is the process that is responsible for maintaining the mitochondrial population of the cell by fusing two mitochondria. In recent studies, it has been elucidated stressful conditions can induce mitochondrial hyper-fusion resulting in increased ATP biogenesis (Panchal & Tiwari, 2019). The homeostasis between mitochondrial fission and fusion must be maintained under normal physiological conditions, but in the case of NDs, this process gets abrupted resulting in defective ATP biogenesis.

Mitochondrial dysfunction and oxidative stress are one of the extensively studied pathogenic mechanisms for PD. The complex mechanisms of mitochondrial functioning are being studied substantially, and locating therapeutic targets within these complex cascades of mechanisms could prove to be useful for the discovery of a cure for PD.

14.5 FRUITS, VEGETABLE, AND SPICE DERIVED NUTRACEUTICALS FOR PD

Plant-derived molecules are often used in pharmaceutical industries to develop therapeutic compounds treating a wide array of diseases. Their easy accessibility and versatile nature make them ideal to work with. The use of plant-derived therapeutics for the treatment of diseases has been observed in many cultures throughout history. Not only plants but the food we eat has been long considered a key element in the prevention and/or treatment of diseases. This is where the concept of functional food (Keservani, 2010a, 2017) and nutraceuticals was developed. Recent studies have shown that early dietary intervention with neuroprotective nutraceuticals is delay the onset of PD. In this section, we will discuss some of the examples of nutraceuticals (Keservani, 2010b, 2020) that have a therapeutic effect against PD.

14.5.1 CAMELLIA SINESIS

- **Potential Neuroprotective Effect:** Tea is an important part of Asian and Mediterranean diets consumed as a beverage. It possesses numerous bioactive compounds with antioxidant and anti-inflammatory properties. The studies have found, the presence of catechins, epicatechin, and epigallocatechin gallate are responsible for its neuroprotective activity. Experimental studies on PD diseased models confirm the neuroprotective effect of green tea.

- **Mechanism of Action:** Several mechanisms elucidating how green tea and bioactive compounds present in them show neuroprotective potential against PD have been discovered. As inflammation and oxidative stress play a major role in the manifestation of PD, inhibition of these pro-inflammatory enzymes by green tea along with its anti-inflammatory and antioxidant properties makes it a great candidate as a preventive measure for PD. In 6-OHDA model of PD, tea extracts have depicted antidepressant activity. Also, mice that received pre-treatment of tea extracts showed a reduced MPTP-induced DA neuronal loss, lipid peroxidation, ROS, nitrite contents, COX-1, COX-2, iNOS, activate tyrosine hydroxylase, the rate-limiting enzyme in dopamine synthesis (Pinto et al., 2015). The study shows that oxidative stress induced by iron accumulation is another pathological factor playing a significant role in the onset of PD (Batista-Nascimento et al., 2012). Evidence has confirmed that the presence of Fe^{3+} can promote the aggregation of α-synuclein by accelerating the rate of fibrillization. The EGCG present in green tea enhances hepcidin-ferroportin expression and can form complex with Fe^{3+} thereby preventing its accumulation in the brain (Malar e al., 2020). In conclusion, green tea polyphenols (Keservani & Sharma, 2014; Keservani et al., 2016a, b) have a wide range of pharmacological potential against PD.

14.5.2 *CURCUMA LONGA L.*

- **Potential Neuroprotective Effect:** Curcumin is the principal polyphenolic flavonoid present in turmeric and has been long used in traditional Asian culinary and ayurvedic practices for centuries. Experimental studies have demonstrated its diverse array of medicinal use, starting from antioxidant, to anti-inflammatory, anticarcinogenic, anti-arthritic, anti-anxiety properties. It is also known for its effects in reducing hyperlipidemia (Hewlings & Kalman, 2017).
- **Mechanism of Action:** A study on *Drosophila melanogaster* by Nguyen et al. (2018) demonstrated neuroprotective properties

of curcumin in environmental (6-OHDA, MPTP, rotenone, and paraquat) and genetic factors (*α-synuclein*, *PINK1*, *DJ-1*, *LRRK2*, and *UCH-L1*) induced models of PD. Curcumin is a multi-target molecule as it stimulates neuronal regeneration by acting on Trk/PI3K signaling cascade and lowering the caspase activity thereby inhibiting apoptosis pathways, inhibits inflammatory molecules such as cytokines, interleukins, chemokines, COX-2 (Ghasemi et al., 2019), and inhibits aggregation of α-synuclein reducing neurotoxicity (Sharma & Nehru, 2018). It also possesses iron-chelating properties to help maintain Fe homeostasis in neurons (Rainey et al., 2019) and helps maintain mitochondrial function (Khatri & Juvekar, 2016).

14.5.3 MORUS ALBA L.

- **Potential Neuroprotective Effect:** Extracts of the edible fruit of the deciduous *Morus alba* L. (mulberry) have demonstrated neuroprotective activity in *in vitro* and *in vivo* PD models, with the efficacy related to higher concentrations of polyphenols, rutin, anthocyanin, tocopherol, and quercetin (Liu et al., 2008).
- **Mechanism of Action:** The extract possesses antioxidant and anti-apoptotic properties in 6-OHDA-mediated neurotoxic cell lines. It was found to maintain the mitochondrial membrane integrity and control the expression of apoptosis-related proteins namely Bax, Bcl-2, and caspase 3 (Pinto et al., 2015). *In vivo* administration of the mulberry extract also reduces bradykinesia and DA neuronal damage in the SN (Kim et al., 2010; Zheng et al., 2008).

14.5.4 CITRUS FRUITS

- **Potential Neuroprotective Effect:** Citrus fruits include the most consumed *Citrus lemon* (lemon), *Citrus sinensis* (orange), and *Citrus paradisi* (grapefruit) rich in flavonoids such as flavonones, polymethoxyflavones, and flavone glycosides (Datla et al., 2001). Polymethoxyflavones present in lemon and orange peel reduce the loss of DA neurons and tyrosine hydroxylase activity in the SN of 6-OHD rats (Zbarsky et al., 2001).

- **Mechanism of Action:** In PD models, the flavonoid naringenin demonstrates similar effects (Rail et al., 1981) along with the suppression of LPS/IFN-induced iNOS expression, p38 mitogen-activated protein kinase (MAPK) activation, and transcription-1 (STAT-1) in glial cells. These responses are linked to astrocyte- and microglial-mediated inflammatory responses (Kanter, 2008).

14.5.5 *NIGELLA SATIVA*

- **Potential Neuroprotective Effect:** *N. sativa* (400 mg/kg body weight/day) given orally was found to prevent neuronal damage in the frontal cortex of rats chronically exposed to toluene via inhalation of 3,000 pm toluene for 8 h/d (Kanter, 2008). The control group demonstrated symptoms such as dilated endoplasmic reticulum cisternae, cytoplasm shrinkage, swollen mitochondria, and breakdown of the nuclear membrane.
- **Mechanism of Action:** Thymoquinone a major bioactive component of the essential oil (EO) isolated from *N. sativa* given orally at 50 mg/kg body weight/day demonstrated protective effects in the hippocampus and absence of neurodegeneration-related histological alterations (Al-Majed et al., 2006). The impact of pretreatment with TQ (5 mg/kg/day p.o.) on transient forebrain ischemia-induced neuronal injury in the rat hippocampus demonstrated reduced death of hippocampal cells when compared to the control group. In addition, an increase in GSH levels, catalase activity, and SOD activity, and a decline in MDA levels were reported (Geng et al., 2007).

14.5.6 *CAFFEINE*

- **Potential Neuroprotective Effect:** Caffeine is one of the most regularly consumed psychomotor stimulants in the world. Coffee is the common source of caffeine consumed for its psychostimulant effect. It is known to stimulate our CNS, by acting on adenosine receptors. Recent studies have shown evidence of caffeine modulating neuroinflammation, and autophagy (Kim et al., 2003).
- **Mechanism of Action:** Caffeine possesses an anti-neuroinflammatory effect, which is beneficial for PD. Several mechanisms are

involved in its anti-neuroinflammatory effect, starting with modulation of lipopolysaccharide (LPS)-induced and microglia-mediated neuroinflammation (Kim et al., 2003; Santini et al., 2007), and prevention of BBB dysregulation (Kim et al., 2012). Some studies on α-synuclein fibril models of PD demonstrated caffeine modulation of α-syn aggregation and neurotoxicity along with how it attenuates the activation of microglia (Ha et al., 2001). Thus, caffeine is involved in several mechanisms providing an overall neuroprotective effect.

14.5.7 OLIVE OIL

- **Potential Neuroprotective Effect:** Olive oil is an essential part of Mediterranean culinary practices. It gained the spotlight worldwide in the past decade because of its healthy nature among other oil alternatives and its medicinal properties including, anti-inflammation, ability to reduce lipids and DNA oxidation, and the modulation blood pressure. Recent studies have elucidated its neuroprotective potential due to the presence of polyphenols and terpenes.
- **Mechanism of Action:** *Olea europaea* or olive tree is rich in polyphenols (phenolic alcohols, phenolic acids, flavonoids, lignans, and secoiridoids) stored mainly in the drupes or leaves. These polyphenols give olive oil its distinct taste and anti-inflammatory potential. Most of the plant-based anti-inflammatory compounds are unable to cross BBB thereby not helping with neuro-inflammation, but oleuropein aglycone can cross BBB and thereby have a better neuro anti-inflammatory effect (Wu et al., 1996). Oleuropein and hydroxytyrosol are the bioactive compounds present in olive oil that provide neuroprotection by scavenging ROS, inhibiting COX, modulating Fe^{2+} homeostasis, inhibiting NO-induced ATP loss, counteracts depolarization of mitochondria, and preventing DA neuronal death by modulating autophagy (Lai et al., 2006). This makes olive oil another multitarget nutraceutical that can deaccelerate the progression of PD by modulating different cellular mechanisms.

14.5.8 GRAPES

- **Potential Neuroprotective Effect:** Grape is part of the berry family and is consumed regularly in the diet. Although it has been part of our diet for centuries, it was not until recently that it gained the interest of the scientific community for its medical properties. The bioactive compounds present in it are known to have antioxidant, anticarcinogenic, immunomodulatory, antidiabetic, anti-atherogenic, neuroprotective, anti-obesity, anti-aging, and anti-infection properties (Zhang et al., 2011). The presence of bioactive compounds like resveratrol (RES) is responsible for their medicinal properties.

- **Mechanism of Action:** Grape seed and the grape skin are the parts of this berry that store the most bioactive compounds. Recent studies on different disease models of PD have shown that bioactive compounds present in grapes can deaccelerate the progression and delay the onset. There are several cellular mechanisms that it can modulate to bring about neuroprotection. A study elucidated that polyphenol-rich extract of grape seed and skin reduces cytotoxicity in 6-OHDA PD model via inhibition of ROS, free radical formation, antioxidant (inhibit phospholipase, COX-2, reduce glutathione (GSH) peroxidase), inflammation (regulation of NF-kB nuclear factor) and apoptosis (inhibit caspase 3 activation). Therefore, a further investigation in grapeseed and skin is needed to determine its true potential as a preventive food supplement to delay the onset of PD (Ahlskog & Muenter, 2001).

14.6 PARASITIC PLANT AND FUNGI DERIVED BIOACTIVES AGAINST PD

14.6.1 CISTANCHES HERBA

- **Potential Neuroprotective Effect:** The neurotoxin MPTP generating the active metabolite 1-methyl-4-phenyl pyridinium (MPP+) in a mouse model was found to enters DA neurons via the dopamine transporter where it inhibits the mitochondrial ETC and causes cell death (Sheng et al., 2002). Bioactives campneoide and tubuloside B from *C. herba* afford protection to neurons MPP+-induced (Chen et al., 2007). Acteoside provides neuroprotection against rotenone-induced damage of SH-SY5Y cells. Similar results were also seen for echinacoside, a phenylethanoid glycoside from *C. salsa* in the prevention and treatment of PD (Wang et al., 2017).
- **Mechanism of Action:** Tubuloside B, Cistanoside K, and acteoside from *C. herba* inhibits the arachidonic acid pathway by decreasing the activities of COX-2 and PGE2, while activating phospholipase A2 and

iNOS expression (Zhou et al., 2015). In addition, inhibition of ROS, NO, TNF, and histamine by macrophages, neutrophils, and mast cells were recorded. This is linked to down-regulation of Ca/nuclear factor in activated T cells and the JNK-MAPK cascade (May et al., 2009).

14.6.2 *GASTRODIA ELATA BLUME*

- **Potential Neuroprotective Effect:** It is one of the most often utilized herbs in Asia for its ability to treat disorders such as ischemia, convulsions, dementia, vertigo, and tremors (Lee et al., 1999). It is known to block apoptosis in neuroblastoma cells induced by glutamate (Kim et al., 2001), cytotoxicity in SH-SY5Y cells induced by MPP+ (Kumar et al., 2013), and neuronal damage induced by kainite (Kim et al., 2003). In addition, it can prevent temporary global ischemic damage in neurons (Santini et al., 2007).
- **Mechanism of Action:** L-dopa-induced dyskinesia (LID) manifested as over-plasticity in an aberrant glutamate receptor causing over-excited signaling in the basal ganglia system was found to be inhibited by pharmacological blockade of ERK1/2 signaling (Kim et al., 2012) in the presence of gastrodin, a phenolic glucoside and major constituent of *G. elata.* This bioactive is also successful in reducing inflammation (Ha et al., 2001), ROS (Wu et al., 1996), PD (Kim et al., 2003), and memory loss (Lai et al., 2006). It suppresses aberrant elevations in ERK1/2, JNK, and p38MAPK expression thereby inhibiting cytokine and MAPK signaling triggered as a result of inflammation. This could also correct over plasticity of dopamine D1-like receptors.

14.6.3 *GANODERMA LUCIDUM*

- **Potential Neuroprotective Effect:** The extract of *Ganoderma lucidum* provides immunomodulatory and anti-inflammatory effects along with cognitive enhancement, free radical scavenging, cholesterol reduction, anti-aging, and mitochondrial energy generation abilities (Zhang et al., 2011). It also reduces neuronal loss following cerebral ischemia. Polysaccharides, such asβ-d-glucans, possess anti-tumor capabilities through immunomodulation and anti-angiogenesis.

- **Mechanism of Action:** *G. lucidum* efficiently protects DA neurons from inflammatory damage caused by microglial activation following exposure to MPP+ and LPS (Ahlskog & Muenter, 2001). This is due to its ability to reduce NO, IL-1, TNF-α, and superoxide generated by microglia via reduction in TNF-α and IL-1 mRNA expression (Ahlskog & Muenter, 2001).

14.7 CURRENT DEVELOPMENTS

Current PD pharmacotherapies do not give patients with the required long-term curative advantages. Natural products have been found as viable adjuvant therapies to conventional pharmacological therapy to minimize the dosage of anti-Parkinson medication and the occurrence of related side effects. Following long-term therapy with levodopa, there is a significant prevalence of dyskinesia and return of Parkinsonian symptoms. Combination treatment with natural herbal remedies has shown to reduce levodopa-related problems significantly (Schulz et al., 2010). However, the long-term effects of such combination therapy have not been studied, and possible interactions with drugs already used in the treatment of PD are unknown. Because PD medication must be taken for the rest of one's life, the dosing regimen is crucial. Adjunctive treatment using natural products may eventually be effective in lowering the dosage of levodopa used to treat PD symptoms.

While many natural products show promise for the treatment of PD and dyskinesia, a worldwide effort to standardize formulations for pharmaceutical uses is required to assure excellent quality and effectiveness. The efficiency of natural formulations is restricted owing to a number of issues including the original species of particular herbs, chemical compositions and indices, herbal product preparations, and indications. Standardization of single and combined natural products in human clinical trials will be a key aspect in transferring preclinical findings to the clinic. To assure excellent repeatability, increase therapeutic benefit, and limit the risk of damage, future clinical studies employing natural products for the treatment of PD must be improved methodologically. In the research design, double-blind trials and the use of placebos should be used. To ensure the openness of research findings and avoid publication bias, protocols must be set prior to the start of clinical trials.

14.8 CONCLUSION

Various natural ingredients have been used as 'food for medicine' by cultures all over the globe prior to the advent of modern medicine. The use of natural items for the prevention and amelioration of illness was judged a relic of the past due to the primacy of biomedical sciences and their reductionist approach. Grandmother's medication was often shown as the concept embodying food for medicine.' Despite decades of reductionist health and medicine, certain growing common diseases, such as PD, seem to be inaccessible to current medical therapies. Interest in natural products has been reignited in this era of paradigm transition. There is a lot of evidence that several natural compounds have heuristic value. This study looked at the effect of herbs, fruits, vegetables, and spices, as well as parasitic plants, and fungi, in the progression of PD. The Natural products may be able to treat the underlying pathogenic processes involved with the development and progression of neurodegenerative disorders in general, and PD in particular, if research advances in this direction.

KEYWORDS

- **abnormal mitochondrial function**
- **bioactives**
- **mitochondrial dysfunction**
- **oxidative stress**
- **Parkinson's disease**
- **pathophysiology**
- **polyphenols**

REFERENCES

Ahlskog, J. E., & Muenter, M. D., (2001). Frequency of levodopa-related dyskinesias and motor fluctuations as estimated from the cumulative literature. *Movement Disorders: Official Journal of the Movement Disorder Society, 16*(3), 448–458.

Al-Majed, A. A., Al-Omar, F. A., & Nagi, M. N., (2006). Neuroprotective effects of thymoquinone against transient forebrain ischemia in the rat hippocampus. *European Journal of Pharmacology, 543*(1–3), 40–47.

Alster, P., Madetko, N., Koziorowski, D., & Friedman, A., (2020). Progressive supranuclear palsy-parkinsonism predominant (PSP-P)-A clinical challenge at the boundaries of PSP and Parkinson's disease (PD). *Frontiers in Neurology, 11*, 180.

Bandres-Ciga, S., Diez-Fairen, M., Kim, J. J., & Singleton, A. B., (2020). Genetics of Parkinson's disease: An introspection of its journey towards precision medicine. *Neurobiology of Disease, 137*, 104782.

Batista-Nascimento, L., Pimentel, C., Menezes, R. A., & Rodrigues-Pousada, C., (2012). Iron and neurodegeneration: From cellular homeostasis to disease. *Oxidative Medicine and Cellular Longevity, 128*647.

Benabid, A. L., Benazzous, A., & Pollak, P., (2002). Mechanisms of deep brain stimulation. *Movement Disorders: Official Journal of the Movement Disorder Society, 17*(Suppl 3), S73–S74.

Benabid, A. L., Chabardes, S., Mitrofanis, J., & Pollak, P., (2019). Deep brain stimulation of the subthalamic nucleus for the treatment of Parkinson's disease. *The Lancet. Neurology, 8*(1), 67–81.

Bjørklund, G., Hofer, T., Nurchi, V. M., & Aaseth, J., (2019). Iron and other metals in the pathogenesis of Parkinson's disease: Toxic effects and possible detoxification. *Journal of Inorganic Biochemistry, 199*, 110717.

Bose, A., Petsko, G. A., & Eliezer, D., (2018). Parkinson's disease and melanoma: Co-occurrence and mechanisms. *Journal of Parkinson's Disease, 8*(3), 385–398.

Brocks, D. R., (1999). Anticholinergic drugs used in Parkinson's disease: An overlooked class of drugs from a pharmacokinetic perspective. *Journal of Pharmacy & Pharmaceutical Sciences: A Publication of the Canadian Society for Pharmaceutical Sciences, Societe Canadienne des Sciences Pharmaceutiques*, *2*(2), 39–46.

Brooks, D. J., (2000). Dopamine agonists: Their role in the treatment of Parkinson's disease. *Journal of Neurology, Neurosurgery, and Psychiatry, 68*(6), 685–689.

Cacabelos, R., (2017). Parkinson's disease: From pathogenesis to pharmacogenomics. *International Journal of Molecular Sciences, 18*(3), 551.

Callaghan, R. C., Cunningham, J. K., Sajeev, G., & Kish, S. J., (2010). Incidence of Parkinson's disease among hospital patients with methamphetamine-use disorders. *Movement Disorders: Official Journal of the Movement Disorder Society, 25*(14), 2333–2339.

Carlssone, A., Lindquist, M., & Magnusson, T., (1957). 3,4-dihydroxyphenylalanine and 5-hydroxytryptophan as reserpine antagonists. *Nature, 180*(4596), 1200.

Caudle, W. M., Richardson, J. R., Delea, K. C., Guillot, T. S., Wang, M., Pennell, K. D., & Miller, G. W., (2006). Polychlorinated biphenyl-induced reduction of dopamine transporter expression as a precursor to Parkinson's disease-associated dopamine toxicity. *Toxicological Sciences: An Official Journal of the Society of Toxicology, 92*(2), 490–499.

Chang, C., & Ramphul, K., (2022). *Amantadine*. In: StatPearls [Internet]. Treasure Island (FL): StatPearls Publishing.

Chen, H., Jing, F. C., Li, C. L., Tu, P. F., Zheng, Q. S., & Wang, Z. H., (2007). Echinacoside prevents the striatal extracellular levels of monoamine neurotransmitters from diminution in 6-hydroxydopamine lesion rats. *Journal of Ethnopharmacology, 114*(3), 285–289.

Cheng, P., Yu, J., Huang, W., Bai, S., Zhu, X., Qi, Z., Shao, W., & Xie, P., (2015). Dietary intake of iron, zinc, copper, and risk of Parkinson's disease: A meta-analysis. *Neurological Sciences: Official Journal of the Italian Neurological Society and of the Italian Society of Clinical Neurophysiology, 36*(12), 2269–2275.

Cheong, J., De Pablo-Fernandez, E., Foltynie, T., & Noyce, A. J., (2020). The association between type 2 diabetes mellitus and Parkinson's disease. *Journal of Parkinsons Disease, 10*(3), 775–789.

Chishti, M. A., Fisher, J. P., & Seegal, R. F., (1996). Aroclors 1254 and 1260 reduce dopamine concentrations in rat striatal slices. *Neurotoxicology, 17*(3, 4), 653–660.

Cotzias, G. C., Van, W. M. H., & Schiffer, L. M., (1967). Aromatic amino acids and modification of parkinsonism. *The New England Journal of Medicine, 276*(7), 374–379.

Datla, K. P., Christidou, M., Widmer, W. W., Rooprai, H. K., & Dexter, D. T., (2001). Tissue distribution and neuroprotective effects of citrus flavonoid tangeretin in a rat model of Parkinson's disease. *Neuroreport, 12*(17), 3871–3875.

Delgado-Morales, R., (2017). Neuroepigenomics in aging and disease. *Advances in Experimental Medicine, and Biology*. doi: 10.1007/978-3-319-53889-1_19. Springer International Publishing AG 363.

Delic, V., Beck, K. D., Pang, K., & Citron, B. A., (2020). Biological links between traumatic brain injury and Parkinson's disease. *Acta Neuropathologica Communications, 8*(1), 45.

Duvoisin, R. C., (1974). Variations in the "on-off" phenomenon. *Advances in Neurology, 5*, 339–340.

Egger, G., Liang, G., Aparicio, A., & Jones, P. A., (2004). Epigenetics in human disease and prospects for epigenetic therapy. *Nature, 429*, 457–463.

Ehringer, H., & Hornykiewicz, O., (1960). *Klinische Wochenschrift, 38*, 1236–1239.

Fields, C. R., Bengoa-Vergniory, N., & Wade-Martins, R., (2019). Targeting alpha-synuclein as a therapy for Parkinson's disease. *Frontiers in Molecular Neuroscience, 12*, 299.

Galaris, D., Barbouti, A., & Pantopoulos, K., (2019). Iron homeostasis and oxidative stress: An intimate relationship. *Biochimica et Biophysica Acta: Molecular Cell Research, 1866*(12), 118535.

GBD 2016 Parkinson's Disease Collaborators, (2018). Global, regional, and national burden of Parkinson's disease, 1990–2016: A systematic analysis for the Global Burden of Disease Study 2016. *The Lancet. Neurology, 17*(11), 939–953.

Geng, X., Tian, X., Tu, P., & Pu, X., (2007). Neuroprotective effects of echinacoside in the mouse MPTP model of Parkinson's disease. *European Journal of Pharmacology, 564*(1–3), 66–74.

Ghasemi, F., Bagheri, H., Barreto, G. E., Read, M. I., & Sahebkar, A., (2019). Effects of curcumin on microglial cells. *Neurotoxicity Research, 36*(1), 12–26.

Granado, N., Ares-Santos, S., & Moratalla, R., (2013). Methamphetamine and Parkinson's disease. *Parkinsons Disease,* 308052.

Groiss, S. J., Wojtecki, L., Südmeyer, M., & Schnitzler, A., (2009). Deep brain stimulation in Parkinson's disease. *Therapeutic Advances in Neurological Disorders, 2*(6), 20–28.

Ha, J. H., Shin, S. M., Lee, S. K., Kim, J. S., Shin, U. S., Huh, K., Kim, J. A., et al., (2001). *In vitro* effects of hydroxybenzaldehydes from *Gastrodia elata* and their analogues on GABAergic neurotransmission, and a structure-activity correlation. *Planta Medica, 67*(9), 877–880.

Hatcher-Martin, J. M., Gearing, M., Steenland, K., Levey, A. I., Miller, G. W., & Pennell, K. D., (2012). Association between polychlorinated biphenyls and Parkinson's disease neuropathology. *Neurotoxicology, 33*(5), 1298–1304.

Hewlings, S. J., & Kalman, D. S., (2017). Curcumin: A review of its effects on human health. *Foods (Basel, Switzerland), 6*(10), 92.

Jankovic, J., & Tan, E. K., (2020). Parkinson's disease: Etiopathogenesis and treatment. *Journal of Neurology, Neurosurgery, and Psychiatry, 91*(8), 795–808.

Jonsson, G., (1983). Chemical lesioning techniques: Monoamine neurotoxins. In: Björklund, A., & Hökfelt, T., (eds.), *Handbook of Chemical Neuroanatomy: Methods in Chemical Neuroanatomy* (1st edn., Vol. 1, pp. 463–507). Amsterdam: Elsevier Science Publishers B.V.

Kalteis, K., Standhardt, H., Kryspin-Exner, I., Brücke, T., Volc, D., & Alesch, F., (2006). Influence of bilateral Stn-stimulation on psychiatric symptoms and psychosocial functioning in patients with Parkinson's disease. *Journal of Neural Transmission, 113*(9), 1191–1206.

Kanter, M., (2008). Nigella sativa and derived thymoquinone prevents hippocampal neurodegeneration after chronic toluene exposure in rats. *Neurochemical Research, 33*(3), 579–588.

Keservani, R. K., & Sharma, A. K., (2014). Flavonoids: Emerging trends and potential health benefits. *Journal of Chinese Pharmaceutical Sciences, 23*(12), 815.

Keservani, R. K., Kesharwani, R. K., Sharma, A. K., Vyas, N., & Chadoker, A., (2010b). Nutritional supplements: An overview. *International Journal of Current Pharmaceutical Review and Research, 1*(1), 59–75.

Keservani, R. K., Kesharwani, R. K., Vyas, N., Jain, S., Raghuvanshi, R., & Sharma, A. K., (2010a). Nutraceutical and functional food as future food: A review. *Der Pharmacia Lettre, 2*(1), 106–116.

Keservani, R. K., Sharma, A. K., & Kesharwani, R. K., (2016a). Nutraceutical and functional foods for cardiovascular health. In: *Food Process Engineering* (pp. 257–278, 291–312). Apple Academic Press, CRC Press. ISBN: 9781771884020.

Keservani, R. K., Sharma, A. K., & Kesharwani, R. K., (2016b). Medicinal effect of nutraceutical fruits for the cognition and brain health. *Scientifica*, 10. Article ID 3109254. doi: 10.1155/2016/3109254.

Keservani, R. K., Sharma, A. K., & Kesharwani, R. K., (2017). An overview and therapeutic applications of nutraceutical and functional foods. *Recent Advances in Drug Delivery Technology*, 160–201. ISBN:9781522507543.

Keservani, R. K., Sharma, A. K., & Kesharwani, R. K., (2020). *Nutraceuticals and Dietary Supplements: Applications in Health Improvement and Disease Management*. CRC Press. ISBN: 9781771888738.

Khatri, D. K., & Juvekar, A. R., (2016). Neuroprotective effect of curcumin as evinced by abrogation of rotenone-induced motor deficits, oxidative and mitochondrial dysfunctions in mouse model of Parkinson's disease. *Pharmacology, Biochemistry, and Behavior, 150, 151*, 39–47.

Kim, B. W., Koppula, S., Kim, J. W., Lim, H. W., Hwang, J. W., Kim, I. S., Park, P. J., & Choi, D. K., (2012). Modulation of LPS-stimulated neuroinflammation in BV-2 microglia by *Gastrodia elata*: 4-hydroxybenzyl alcohol is the bioactive candidate. *Journal of Ethnopharmacology, 139*(2), 549–557.

Kim, H. G., Ju, M. S., Shim, J. S., Kim, M. C., Lee, S. H., Huh, Y., Kim, S. Y., & Oh, M. S., (2010). Mulberry fruit protects dopaminergic neurons in toxin-induced Parkinson's disease models. *The British Journal of Nutrition, 104*(1), 8–16.

Kim, H. J., Lee, S. R., & Moon, K. D., (2003). Ether fraction of methanol extracts of *Gastrodia elata*, medicinal herb protects against neuronal cell damage after transient global ischemia in gerbils. *Phytotherapy Research: PTR, 17*(8), 909–912.

Kim, H. J., Moon, K. D., Oh, S. Y., Kim, S. P., & Lee, S. R., (2001). Ether fraction of methanol extracts of *Gastrodia elata*, a traditional medicinal herb, protects against kainic acid-induced neuronal damage in the mouse hippocampus. *Neuroscience Letters, 314*(1, 2), 65–68.

Kim, W., Lee, Y., McKenna, N. D., Yi, M., Simunovic, F., Wang, Y., Kong, B., et al., (2014). miR-126 contributes to Parkinson's disease by dysregulating the insulin-like growth factor/phosphoinositide 3-kinase signaling. *Neurobiology of Aging, 35*(7), 1712–1721.

Klein, C., & Westenberger, A., (2012). Genetics of Parkinson's disease. *Cold Spring Harbor Perspectives in Medicine, 2*(1), a008888.

Kouli, A., Torsney, K. M., & Kuan, W. L., (2018). Parkinson's disease: Etiology, neuropathology, and pathogenesis. In: Stoker, T. B., et. al., (eds.), *Parkinson's Disease: Pathogenesis and Clinical Aspects*. Codon Publications.

Krack, P., Batir, A., Van, B. N., Chabardes, S., Fraix, V., Ardouin, C., et al., (2003). Five-year follow-up of bilateral stimulation of the subthalamic nucleus in advanced Parkinson's disease. *The New England Journal of Medicine, 349*(20), 1925–1934.

Kumar, H., Kim, I. S., More, S. V., Kim, B. W., Bahk, Y. Y., & Choi, D. K., (2013). Gastrodin protects apoptotic dopaminergic neurons in a toxin-induced Parkinson's disease model. *Evidence-Based Complementary and Alternative Medicine: ECAM*, 514095.

Lai, K. N., Chan, L. Y., Tang, S. C., & Leung, J. C., (2006). Ganoderma extract prevents albumin-induced oxidative damage and chemokines synthesis in cultured human proximal tubular epithelial cells. *Nephrology, Dialysis, Transplantation: Official Publication of the European Dialysis and Transplant Association – European Renal Association, 21*(5), 1188–1197.

Langston, J. W., Ballard, P., Tetrud, J. W., & Irwin, I., (1983). Chronic parkinsonism in humans due to a product of meperidine-analog synthesis. *Science (New York, N.Y.), 219*(4587), 979, 980.

Lappin, J. M., Darke, S., & Farrell, M., (2018). Methamphetamine use and future risk for Parkinson's disease: Evidence and clinical implications. *Drug and Alcohol Dependence, 187*, 134–140.

Lee, Y. S., Ha, J. H., Yong, C. S., Lee, D. U., Huh, K., Kang, Y. S., Lee, S. H., et al., (1999). Inhibitory effects of constituents of *Gastrodia elata* Bl. on glutamate-induced apoptosis in IMR-32 human neuroblastoma cells. *Archives of Pharmacal Research, 22*(4), 404–409.

Lees, A. J., (1989). The on-off phenomenon. *Journal of Neurology, Neurosurgery, and Psychiatry*, 29–37.

Li, S., & Le, W., (2017). Milestones of Parkinson's disease research: 200 years of history and BEYOND. *Neuroscience Bulletin, 33*(5), 598–602.

Limphaibool, N., Iwanowski, P., Holstad, M., Kobylarek, D., & Kozubski, W., (2019). Infectious etiologies of parkinsonism: Pathomechanisms and clinical implications. *Frontiers in Neurology, 10*, 652.

Liu, L. K., Lee, H. J., Shih, Y. W., Chyau, C. C., & Wang, C. J., (2008). Mulberry anthocyanin extracts inhibit LDL oxidation and macrophage-derived foam cell formation induced by oxidative LDL. *Journal of Food Science, 73*(6): H113–H121.

Malar, D. S., Prasanth, M. I., Brimson, J. M., Sharika, R., Sivamaruthi, B. S., Chaiyasut, C., & Tencomnao, T., (2020). Neuroprotective properties of green tea (*Camellia sinensis*) in Parkinson's disease: A review. *Molecules (Basel, Switzerland), 25*(17), 3926.

Manthey, J. A., Grohmann, K., & Guthrie, N., (2001). Biological properties of citrus flavonoids pertaining to cancer and inflammation. *Current Medicinal Chemistry, 8*(2), 135–153.

Marsden, C. D., (1994). Problems with long-term levodopa therapy for Parkinson's disease. *Clinical Neuropharmacology, 17*(Suppl 2), S32–S44.

Martin, I., Kim, J. W., Dawson, V. L., & Dawson, T. M., (2014). LRRK2 pathobiology in Parkinson's disease. *Journal of Neurochemistry, 131*(5), 554–565.

May, B. H., Lit, M., Xue, C. C., Yang, A. W., Zhang, A. L., Owens, M. D., Head, R., et al., (2009). Herbal medicine for dementia: A systematic review. *Phytotherapy Research: PTR, 23*(4), 447–459.

McCormack, A. L., Thiruchelvam, M., Manning-Bog, A. B., Thiffault, C., Langston, J. W., Cory-Slechta, D. A., & Di Monte, D. A., (2002). Environmental risk factors and Parkinson's disease: Selective degeneration of nigral dopaminergic neurons caused by the herbicide paraquat. *Neurobiology of Disease, 10*(2), 119–127.

Miranda-Morales, E., Meier, K., Sandoval-Carrillo, A., Salas-Pacheco, J., Vázquez-Cárdenas, P., & Arias-Carrión, O., (2017). Implications of DNA methylation in Parkinson's disease. *Frontiers in Molecular Neuroscience, 10*, 225.

Moon, H. E., & Paek, S. H., (2015). Mitochondrial dysfunction in Parkinson's disease. *Experimental Neurobiology, 24*(2), 103–116.

Mutter, J., (2011). Is dental amalgam safe for humans? The opinion of the scientific committee of the European commission. *Journal of Occupational Medicine and Toxicology, 6*(1), 2.

Network Management, (2021). NobelPrize.org. Nobel Media AB. https://www.nobelprize.org/prizes/medicines/1932/speedread/ (accessed on 15 November 2023).

Ngim, C. H., & Devathasan, G., (1989). Epidemiologic study on the association between body burden mercury level and idiopathic Parkinson's disease. *Neuroepidemiology, 8*(3), 128–141.

Nguyen, T. T., Vuu, M. D., Huynh, M. A., Yamaguchi, M., Tran, L. T., & Dang, T., (2018). Curcumin effectively rescued Parkinson's disease-like phenotypes in a novel *Drosophila melanogaster* model with Duch knockdown. *Oxidative Medicine and Cellular Longevity, 20*38267.

Panchal, K., & Tiwari, A. K., (2019). Mitochondrial dynamics, a key executioner in neurodegenerative diseases. *Mitochondrion, 47*, 151–173.

Panenka, W. J., Procyshyn, R. M., Lecomte, T., MacEwan, G. W., Flynn, S. W., Honer, W. G., & Barr, A. M., (2013). Methamphetamine use: A comprehensive review of molecular, preclinical and clinical findings. *Drug and Alcohol Dependence, 129*(3), 167–179.

Parkinson, J., (2002). An essay on the shaking palsy. 1817. *The Journal of Neuropsychiatry and Clinical Neurosciences, 14*(2), 223–222.

Petelin, L. S., Vartanian, K. Z., & Romenskaia, L., (1977). Dlitel'naia terapiia l-DOFA pri parkinsonizme i ee patokhimicheskiĭ aspekt [long-term l-DOPA therapy for parkinsonism and its pathochemical aspect]. *Zhurnal Nevropatologii I Psikhiatrii Imeni S.S. Korsakova (Moscow, Russia: 1952), 77*(12), 1810–1813.

Pinto, N., Da Silva, A. B., Neves, K. R., Silva, A. H., Leal, L. K., & Viana, G. S., (2015). Neuroprotective properties of the standardized extract from camellia sinensis (green tea) and its main bioactive components, epicatechin and epigallocatechin gallate, in the 6-OHDA model of Parkinson's disease. *Evidence-Based Complementary and Alternative Medicine: ECAM*, 161092.

Prusiner, S. B., Woerman, A. L., Mordes, D. A., Watts, J. C., Rampersaud, R., Berry, D. B., Patel, S., et al., (2015). Evidence for α-synuclein prions causing multiple system atrophy in humans with parkinsonism. *Proceedings of the National Academy of Sciences of the United States of America, 112*(38), E5308–E5317.

Przedborski, S., & Ischiropoulos, H., (2005). Reactive oxygen and nitrogen species: Weapons of neuronal destruction in models of Parkinson's disease. *Antioxidants & Redox Signaling, 7*(5, 6), 685–693.

Przedborski, S., Vila, M., & Jackson-Lewis, V., (2003). Neurodegeneration: What is it and where are we? *The Journal of Clinical Investigation, 111*(1), 3–10.

Qureshi, I. A., & Mehler, M. F., (2011). Advances in epigenetics and epigenomics for neurodegenerative diseases. *Current Neurology and Neuroscience Reports, 11*(5), 464–473.

Rail, D., Scholtz, C., & Swash, M., (1981). Post-encephalitic Parkinsonism: Current experience. *Journal of Neurology, Neurosurgery, and Psychiatry, 44*(8), 670–676.

Rainey, N. E., Moustapha, A., Saric, A., Nicolas, G., Sureau, F., & Petit, P. X., (2019). Iron chelation by curcumin suppresses both curcumin-induced autophagy and cell death together with iron overload neoplastic transformation. *Cell Death Discovery, 5,* 150.

Richardson, J. R., & Miller, G. W., (2004). Acute exposure to aroclor 1016 or 1260 differentially affects dopamine transporter and vesicular monoamine transporter 2 levels. *Toxicology Letters, 148*(1, 2), 29–40.

Rietdijk, C. D., Perez-Pardo, P., Garssen, J., Van, W. R. J., & Kraneveld, A. D., (2017). Exploring Braak's hypothesis of Parkinson's disease. *Frontiers in Neurology, 8*, 37.

Rivest, J., Barclay, C. L., & Suchowersky, O., (1999). COMT inhibitors in Parkinson's disease. *The Canadian Journal of Neurological Sciences: Le Journal Canadien Des Sciences Neurologiques, 26*(Suppl 2), S34–S38.

Rocha, E. M., De Miranda, B., & Sanders, L. H., (2018). Alpha-synuclein: Pathology, mitochondrial dysfunction and neuroinflammation in Parkinson's disease. *Neurobiology of Disease, 109*(Pt B), 249–257.

Santini, E., Valjent, E., Usiello, A., Carta, M., Borgkvist, A., Girault, J. A., Hervé, D., et al., (2007). Critical involvement of cAMP/DARPP-32 and extracellular signal-regulated protein kinase signaling in L-DOPA-induced dyskinesia. *The Journal of Neuroscience: The Official Journal of the Society for Neuroscience, 27*(26), 6995–7005.

Schuler, F., & Casida, J. E., (2001). Functional coupling of PSST and ND1 subunits in NADH: Ubiquinone oxidoreductase established by photoaffinity labeling. *Biochimica et Biophysica Acta, 1506*(1), 79–87.

Schulz, K. F., Altman, D. G., Moher, D., & CONSORT Group, (2010). CONSORT 2010 statement: Updated guidelines for reporting parallel group randomized trials. *BMC Medicine, 8*, 18.

Seegal, R. F., Brosch, K. O., & Shain, W., (1990). Lightly chlorinated ortho-substituted PCB congeners decrease dopamine in nonhuman primate brain and in tissue culture. *Toxicology and Applied Pharmacology, 106*(1), 136–144.

Sharma, N., & Nehru, B., (2018). Curcumin affords neuroprotection and inhibits α-synuclein aggregation in lipopolysaccharide-induced Parkinson's disease model. *Inflammopharmacology, 26*, 349–360.

Sheng, G., Pu, X., Lei, L., Tu, P., & Li, C., (2002). Tubuloside B from *Cistanche salsa* rescues the PC12 neuronal cells from 1-methyl-4-phenylpyridinium ion-induced apoptosis and oxidative stress. *Planta Medica, 68*(11), 966–970.

Shimizu, K., Ohtaki, K., Matsubara, K., Aoyama, K., Uezono, T., Saito, O., Suno, M., et al., (2001). Carrier-mediated processes in blood--brain barrier penetration and neural uptake of paraquat. *Brain Research, 906*(1, 2), 135–142.

Siddiqui, I. J., Pervaiz, N., & Abbasi, A. A., (2016). The Parkinson disease gene SNCA: Evolutionary and structural insights with pathological implication. *Scientific Reports, 6*, 24475.

Singh, R., Gautam, N., Mishra, A., & Gupta, R., (2011). Heavy metals and living systems: An overview. *Indian Journal of Pharmacology, 43*(3), 246–253.

Smeyne, R. J., Noyce, A. J., Byrne, M., Savica, R., & Marras, C., (2021). Infection and risk of Parkinson's disease. *Journal of Parkinson's Disease, 11*(1), 31–43.

Sposato, L. A., (2014). *Handbook of Clinical Neurology: Neurologic Aspects of Systemic Disease, Part III, 121,* 1635–1671.

Stoker, T. B., & Greenland, J. C., (2018). *Parkinson's Disease: Pathogenesis and Clinical Aspects*. Codon Publications.

Tanner, C. M., & Goldman, S. M., (1996). Epidemiology of Parkinson's disease. *Neurologic Clinics, 14*(2), 317–335.

Thanvi, B. R., & Lo, T. C., (2004). Long term motor complications of levodopa: Clinical features, mechanisms, and management strategies. *Postgraduate Medical Journal, 80*(946), 452–458.

Thiruchelvam, M., Richfield, E. K., Baggs, R. B., Tank, A. W., & Cory-Slechta, D. A., (2000). The nigrostriatal dopaminergic system as a preferential target of repeated exposures to combined paraquat and maneb: Implications for Parkinson's disease. *The Journal of Neuroscience: The Official Journal of the Society for Neuroscience, 20*(24), 9207–9214.

Triarhou, L. C., (2013). Dopamine and Parkinson's disease. In: *Madame Curie Bioscience Database [Internet].* Austin (TX): Landes Bioscience.

Ullah, I., Zhao, L., Hai, Y., Fahim, M., Alwayli, D., Wang, X., & Li, H., (2021). Metal elements and pesticides as risk factors for Parkinson's disease: A review. *Toxicology Reports, 8*, 607–616.

Vafeiadou, K., Vauzour, D., Lee, H. Y., Rodriguez-Mateos, A., Williams, R. J., & Spencer, J. P., (2009). The citrus flavanone naringenin inhibits inflammatory signaling in glial cells and protects against neuroinflammatory injury. *Archives of Biochemistry and Biophysics, 484*(1), 100–109.

Veeriah, S., Taylor, B. S., Meng, S., Fang, F., Yilmaz, E., Vivanco, I., Janakiraman, M., et al., (2010). Somatic mutations of the Parkinson's disease-associated gene PARK2 in glioblastoma and other human malignancies. *Nature Genetics, 42*(1), 77–82.

Wang, N., Ji, S., Zhang, H., Mei, S., Qiao, L., & Jin, X., (2017). Herba Cistanches: Anti-aging. *Aging and Disease, 8*(6), 740–759.

Wu, C. R., Hsieh, M. T., Huang, S. C., Peng, W. H., Chang, Y. S., & Chen, C. F., (1996). Effects of *Gastrodia elata* and its active constituents on scopolamine-induced amnesia in rats. *Planta Medica, 62*(4), 317–321.

Zbarsky, V., Datla, K. P., Parkar, S., Rai, D. K., Aruoma, O. I., & Dexter, D. T., (2005). Neuroprotective properties of the natural phenolic antioxidants curcumin and naringenin but not quercetin and fisetin in a 6-OHDA model of Parkinson's disease. *Free Radical Research, 39*(10), 1119–1125.

Zhang, R., Xu, S., Cai, Y., Zhou, M., Zuo, X., & Chan, P., (2011). *Ganoderma lucidum* protects dopaminergic neuron degeneration through inhibition of microglial activation. *Evidence-Based Complementary and Alternative Medicine: eCAM*, 156810.

Zheng, Z. P., Cheng, K. W., Chao, J. F., Wu, J. J., & Wang, M. F., (2008). Tyrosinase inhibitors from paper mulberry (*Broussonetia papyrifera*) *Food Chemistry, 106*(2), 529–535.

Zhou, Y. Y., Li, Y., Jiang, W. Q., & Zhou, L. F., (2015). MAPK/JNK signaling: A potential autophagy regulation pathway. *Bioscience Reports, 35*(3), e00199.

INDEX

E

F

G

N

Q

R

T